The sea-surface microlayer has often been operationally defined as roughly the top 1000 micrometres of the ocean surface. Considerable research over the past 10 years has led to increased understanding of this vitally important interface between the ocean and the atmosphere and how it may interact with global change processes. This book offers the first comprehensive review of the physics, chemistry and biology of the surface microlayer in a decade. In addition to a review of these fundamental features, the authors address the potential global marine impacts at the air–sea interface of such phenomena as large-scale atmospheric ozone depletion, climate change and industrial pollution. Environmental scientists, oceanographers and atmospheric scientists interested in global change will welcome this authoritative reference work, at graduate or research level.

THE SEA SURFACE AND GLOBAL CHANGE

THE SEA SURFACE
AND GLOBAL CHANGE

Edited by

PETER S. LISS

School of Environmental Sciences,
University of East Anglia
Norwich, UK

and

ROBERT A. DUCE

College of Geosciences & Maritime Studies,
Texas A & M University,
College Station
Texas, USA

CAMBRIDGE
UNIVERSITY PRESS

PUBLISHED BY THE PRESS SYNDICATE OF THE UNIVERSITY OF CAMBRIDGE
The Pitt Building, Trumpington Street, Cambridge CB2 1RP, United Kingdom

CAMBRIDGE UNIVERSITY PRESS
The Edinburgh Building, Cambridge CB2 2RU, United Kingdom
40 West 20th Street, New York, NY 10011–4211, USA
10 Stamford Road, Oakleigh, Melbourne 3166, Australia

First published 1997

Printed in the United Kingdom at the University Press, Cambridge

Typeset in 10/13pt Linotron Times

A catalogue record for this book is available from the British Library

Library of Congress Cataloguing in Publication data

The sea surface and global change / edited by Peter S. Liss and Robert A. Duce
p. cm.
Includes index.
ISBN 0 521 56273 2
1. ocean–atmosphere interaction. 2. Climatic changes. I. Liss, P. S. II. Duce,
Robert A., 1935– .
GC190.2.S43 1997
551,46'01–dc20 96-9377 CIP

ISBN 0 521 56273 2 hardback

Contents

Contributors

William Asher
Joint Institute for the Study of the Atmosphere and the Ocean, Box 354235, University of Washington, Seattle, Washington, 98195, USA

Neil Blough
Department of Chemistry and Biochemistry, University of Maryland, College Park, Maryland, 70742, USA

Erik Bock
Department of Ocean Engineering, Woods Hole Oceanographic Institution, Woods Hole, Massachusetts, 02543, USA

Dominique Calmet
Institut de Protection et de Surete Nucleaire, Department de Protection de l'Environnement et des Installations, Batiment 601, Bois des Rames, Orsay, France

John Cleary
Plymouth Marine Laboratory, Prospect Place, West Hoe, Plymouth, PL1 3DH, United Kingdom

Robert Duce
Room 204, O & M Building, College of Geosciences and Maritime Studies, Texas A & M University, College Station, Texas, 77843–3148, USA

Manfred Ehrhardt
Abteilung Meereschemie, Institut fuer Meereskunde, Universitaet Kiel, Duesternbrooker Weg 20, D 24105 Kiel, Germany

Thomas Forbes*
Department of Marine Ecology and Microbiology, Frederiksborgvej 399, P.O. Box 358, DK-4000, Roskilde, Denmark

Nelson Frew
Department of Marine Chemistry and Geochemistry, Woods Hole Oceanographic Institution, Woods Hole, Massachusetts, 02543, USA

Michail Gladyshev
Institute of Biophysics, Siberian Branch, Russian Academy of Sciences, Akademgorodok, Krasnoyarsk, 660036, Russia

Gareth Harding
Department of Fisheries and Oceans, Bedford Institute of Oceanography, Box 1006, Dartmouth, Nova Scotia, B2Y 4A2, Canada

John Hardy
Huxley College of Environmental Studies, Western Washington University, Bellingham, Washington, 98225, USA

Lutz Hasse
Abteilung Maritime Meteorologie, Institut fuer Meereskunde, Universitaet Kiel, Duesternbrooker Weg 20, D 24105 Kiel, Germany

Keith Hunter
Department of Chemistry, University of Otago, Box 56, Dunedin, New Zealand

Bernd Jähne
Physical Oceanography Research Division, 0230, Scripps Institution of Oceanography, University of California, San Diego, La Jolla, California, 92093, USA

Gerald Korenowski
Department of Chemistry, Rensselaer Polytechnic Institute, Troy, New York, 12180–3590, USA

Peter Liss
School of Environmental Sciences, University of East Anglia, Norwich, NR4 7TJ, United Kingdom

Liliane Merlivat
Laboratoire d'Oceanographie Dynamique et de Climatologie, Tour 14, 2e etage, Universite Pierre et Marie Curie, Place Jussieu, Paris, France

Leon Phillips
Department of Chemistry, University of Canterbury, Christchurch, New Zealand

John Plane
School of Environmental Sciences, University of East Anglia, Norwich, NR4 7TJ, United Kingdom

Ian Robinson*
Department of Oceanography, University of Southampton, Highfield, Southampton, SO17 1BJ, United Kingdom

Peter Schluessel
Meteorologisches Institut, Bundesstrasse 55, University of Hamburg, Hamburg, Germany

Jon Shenker
Department of Biological Sciences, Florida Institute of Technology, West University Boulevard, Melbourne, Florida, 32901–6988, USA

Alexander Soudine
World Meteorological Organization, 41 Avenue Giuseppe Motta, Case Postale No. 2300, CH-1211, Geneva 2, Switzerland

Paul Tratnyek
Department of Environmental Science and Engineering, Oregon Graduate Institute of Science and Technology, NW Walker Road, Box 91000, Portland, Oregon, 97291–1000, USA

Kirk Waters
JIMAR, MSB Room 312, Pope Road, University of Hawaii, Honolulu, Hawaii, 96822, USA

Andrew Watson
School of Environmental Sciences, University of East Anglia, Norwich, NR4 7TJ, United Kingdom

David Woolf
Department of Oceanography, University of Southampton, Southampton Oceanography Centre, Waterfront Campus, European Way, Southampton SO14 3ZH, United Kingdom

Yuvenaly Zaitsev
Odessa Branch, Institute of Biology of the Southern Seas, Academy of Sciences of Ukraine, Pushkinskaya Street, 270011 Odessa, Ukraine

Richard Zepp
EPA Environmental Research Laboratory, College Station Road, Athens, Georgia, 30605–2720, USA

Rod Zika
Marine and Atmospheric Chemistry, RSMAS, University of Miami, Rickenbacker Causeway, Miami, Florida, 33149–1098, USA

* Did not attend the workshop at University of Rhode Island, February, 1994.

Preface

The sea-surface microlayer has often been operationally defined as roughly the top 1 to 1000 micrometres of the ocean surface. There has been considerable new research in this area over the past 5–10 years. The microlayer is known to concentrate, to varying degrees, many chemical substances, particularly those that are surface active, and many organisms live and/or find food there. It is clearly the interface through which all gaseous, liquid and particulate material must pass when exchanging between the ocean and the atmosphere. It also plays a vital role in the transfer of various forms of energy (momentum, heat) between the two media.

It is now recognized that important physical, chemical, and biological processes near the air–sea interface are not restricted to what has been traditionally referred to as the 'microlayer', but rather occur over gradients of varying thickness. Above the interface is an atmospheric boundary layer of 50–500 μm, where atmospheric turbulence is much reduced. Below the air-water interface the aquatic surface layer contains a series of sublayers (as described by Jack Hardy in Chapter 11). In this book, we use the term 'microlayer' in its operational meaning to refer to roughly the uppermost millimetre of the oceans, where properties are most altered relative to deeper waters. We also utilize the following terminology: a 'film' refers to a surfactant-influenced surface and a 'slick' refers to a visibly surfactant-influenced surface. The viscous and thermal sublayers are roughly the top 1000 and 300 μm of the water surface, respectively, while the diffusion sublayer refers to the top 50 μm.

Natural surface-active substances are often enriched in the sea surface compared with sub-surface water. Amino acids, proteins, fatty acids, lipids, phenols, and a great variety of other organic compounds collect on the surface, especially in the particulate phase. The biota of the water column below are the source for most of the enrichment of natural (non-pollutant) chemicals. Plankton produce dissolved compounds as products of their metabolism. Air bubbles, rising through the water column, scavenge these organic materials and bring them to the surface. Also, as plankton die and disintegrate some particles and many of the breakdown products (oils, fats, proteins, etc) float or are transported by active processes to the surface.

The accumulation of natural organic chemicals modifies the physical and optical properties of the sea surface. Thin organic films, invisible to the naked eye, are ubiquitous in aquatic systems. In areas where currents converge, films become more concentrated. Under light to moderate wind conditions, areas of accumulated film dampen small waves and become visible as 'surface slicks'. Strong surface tension forces exist, creating a boundary region where turbulent mixing is much reduced.

Of particular concern are processes occurring at the sea-surface microlayer that either affect or are affected by global change. For example, growing population and industrialization have resulted in increasing atmospheric transport of pollutant materials over the ocean. Atmospheric deposition of this material and of naturally occurring substances represents an important source of inorganic and organic chemicals to the sea-surface microlayer. Many of these substances are surface active and contribute to increased concentrations in the surface microlayer. They could result in increasing incidence of coherent films or slicks as well as the concentration of some surface-active substances at the ocean surface in both coastal and open ocean regions. High concentrations of toxic chemicals have been reported in the surface microlayer compared with the sub-surface bulk-water in some coastal environments.

Global decreases in stratospheric ozone resulting from CFC and halon releases have led to increased levels of solar ultraviolet-B (UV-B) radiation (280 to 315 nm) reaching the earth's surface. Because of the long residence times of different CFC compounds in the atmosphere (8 to 380 years), stratospheric ozone depletion and increases in UV-B radiation are expected to continue well into the twenty-first century.

These changes, or potential changes, have raised several important and inter-related questions concerning global marine impacts, including:

(a) Could continuing or increased deposition of toxic chemicals and surface-active agents, and/or increased UV-B radiation, alter either physically or biologically mediated fluxes of radiatively active and atmospheric chemically important trace gases between the atmosphere and ocean and vice versa?
(b) What is the likely impact of chemical enrichment of the sea surface, along with increased UV-B radiation, on the health of biological communities inhabiting the sea surface, including the egg and larval stages of many commercially important fish species?

To address adequately these and related issues it is timely to review our current understanding of the fundamental physical, chemical and biological processes in the surface of the oceans that may affect or be affected by global changes. Much new information has been generated in recent years about the chemical composition and structure of the surface layer and the types and rates of reaction occurring there (particularly photochemical reactions, but not explicitly in the microlayer). This new information provides an important foundation, but there are many processes which may be of global importance for which we have incomplete or virtually no information. In addition, there has been no comprehensive review of our understanding of the surface microlayer for almost a decade. This lack of comprehensive evaluation of the micro-layer, its potential future changes, and its impact on global change processes is the primary motivation for the production of this book.

A vital precursor to the book was a meeting of the GESAMP (Group of Experts on the Scientific Aspects of Marine Environmental Protection) Working Group No. 34 held in the form of a Workshop on 'The sea-surface microlayer and its potential role in global change' at the W. Alton Jones Campus of the University of Rhode Island, USA, on 20–24 February 1994. Most of the contributors to this book attended that workshop. At the workshop three group reports were prepared and appear at the beginning of the book. (These group reports are also published in a somewhat amended form in the *GESAMP Reports and Studies* Series, No. 59 (1995).) They are followed by 13 individually authored chapters dealing with different aspects of the sea-surface micro-layer.

We thank Jack Hardy, John Plane and Andrew Watson for their help in planning and leading the workshop. Financial support for the meeting and subsequent work was provided by the following agencies contributing to GESAMP: WMO, UNEP, IMO, IOC, IAEA, and we thank Alexander Soudine of WMO who was instrumental in organising the funding and participated in the workshop. We are very pleased to acknowledge the assistance of Van Chisholm and Stephen Brown in providing logistic support during the workshop. Finally, we thank Phillip Williamson for his invaluable editorial help with several of the chapters.

Peter S. Liss,
Norwich, UK
Robert A. Duce,
College Station, USA
July 1996

1

Report Group 1 – Physical processes in the microlayer and the air–sea exchange of trace gases

P. S. LISS and A. J. WATSON (*Chairs*), E. J. BOCK
and B. JÄHNE (*Rapporteurs*), W. E. ASHER, N. M. FREW,
L. HASSE, G. M. KORENOWSKI, L. MERLIVAT, L. F.
PHILLIPS, P. SCHLUESSEL and D. K. WOOLF

Surface films

Sources, sinks, and properties of surface films

Sea-surface films are derived from multiple sources, both in the sea and on land. Even in oligotrophic waters where biological productivity is low, the concentration of dissolved organic matter, including degraded bio-polymers and geopolymeric materials, is sufficient to produce surface enrichments of organic matter under favourable physical conditions. Specific inputs from phytoplankton blooms and from neuston indigenous to the microlayer also contribute to the enrichment of surface-active matter at the interface. Terrestrial sources include both natural and anthropogenic contributions. Terrestrial plant-derived materials are released directly to the atmosphere and introduced via dry and wet deposition, or enter the ocean environment via riverine inputs as decay products of vegetation. Anthropogenic contributions include point sources related to industrial processes, agricultural runoff, and spills of petroleum products (catastrophic and chronic). In addition, municipal wastewater discharges are frequently highly enriched in surfactants that ultimately enter coastal seas and sediments. In shallow coastal environments, resuspension of sediments and release of sediment pore-water materials are other potential sources of surfactants.

The relative importance of these sources is not known. A major source of surfactants is thought to be production by phytoplankton, which exude natural surfactants as metabolic by-products (Zutic et al., 1981). Exudates of many types of marine phytoplankton are rich in complex, high-molecular-weight homo- and hetero-polysaccharides (Allan et al., 1972; Ramus, 1972; Smestad et al., 1975; Percival et al., 1980; Mykelstad, 1974; Mykelstad et al., 1982), and these compounds are commonly found in surface waters during phytoplankton blooms (Sakugawa and Handa, 1985a, b). A significant fraction of these polysaccharides is conjugated to hydrophobic moieties that make them at least weakly surface active (Frew *et al.*, 1990), as is the case with most macromolecular materials (Leenheer, 1985). Proteinaceous and lipid materials are less abundant than carbohydrates in phytoplankton exudates (Hellebust, 1974), but their influence on the surface physical properties of sea-surface films can be significant despite their low concentrations (Van Vleet and Williams, 1983).

Indirect production of surface-active materials from plankton results from degradative processes, and synthesis from low-molecular-weight metabolic products. Chemical and microbiological transformations of dead organisms lead to a variety of compounds whose molecular structures contain hydrophobic and hydrophilic moieties. Condensation of relatively low molecular weight exudates to form complex surface-active macromolecular structures has been suggested as a possible production pathway (Nissenbaum, 1974; Zutic *et al.*, 1981). Surface-active marine gelbstoff or humic materials (Rashid, 1985; Gillam and Wilson, 1985) are thought to be derived from phytoplankton exudates by such transformation processes and may be important in direct production of surfactants as a source of microlayer films.

Surfactants are concentrated at the air–sea interface over broad areas of the ocean by a number of physical processes including diffusion, turbulent mixing, scavenging and transport by bubbles and buoyant particles. Direct atmospheric deposition is also likely to contribute to films over broad areas and will be relatively more important at long distances from terrestrial sources. Convergent circulations driven by wind, tidal forces, current shear, upwelling and internal waves lead to localized concentrations of surface-active materials on various spatial scales, ranging from features a few metres (i.e., internal waves, current shear zones) to kilometres (i.e., large-scale eddies) in size.

Organized surfactant films, representing high surface concentrations of organic material, are prevalent on only a small fraction of the global oceans. Based on measurements of surface tension reduction (surface pressure) in the open ocean, surface concentrations of surfactants are quite low. This may not be an accurate reflection of their importance in modulating wave spectra or gas transfer since natural surfactants have been shown in the laboratory to exhibit substantial dynamic elasticities. It is this quantity that is relevant in controlling dynamical processes related to surface tangential straining of the ocean surface. Recent laboratory studies have demonstrated that natural surfactants reduce gas transfer at low surface pressures and surface concentrations (Goldman *et al.*, 1988; Frew *et al.*, 1990).

The composition and concentration of the surfactant films at the surface of the ocean are subject to dynamic changes. Various biological, chemical and physical processes lead to alteration of film chemical composition, surface physical properties, surface concentration and spatial distributions. The chemical composition of microlayer films is believed to be controlled mainly by source contributions, but may also be modified by physical processes (Bock and Frew, 1993).

The composition and molecular arrangement of microlayer films can vary with changes in physical forcing, such as compressional–dilational forcing by wave motion, and wind and water circulation. Due to their complex composition and the possibility for relaxation processes to occur on a variety of timescales (van den Tempel and Lucassen-Reynders, 1983; Lucassen-Reynders, 1986), microlayer films exhibit a complex stress–strain relation in response to compression and dilation that is purely elastic (Bock and Frew, 1993). This response is quantified by the dilational viscoelastic modulus, a complex number that is useful in deriving dynamical relations at the air–sea interface (Levich, 1962; Hansen and Mann, 1964; Lucassen-Reynders and Lucassen, 1969; Bock and Mann, 1989; Fernandez *et al.*, 1992). This modulus is determined by chemical composition and concentration, and by relaxation processes. These relaxations take place as the result of shift of the local surface composition away from equilibrium, and include adsorption, desorption, micelle and solid phase formation, and molecular reorientation. In the case of surfactants with appreciable solubility in the bulk phase, the surfactant is capable of diffusional exchange between the interface and bulk phase. For these molecules, the dilational viscosity may be appreciable. Surfactant films comprising many different molecular species, when subject to long-time compression, relax differentially by the desorption of the more soluble species. The remaining film materials (the more hydrophobic ones) dominate viscoelastic response and generally exhibit enhanced damping capabilities (Bock and Frew, 1993).

As with the source terms, only general statements can be made concerning the sink terms leading to dissipation of surface films. Sink terms that represent loss of material at the surface include microbial degradation, photooxidation, micellization, or loss due to adsorption onto particulates. Some estimates of microbial turnover rates for amino acids have been reported (Williams *et al.*, 1986). Photooxidation effects can be estimated for coloured organic matter in the microlayer and are discussed in Report Group 3 (Chapter 3). However, overall residence times of material in the surface microlayer are still poorly understood because of the lack of knowledge about the magnitude of the source and sink terms.

Surface films and gas exchange

Surfactant films are effective in reducing the amplitude of short (capillary, ~1 cm wavelength) waves. Wind-waves, the result of wind stress, are thought to be generated in a rather limited range of frequencies (Hasselmann, 1960, 1962, 1963a,b; Hasselmann and Hasselmann, 1985), and nonlinear interactions and dissipative mechanisms redistribute wave energy into the resultant wave spectrum. The interaction between surfactants at the ocean surface and the wind stress result in the existence of a critical wind speed, below which waves either do not form or grow at a small rate. In the absence of surfactants, in carefully controlled laboratory conditions, waves have been observed to grow at wind speeds as low as 0.5 m s^{-1}. When surfactants are introduced into these clean systems, a critical wind speed is established. Figure 1.1 shows the mean square slope (a measure of waviness) as a function of wind speed. In the response of the surface to the onset of waves for the systems containing 0.03 ppm and 1.0 ppm concentrations of a water-soluble surfactant, an abrupt knee demonstrates the existence of a critical wind speed. For wind speeds below this critical value, waves increase at a very low rate in response to wind. Above this speed, waves increase more rapidly, but do not attain the same level as they did in the case of a clean surface (at least not in the range of these experiments). Measurements of the transfer velocity (the term which quantifies the kinetics of air–sea gas transfer) indicate that it too is lower than for the clean water case. It appears that the presence of surfactant produces an energy barrier to the formation of a robust wind–wave field.

On the ocean, variability in the wind field makes assessment of the situation more difficult. However, qualitative observations have been made that are consistent with the laboratory results. At low wind speeds, slicked areas are prevalent, particularly in coastal areas where biological productivity is high. As winds increase, these slicks rapidly become developing waves, often at wind speeds of ~3–4 m s^{-1}. It is suspected that this speed represents the critical wind speed. In this respect, the natural ocean appears to behave similarly to samples tested in tank studies. As

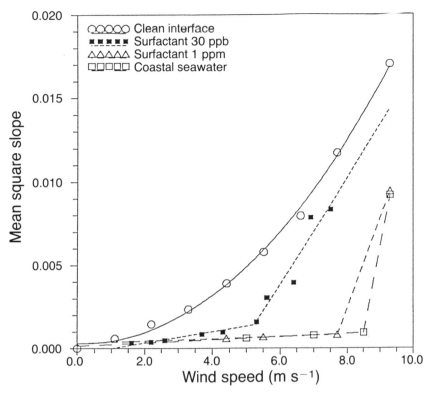

Figure 1.1. Mean square wave slope as a function of wind speed for clean and surfactant-influenced water surfaces (N. M. Frew, pers. commun.).

the wind continues to increase, wave breaking begins and whitecaps form. It is unclear what is happening to the surfactant material at the surface under these conditions, since mechanical continuity of the surface is broken in areas of plunging waves. More insoluble components may interact to produce micelles that are transported away from the surface, but very soluble components (already in solution and capable of read-sorption) may rapidly reestablish both the surface enrichment and a finite viscoelasticity. The bubble plumes generated by breaking waves may act as scavengers for the more soluble surfactants and help to promote their transport to the surface. Owing to these mechanisms of film renewal, it is likely that the more soluble components continue to influence gas exchange rates at wind speeds above the critical wind speed, as in the laboratory studies.

Surface films and bulk material

The material found in the sea-surface microlayer has properties which indicate that it contains only a small percentage of highly surface-active material, such as straight-chain lipids. The majority appears to be less surface active, with the molecules having both water-soluble and surface-active groups arranged on convoluted polymeric chains. In this it is structurally similar to much of the organic material found in bulk sea-water. There is certain to be continuous interchange of this material between the microlayer and bulk, under the action of turbulent mixing processes induced by wind, waves, currents, bubbles and tides. Support for this concept comes from the finding that if the surface material is removed, then the microlayer is reformed by material coming to the surface from the bulk. The speed of the renewal process is dependent on the degree of mixing in the water; the timescale being a few hours for stagnant conditions but a matter of minutes when the water is bubbled. In view of the existence in the bulk of material similar to that found near the surface, care needs to be exercised in interpreting results obtained from microlayer samplers which come into contact with sub-surface water as well as microlayer material.

Extent and viscoelastic properties of surface films

The extent of surface films on the ocean is difficult to quantify. Observations from space, in the form of the now well-known Scully-Power photographs (Scully-Power, 1986), have shown the widespread occurrence of filamentous and eddy-like structures representing differences in roughness of the ocean surface. These structures are believed to be slicked zones that result from local convergence areas that pack surfactants to concentrations large enough to suppress waves, making areas of the ocean reflect light specularly (i.e., as if they were perfectly smooth).

This effect was once considered to be restricted to areas in which oil production is high, and a direct result of seepage and production losses. The observation of these phenomena in areas like the New York Bight renewed interest in natural slicks and revitalized research in the surface chemistry of the world's oceans (e.g., *Sea Surface Microlayer* issue of *J. Geophys. Res.* 97:C4, 1992). No evidence exists that organized structures are common in open-ocean environments, but the conclusions reached above, that soluble surfactants quickly reestablish at the air–sea interface, suggest this possibility.

The probable range of the viscoelastic moduli on the ocean surface is between zero (for a clean, surfactant-free surface) and 50 mN m^{-1} (milliNewtons per metre). In visible slicks, spreading pressures can range from a few mN m^{-1} to tens of mN m^{-1}. In these cases, moduli can reach their highest values, but because of the relation between spreading pressure and viscoelastic modulus, higher pressure does not necessarily cause higher viscoelasticity. Large moduli (based on laboratory measurements and estimates obtained by Lombardini *et al.*, 1989) have values not exceeding 50 mN m^{-1}. To contrast this, areas not visibly slicked can have spreading pressures of the order of a few tenths of a mN m^{-1}, yet the viscoelasticity may be large enough to reduce significantly waves and gas transfer. The probability that regions of the ocean can actually have zero viscoelasticity is extremely low. It can be argued that the observed values measured globally for dissolved organic carbon imply that sufficient source material is generally available to obtain a surfactant-influenced surface.

Presently, it is not known how long-term environmental changes will affect the concentration and composition of the global ocean surface. It is reasonable to assume that the prevalence of oceanic films will closely track biological productivity, since this is a large source term. It may prove that the extent of film coverage on the ocean is a sensitive measure of productivity, including global mass flux. Because of this, it may be beneficial to develop the remote sensing capabilities described later which use multiple band backscatter, along with ground truthing methodologies, to create an ocean surfactant monitoring capability. This should be a realistic long-term goal.

Physical processes in the microlayer

Comparison of surface-renewal and boundary-layer models of near-surface transfer

As the sea-surface interface is approached, turbulence is damped. Within the diffusion sublayer, molecular transport dominates, while in the bulk of the water turbulence largely controls the transport. The concept of the viscous boundary layer explicitly describes the cooperation of molecular and turbulent transport in the transition between the two regimes, based on continuity arguments and observations. The surface-renewal model assumes that at regular or irregular intervals part of the surface layer is removed and replaced by water from the bulk. The significance of surface renewal can be seen by the exponent, n, of the Schmidt number (Sc, the ratio of viscosity to molecular diffusivity) in the equation for the transfer velocity (k). The value of n is $-1/2$ in the surface-renewal model and $-2/3$ in the boundary-layer model.

In wind–wave facilities, a clear correlation between the Schmidt number exponent and the mean square slope of the waves has been established (Jähne, 1985). With increasing mean square wave slope, the exponent gradually decreases from $-2/3$ to $-1/2$. However, no direct determination of the Schmidt number exponent from oceanic field experiments is available. For lakes, Watson et al. (1991) showed that $n = -1/2$. For practical purposes, it has been widely assumed that n is $-1/2$ for moderate and higher wind speeds and $-2/3$ for low wind speeds. As the laboratory measurements indicate, this simple assumption might be wrong in the case of a surfactant-influenced surface, since n seems to be closely related to the presence of waves that are suppressed by surfactants. The uncertainty in n is most critical – up to a factor of about two (see later) for typical gases – if gas transfer rates are determined from measurements of heat transfer.

To eliminate this uncertainty, it is necessary to estimate the amount of surface renewal occurring in the oceanic surface microlayer. By dyeing the sea surface, Gemmrich and Hasse (1992) were able to identify certain types of surface streaming under natural conditions that would induce surface renewal. The uncertainty in n would also be eliminated if a satisfactory theory of gas exchange were available.

Short capillary waves

One of the most striking phenomena observed in wind–wave facilities is the large enhancement of gas exchange with the onset of waves (Broecker et al., 1978; Jähne et al., 1979, 1984; McCready and Hanratty, 1984). Based on simple scale considerations, this effect was first attributed to short waves (Hasse and Liss, 1980; Coantic, 1986; Back and McCready, 1988). However, the experimental data from different wind–wave facilities taken by Jähne et al. (1987), do not support this hypothesis, but rather suggest that the gas exchange rate is better correlated to the mean square slope of the waves. Laboratory experiments from Frew (Chapter 5, this volume) show that this correlation still holds when various surfactants are present at the surface.

The strong correlation between mean square slope and the gas exchange rate is due to the fact that the mean square slope is a general measure of the nonlinearity, and thus instability, of the wave field. The detailed mechanisms are still not understood and require further detailed laboratory investigations, including simultaneous gas exchange, wave and turbulence measurements. Although there is no direct proof of the influence of waves on the air–sea gas exchange rate from direct field measurements, the effect manifests itself in the data indirectly. First, the field transfer velocities reach similar values at the same friction velocities as in wind–wave facilities. Such high transfer rates can only be brought into agreement with theoretical models if a higher turbulence level than for flow at a rough or smooth solid wall is assumed (Jähne, 1985). Second, the large variability of gas transfer rates found in the field is a clear indication that parameters other than the direct effect of waves are of importance. Variability of the wave field, which depends significantly on surface films and other parameters, is one possible explanation.

The relation between heat and mass transfer

Transfer of a gas such as CO_2 through the air–sea interface is affected by heat transfer in two distinct ways. First, the driving force for gas flux across the air layer adjacent to the surface contains a factor $\{(Q^*/RT)T'/T+P'/P\}$, where R is the ideal gas constant, T' and P' are the gradients of temperature and pressure across this air layer and Q^* is the heat of transport of the gas. The need for the term involving Q^* has been demonstrated for the wind-tunnel measurements of oxygen exchange by Liss *et al.* (1981). (In these calculations, the oxygen flux in the liquid was scaled to the total heat flux through the ratio of molecular and thermal diffusivities – another link between heat and matter transfer.) At a temperature of 290 K, the factor Q^*/RT varies from -2.2 for helium, through 4.0 for CO_2, to 14.2 for water vapour. The term involving Q^* becomes important when Q^*/RT is large and there is a significant temperature gradient at the interface. However, because over the majority of the oceans the difference between the bulk air and sea temperatures is generally small, this effect is invariably neglected by oceanographers.

The second way that mass transfer is affected by heat transfer is through the combined effect of the sensible, latent, and radiative heat fluxes on the surface temperature of the liquid. Matter transfer through the liquid layer immediately below the surface is controlled by the difference between the surface and bulk concentrations and by the liquid-phase diffusion coefficient, whose temperature dependence is small. The surface concentration is given by the product of the gas partial pressure adjacent to the surface and the gas solubility at the surface temperature. Further, correction of the air–sea flux of CO_2 for change in its solubility caused by this skin temperature deviation appears large enough to be globally significant.

In common with almost all gases, CO_2 has a solubility which decreases with increasing temperature. For CO_2 at 20 °C the solubility decreases by about 2.8% per degree. Typically, the surface p_{CO_2} is about 35 Pa and the deviation of the skin temperature from the bulk temperature is of the order of -0.3 °C, so surface p_{CO_2} values have to be adjusted by $35 \times 0.028 \times -0.3 = -0.3$ Pa. If applied to the ocean p_{CO_2} field globally, such a shift would be sufficient to increase the global air-to-sea flux by around 0.6 Gt $(10^{15}$ g) C y^{-1}, which is quite significant given that the global net flux is of the order of 2.0 Gt C y^{-1}.

A recognition of this correction does not contribute directly to solving the 'missing carbon' problem: the accepted figure for ocean uptake of 2 Gt C is based on ocean model uptake and is not sensitive to measured values of air–sea CO_2 fluxes. However, it does help to explain why attempts to calculate directly the air–sea flux over the globe by integrating the gas transfer flux always give numbers which are low compared with these models (Tans *et al.*, 1990). It is also important for identifying the distribution of sources and sinks of CO_2 in the ocean, which in turn is necessary in order to interpret correctly the distribution of atmospheric CO_2 measurements in terms of sources and sinks. This is currently the primary method for placing limits on the 'missing carbon' problem and in particular for deducing the size of the enigmatic 'land sink'.

The importance of the skin temperature deviation for the global CO_2 budget has been pointed out by Robertson and Watson (1992), who made a more detailed spatial calculation of the effect using the parameterization of Hasse (1971) combined with monthly mean data on global heat fluxes and wind speeds. The largest changes to the air–sea flux were found at high latitudes in the winter season and in regions of especially high sea-to-air heat flux, such as the Gulf Stream and Kuroshio Current. Globally the effect integrated to 0.7 Gt C per year. Refinement of this calculation using a more accurate representation of skin temperature deviation, in particular at high wind speeds, needs to be attempted.

Wave breaking and bubbles

The role of bubbles in heat flux

A latent heat flux between the atmosphere and ocean is associated with marine aerosol droplets. Most atmospheric seawater droplets are film drops or jet drops produced by bubbles bursting at the sea surface (Blanchard, 1963, 1983; Monahan, 1986; Woolf *et al.*, 1987). Large jet drops and large drops torn directly from the sea surface are most effective in humidity transfer. The influence of droplets has been estimated to be very large by some investigators (Ling and Kao, 1976). Participants in HEXOS, a comprehensive experimental programme directed at answering this question, concluded, however, that the net influence on humidity and latent heat transfer was very small. There remains considerable interest in this issue and some still argue that spray is significant (e.g., Ocampo-Torres and Donelan, 1994).

The role of bubbles in gas exchange

The entrainment of bubbles by breaking waves contributes to gas exchange at high wind speeds in wind–wave tunnels (Memery and Merlivat, 1983; Broecker and Siems, 1984). Asher *et al.* (1992) have shown in the laboratory that breaking waves greatly enhance gas transfer. It is well-established that deep bubble clouds are capable of significantly supersaturating very poorly soluble gases such as oxygen (Thorpe, 1982, 1986; Woolf and Thorpe, 1991; Wallace and Wirick, 1992). Farmer *et al.* (1993) have found very high rates of gas transfer associated with bubble clouds. Two numerical models (Keeling, 1993; Woolf, 1993) predict a considerable contribution of large, near-surface bubbles to the air–sea transfer velocity of gases, but uncertainties in the value for any gas are still large. Measurements of the distribution of bubbles very near the sea surface and of the mass transfer coefficients of individual bubbles in nature are necessary to improve the models. In addition to transfer via bubbles (bubble-mediated exchange) the generation of turbulence by surfacing bubbles may also enhance exchange directly across the sea surface (Monahan and Spillane, 1984).

The effect of surfactants on bubble-mediated gas exchange

A bubble may carry surfactant from the sea surface on formation, and it will scavenge material from the bulk as it rises (Scott, 1975). A bubble may be covered with material after rising only a few centimetres (Blanchard, 1983). A coating of material (changing the bubble from hydrodynamically 'clean' to 'dirty') has a very great effect on the transfer rate of gas between the bubble and the surrounding water, particularly for large bubbles. Woolf and Thorpe (1991) assumed that small bubbles would usually be dirty. On the other hand, Keeling (1993) and Woolf (1993) have argued that formulae for clean bubbles are more appropriate for large near-surface bubbles. This assumption is quite uncertain. Woolf (1993) estimated a contribution of 8.5 cm h^{-1} to the mean global transfer velocity of carbon dioxide from bubbles if they were clean, but only 2.6 cm h^{-1} if the bubbles were dirty. Thus, the contribution of bubbles to air–sea gas exchange is sensitive to surfactants, and might respond significantly to changes in the concentration of these materials.

The effects of wave breaking on microlayer composition

The composition of the surface microlayer must be partly determined by the efficiency with which physical processes remove various materials from the sea surface and likewise supply new materials (MacIntyre, 1974). Simple turbulence cannot remove buoyant or bound material from the surface. Removal must depend on surface renewal (which is related to microscale breaking) and the especially vigorous overturning associated with wave breaking. The most highly surface-active materials will be the most difficult to remove. The production of bubbles involves the folding of part of the sea surface into the surface of bubbles and is presumably an irresistible removal process. Most of the larger bubbles will eventually return to the sea surface and burst (Woolf and Monahan, 1988). While below the surface, a bubble will scavenge material from the surrounding water (Blanchard, 1975; Wallace and Duce, 1978). The transfer of particulate material between a bubble and the surrounding water depends on the hydrodynamic flow around the bubble, the size of the particle and its adsorption properties. A bubble may become saturated with material, in which case globules of material will fall from the bottom of the bubble. This should be a fractionating process, with the most surface-active materials remaining on the bubble. When a bubble bursts, material originally on the bubble or on the sea surface may be ejected on the aerosol droplets generated (see below). Some of the aerosol droplets will quickly fall back to the sea surface. The paradigm of the effect of wave breaking on microlayer composition should be one of renewal rather than simply destruction. In addition to vertical movement of material, wave breaking will also affect the surface dispersion of materials in the microlayer. The compaction of materials associated with strong surface convergences will be superimposed on that associated with Langmuir cells (Sutcliffe *et al.*, 1963) and small-scale surface streaming.

Bubble floatation and aerosol formation processes

When bubbles break at the surface of the ocean they skim off part of the surface microlayer to produce atmospheric film and jet droplets. In addition to providing an important mechanism for charge separation and electrification of the atmosphere (Blanchard, 1963), this process results in the injection of material enriched by contributions from both the sea-surface microlayer before the bubble arrives at the surface and material from the bubble skin itself. Thus, aerosol particles often have concentrations significantly higher than the bulk composition of the seawater from which they were generated. This process enables materials, including pollutants, in one region of the ocean to be effectively transported horizontally via the atmosphere far from their marine source region. For example, a number of studies have shown that microorganisms such as bacteria and viruses can be enriched on aerosol particles produced by bursting bubbles, with subsequent atmospheric transport significant distances from the source area (Baylor *et al.*, 1977; Blanchard and Syzdek, 1974, 1982; Aubert, 1974; Gruft *et al.*, 1981). Similar results have been found for radionuclides (Walker *et al.*, 1986), many organic substances, trace metals (Weisel *et al.*, 1984), and irritants derived from dinoflagellate (red tide) blooms (Woodcock, 1948).

The effect of rain on exchange processes

Rain spectra show an exponential form, with an abundance of small droplets and a few large drops. Most of the mass flux is with the typical rain drops of 1 mm to 3 mm diameter – depending on the type of rain. Typical rain drops penetrate the microlayer. The formation of a stable layer of reduced salinity in the uppermost few metres of the ocean mixed layer has been reported. Rain is also known to calm the seas. It has been observed that primarily the smaller waves are damped by rain, producing a smoother appearance at the surface. It is also well known in hydrodynamics that rain drops striking the water surface at rest produce circular waves propagating outwards, as well as splash drops. Rain drops are expected to have a temperature equal to the wet bulb temperature of the air in the atmospheric boundary layer. Rain is known to be an important pathway for material transport from the atmosphere to the oceans (wet deposition, GESAMP, 1980), but this process is not dealt with in this report.

Since we lack direct measurements of the effect of rain on the microlayer, we must work from known features. The slowly falling fraction of small rain droplets below ~1 mm could form a thin film of lower density on top of the microlayer that is slowly mixed with the saline water below. Because of the usually cooler skin present on the ocean surface, the cooling effect of raindrops being added to the surface layer would not drastically change the temperature. The influence of salinity change on the properties of surfactants in the microlayer is deemed less important, but needs experimental verification. For the transport of chemicals from the atmosphere to the ocean it is probably less important whether these materials are deposited directly with the small droplet fraction of rain on top of the microlayer or with the primary mass flux by rain in the upper part of the mixed layer.

The primary dynamic effect of rain on the microlayer is the production of turbulence or secondary motions. The passage of rain drops through the microlayer disturbs the mean flow and leads to secondary motions that may well be a local source of turbulence in the microlayer. However, the mechanism of turbulence formation has not yet been identified. Whether secondary motion or turbulence, both effects would lead to localized surface renewal and similarly enhanced gas transfer. Rain drops cover only a small fraction of the surface area at a given time during rain, and at any single location it is raining only a small fraction of the time. Thus, the net effect of rain on gas exchange is usually believed to be negligible, but this requires confirmation.

Damping of small waves by heavy rain has been known for centuries. This effect need not influence gas exchange directly. However, it is likely to reduce the rate of wave dissipation through wave breaking. This implies less secondary motion, less surface renewal and therefore less effective gas exchange. Again, the direct effect on gas exchange will probably not be important. There is an indirect effect of small-scale wave damping by heavy rain implied by the parameterization of gas exchange in terms of mean square wave slope (see earlier). The error in global estimates again is likely to be small, and satellite microwave remote sensing during heavy rain is doubtful anyway. There are other physical effects of rain at the sea surface (e.g., momentum transfer) but they are not expected to have noticeable influence on gas transfer.

Horizontal transport and deposition of surface slicks in coastal zones

Surface slicks containing enhanced levels of organic and inorganic substances and biota can be moved by currents and winds such that they become grounded on tidal flats and other shallow water areas. This route is clearly one by which such areas could receive additional amounts of a variety of substances. The relative importance of this route will depend on what other sources (e.g., from adjacent rivers and the atmosphere) of the substances exist, and this is likely to be site specific. Thus, it is difficult to make other than very general statements relative to the possible importance of this transport. Nonetheless, the scums and foams formed by wind effects on microlayer films are very conspicuous in littoral areas and are often identified as 'pollution' by the general public.

Review of experimental data on gas transfer

Gas transfer velocities

Gas transfer velocities at the surface of the ocean are derived from the application of three methods:

(i) inventories of natural and bomb-produced $^{14}CO_2$ in the ocean,
(ii) radon deficiencies in the surface-layer,
(iii) the dual-tracer technique.

Figure 1.2 shows all oceanic gas exchange data published before mid 1993. It includes two values utilizing $^{14}CO_2$, 20 utilizing radon, and four utilizing the dual-tracer method. Arguably, the ^{14}C data points should not be compared directly with the radon or SF_6–^3He data as they correspond to processes which take place on much longer timescales (order of several years), whereas the radon and SF_6–^3He results are adapted to timescales of the same order (i.e., several days). Line 1 represents the Liss and Merlivat (1986) relationship which, initially normalized for lake studies, appears to fit the oceanic measurements made with radon and SF_6–^3He. Line 2 represents the same relationship after normalization to ^{14}C and accounting for the nonlinear nature of the transfer velocity/wind speed relationship. There is a clear discrepancy between the two curves. A scaling factor of 1.6 must be introduced if the two curves are to be reconciled (Merlivat *et al.*, 1993).

Line 3 represents the equation proposed by Wanninkhof (1992) for steady winds (i.e., valid for ^{14}C data points). Lines 2 and 3 necessarily coincide for $U = 7.6$ m s^{-1}, as they go through the same ^{14}C data point. Wanninkhof has preferred for the sake of simplicity to use a quadratic relationship for transfer velocity–wind speed. This artificially introduces an increase in the transfer velocities at high wind speeds.

Nightingale *et al.* (1996) have shown 11 new results from experiments

Report Group 1

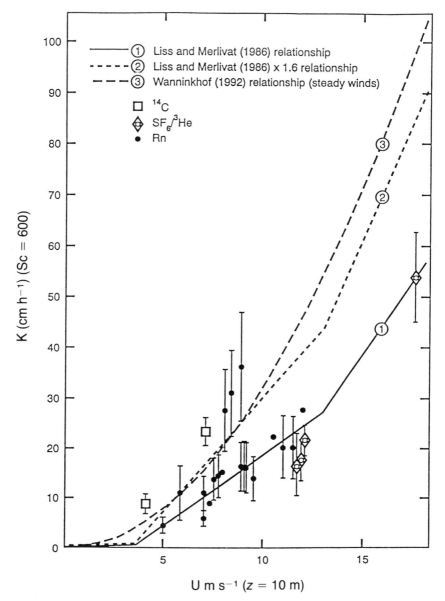

Figure 1.2. Gas exchange measurements in the ocean and transfer velocity (K)/ wind speed $(U$, at a reference height (z) of 10 cm) relationships (after L. M. Merlivat, pers. commun.).

made using the dual-tracer technique in the North Sea. They confirm their previous data and fit the Liss and Merlivat relationship. Wanninkhof *et al.* (1993) have reported four results measured by the dual-tracer technique on Georges Bank. In their analysis the results are above line 3. However, reanalysis of their raw data brings the results closer to the Liss and Merlivat relationship. This point is described in more detail on page 19.

Boutin and Etcheto (1994, 1996) have recently made detailed comparisons of colocated Special Sensor Microwave/Imager (SSM/I) wind speeds and those measured on board ship. They identified large biases (underestimations) for the SSM/I under high winds. After applying a correction to suppress this bias, they have recalculated the mean global exchange coefficient using the Liss and Merlivat relationship. Using their new value, the ratio with the ^{14}C data is now 1.4 instead of 1.6.

Discrepancy between ^{14}C-based and inert gas exchange rates

The entire set of experimental results reported in the literature since 1991 allows us to conclude that there is a discrepancy between exchange rates based on ^{14}C and those measured using inert gases of the order of 40%. This should not be confused with scatter in the experimental data; there appears to be a systematic difference between the results for ^{14}C and other gases. However, it should be noted that the discrepancy may not be as large as 40%, since a recent report (Hesshaimer *et al.*, 1994) suggests that the previous estimates of the ocean uptake of ^{14}C may have to be revised downwards.

Taking into consideration all the knowledge that has been acquired in making the measurements at sea, there is a high priority to design experiments dedicated to understanding the behaviour of these different gases. These experiments could involve either laboratory measurements using natural seawater or field experiments. We have not discussed the possible effect of the surfactant-influenced sea surface; for example could it affect CO_2, SF_6, [3]He and radon differently? This effect, if important, must be identified in the existing data sets.

Do catalysts exist in the microlayer which may enhance CO_2 gas exchange?

A possibility not yet considered here is that the above discrepancy between CO_2 and inert gases is due to specific reactions of CO_2 occurring in the microlayer. Carbon dioxide reacts with water molecules on a timescale of the order of 1 minute in seawater to form H_2CO_3, which in turn very rapidly dissociates into bicarbonate. Because of the rather slow hydration of CO_2, it is normally possible to treat dissolved CO_2 as an inert gas for the process of gas exchange, and to ignore the presence of the bicarbonate in the microlayer. However, the enzyme carbonic anhydrase catalyses the hydration of CO_2; if this were present in sufficient quantities in the microlayer, the effect would be to increase the effective rate of gas exchange by making some of the bicarbonate available to transport CO_2 across the diffusion sublayer.

This idea was first suggested by Berger and Libby (1969). Subsequent studies (Goldman and Dennett, 1983) found no evidence for the presence of sufficient quantities of carbonic anhydrase in bulk seawater, but to our knowledge there have been no studies which have looked for this substance specifically in the microlayer. Given the large enrichments of biological activity sometimes observed there, it remains a possibility that warrants investigation.

Review of experimental techniques

'New' techniques for measuring air–sea gas exchange rates

Until quite recently methods used at sea for measuring gas exchange rates have been very crude. For example, the enclosure of a small area of sea surface by a cover and measurement of changes in gas concentration in it over a period of time suffers from the major problem that air and water motions are severely disrupted. More satisfactory methods have been tried recently, and further developments of them are on the horizon.

Dual tracer technique

One group of techniques uses micrometeorological methods such as eddy correlation (which was developed for the measurement of terrestrial fluxes) and applies them to the generally smaller fluxes found over the sea. To date the results have been disappointing in the sense that the fluxes appear at least an order of magnitude larger than are expected or explicable given present understanding. The primary problem is that the flux being measured (generally CO_2) is at the sensitivity limit of the equipment, because a large correction has to be applied for the interfering effects of water vapour. Current developments of the technique involving conditional sampling and eddy accumulation approaches hold promise for an improvement in this situation.

Another technique for measuring gas exchange rates is a recently introduced experimental procedure in which two inert gas tracers are released into a region of the sea (Watson *et al.*, 1991). Measurements of the ratio of the two tracers are then used to deduce the gas exchange rate as a function of time. The advantage of the technique is that measurements can be made at sea over times which are comparatively short (hours to days) so that it is unnecessary to average over widely different environmental conditions.

Ideally, one of the tracers should be completely non-volatile, in which case the measurements would enable the direct calculation of the gas exchange rate of the other one. In practice, there presently is no adequate non-volatile tracer (but a purposely added strain of bacteria may be a possible candidate – Nightingale *et al.*, 1996) so the two tracers used have been ^{3}He and sulphur hexafluoride, both of which are volatile though with very different diffusivities. When using two volatile tracers, the derivation of absolute values for gas transfer rates requires that the ratio of the transfer rates of the two gases be known. In the unbroken clean surface regime there is general agreement that this ratio can be calculated using an Sc^n where $n = -0.5$. However, when spray and bubbles are present there is reason to expect that the component of the gas transfer carried by bubbles will not follow the same rule, but as yet there is no general agreement on what relationship they will follow. For example, Watson *et al.* (1991) reduced their data using a fixed ratio between the transfer velocities of the two tracers corresponding to Sc^n with $n = -0.5$, whereas Wanninkhof *et al.* (1993) used a variable-n formulation based on measurements made by Asher *et al.* (1992) in a whitecap simulation tank.

If the $n = -0.5$ relationship breaks down due to the effect of bubbles, two types of error are potentially introduced into the procedure (Woolf, 1993). First, the reduction to a Schmidt number of 600 (corresponding to CO_2 at 20 °C), commonly used to display and compare results, would be incorrect. If the effect of bubbles is to reduce n, the effect would be to make the results too low if the $n = -0.5$ relationship is used. However, the application of the results to more soluble gases such as CO_2 would additionally be incorrect because the contribution of bubbles to gas exchange is greater for less soluble gases, so the effect of bubbles would be overestimated. The net effect of these two errors appears to be that CO_2 exchange rates tend to be overestimated by the method (Merlivat *et al.*, 1993). In fact these difficulties are not exclusive to the dual tracer method. They apply to the derivation of CO_2 exchange rates from those of other gases regardless of the experimental method, and are equally relevant to the results from the radon deficit technique, for example.

Asher *et al.* (unpublished results) have made observations of the effect of solubility on gas exchange in the presence of foam, using the same tank apparatus mentioned above. The experimental effect of solubility appears to be considerably less than the model of Memery and Merlivat (1983) predicts. Clearly, gas exchange measurements made using inert gases must for the present be interpreted with caution, particularly if they are to be used to predict CO_2 exchange. Further investigation of the effect of whitecapping on gas exchange, both in laboratory and field situations, is required. However, in the absence of a reliable method for directly measuring the exchange of CO_2, the dual tracer technique remains one of the few ways of obtaining field data with which to constrain models of gas exchange at high wind speeds.

How can we deal with the large spatial variability of p_{CO_2}?

The gas flux through the interface is given as the product of the transfer velocity and the pressure difference. The values of the air–sea p_{CO_2} concentration difference, Δp_{CO_2}, at the surface of the ocean show a very high spatial and temporal variability. This is documented either from ship results sailing across the oceans or a few time series such as those at station P or the North Atlantic JGOFS site (Wong, 1991; Robertson *et al.* 1993).

The problem is to know how to average instantaneous values of the transfer velocity and Δp_{CO_2} to obtain their corresponding mean values. This is important as the product of the means is not equal to the mean of the products. All existing studies have shown that at a given site, the most variable term is the wind speed driven transfer velocity. For this reason, the time to consider is imposed by the temporal variability of the concentration difference. However, it must be remembered that the exchange coefficient must be computed from non-averaged wind speed data due to the nonlinear character of the transfer velocity/wind speed relationship.

Controlled flux technique

Another very promising technique uses heat as a proxy tracer and is based on an entirely new concept. While conventional mass balance methods derive the flux density from a temporal change of the concentration of dissolved gases, this new technique forces a controlled flux density across the interface and infers the transfer velocity by simply measuring the surface concentration difference related to a certain change in the heat flux density. This technique has been successfully used in wind–wave facilities (Jähne *et al.*, 1989) using a mechanically chopped IR radiator to force a heat flux density across the interface and an IR radiometer to measure the skin temperature variations. In the field, this technique was used for the first time on the west coast of the United States during the US Marine Boundary Layer Project in spring 1995, when a CO_2 laser was used as a heat source with an IR camera to image and track the heated spots. First results from this study can be found in Haussecker and Jähne (1995).

This new technique has the tremendous advantage that it gives a local measurement with a high temporal resolution of only several minutes. Thus, it is possible to achieve wide coverage with the technique, e.g., to traverse slicks, and to perform a direct parameterization with friction velocity, wave parameters, and properties of the surfactant-influenced surface such as the viscoelastic modulus. It would be especially valuable to use the technique together with the dual-tracer technique. While the latter appears to overestimate the CO_2 gas exchange rate (Merlivat *et al.*, 1993), the controlled flux technique underestimates it by not including bubble-mediated transfer.

Are direct measurements of the heat flux possible?

Turbulent and longwave radiative heat fluxes have to be taken over by molecular and turbulent heat conduction through the oceanic thermal boundary layer. Thus, a temperature gradient across the thermal sub-layer is established. Right at the surface, the gradient is given by molecular conduction only. Thus, the heat flux can be estimated from the temperature gradient at the surface. McAlister and McLeish (1969) proposed a method utilizing multispectral infrared measurements that retrieve information from different depths of the thermal sublayer. The penetration depths of radiation at wavelengths of 3.7, 4.0, 8.3, 10, and 20 micrometres amount to 80, 65, 20, 15, and 4 micrometres, respectively (e-folding depths).

Together with an accurate calibration that accounts for the non-blackness of the sea surface (Grassl and Hinzpeter, 1975; Schlüssel *et al.*, 1990), it appears to be feasible to measure temperature gradients and thus the heat flux through the thermal sublayer. Given the significant progress in infrared technology, utilization of this technique appears worthwhile. Mammen and von Bosse (1990) utilized a buoyant pop-up profiler carrying a fast thermometer that measures the temperature profile just beneath the sea surface. While resolving parts of the entire depth of the microlayer, their technique could be used to elucidate heat transfer processes in the thermal sublayer in greater detail.

Novel ocean-surface sampling techniques

State-of-the-art microlayer sampling is defined by the rotating drum sampler (Hardy *et al.*, 1988; Carlson *et al.*, 1988). Microlayer samples are lifted from the ocean surface as an aqueous layer adhering to the surface of a rotating, partially submerged drum. Samples are then removed for *ex situ* study and analysis. Current understanding of microlayer chemical structure is largely based on samples obtained with this system. Questions exist, however, about the method. Does bulk water adhering to the drum dilute the microlayer sample and are there bulk material contributions? In addition, rotating drum samplers may not exhibit the real-time concentration sensitivity necessary to characterize chemically the complete range of surfactant-influenced surfaces. (Surfactant-influenced surfaces range from heavily slicked areas of ocean where surfactants can be highly compacted surface films to surfaces on which surfactants might be compared with highly expanded monolayers.) Based on existing knowledge of microlayer composition (Frew and Nelson, 1992a, b), some *ex situ* separation and analysis may always be necessary in surfactant characterization.

A solution is the application of nonintrusive *in situ* measurement methods which could be used in conjunction with existing sampling techniques. One group of possible techniques is optical sampling methods, including nonlinear laser spectroscopy. Surface selective methods such as reflected second harmonic generation and reflected sum frequency generation have gained acceptance as laboratory surface probes (Shen, 1989). These interfacial spectroscopic probes have also been demonstrated to function as *in situ* probes for studying surfactant-influenced ocean surfaces (Korenowski *et al.* 1989, 1993; Frysinger *et al.*, 1992). Korenowski *et al.* (1989) and Frysinger *et al.* (1992) demonstrated that it may be possible to use second harmonic generation as a true remote sensing probe of sea-surface chemistry. Other laser-based methods such as reflected four-wave mixing spectroscopy are yet to be explored as *in situ* sea-surface probes. Development and research into these methods could result in a host of *in situ* surface spectroscopic measurement techniques which are nonintrusive and sensitive to chemical composition (through functional groups) and physical structure.

Parameterization of air–sea exchange processes

Best parameterization for momentum, heat, and material fluxes

In the air–sea interaction community, parameterizations for momentum, heat and material fluxes are commonly formulated in terms of surface layer variables using the so-called bulk aerodynamic method and bulk transfer coefficients (drag coefficients, Stanton and Dalton numbers). Different formulations of the variation of the neutral drag coefficient with wind speed are available from different experiments but, for practical purposes, the parameterization for the drag coefficient by Large and Pond (1981) is recommended. Parameterization for water vapour flux is more difficult. The best known parameterization is given by DeCosmo *et al.* (1996). The parameterization for sensible heat is less well established. It is reasonable to use the bulk transfer coefficients for water vapour for heat too, or a slightly smaller coefficient (say by 8%).

The effect of surfactants on the transfer of momentum, sensible heat and water vapour is not explicitly formulated in these parameterizations. However, it is understood that the recommended parameterizations are obtained from measurements in the oceans under natural conditions. Hence, they should contain a possible influence of surfactants in the natural mix found in the sea. Part of the experimental scatter in the determination of bulk transfer coefficients could be from the varying influence of surfactants. At present, the accuracy of direct flux measurements is not good enough to determine such an influence.

Parameterizations including viscoelasticity of the air–sea interface

Capillary and capillary–gravity wave spectra are attenuated by the presence of surfactants at the air–sea interface (Lombardini *et al.*, 1989; Bock and Frew, 1993). The mechanism by which surfactants damp waves is quantified through a description of the surface response function. Included in this description is the modulus of dilational viscoelasticity, which is a complex quantity that relates the stress and strain exhibited at air–liquid interfaces for small expansions and compressions of surface area. While experimental measurements can be obtained from wave propagation characteristics, quantification of the viscoelastic modulus is more difficult from *in situ* observation. An explanation of the equilibrium wave fields that exist as a result of wind stress has been put forward (Hasselmann, 1960, 1962, 1963a,b; Hasselmann and Hasselmann, 1985). In this formulation, the resulting equilibrium wave spectrum is the result of three competing mechanisms, namely energy input into waves from wind, energy distribution between wave numbers by nonlinear inter-actions, and dissipation of energy by linear wave damping and other dissipative processes. Hühnerfuss *et al.* (1987) have argued that the influence of surface films is limited to the damping enhancements described above. They suggested that an estimate of the viscoelastic modulus can be obtained from *in situ* measurements of wave spectra made in the presence and absence of surfactants under the same wind forcing. The ratio of these spectra is supposed to represent the damping ratio (the computed damping coefficient for a surfactant-influenced sea surface divided by the computed damping coefficient for a clean sea surface). By performing a fit of experimental spectra ratios to computed damping ratios, an appropriate estimate of the viscoelastic modulus can be attained. Lombardini *et al.* (1989) report experimental results obtained with a microwave height gauge for wave height spectra. Their findings indicate moduli representative of both soluble and insoluble surfaces, but the scatter in their data is wide. Some of this scatter is probably a result of environmental variability, but some of it is undoubtedly due to the lack of rigour in the underlying assumption.

Recent experiments (Bock *et al.*, 1996) have demonstrated that the ratio of slope spectra of wind driven waves with and without the presence of surfactants is not the same as the damping ratio, obtained from mechanically generated waves. This finding confirms that the effects of surface films influence not only the dissipation mechanisms of the Hasselman balance, but contribute to the other terms as well. This result indicates that no simple technique exists to derive the viscoelastic modulus from spectral data. Despite this, a parameterization that interrelates remotely sensed radar backscatter, transfer velocity, spectral levels, and viscoelastic modulus would be helpful for inclusion into global geochemical models. Such a parameterization would be facilitated by direct *in situ* measurements of capillary and capillary–gravity wave spectra and transfer velocity, along with images of radar backscatter obtained simultaneously. These data, compared with wave spectra obtained under controlled conditions, with coincident measurement of the viscoelastic modulus from mechanical wave experiments, may allow the formulation of an empirical descriptor of short-wave spectra. This formulation would result in a functional relation of spectral variance to both wind stress and viscoelastic modulus. Since Bragg scattering theory has been shown to work well for intermediate angles of incidence, remote sensing from space-borne satellites can be used to infer spectral levels of short waves. Using multiple wavelength radar, it may be possible to infer the shape of the spectrum. If this turns out to be the case, intercomparison of spectra obtained in the field with spectra on surfaces that have been characterized under controlled conditions (as described above) and with remotely sensed spectra estimates will allow parameterization of a functional model of transfer velocity as a function of viscoelasticity. Laboratory wind–wave tank results have indicated a strong linear correlation between mean square slope and transfer velocity (Jähne *et al.*, 1987; Bock and Frew, 1993). If this correlation applies to *in situ* wave fields, a direct estimate of transfer velocity using remotely sensed backscatter should be possible, owing to (among other validations obtained in wind-wave tanks) the recent *in situ* validation of Bragg scattering by direct wave spectrum measurements (Bock and Hara, 1993; Hara *et al.*, 1994).

A correlation between backscatter and transfer velocity using data collected in a wind–wave tunnel has been developed by Wanninkhof and Bliven (1991). This demonstrates, in principle, for one surface condition that the backscatter/transfer velocity correlation proposed here is possible.

A method for measuring transfer velocities at high wind speeds from a satellite-measurable parameter could be possible through passive microwave radiometry. Asher *et al.* (1995a,b) have developed a parameterization of the transfer velocity in terms of the microwave brightness temperature of the ocean surface. They showed that the increase in brightness temperature correlated with a 10% increase in the transfer velocity could be resolved by currently available microwave radiometers. This would provide a method for estimating transfer velocities at high wind speeds.

Conclusions and recommendations

There is increasing evidence for the importance of surface films in the transfer of mass, heat and momentum across the air–sea interface. However, in order to aid interpretation of the results of field experiments and to allow extrapolation of laboratory data to describe real ocean phenomena, we need to have a better understanding of the physics of near-surface transport mechanisms and the effect of films on them. The viscoelastic modulus appears to be the most relevant parameter to characterize the films' ability to modulate these transport processes.

The present global distribution of surfactants capable of affecting exchange processes is largely unknown, as are the factors controlling future distributions. Further, it is important to ascertain what fraction of the materials are of anthropogenic versus biogenic origin, both with respect to coastal and open ocean areas. For relatively unpolluted areas, it is reasonable to assume that marine biological productivity will give a good first-order estimate of the extent of films, since it is clearly a large source term. If this is so, then remote sensing techniques, such as multiple band backscatter, may be useful in establishing a monitoring capability for ocean surfactants

With respect to air–sea gas exchange, effort needs to be devoted to several important uncertainties related to the microlayer. These include the importance of the cool skin effect particularly for CO_2 exchange, the role of bubbles as gas exchangers and more specifically differences between the exchange properties of clean versus dirty bubbles, and the possible role of carbonic anhydrase in enhancing uptake of CO_2. Direct methods, adapted from existing micrometerological or other techniques, capable of measuring air–sea gas fluxes *in situ* are urgently needed. Due to lack of knowledge concerning surfactant distributions, attempts at estimating the effect of films on current global gas fluxes across the sea surface are subject to great uncertainty, as discussed by Asher (Chapter 8, this volume).

The research in the above directions should be pursued via both field and laboratory experiments. Future field experiments need to be performed that include integrated measurements of, *inter alia*, wind stress, heat flux, bubble concentrations, wave spectra and breaking wave statistics, viscoelasticity of films and their biological properties, multi-tracer gas transfer, and a variety of remote sensing approaches. Laboratory experiments are necessary to elucidate fundamental physical and chemical phenomena, and they are necessary for interpretation of results obtained in the field.

References

Allan, G. G., Lewin, L. and Johnson, P. G. (1972). Marine polymers. IV. Diatom polysaccharides. *Bot. Mar.*, **25**, 102–8.

Asher, W. E., Farley, P. J., Wanninkhof, R., Monahan, E. C. and Bates, T. S. (1992). Laboratory and field measurements concerning the correlation of fractional area whitecap coverage with air/sea gas transport. In *Precipitation Scavenging and Atmosphere–Surface Exchange. Vol. 2. The Semonin Volume: Atmosphere–Surface Exchange Processes*, ed. S. E. Schwartz and W. G. N. Slinn, pp. 815–28. Washington DC: Hemisphere.

Asher, W. E., Wanninkhof, R. and Monahan, E. C. (1995a). Correlating whitecap coverage with air-sea gas transfer velocities. *J. Geophys. Res.* (in press)

Asher, W. E., Smith, P. M., Monahan, E. C. and Wang, Q. (1995b). Estimation of air–sea gas transfer velocities from apparent microwave brightness temperature. In *Proceedings of the Third Thematic Conference: Remote Sensing for Marine and Coastal Environments, Volume 1*, pp. 104–115. Ann Arbor MI: Environmental Research Institute of Michigan.

Aubert, J. (1974). Les aerosols marins, vecteurs de microorganismes. *J. Rech. Atmos.*, **8**, 541–54.

Back, D. D. and McCready, M. J. (1988). Effect of small-wavelength waves on gas transfer across the ocean surface. *J. Geophys. Res.*, **93**, 5143–52.

Baylor, E. R., Baylor, M. B., Blanchard, D. C., Syzdek, L. D. and Appel, C. (1977). Virus transfer from surf to wind. *Science*, **198**, 575–80.

Berger, R. and Libby, W. F. (1969). Equilibration of atmospheric carbon dioxide with seawater: possible enzymatic control of the rate. *Science*, **164**, 1395–7.

Blanchard, D. C. (1963). The electrification of the atmosphere by particles from bubbles in the sea. *Prog. Oceanogr.*, **1**, 71–202.

Blanchard, D. C. (1975). Bubble scavenging and the water-to-air transfer of organic material in the sea. In *Applied Chemistry of Protein Interfaces*, ed. R. E. Baier, pp. 360–87. ACS Adv. Chem. Series, **145**.

Blanchard, D. C. (1983). The production, distribution and bacterial enrichment of the sea-salt aerosol. In *Air–Sea Exchange of Gases and Particles*, ed. P. S. Liss and W. G. N. Slinn, pp. 407–54. Dordrecht, Holland: Kluwer Academic Publishers.

Blanchard, D. C. and Syzdek, L. D. (1974). Importance of bubble scavenging in the water-to-air transfer of organic material and bacteria. *J. Rech. Atmos.*, **8**, 529–40.

Blanchard, D. C. and Syzdek, L. D. (1982). Water-to-air transfer and enrichment of bacteria in drops from bursting bubbles. *Appl. Environ. Microbiol.*, **43**, 1001–5.

Bock, E. J. and Frew, N. M. (1993). Static and dynamic response of natural multicomponent oceanic surface films to compression and dilation: laboratory and field observations. *J. Geophys. Res.*, **98**, 14 599–617.

Bock, E. J. and Hara, T. (1993). Optical measurements of ripples using a scanning laser slope gauge. *J. Atmos. Ocean. Technol.*, **12**, 395–403.

Bock, E. J. and Mann, J. A. Jr (1989). On ripple dynamics. II. A corrected dispersion relation for surface waves in the presence of surface elasticity. *J. Colloid Interface Sci.*, **129**, 501–5.

Bock, E. J., Hara, T. and Donelan, M. (1996). Wind tunnel measurements of spatial wave spectra as a function of wind stress and visco-elastic modulus. *J. Geophys. Res.*, in preparation.

Boutin, J. and Etcheto, J. (1994). Comparison of ERS1 and ship wind speed. In *Proceedings of 2nd ERS1 Workshop, IFREMER, Brest, March 1994. Tech. Rep. 94–06*, ed. K. B. Katsaros, pp. 145–52. Brest: IFREMER.

Boutin, J. and Etcheto, J. (1996). Consistency of GEOSAT, SSM/I and ERS1 global surface wind speeds; comparison with in-situ data. *J. Atmos. Oceanic Technol.*, **13**, 183–97.

Broecker, H. C. and Siems, W. (1984). The role of bubbles for gas transfer from water to air at higher windspeeds: experiments in the wind–wave facility in Hamburg. In *Gas Transfer at Water Surfaces*, ed. W. Brutsaert and G. H. Jirka, pp. 229–36. Dordrecht, Holland: Kluwer Academic Publishers.

Broecker, H. C., Petermann, J. and Siems, W. (1978). The influence of wind on CO_2-exchange in a wind–wave tunnel, including the effects of mono-layers. *J. Mar. Res.*, **36**, 595–610.

Carlson, D. J., Cantey, L. L. and Cullen, J. J. (1988). Description of and results from a new surface microlayer sampling device. *Deep-Sea Res.*, **35A**, 1205–13.

Coantic, M. (1986). A model of gas transfer across air-water interfaces with capillary waves. *J. Geophys. Res.*, **91**, 3925–43.

DeCosmo, J., Katsaros, K. B., Smith, S. D., Anderson, R. J., Oost, W. A., Bumke, K. and Chadwick, K. (1996). Air–sea exchange of water vapor and sensible heat: the humidity exchange over the sea. *J. Geophys. Res.*, **101**, 12001–12016.

Farmer, D. M., McNeil, C. L. and Johnson, B. D. (1993). Evidence for the importance of bubbles in increasing air–sea gas flux. *Nature*, **361**, 620–3.

Fernandez, D. M., Vesecky, J. F., Napolitano, D. J., Khuri-Yakub, B. T. and Mann, J. A. (1992). Computation of ripple wave parameters: a comparison of methods. *J. Geophys. Res.*, **97**, 5207–13.

Frew, N. M. and Nelson, R. K. (1992a). Isolation of marine microlayer film surfactants for *ex situ* study of their surface physical and chemical pro-perties. *J. Geophys. Res.*, **97**, 5281–90.

Frew, N. M. and Nelson, R. K. (1992b). Scaling of marine microlayer film surface pressure–area isotherms using chemical attributes. *J. Geophys. Res.*, **97**, 5291–300.

Frew, N. M., Goldman, J. C., Dennett, M. R. and Johnson, A. S. (1990). Impact of phytoplankton-generated surfactants on air–sea gas exchange. *J. Geophys. Res.*, **95**, 3337–52.

Frysinger, G. S., Asher, W. E., Korenowski, G. M., Barger, W. R., Klusty, M. A., Frew, N. M. and Nelson, R. K. (1992). Study of ocean slicks by nonlinear laser processes. 1. Second harmonic generation. *J. Geophys. Res.*, **97**, 5253–69.

Gemmrich, J. and Hasse, L. (1992). Small-scale surface streaming under natural conditions as effective in air–sea gas exchange. *Tellus*, **44B**, 150–9.

GESAMP (1980). *Interchange of Pollutants between the Atmosphere and the Oceans*. Report and Studies Series No. 13, 55 pp. Geneva: WMO.

Gillam, A. H. and Wilson, M. A. (1985). Pyrolysis-GC-MS and NMR studies of dissolved seawater humic substances and isolates of a marine diatom. *Org. Geochem.*, **8**, 15–25.

Goldman, J. C. and Dennett, M. R. (1983). Carbon dioxide exchange between air and sea-water: no evidence for rate catalysis. *Science*, **220**, 199–201.

Goldman, J. C., Dennett, M. R. and Frew, N. M. (1988). Surfactant effects on air–sea gas exchange under turbulent conditions. *Deep-Sea Res.*, **35**, 1953–70.

Grassl, H. and Hinzpeter, H. (1975). The cool skin of the ocean. *GATE Report*, **14 I**, 229–36. Geneva: WMO/ICSU.

Gruft, H., Falkinam, J. O. III and Parker, B. C. (1981). Recent experience in the epidemiology of disease caused by atypical mycobacteria. *Rev. Infect. Dis.*, **3**, 990–6.

Hansen, R. S. and Mann, J. A. Jr (1964). Propagation characteristics of capillary ripples. I. The theory of velocity dispersion and amplitude attenuation of plane capillary waves on viscoelastic films. *J. Appl. Phys.*, **35**, 152–61.

Hara, T., Bock, E. J. and Lyzenga, D. (1994). *In situ* measurements of capillary-gravity wave spectra using a scanning laser slope gauge and microwave radars. *J. Geophys. Res.*, **99**, 12 593–602.

Hardy, J. T., Coley, J. A., Antrim, L. D. and Kiesser, S. L. (1988). A hydrophobic large-volume sampler for collecting aquatic surface microlayers: characterization and comparison with glass plate method. *Can. J. Fish Aquat. Sci.*, **45**, 822–6.

Hasse, L. (1971). The sea surface temperature deviation and the heat flow at the sea–air-interface. *Boundary-Layer Meteorol.*, **1**, 368–79.

Hasse, L. and Liss, P. S. (1980). Gas exchange across the air–sea interface. *Tellus*, **32**, 470–81.

Hasselmann, K. (1960). Grundgleichungen der Seegangsvorhersage. *Schifftech*, **7**, 191–5.

Hasselmann, K. (1962). On the nonlinear energy transfer in a gravity wave spectrum. 1. General theory. *J. Fluid Mech.*, **12**, 481–500.

Hasselmann, K. (1963a). On the nonlinear energy transfer in a gravity wave spectrum. 2. Conservation theorems, wave-particle analogy, irreversibility. *J. Fluid Mech.*, **15**, 273–81.

Hasselmann, K. (1963b). On the nonlinear energy transfer in a gravity wave spectrum. 3. Evaluation of the energy flux and swell–sea interactions for a Neumann spectrum. *J. Fluid Mech.*, **15**, 385–98.

Hasselmann, S. and Hasselmann, K. (1985). Computations and parameterization of the nonlinear energy transfer in a gravity-wave spectrum. 1. A new method for efficient computations of the exact nonlinear transfer integral. *J. Phys. Oceanogr.*, **15**, 1369–77.

Haussecker, H. and Jähne, B. (1995). In situ measurements of the air–sea gas transfer rate during the MBL/CoOP West Coast Experiment. In *Air–Water Gas Transfer*, ed. B. Jähne and E. C. Monahan, pp. 775–84. Hanau, Germany: AEON Verlag & Studio.

Hellebust, J. A. (1974). Extracellular products. In *Algal Physiology and Biochemistry*, ed. W. D. P. Stewart, pp. 838–63. Berkeley: University of California Press.

Hesshaimer, V., Heimann, M. and Levin, I. (1994). Radiocarbon evidence for a smaller oceanic carbon dioxide sink than previously believed. *Nature*, **370**, 201–3.

Hühnerfuss, H., Walter, W., Lange, P. and Alpers, W. (1987). Attenuation of wind-waves by monomolecular sea slicks by the marangoni effect. *J. Geophys. Res.*, **92**, 3961–3.

Jähne, B. (1985). Transfer processes across the free water surface. Habilitation Thesis, University of Heidelberg.

Jähne, B., Münnich, K. O. and Siegenthaler, U. (1979). Measurements of gas exchange and momentum transfer in a circular wind–water tunnel. *Tellus*, **31**, 321–9.

Jähne, B., Huber, W., Dutzi, A., Wais, T. and Ilmberger, J. (1984). Wind/ wave tunnel experiments on the Schmidt number and wave field dependence of air–water gas exchange. In *Gas Transfer at Water Surfaces*, ed. W. Brutsaert and G. H. Jirka, pp. 303–9. Dordrecht: Reidel.

Jähne, B., Münnich, K. O., Bösinger, R., Dutzi, A., Huber, W. and Libner, P. (1987). On the parameterization of air–water gas exchange. *J. Geophys. Res.*, **92**, 1937–49.

Jähne, B., Libner, P., Fischer, R., Billenand, T. and Plate, E. J. (1989). Investigating the transfer processes across the free aqueous viscous boundary layer by the controlled flux method. *Tellus*, **41B**, 177–95.

Keeling, R. F. (1993). On the role of large bubbles in air-sea gas exchange and supersaturation in the ocean. *J. Mar. Res.*, **51**, 237–71.

Korenowski, G. M., Frysinger, G. S., Asher, W. E., Barger, W. R. and Klusty, M. A. (1989). Laser-based nonlinear optical measurements of organic surfactant concentration variations at the air/sea interface. In *Proceedings of the International Geoscience and Remote Sensing Symposium, IEEE 89CH2768-0*, **3**, 1506–9.

Korenowski, G. M., Frysinger, G. S. and Asher, W. E. (1993). Noninvasive probing of the ocean surface using a nonlinear optical methods. *Photogramm. Engng. Remote Sensing*, **59**, 363–9.

Large, W. G. and Pond, S. (1981). Open ocean momentum fluxes in moderate to strong winds. *J. Phys. Oceanogr.*, **11**, 324–36.

Leenheer, J. A. (1985). Fractionation techniques for aquatic humic substances. In *Humic Substances in Soil, Sediment and Water*, ed. G. R. Aiken *et al.*, pp. 409–29. New York: John Wiley and Sons.

Levich, V. G. (1962). *Physicochemical Hydrodynamics*. Englewood Cliffs, NJ: Prentice-Hall.

Ling, S. C. and Kao, T. W. (1976). Parameterization of the moisture and heat transfer process over the ocean under whitecap sea states. *J. Phys. Oceanogr.*, **6**, 306–15.

Liss, P. S. and Merlivat, L. (1986). Air–sea gas exchange rates: introduction and synthesis. In *The Role of Air–Sea Exchange in Geochemical Cycling*, ed. P. Buat-Menard, pp. 113–27. Dordrecht, Holland: Reidel.

Liss, P. S., Balls, P. W., Martinelli, F. N. and Coantic, M. (1981). The effect of evaporation and condensation on gas transfer across an air–water interface. *Oceanol. Acta*, **4**, 129–38.

Lombardini, P. P., Fiscella, B., Trivero, P., Cappa, C. and Garrett, W. D. (1989). Modulation of the spectra of short gravity waves by sea surface films: slick detection and characterization with a microwave probe. *J. Atmos. Oceanic Technol.*, **6**, 882–90.

Lucassen-Reynders, E. H. (1986). Dynamic properties of film-covered surfaces. In *ONRL Workshop Proceedings: Role of Surfactant Films on the Interfacial Properties of the Sea Surface*, ed. F. Herr and J. Williams, pp. 175–86. Washington, DC: Office of Naval Research Report, C-11–8.

Lucassen-Reynders, E. H. and Lucassen, J. (1969). Properties of capillary waves. *Adv. Colloid Interface Sci.*, **2**, 347–95.

MacIntyre, F. (1974). Chemical fractionation and sea-surface microlayer processes. In *The Sea (Marine Chemistry)*, Vol. 5, ed. E. D. Goldberg, pp. 245–99. New York: John Wiley and Sons.

Mammen, T. C. and von Bosse, N. (1990). STEP: a temperature profiler for measuring the oceanic thermal boundary layer at the ocean–air interface. *J. Atmos. Oceanic Technol.*, **7**, 312–22.

McAlister, E. D. and McLeish, W. (1969). Heat transfer in the top millimeter of the ocean. *J. Geophys. Res.*, **74**, 3408–14.

McCready, M. J. and Hanratty, T. J. (1984). A comparison of turbulent mass transfer at gas–liquid and gas–solid interfaces. In *Gas Transfer at Water Surfaces*, ed. W. Brutsaert and G. H. Jirka, pp. 283–92. Dordrecht, Holland: Reidel.

Memery, L. and Merlivat, L. (1983). Gas exchange across an air–water interface: experimental results and modeling of bubble contribution to transfer. *J. Geophys. Res.*, **88**, 707–24.

Merlivat, L., Memery, L. and Boutin, J. (1993). Gas exchange at the air–sea interface: present status. Case of CO_2. In *Abstract Volume, 4th International Conference on CO_2*, Carquerainne, France, September 13–17.

Monahan, E. C. (1986). The ocean as a source of atmospheric particles. In *The Role of Air–Sea Exchange in Geochemical Cycling*, ed. P. Buat-Menard, pp. 129–63. Dordrecht, Holland: Kluwer Academic Publishers.

Monahan, E. C. and Spillane, M. C. (1984). The role of oceanic whitecaps in air–sea gas exchange. In *Gas Transfer at Water Surfaces*, ed. W. Brutsaert and G. H. Jirka, pp. 495–503. Dordrecht, Holland: Kluwer Academic Publishers.

Mykelstad, S. (1974). Production of carbohydrates by marine planktonic diatoms, I. Comparison of nine different species in culture. *J. Exp. Mar. Biol. Ecol.*, **15**, 261–74.

Mykelstad, S., Djurhuus, R. and Mohus, A. (1982). Determination of exo-(β-1,3)-D-glucanase activity in some planktonic diatoms. *J. Exp. Mar. Biol. Ecol.*, **56**, 205–11.

Nightingale, P. D., Watson, A. J., Malin, G., Liss, P. S., Upstill-Goddard, R. C. and Liddicoat, M. I. (1996). Gas exchange at sea measured by dual and triple tracer releases, in preparation.

Nissenbaum, A. (1974). The organic geochemistry of marine and terrestrial humic substances. In *Advances in Organic Geochemistry 1973*, ed. F. Biernner and B. Tissot, pp. 39–52. Paris: Editions Technip.

Ocampo-Torres, F. J. and Donclan, M. A. (1994). Laboratory measurements of mass transfer of CO_2 and H_2O vapour for smooth and rough flow conditions. *Tellus*, **46**, 16–32.

Percival, E., Rahman, M. A. and Weigel, H. (1980). Chemistry of the polysaccharides of the diatom *Coscinodiscus nobilis*. *Phytochemistry*, **19**, 809–811.

Ramus, J. (1972). The production of extracellular polysaccharides by the unicellular red alga *Porphyridium aerugineum*. *J. Phycol.*, **8**, 97–111.

Rashid, M. A. (1985). *Geochemistry of Marine Humic Compounds*. New York: Springer-Verlag.

Robertson, J. E. and Watson, A. J. (1992). Thermal skin effect of the surface ocean and its implications for CO_2 uptake. *Nature*, **358**, 738–40.

Robertson, J. E., Watson, A. J., Langdon, C., Ling, R. D. and Cooper, D. J. (1993). Diurnal variations in surface pCO_2 and oxygen at 60-N, 20-W in the NE Atlantic. *Deep-Sea Res.*, **40**, 409–22.

Sakugawa, H. and Handa, N. (1985a). Isolation and chemical characterization of dissolved and particulate polysaccharides in Mikawa Bay. *Geochim. Cosmochim. Acta*, **49**, 1185–93.

Sakugawa, H. and Handa, N. (1985b). Chemical studies on dissolved carbohydrates in water samples collected from the North Pacific and Bering Sea. *Oceanol. Acta*, **8**, 185–96.

Schlüssel, P., Emery, W. J., Grassl, H. and Mammen, T. C. (1990). On the bulk-skin temperature difference and its impact on satellite remote sensing of the sea surface temperature. *J. Geophys. Res.*, **95**, 13 341–56.

Scott, J. C. (1975). The preparation of water for surface-clean fluid mechanics. *J. Fluid Mech.*, **69**, 339–51.

Scully-Power, P. (1986). Navy oceanographer shuttle observations, Mission Report. *Rep. STSS 41-G, NUSC Tech. Doc. 7611*. Newport, RI: Naval Underwater Systems Center 2.2, 5.3.

Shen, Y. R. (1989). Surface properties probed by second-harmonic and sum-frequency generation. *Nature*, **337**, 519–25.

Smestad, B., Haug, A. and Myklestad, S. (1975). Structural studies of the extracellular polysaccharide produced by the diatom *Chaetoceros curvisetus* Cleve. *Acta Chem. Scand.*, **29**, 337–40.

Sutcliffe, W. H. Jr, Baylor, E. R. and Menzel, D. W. (1963). Sea surface chemistry and Langmuir circulation. *Deep-Sea Res.*, **10**, 233–43.

Tans, P., Fung, I. Y. and Takahashi, T. (1990). Observational constraints on the global atmospheric CO_2 budget. *Science*, **247**, 1431–8.

Thorpe, S. A. (1982). On the clouds of bubbles formed by breaking wind–waves in deep water, and their role in air–sea gas transfer. *Phil. Trans. R. Soc. London*, **A304**, 155–210.

Thorpe, S. A. (1986). Measurements with an automatically recording inverted echo sounder; ARIES and the bubble clouds. *J. Phys. Oceanogr.*, **16**, 1462–78.

van den Tempel, M. and Lucassen-Reynders, E. H. (1983). Relaxation processes at fluid interfaces. *Adv. Colloid Interface Sci.*, **18**, 281–301.

Van Vleet, E. S. and Williams, P. M. (1983). Surface potential and film pressure measurements in seawater systems. *Limnol. Oceanogr.*, **28**, 401–14.

Walker, M. I., McKay, W. A., Pattenden, N. J. and Liss, P. S. (1986). Actinide enrichment in marine aerosols. *Nature*, **323**, 141–3.

Wallace, D. W. R. and Wirick, C. D. (1992). Large air–sea gas fluxes associated with breaking waves. *Nature*, **356**, 694–6.

Wallace, G. T. Jr and Duce, R. A. (1978). Transport of particulate organic matter by bubbles in marine waters. *Limnol. Oceanogr.*, **23**, 1155–67.

Wanninkhof, R. (1992). Relationship between wind speed and gas exchange over the ocean. *J. Geophys. Res.*, **97**, 7373–82.

Wanninkhof, R. and Bliven, L. F. (1991). Relationship between gas exchange, wind speed, and radar backscatter in a large wind–wave tank. *J. Geophys. Res.*, **96**, 2785–96.

Wanninkhof, R., Asher, W. E., Weppernig, R., Chen, H., Schlosser, P., Langdom, C. and Sambrotto, R. (1993). Gas transfer experiment on Georges Bank using two volatile deliberate tracers. *J. Geophys. Res.*, **98**, 20 237–48.

Watson, A. J., Upstill-Goddard, R. C. and Liss, P. S. (1991). Air–sea gas exchange in rough and stormy seas measured by a dual tracer technique. *Nature*, **349**, 145–7.

Weisel, C. P., Duce, R. A., Fasching, J. L. and Heaton, R. W. (1984). Estimates of the transport of trace metals from the ocean to the atmosphere. *J. Geophys. Res.*, **89**, 11 607–18.

Williams, P. M., Carlucci, A. F., Henrichs, S. M., Van Vleet, E. S., Horrigan,

S. G., Reid, F. M. H. and Robertson, K. J. (1986). Chemical and microbiological studies of sea-surface films in the southern Gulf of California and
off the west coast of Baja California. *Mar. Chem.*, **19**, 17–98.

Wong, C. S. (1991). Temporal variations in the partial pressure and flux of CO_2 at
Ocean Station P in the subarctic NE Pacific Ocean. *Tellus*, **43B**, 206–23.

Woodcock, A. H. (1948). Note concerning human respiratory irritation
associated with high concentrations of plankton and mass mortality of
marine organisms. *J. Mar. Res.* **7**, 56–62.

Woolf, D. K. (1993). Bubbles and the air–sea transfer velocity of gases.
Atmosphere–Ocean, **31**, 517–40.

Woolf, D. K. and Monahan, E. C. (1988). Laboratory investigations of the
influence on marine aerosol production of the interaction of oceanic
whitecaps and surface-active material. In *Aerosols and Climate*, ed. P. V.
Hobbs and M. P. McCormick, pp. 1–8. Hampton, VA: A. Deepak
Publishers.

Woolf, D. K. and Thorpe, S. A. (1991). Bubbles and the air–sea exchange of gases
in near-saturation conditions. *J. Mar. Res.*, **49**, 435–66.

Woolf, D. K., Bowyer, P. A. and Monahan, E. C. (1987). Discriminating
between the film drops and jet drops produced by a simulated whitecap. *J.
Geophys. Res.*, **92**, 5142–50.

Zutic, V. B., Cosovic, B., Marcenko, E. and Bihari, N. (1981). Surfactant
production by marine phytoplankton. *Mar. Chem.*, **10**, 505–20.

2

Report Group 2 – Biological effects of chemical and radiative change in the sea surface

J. T. HARDY (*Chair*), K. A. HUNTER (*Rapporteur*),
D. CALMET, J. J. CLEARY, R. A. DUCE, T. L. FORBES,
M. L. GLADYSHEV, G. HARDING, J. M. SHENKER,
P. TRATNYEK and Y. ZAITSEV

Introduction

The physics, chemistry and biology of the sea surface are closely inter-related. Plankton in the water column produce an abundance of particulate and dissolved organic material, some of which is transported to the surface either passively by floatation or actively by bubble transport. Atmospheric deposition also enriches the sea surface with natural and anthropogenic compounds, which often accumulate there in relatively high concentrations compared with those in the water column. The abundance of organic matter at the sea surface provides a substrate for the growth of organisms that inhabit the sea surface microlayer: the neuston. Most studies suggest that the sea surface represents a highly productive, metabolically active interface. Organisms from most major divisions of the plant and animal kingdoms either live, reproduce or feed in the surface layers. Of particular interest are the microneuston, which may be involved in biogeochemical cycling, and neustonic eggs and larvae of commercially important fish and shellfish.

The quantities and types of anthropogenic chemicals entering the earth's atmosphere continue to grow. Many of these chemicals, some of which are highly toxic, are now globally distributed in the atmosphere and deposit to the sea surface even in remote areas. Due to stratospheric ozone depletion, ultraviolet-b (UV-B) radiation reaching the sea surface is increasing annually. These global changes represent a potential threat to the marine environment. Neuston, living in the sea surface, will experience the greatest increases in chemical and radiative change. The potential impact of global changes on this unique community should be assessed.

The condition of the neuston community could possibly be used as a 'sentinel' or early indicator of changes in the marine environment occurring on a regional and global scale. For example, evidence suggests that in the Black Sea over the past 30 years eutrophication and other pollution has produced major alterations in the plankton and neuston communities (Zaitsev, 1992).

Chemicals and radionuclides that partition and concentrate at the surface may represent relatively recent inputs from both the water column and atmosphere and, thus, may be useful for source identification and 'fingerprinting'. For example, ratios of certain specific hydrocarbons in microlayer samples can suggest different types of source, such as fossil fuel combustion, petroleum spills, etc. (Hardy *et al.*, 1990).

Samplers and sampling techniques

A variety of distinct methods have been used to sample the sea surface for chemical and biological analysis. These are distinguished from each other both by the physical manner in which they collect surface material and by the thickness of the sample collected.

The prism-dipping technique

The prism-dipping technique, devised by Baier (1972), samples mono-molecular films of organic material adsorbed at the air–sea interface by adsorption onto a small germanium prism. This method is based on the technique first described by Blodgett (1934) for recovering fatty acid monolayers from a water surface. The recovered organic films may subsequently be examined by several techniques, including internal reflection infrared (IR) spectroscopy to determine major functional groups, ellipsometry for measurement of film thickness and refractive index, surface potential and contact angle. Since these properties relate to general physical and chemical properties of the film, it is difficult to compare the results with specific chemical analyses conducted on surface samples collected by other methods.

The screen sampler

With the screen sampler (Garrett, 1965), small rectangular cells of water from a layer of the sea surface are captured in the interstitial spaces of a wire or plastic mesh by means of surface tension forces. The physical thickness of the microlayer sample collected by the screen is calculated from the void area of the screen and the volume of seawater collected and is typically the upper 200–400 μm. This thickness is determined primarily by the diameter of the screen filaments.

Plate and drum samplers

These sampling devices collect a discrete slice of the water surface, including the air–water interface, by means of viscous adhesion to a solid surface, and superficially they resemble the screen sampler. The simplest type is the plate sampler in which a viscous film of water from the sea surface adheres to a glass plate that is submerged below the water surface and then withdrawn vertically at a velocity of about 20 cm s^{-1} (Harvey and Burzell, 1972; Hardy *et al.*, 1985). The film is removed by using a wiper blade. More sophisticated variants use a rotating drum of ceramic, glass or Teflon that is pushed slowly forward while mounted on a catamaran (Harvey, 1966; Hardy *et al.*, 1988; Carlson *et al.*, 1988). The film adhering to the drum is automatically wiped off into a sample container.

The physical thickness of the sample obtained with these devices is calculated from the area of surface sampled and the volume of seawater collected, and varies from 20 to 100 μm for both plate and drum samplers. The thickness depends on a number of factors including water temperature, the presence and density of surface slicks and, in the case of drums, speed of rotation.

Techniques for sampling neuston

Microneuston may be sampled using a membrane filter (Hardy and Apts, 1984), glass plate (Hardy *et al.*, 1985), or rotating drum (Hardy *et al.*, 1988). Macroneuston can be sampled using a variety of nets (e.g., Zaitsev, 1971; Brown and Cheng, 1981; Shenker, 1988). Care must be used to select the appropriate mesh size and net mouth size for different target species and life stages. Comparisons among sampling programmes must, therefore, consider the types of gear used to collect the samples.

Characteristics of the surface microlayer

Enrichment factors

The degree to which microlayer samples collected by screen, plate and drum samples are enriched in chemical or biological species is usually assessed by calculating the enrichment factor, EF, defined as the ratio of the concentration found in the microlayer to that of a sub-surface water sample (usually collected 10–20 cm below the surface):

$$EF = [X]_\mu/[X]_b \qquad (2.1)$$

where $[X]_\mu$ is the microlayer concentration of a species X, and $[X]_b$ is the concentration of X in bulk (sub-surface) water. EF values greater than unity indicate enrichment in the microlayer, while EF values less than 1 indicate depletion. Since the screen, plate and drum samplers collect mostly water along with materials adsorbed at the air–water interface, the possibility exists that some, or even most, of the sample collected may be sub-surface water not enriched in surface-active species. As discussed below, this can lead to dilution of actual surface concentrations and underestimation of EFs.

Surface excess concentrations

Because the sample thicknesses of the various sampling methods are different, the results of chemical or biological analysis cannot be directly compared in terms of either microlayer concentrations or enrichment factors. However, it is possible to compare results for different sampler thicknesses by calculating the surface excess concentration (SE) defined as:

$$SE = ([X]_\mu - [X]_b) \cdot d \qquad (2.2)$$

where d is the sample thickness (Hunter and Liss, 1981a). This relationship may be used to calculate total excess concentration through a cross-section of the surface. The results of such comparisons show only fair agreement between screen and plate samplers, consistent with their different modes of action (Carlson, 1982b). However, plate and drum samplers seem to operate similarly when results are corrected for sample thickness using the surface excess concentration.

Variability of organisms and contaminants

Distributions of sea-surface chemical contaminants and organisms are patchy, but both tend to become concentrated in the same areas. One of the unusual features of the surface microlayer, compared with sub-surface waters, is the coincident concentration of contaminants and biota. Processes such as particle adsorption, organic complexation, and bubble scavenging concentrate natural organics and nutrients in the microlayer creating conditions for enhanced biological activity. Contaminants tend to concentrate in the microlayer due to the same processes that concentrate the organic materials and organisms. Therefore, the exposure hazard for neuston, i.e., occurrence of organisms in areas of contamination, may be higher than in the water column. This situation parallels that in the sediments.

The development of new chemical sensors for nutrients, trace elements and other chemical components in the near future, coupled with drum samplers, should provide much more detailed information on microlayer variability. More detailed, larger-scale information on microlayer spatial and temporal variability relevant to the issue of capillary wave damping effects on gas exchange is likely to arise from future applications of remote sensing techniques now being developed, e.g., the infrared radiometer, microwave radar, laser scattering and SHG (single harmonic generation) techniques. These methods will need 'ground truthing' using more systematic chemical measurements of the microlayer, particularly of organic enrichments and film pressures.

Thickness of sea-surface films

Because the sea-surface microlayer is a complex phenomenon involving physical, chemical and biological processes operating over widely different physical dimensions, the question of its thickness has no simple answer. It is important, however, to consider how the methods used for sampling the microlayer for chemical or biological analysis are related to dimensions of the physical regions within which chemical and biological materials are accumulated. In particular, if the microlayer, seen as a region of concentrated contaminants, is significantly smaller in thickness than that sampled for chemical analysis, then microlayer organisms may be exposed to much higher concentrations of contaminants than are indicated by chemical studies. On the other hand, a much thinner region of enriched contaminants may mean that many neustonic organisms are not actually in contact with the enriched region and thus not exposed to the toxicant, particularly if they are not attached to the air–water interface.

Conventional wisdom from surface chemistry would suggest that the natural organic matter in the microlayer consists of material adsorbed at the air–water interface, making up a film of molecular thickness. This view is certainly supported by the visual appearance of surface slicks.

However, visible surface slicks are relatively rare on the open oceans, although in coastal seas they are more prevalent. For example, Garabetian *et al.* (1993) report that in the Mediterranean slicks were present 30% of the time and covered >50% of the photographed area. General measures of organic concentration, such as dissolved organic carbon (DOC), show microlayer enrichment even when slicks are absent. EF values usually do not increase markedly even when slicks are present, suggesting that the appearance of surface films and enrichment of natural organic material in the microlayer may not be related to each other in a simple, linear fashion. However, in contaminated areas, EF values for contaminants are generally greater in slicks than outside (Hardy *et al.*, 1988).

The compiled results for DOC enrichments (Carlson, 1982a) and other quantities often show depletion in microlayer samples (EF < 1). Examples for UV absorbance and chlorophyll fluorescence are shown in Figure 2.1. If the region of enrichment is very near to the air–water interface, e.g., within 1 nm, then the much thicker plate, drum or screen microlayer samples would comprise mostly sub-surface water. In this case, it is very difficult to see how microlayer depletion can arise. On the other hand, if most of the enriched organic material in the microlayer region consists of very expanded, hydrated material extending over a thickness comparable to those of sampling devices (i.e., 50–200 μm), more hydrophilic fractions of the dissolved organic matter (DOM) could be volume-excluded from the microlayer region, thus giving rise to microlayer depletion.

Other evidence points to the fact that microlayer film materials can be relatively thick. Ellipsometric measurements of the thickness of organic layers collected using the prism-dipping technique show that the dried material can be as thick as 1 μm (Baier *et al.*, 1974). This suggests that once re-hydrated, the organic material is likely to occupy a much greater thickness. When the drum sampler is used to collect from slick-covered waters, the volume of sample obtained per unit time increases over that obtained in clean areas. This implies that the adsorbed film material is able to influence the viscous properties of the water film (thickness 20–100 μm) collected by the drum. The thickness of microlayer samples collected with the glass plate increases exponentially from 30 to 55 μm as the surface pressure (an indicator of the amount of organic film present) increases from 0 to >20 dyne cm^{-1} (Hardy *et al.*, 1985).

The only evidence that the microlayer might be significantly thinner than that sampled conventionally comes from the use of 'bubble microtome' samplers that are based on air bubbles bursting at the air–water

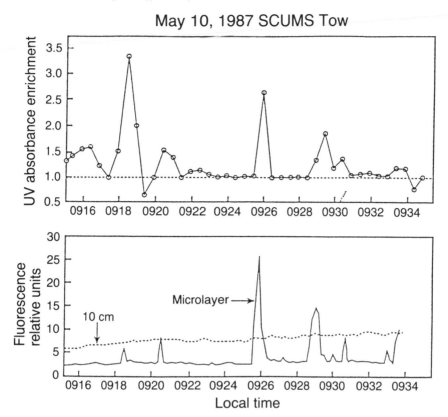

Figure 2.1. Results from a SCUMS (Self-Contained Underway Microlayer Sampler) tow (discussed by Hunter, Chapter 9, this volume) in Damariscotta Estuary, Maine, USA. The upper panel shows microlayer UV absorbance enrichment. The lower panel shows surface microlayer and sub-surface chlorophyll fluorescence, in relative fluorescence units. From Carlson *et al.*, 1988.

interface (e.g., Fasching *et al.*, 1974). However, the greater enrichments observed with this type of sampler compared with conventional methods most likely result from scavenging of surface-active material in sub-surface waters by rising air bubbles rather than a simple, very thin sample taken only from the air–water interface itself. This view is supported by studies on the scavenging of bacteria by rising bubbles (Blanchard, 1975) and by the considerable body of scientific knowledge concerned with mineral floatation processes.

The above considerations suggest that a simple monomolecular film, e.g., of a simple surfactant like oleic acid, is a poor analogue of natural microlayers. A more accurate picture is that of a gel-like structure occupying a significant thickness comparable with that sampled by screen, plate and drum samplers (50–200 μm).

Biology of the sea surface

The neustonic realm is a vast habitat covering 70% of the earth's surface. The unique physical and chemical characteristics of the sea surface have contributed to the evolution of a highly diverse and abundant assemblage of species, including many that are of commercial and ecological importance (Zaitsev, 1971) (Figure 2.2). Permanent inhabitants of the surface layer, such as bacterioneuston, phytoneuston, zooneuston and ichthyoneuston, often reach much higher densities than similar organisms found in sub-surface waters. Temporary inhabitants of the neuston, particularly the eggs and larvae of a great number of fish and invertebrate species, utilize the surface during a portion of their embryonic and larval development. Temperature may be increased and permanent neustonic organisms may be concentrated by meteorological and oceanographic processes that result in convergences within the surface layer. Some neuston can remain in the upper milli- or centi-layer until turbulence created by winds exceeding 10–15 m s^{-1} disperses them (Zaitsev, 1971). Many organisms become strongly attached to the air–water interface because of surface tension forces. Neuston, like plankton and nekton, may be classified for purposes of study according to their size.

Piconeuston

Piconeuston (<2 μm), like picoplankton (Sieburth *et al.*, 1976), are being increasingly recognized as important links in the recycling of organic matter in aquatic ecosystems. High densities of metabolically active bacterioneuston are found in the surface microlayer (Sieburth, 1971; Bezdek and Carlucci, 1972; Sieburth *et al.*, 1976; Carlucci *et al.*, 1985). For example, bacterioneuston off the coast of Baja California accounted for 1.4 to 5.9% of the total microbial carbon biomass and a similar percentage (1.9 to 5.1%) of the microbial carbon production in the surface microlayer. The enrichment of bacteria in the surface microlayer results, at least in part, from the greater degree of hydrophobicity of bacterioneuston compared with bacterioplankton and, thus, their adhesion to organically enriched surface microlayers (Dahlback *et al.*, 1981). Bacteria also adhere to the air–water interface of bubbles and may

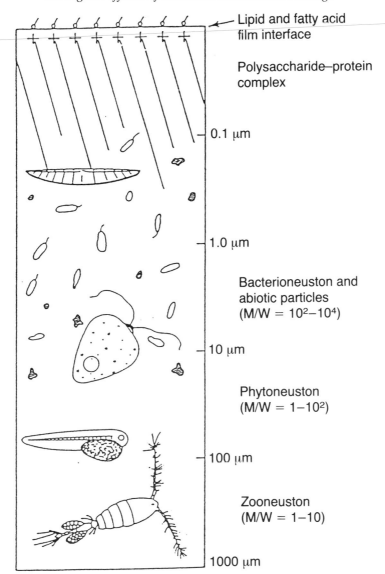

Figure 2.2. Conceptual model of the sea-surface microlayer ecosystem. M/W = typical microlayer to water concentration ratios based on a number of studies. (From Hardy, 1982.)

be injected into the atmosphere as part of sea-salt aerosols (Blanchard, 1983). Bacterioncuston probably play an important role in degrading not only natural organics, but also anthropogenic chemicals collected at the sea surface.

The importance of autotrophic picoplankton is also being increasingly recognized (Johnson *et al.*, 1982). In the water column in the subarctic Pacific, 28% of the biomass (as carbon) was in the <2 μm size fraction and was almost exclusively the blue-green autotroph *Synechococcus* spp. (Booth *et al.*, 1988). Some early data for the enrichments of bacteria in microlayer samples may have given anomalously high enrichment factors because many allochthanous species grow readily on culture plates and may become over-represented in the final counts made by this method. Direct counting, e.g., using epifluorescence spectroscopy, autoradiography or adenosine triphosphate (ATP) biomass determination, is preferred for measurement of bacterioneuston. With direct counts of membrane-collected samples, bacterioneuston/bacterioplankton enrichments in Puget Sound are 10^2 to 10^6 (Hardy and Apts, 1984). However, few studies of open ocean samples have been made and the global distribution of bacterioneuston is poorly known.

Nano- and microneuston

Nano- and microneuston (2 to 200 μm in size), including dozens of species of microalgae (phytoneuston), occur in great abundance in the microlayer (Hardy 1973; Hardy and Valett, 1981; Hattori *et al.*, 1983; Hardy and Apts, 1984; De Souza-Lima and Chretiennot-Dinet, 1984). Enrichment ratios (number of individuals per volume in the surface microlayer/ numbers per volume in the sub-surface water) are generally 10 to 1000.

The blue-green alga (cyanobacterium) *Trichodesmium* is a common phytoneustonic organism in tropical open ocean waters. In the tropical North Atlantic Ocean it is the most important primary producer (165 mg C m^{-2} d^{-1}) and is also responsible, through nitrogen fixation, for the largest fraction of new nitrogen to the euphotic zone (30 mg m^{-2} d^{-1}) (Carpenter and Romans, 1991). Although maximum densities of *Trichodesmium* sp. occurred at near-surface (15 m) depths (Carpenter and Romans, 1991), this may simply reflect the absence of microlayer sample collection, which might otherwise show the buoyant filaments even more concentrated in the microlayer. For example, in the Atlantic Ocean 150 km off the Florida coast, a surface microlayer enrichment ratio for total phytoneuston (dominated by *Trichodesmium* sp. and microflagellates) of 400 was found, as well as a phytoneuston species composition very different from that of the sub-surface phytoplankton community (Hardy *et al.*, 1988).

Certain dinoflagellate species are often found in the surface microlayer. For example, in the North Sea, near the island of Sylt, a bloom of the dinoflagellate *Prorocentrum micans*, and several other phytoneuston species, contributed to the formation of visible surface slicks. Microlayer enrichment ratios for *P. micans* were >1000 (Brockmann *et al.*, 1976). The phytoneuston community off Baja California is dominated by dinoflagellates and differs greatly from the phytoplankton in species composition (Williams *et al.*, 1986). Also, in the Black Sea continuing eutrophication has led to frequent algal blooms (including dinoflagellates) and screen-collected microlayer phytoneuston/phytoplankton enrichment ratios of 100 to 10000 (Nestrova, 1980).

There have been only a few measurements of photosynthetic carbon reduction in the surface microlayer. Most have focused on nearshore, productive environments: an enclosed lagoon (Hardy, 1973), a salt marsh (Gallagher, 1975), a river estuary (Albright, 1980) and the Mediterranean Sea near Marseilles (De Souza-Lima and Chretiennot-Dinet, 1984). One offshore study was conducted near Baja California (Williams *et al.*, 1986). All these investigators enclosed collected microlayers and their communities (microneuston) in bottles, thus disrupting the natural surface-film habitat.

Hardy and Apts (1989) determined microalgal standing stocks and activities in the surface microlayer and sub-surface waters in areas with (S) and without (NS) visible natural surface slicks. Phytoneuston photosynthesis was measured without enclosing samples in bottles. Microlayer and sub-surface water were enclosed in open-top microcosms and incubated *in situ*. Enrichment ratios (surface microlayer/sub-surface water concentration) were: phytoneuston population abundance, 37 (NS) to 154 (S); total chlorophyll, 1.3 (NS) to 18 (S); particulate carbon fixation, 2 (NS) to 52 (S) and dissolved carbon excretion by autotrophs, 17 (NS) to 63 (S). As in other studies, the species composition of the phytoneuston was distinctly different from that of the phytoplankton.

Winter and spring carbon reduction rates at one coastal marine site were 20 to 150 times greater in the surface microlayer than in the sub-surface water (Hardy and Apts, 1984). However, in high summer light intensities, photoinhibition of 36 to 89% occurred in phytoneuston. Williams *et al.* (1986) also found evidence for photoinhibition in phytoneuston; i.e., photosynthesis increased 179% in neuston incubated at 1 m depth compared with incubation at the surface.

Microzooplankton (the heterogenous group of organisms between 2 and 200 μm) form a vital link in the planktonic food web between small bacteria, flagellates, and larger metazoans including fish. They include protozoa and the developmental stages of many pelagic and benthic animals, but few adult metazoans. Their feeding modes include particle filtration, phagotrophy, and pinocytosis (Conover, 1982). From 4% to greater than 70% of primary production is consumed by micro-zooplankton (Conover, 1982). Small (<20 μm) ciliates at times make up 4% to 57% of the total biomass of heterotrophic (apochlorotic) nanoplankton in diverse marine systems (Sherr *et al.*, 1986). Nano- and microheterotrophs release dissolved organics that form substrates (along with organics from phototrophs) for the growth of bacteria on which they feed. They can be thought of as 'microbial gardeners' (Davis and Sieburth, 1982). Despite their probable importance, few quantitative data are available on their abundance or trophic relationships. 'To be able to attach reliable rates and biomass to just one of the pathways would be a major achievement' (Conover, 1982).

Compared with the microzooplankton, even less is known about the species composition, abundance or dynamics of microzooneuston. Based on a limited number of investigations, heterotrophic microneuston in-habiting the surface microlayer include small ciliates (Zaitsev, 1971), protozoa (Norris, 1965) and yeasts and moulds (Kjelleberg and Hakansson, 1977). Tintinnids often form an important component of the marine neuston, probably feeding on the high densities of bacterio-neuston (Hardy, 1971; Zaitsev, 1971). Also, the heterotrophic dino-flagellates *Noctiluca scintillans* and *Oxyrrhis marina* are often abundant in neuston samples (Hardy, 1971; Zaitsev, 1971).

Mesozooneuston

Some mesozooneuston (0.2 to 20 mm in size) feed on the high densities of microneuston (Zaitsev, 1971). Samples from many areas of the world's oceans indicate that copepods, often dominated by pontellid species, are abundant components of the surface microlayer (Zaitsev, 1971; Hardy, 1982). In Galway Bay (Tully and O'Ceidigh, 1986) and many other locations (Zaitsev, 1971), neustonic isopods and amphipods are abund-ant, often associated with floating driftweed and debris. Numerous species of fish, including cod, sole, flounder, hake, menhaden, anchovy, mullet, flying fish, greenling, saury, rockfish, halibut, and many others have surface-dwelling egg and/or larval stages (Zaitsev, 1971; Ahlstrom and Stevens, 1975; Brewer, 1981; Safronov, 1981; Kendall and Clark, 1982; Shenker, 1988). Some crab and lobster larvae in estuarine, coastal, and shelf areas concentrate in the surface film during midday as a result of positive phototaxis (Smyth, 1980; Jacoby, 1982; Provenzano *et al.*, 1983; Tully and O'Ceidigh, 1986).

In Puget Sound, English sole (*Parophrys vetulus*) and sand sole (*Psettichthys melanostictus*) spawn between January and April, releasing trillions of eggs that collect on the water surface. The embryos float until hatching occurs, generally 6 to 7 days after fertilization. Because of the buoyancy of their large yolk sacks, newly hatched larvae of both species often float upside-down at the surface of the water (Budd, 1940).

Although the micro-vertical distribution of zooneuston in the upper 0 to 50 cm can be disturbed by wind mixing, the effect seems to disappear quickly once the wind subsides (Champalbert, 1977). As Zaitsev (1971) noted: '. . . most of the eggs must rise to the surface film and remain suspended there. Although the specific weight of the egg increases during development, it remains low enough to hold the egg close to the surface film . . . and surface enrichments of organisms were found to be stable even at sea states of 5 to 6 Beaufort.' Fish eggs, concentrated at the sea surface, are dispersed at wave heights exceeding 1 to 2 m, while larvae and fry remain there even when waves reach heights of 3 to 4 m (Zaitsev, 1971).

Neuston net tows, from widely dispersed areas of the global ocean, indicate that floating tar and plastic debris is important as a habitat colonized by dozens of species of marine invertebrates. The distributional abundance of many neuston, including attached eggs, is positively correlated with the distribution of pelagic tar. While the tar acts as a habitat for some species, it may exert a detrimental effect on the majority of neuston (Holdway and Maddock, 1983).

Macroneuston

The larger (>2 cm) neuston organisms have been referred to as 'pleuston', although a more consistent term might be 'macroneuston'. These organisms, of which there are perhaps 100 species, inhabit the surface-layer (upper metre) generally by floatation or by association with floatable seaweed (Cheng, 1975). Coelenterates include the jellyfish *Physalia* spp., which feed on small fish, and *Velella* spp., which feed on copepods and fish and invertebrate eggs. Common gastropods in the open ocean surface layer are species of prosobranchiata and nudibranchiata. Oceanic crabs and many other organisms are found associated with the floating pelagic seaweed *Sargassum* spp. These macroneuston undoubtedly form part of the food web linking the sea-surface and sub-surface layer, but the details of this food web remain unclear.

Freshwater neuston

Neustonic populations in freshwaters, although represented by different species, fill analogous niches to their marine counterparts. Freshwater eutrophication (artificial and natural) often leads to increasing formation of surface materials and 'scums' – thick organically enriched films. One important difference in freshwaters is the absence of fish and benthic invertebrate species with neustonic eggs or larvae. In many high-latitude lakes, winter ice cover certainly alters the lake surface and would affect chemical and radiative exposures. Ice could trap surface materials and nutrients and release them again during the spring thaw, rapidly enriching the surface layer. Algal blooms on the underside of ice form a unique surface community.

Chemistry in the sea surface

Organic components of natural origin

Organic compounds of natural (non-anthropogenic) origin are the principal film-forming components in aquatic systems and generally influence the presence of other microlayer constituents, including trace elements, radionuclides and particulate matter. The infrared spectra of marine films obtained using the prism-dipping technique show a functional group chemistry dominated by glycoprotein, with lipid components important only in samples collected from grossly polluted waters (Baier *et al.*, 1974).

This analysis is confirmed by extensive measurements of general organic chemical quantities such as dissolved organic carbon (DOC) and nitrogen (DON), or specific organic compound classes such as protein, polysaccharide, lipid and hydrocarbons. Williams *et al.* (1986) have reported a very comprehensive study of microlayer samples from the Gulf of California and Baja California, the results of which are typical of most studies. Mean enrichment factors (screen sampler) were 1.1–2.4 for DOC, DON, urea, lipids, inorganic nutrients; 1.3–2.0 for ATP, chlorophyll-*a*, microplankton and bacteria and 1.1–3.7 for POC, PON, and protein. The authors report that systematic correlations between the measured variables were few. The results demonstrate that the bulk of microlayer organic material is a complex mixture of biologically derived compounds dominated by polymeric proteins, carbohydrates and their condensation/degradation products, much like the dissolved organic matter (DOM) in bulk seawater from which microlayer materials are derived.

Fatty acids and other lipid components of natural origin have also been widely studied. Although they comprise only a few per cent of the DOM in microlayer samples (Hunter and Liss, 1977; Hardy, 1982), these compound types are valuable as specific markers of biological and other sources. Marty *et al.* (1988) used lipids as specific markers for atmospheric input, floatation from sub-surface algae and input within the surface layer from neuston. Lipids are generally enriched in the microlayer to a similar extent as the major organic compound classes.

Indirect measures of the overall organic content of microlayer samples, for example UV absorbance, fluorescence and surfactant activities based on electrochemical methods (Cosovic and Vojvodic, 1982; 1989; Hunter and Liss, 1981b) have been widely applied to microlayer samples. The results obtained with these methods differ from each other since each measures, or responds to, a different fraction of the DOM that is enriched in microlayers. Carlson and Mayer (1982) consider that UV absorbance (280 nm), which is almost consistently higher in microlayer samples than in those of sub-surface water, is therefore more representative of the presence of surface slicks than either fluorescence or DOC measurements. They also argue that UV absorbance results largely from phenolic materials rather than microalgal exudates. However, these conclusions are restricted to coastal waters in which terrestrial phenolic materials are present. The situation for the open ocean is not well understood.

The surfactant activity method based on suppression of polarographic streaming is well correlated with DOC and gives very similar microlayer enrichments to those of DOC (Hunter and Liss, 1981b). However, the techniques based on the capacity of the mercury electrode double layer are more closely related to surface film enrichments and UV absorbance (Marty *et al.*, 1988).

Spatial and temporal variability in general microlayer properties with a resolution on the scale of metres and seconds can be attained by coupling continuous sensors of UV absorbance or fluorescence to drum-type microlayer samplers (e.g., Carlson *et al.*, 1988). Results of this approach show that enrichments of UV absorbance or fluorescence vary significantly over short time and distance scales. Data from Carlson *et al.* (1988) show that UV absorbance is consistently higher in the microlayer relative to sub-surface water, with absorbance peaks corresponding to areas covered with obvious surface slicks. By contrast, chlorophyll fluorescence is consistently lower in the microlayer than in sub-surface water in unslicked areas, but much higher in the slick-affected areas. The results shown in Figure 2.1 emphasize the natural variability of the sea surface and the consequent need for adequate sample coverage to obtain an accurate picture of the surface state on a regional or global scale.

Microlayer enrichment factors for DOC rarely exceed 1.5 (50 μm plate sampler) (Carlson, 1982a; Savenko, 1990). This implies that most of the DOM pool is hydrophilic in nature with a relatively low affinity for the air–water interface, a conclusion supported by studies using electrochemical surfactant activity measurements (Hunter and Liss, 1981b; Marty *et al.*, 1988).

Organic components of anthropogenic origin

Trace organic compounds, primarily of anthropogenic origin, including polychlorinated biphenyls (PCBs), chlorinated pesticides, and hydrocarbons, have been commonly measured in nearshore microlayer samples, but there is a paucity of open ocean data. Recent improvements in methodology include much better control of sample blanks and contamination, improved detection limits and improved ability to identify individual compounds through the use of gas chromatography – mass spectrometry (GC–MS) identification.

Early studies indicated microlayer/bulkwater enrichments for PCBs, pesticides, and polynuclear aromatic hydrocarbons (PAHs) of 10^1 to 10^4 or more (Hardy, 1982). In Puget Sound, microlayer samples frequently contain relatively high concentrations of pesticides, PCBs, and aromatic hydrocarbons, while these compounds are generally undetectable in bulk water samples from the same sites (Hardy *et al.*, 1987b). Very few samples have been analyzed from open ocean microlayers. A recent study (Sauer *et al.*, 1989) found few detectable PCBs and chlorinated insecticides in either microlayer or sub-surface water samples collected in open ocean waters off the eastern US coast and in the Gulf of Mexico.

Trace elements

Hunter (1980) reviewed the mechanisms through which particulate species are enriched in the microlayer. Both atmospheric deposition and floatation by rising bubbles are important sources of particulate material for the microlayer. The residence times of particles at the surface depend on interfacial tension forces and can range from a few seconds for wettable, high-energy particles to tens of minutes for small hydrophobic, non-wettable particles (Hunter, 1980). Experimental data, using natural urban air particles deposited on the surface of seawater microcosms, suggest microlayer residence times for six different metals between 1.5 and 15 hours (Hardy *et al.*, 1985).

For dissolved trace elements (particularly Pb, Zn and Fe) measurements of offshore samples prior to the early 1980s must be regarded with considerable caution, and it is likely that few, if any, uncontaminated microlayer samples have ever been collected from remote waters. Since the pioneering work of Patterson and others (e.g., Patterson and Settle, 1976; Bruland, 1980), it is now clear that extensive precautions must be taken to avoid spurious contamination artifacts during the collection, handling and analysis of environmental samples for trace elements. Thus, although enrichment of dissolved trace metals in microlayer samples in the remote ocean has not been reliably demonstrated, it almost certainly occurs to some extent. For particulate trace elements, problems of contamination artifacts are much less severe, and reliable particulate trace metal enrichments have been reported by a number of workers (e.g., Hoffman *et al.*, 1974; Hardy *et al.*, 1988). In coastal waters, especially those with elevated trace metal concentrations as a result of human activities, results are probably more reliable. In contaminated coastal waters, enrichment factors of 10 or more have been reported (e.g., Hardy, 1982; Hardy *et al.*, 1985). The enrichment of dissolved trace metals in microlayers is considered to be mediated by association with the primary surface-active organic components of the sea surface (Hunter and Liss, 1981c).

The physical nature of microlayer sampling devices involves the large surface area of the collector making contact with the microlayer water sample. This aspect in particular makes it very difficult to ensure freedom from contamination artifacts. One method to ensure greater confidence in trace element data is to use several different microlayer sampler types in parallel collections. It is not likely that contamination effects will be identical for quite different samplers, such as the screen and the rotating drum. However, this approach cannot replace properly conducted inter-laboratory calibration exercises as a means for validating sample handling techniques. Few, if any, such exercises have been conducted for micro-layer samples.

Organotin compounds

Organotin compounds are in widespread use as biocides in anti-fouling paints, fungicides in agriculture and miticides in fruit crop culture. Tributyltin compounds (TBT) are used primarily as anti-fouling agents in paints used on boats, ships, locks, buoys, etc., and they are among the most toxic anthropogenic pollutants introduced to marine and fresh waters (Goldberg, 1986). TBT concentrations as low as two ng 1^{-1} can have serious effects on marine biota causing shell chambering and gel formation in *Crassostrea gigas* (Alzieu *et al.*, 1989) and imposex in *Nucella lapillus* and *Ilyanassa obsoleta* (Bryan *et al.*, 1989).

Because of their hydrophobic nature and high octanol–water partition coefficient (K_{ow}) ranging from 5500 to 7000 (Laughlin *et al.*, 1986), organotins favour partition into the lipid phase and should therefore readily accumulate in the surface microlayer. Data from the United Kingdom, Canada, and the United States not only confirm the ubiquitous distribution of organotins in both freshwater and marine environments, but also the common occurrence of such surface microlayer enrichment. Studies of Canadian freshwater regions by Maguire and Tkacz (1987) found that microlayer EF values for TBT ranged from 41 to 47 300. In marine waters enhancement was generally lower, with maximum EF values of 35 in Chesapeake Bay (Hall *et al.*, 1987) and 27 in coastal waters of southwest England (Cleary and Stebbing, 1987). Other studies have measured EF values of 135 in the Great Bay estuary in the United States (Donard *et al.*, 1986) and 34 in Hamilton Harbor, Bermuda (Stebbing *et al.*, 1990). The only reported offshore data are from the North Sea, where TBT EF values ranged from 2.0 to 16.5 (Hardy and Cleary, 1992) and total organotin concentrations were significantly correlated with clam larval mortality using cryopreserved Manila clam (*Tapes philippinarum*) bioassay (Cleary *et al.*, 1993).

Radionuclides

The distributions of natural and anthropogenic radionuclides in various marine compartments, including the surface microlayer, have been the subject of extensive research. Natural radionuclides such as ^{210}Po, ^{210}Pb and ^{14}C have been used as tracers of marine processes or differential time markers. Artificial radionuclides such as ^{137}Cs and isotopes of Pu, considered as anthropogenic contaminants of the marine environment, have been studied to assess potential impacts through their vertical and horizontal distributions. The results of these studies show a predominant downward vertical flux of radionuclides from the surface waters to deep water masses and the sediment. However, few field studies have explored the potentially significant accumulation of radionuclides in the surface microlayer. Although bomb-derived Pu isotopes are widely distributed through the upper water column in the global ocean, the open sea distribution in the microlayer is not known.

Indirect indications of artificial radionuclide accumulation in the surface microlayer were proposed in studies of aerosol deposition on land along the coastal areas of the Irish Sea subjected to direct low-level radioactive liquid effluent discharges from the reprocessing plant of Sellafield (Eakins and Lally, 1984). Studies, based on radionuclide measurements of top soil and aerosols, demonstrated that Pu and Am isotopes (primarily adsorbed on particles) had higher concentrations in marine aerosols collected on land than in the bulk seawater of the same area. The observed enrichment factors were at least 5 for these radio-nuclides. However, these studies did not indicate the origin of the aerosols, e.g., surface film, foam, etc.

Two other studies performed in the western Mediterranean Sea (Marseilles) used a rotating drum microlayer sampler to collect the surface microlayer (Badie *et al.*, 1987, Calmet and Fernandez, 1990). Samples between 18 and 37 litres were collected whereas 100 litres were sampled for the surface water (0.5 m). The samples were collected on a 0.45 µm membrane filter. ^{137}Cs microlayer enrichment factors were between 4 and 15. The fraction greater than 0.45 µm gave the highest enrichment factors for both ^{137}Cs and ^{106}Ru. These differences can be explained by the presence of phytoplankton cells that are characterized by concentration factors of 20 (IAEA, 1985).

Systematic studies of natural radionuclides in the sea-surface microlayer should be very valuable as a benchmark for elucidating trace element uptake mechanisms. Unlike stable elements, the measurement of natural radionuclides is usually free from contamination artifacts. To date, natural radionuclide concentrations in the surface microlayer are only known for ^{210}Pb and ^{210}Po. Bacon and Elzerman (1980) reported enrichment factors for these nuclides up to a factor of 7 in screen-collected microlayer samples, with much greater enrichments observed in a surface foam sample from a coastal pond. From a consideration of possible source terms for the radionuclide pair, they concluded that in offshore areas a significant fraction of ^{210}Po and ^{210}Pb in the microlayer was derived from sub-surface waters, whereas in coastal waters this source was relatively less important than deposition of the nuclides from the atmosphere. Heyraud and Cherry (1985) demonstrated that ^{210}Po can accumulate in neuston.

Freshwater chemistry

The general chemical and physical features of the freshwater surface layer are similar to those of the marine surface layer (Gladyshev, 1986). Most studies have focused on freshwaters near pollution sources. In the Great Lakes, for example, metal, pesticide and PCB contamination has historically been great (Rapaport and Eisenreich, 1988). However, studies suggest that the freshwater surfaces in even remote areas are contaminated by chlorinated organics which may originate from long-range atmospheric transport and deposition (Ofstad *et al.*, 1979).

Effects of ultraviolet radiation

Anthropogenic release of chlorofluorocarbons is leading to global decreases in stratospheric ozone (Stolarski *et al.*, 1992) and increases in ultraviolet-B radiation (UV-B) (Lubin *et al.*, 1989; Blumthaler and Ambach, 1990). What are the likely implications of this trend for sea-surface biota? Depletion of stratospheric ozone results in a proportionately larger increase in short UV-B radiation wavelengths compared with longer UV-B radiation wavelengths (Green *et al.*, 1980). Because biological damage generally increases exponentially with decreasing wavelengths within the UV-B radiation band, small decreases in stratospheric ozone translate into rather large increases in biologically damaging radiation. Thus, biologists apply an action spectra which weights the wavelength-specific damage (Behrenfeld *et al.*, 1993a,b).

Based on a time linear extrapolation of the ozone trend data (Stolarski *et al.*, 1992) and spatial and temporal differences in the atmospheric attenuation of UV-B radiation (Green *et al.*, 1980), it appears that noontime UV-B irradiance at the sea surface at mid-latitudes in the Southern Hemisphere could increase about 17% between the years 1979 and 2009. As shown in Figure 2.3, this increase in incident UV-B radiation would translate into a significant change in the biologically effective wavelength region below 310 nm. The sea surface is the region of maximum UV-B radiation exposure hazard, particularly since UV-B radiation may be disproportionately high in this layer because of photon backscatter from neuston and particulates as well as multiple reflections caused by wave action (Regan *et al.*, 1992).

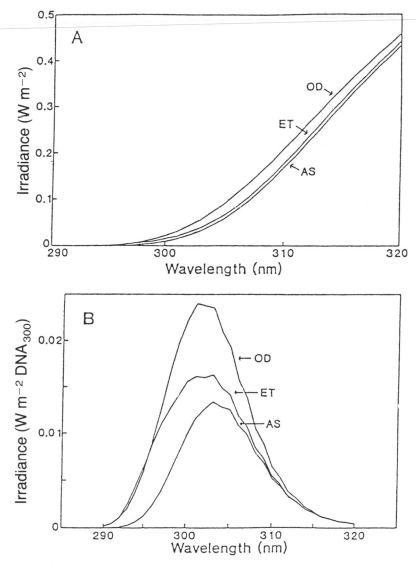

Figure 2.3. UV-B irradiance at noon at 45° S in March. AS = ambient solar for 1979; ET = UV-B radiation in enhanced experimental treatment; OD = UV-B radiation estimated for stratospheric ozone depletion by 2009. (A) Unweighted UV-B radiation, biologically effective UV-B radiation weighted by the DNA_{300} action spectrum. From Behrenfeld *et al.* (1993b).

Considerable evidence indicates that relatively small increases in UV-B radiation can inhibit photosynthesis, growth or reproduction in a variety of marine species (Hardy and Gucinski, 1989). However, neuston occur at the sea surface where the intensity of UV-B radiation is greater than in the water column. Therefore, they should be adapted to high levels of UV-B radiation. Anecdotal evidence suggests that this may be the case. Many neustonic species are highly pigmented and this is thought to be protective in screening out UV-B radiation, while others are transparent at least to visible light. DNA radiation repair mechanisms exist in many organisms to different degrees, but the relative efficiency of repair in neustonic organisms has not been measured. Microsporine-like amino acids occur in some plankton and invertebrates and appear to be effective filters of UV-B radiation, but their presence in neuston has not been investigated (Hardy and Gucinski, 1989).

Many neustonic animals and plants exist near the upper levels of their demonstrated tolerance to UV-B radiation (Hardy and Gucinski, 1989). For example, phytoplankton primary production in temperate to tropical oceans, at least in summer, is already inhibited by UV-B radiation (Behrenfeld, *et al.*, 1993a,b). Many marine fish and benthic organisms, including commercial species, have neustonic eggs and/or larvae that may be threatened by increases in microlayer chemical contamination and UV-B radiation. However, estimating the long-term effects of UV-B radiation on deeper water column organisms (plankton) is extremely difficult and is complicated by such factors as vertical mixing, photo-repair, and photoadaptation (Behrenfeld *et al.*, 1993a,b).

Effects of chemical contamination

The toxic effects of contaminants in the marine environment are affected by the *exposure* of organisms to contaminants, by the *bioavailability* of particular contaminants and by the *sensitivity and response* of different organisms. Any toxic effects of the microlayer that are different from sub-surface waters are likely to result from changes in exposure or bio-availability provided by the unusual conditions of the microlayer.

There is no evidence that microlayer organisms have significantly different sensitivities towards contaminants than the same life stages of the same taxa in sub-surface waters. However, changes in neuston photosynthesis or respiratory quotient could be important for CO_2 transport through the microlayer.

The fact that toxic materials such as trace metals or organics are enriched in the microlayer does not, by itself, imply a significant increase in exposure. Some microlayer organisms spend a significant fraction of time attached to, or floating in contact with, the air–water interface because of surface tension forces. However, some neuston remain free of the interface and enter the microlayer region on an irregular basis, e.g., for opportunistic reasons such as grazing or diurnal changes. For the purposes of a hazard assessment, a 'worst-case' scenario can be developed by considering that neustonic organisms spend long enough in the microlayer that their exposure to contaminants is the same as if the organism were immersed in bulk seawater having the same composition as the microlayer. In this simplified scenario, exposure is proportional to the enrichment factor of the contaminant.

Usually, the microlayer enrichment of toxic materials is less than an order of magnitude. In this case, parallel toxic effects in both the microlayer and sub-surface waters can often be expected, and the toxicities of microlayer and sub-surface waters at the same location should be correlated with each other (Thain, 1992). The greatest exposure for many organisms will be from the sub-surface water. Enhanced microlayer effects would only be expected for organisms that experience some degree of contact with the air–sea interface, including indirect effects due to feeding on organisms or particulate matter located there.

However, in some situations microlayer enrichment of contaminants can be very large, in which case toxic concentrations can occur in the microlayer when the sub-surface water at the same site is non-toxic (Hardy and Cleary, 1992; Hardy *et al.*, 1987a).

Only general principles can be stated concerning the factors affecting bioavailability of contaminants to microlayer organisms. Many are common to bioavailability issues in the water column and/or sediments, e.g., the effects of physical form (particulate, colloidal, dissolved), chemical speciation (aquo ion versus metal–ligand complex), and the specific nature of different organisms (e.g., filter-feeders). Other factors will be specific to the microlayer. As already mentioned, dissolved trace metals are considered to be enriched in the microlayer because of their association with surface-active ligands or colloids. This association may reduce their toxicity toward many organisms. Similarly, the attachment of particles to the air–sea interface involves very strong surface tension forces (Hunter, 1980). Whether this affects their availability to filter-feeders is not known.

One of the most likely cases in which microlayer toxic effects would be expected is that of tributyl tin (TBT), because of its very high toxicity and lipophilic nature. In the absence of biological water quality measurements on samples where microlayer contaminant concentration is known, some interpretation of toxicity may be made from toxicity threshold data and water quality standard criteria. For example, this approach was used to interpret UK organotin microlayer data in relation to toxicity threshold values for neustonic and littoral organisms. Toxicity threshold levels for mussels were shown to range from 100–230 ng TBT 1^{-1}, while lower concentrations of TBT resulted in other toxic effects in oysters at concentrations from 2.5–10 ng TBT 1^{-1} (Cleary, 1992). Concentrations of 2.5 ng TBT 1^{-1} have been demonstrated to induce imposex in gastropods. Microlayer TBT concentrations for Canadian freshwaters at all locations studied exceeded the 12-day LC-100 value of 4.5 ng TBT 1^{-1} for rainbow trout yolk sac fry, and in the most contaminated microlayer the TBT concentration was 262 times this value (Maguire and Tkacz, 1987). Such relationships, while not defining microlayer toxicity, do give cause for concern with regard to chronic toxicity and effects on sensitive organisms.

These toxicity threshold concentrations are similar to, or exceeded by, TBT concentrations reported for marine sub-surface waters, e.g. 20–1000 ng TBT 1^{-1} in coastal UK waters (Cleary and Stebbing, 1987) and 10–80 ng TBT 1^{-1} in the North Sea (Cleary *et al.*, 1993) and the German Bight (Hardy and Cleary, 1992). Since TBT can be enriched in the microlayer by an order of magnitude over these bulk water concentrations, it seems likely that significant toxic effects on microlayer organisms should result. Measurements of TBT from open ocean waters are not presently available. The overall ecological importance of any potential effects of TBT on microlayer organisms in coastal waters must be viewed in the context of the distribution of TBT. For example, in coastal environments the present state of ecological knowledge does not allow one to state, a priori, that the ecological impact of TBT contamination necessarily will be greater in the microlayer than in the sediment. The use of TBT on small boats in enclosed coastal waters is being phased out in many regions.

This situation is even less clear cut for other microlayer contaminants. As discussed on page 305 few reliable microlayer measurements exist for most trace metals, particularly for the open oceans, and much of the data that have been used to argue for the enhanced toxicity of the microlayer with respect to these components are unreliable. However, since enrichment of trace metals in the microlayer is almost certainly mediated by complexing with surface-active organic matter and association with particulate matter (Hunter and Liss, 1981c), trace metal enrichments in the microlayer are unlikely to exceed those of the primary dissolved and particulate organic components. On this basis, microlayer concentrations of dissolved trace metals are likely to be only two to three times greater than in sub-surface waters. This is not a large increase in potential exposure of microlayer organisms compared with that experienced in sub-surface waters. Furthermore, the complexing of metal ions necessary to produce enrichment is expected to reduce the toxicity of most metal ions compared with that of the free aquo ion in sub-surface waters.

For example, the lowest observable effect concentration (LOEC) for Cd in marine organisms is 5000 ng 1^{-1}, and a range of 100–1000 ng 1^{-1} is recommended as one above which toxic effects are likely (OSPAR, 1993). Cd concentrations in uncontaminated surface waters rarely exceed 40 ng 1^{-1}, although levels above this have been reported for contaminated waters. Hardy *et al.* (1990) reported microlayer dissolved Cd concentrations of 41–93 ng 1^{-1} for Chesapeake Bay, which are just below the above threshold for toxic effects.

In contrast, some authors have argued that in contaminated inshore areas toxic effects may occur. For example, microlayer concentrations of PAHs, pesticides and metals such as Pb and Cu many times greater than known toxic levels or US EPA water quality standards have been found in water off southern California (Cross *et al.*, 1987), Puget Sound (Hardy *et al.*, 1987a,b) and the North Sea (Hardy and Cleary, 1992). Further, microlayer PCB concentrations at many sites in Puget Sound exceeded water quality criteria by 130 times. Concentrations of metals and aromatic hydrocarbons at about half the stations in Chesapeake Bay were equal to or greater than concentrations proven toxic to developing fish embryos (and EFs for Cu and Pb were 8.5 and 43.3, respectively, Hardy, *et al.*, 1990). At most stations sampled in Puget Sound and Chesapeake Bay microlayer concentrations of Cu and Pb greatly exceeded the EPA Chronic Water Quality Criteria (2.9 µg 1^{-1} for Cu and 5.6 µg 1^{-1} for Pb).

Little is known about the quantitative importance of biological effects caused by particulate trace metals in the microlayer. Metals in particulate phases can attach readily to the air–sea interface (Hunter, 1980). However, the bioavailability of particulate metals varies greatly depending on their chemical and physical state. Metals bound in mineral phases are unlikely to be available, whereas those associated with organic particles and algae will be more readily available to filter feeders or grazers located at the air–water interface, or to organisms feeding on material deposited in intertidal habitats by surface films (Gardiner,

1992). The overall impact will depend on the ecological importance of organisms affected by these sources.

The chemical properties relevant to the microlayer and bioavailability of many toxic organic compounds (organochlorine insecticides, PCBs, PAHs) are considered similar to those of TBT. Many individual compounds in this group exhibit similarly high toxicity towards marine organisms and have the same high lipophilicity as TBT. For example, provisional ecotoxicological reference values at which toxic effects are expected in seawater solution are 50–500 ng 1^{-1} for fluoranthene and 10–100 ng 1^{-1} for benz[a]pyrene (OSPAR, 1993). For comparison, microlayer samples from Chesapeake Bay contain up to 928 ng 1^{-1} of fluoranthrene (median 67 ng 1^{-1}) and up to 440 ng 1^{-1} of benz[a]pyrene (median 35 ng 1^{-1}) (Hardy *et al.*, 1990). In this case, sub-surface concentrations of both PAHs were extremely low (0.008, < 0.002 ng 1^{-1} respectively).

Toxicity tests have been conducted using microlayer samples after collection by standard screen or drum methods (e.g., Williams, 1992). These experiments generally show toxic effects of different types, e.g., growth inhibition, increased mortality, and chromosomal and developmental abnormalities in neustonic fish eggs and larvae (Dethelfson *et al.*, 1985; Cross *et al.*, 1987; Hardy *et al.*, 1987a; Gardiner, 1992). However, as stated above, such results must be treated with caution because they may not properly reproduce the *in situ* exposure of the organism to the contaminants. On the other hand, the laboratory toxicity studies are supported by some limited *in situ* data. Fertilised floating fish eggs incubated *in situ* displayed marked reduction in hatching success at contaminated compared with uncontaminated field sites (Hardy *et al.*, 1987a), and increased incidences of morphological and chromosomal abnormalities are associated with neustonic fish embryos collected from contaminated areas (Longwell and Hughes, 1980; Cameron and Berg, 1992).

Selection of the organisms for toxicity tests is also important. There is a need for more detailed investigations on the bioavailability of contaminants in the microlayer and exposure of resident organisms under realistic conditions. This will allow the development and interpretation of better bioassays. Other possibilities include the enzymatic-level Microtox® test (luciferin–luciferase) or algal growth tests. It would be valuable to develop tests that can be conducted at sea as part of multidisciplinary studies of the microlayer. Cryo-preserved embryos may be useful as model test organisms for overcoming seasonal availability of microlayer species (Karbe, 1992).

Study of the microlayer can be valuable in ascertaining chemical and biological conditions in the water column as a whole. For example, free fatty acids produced in the water column may indicate the diversity and physiological state of underlying bacterio-, phyto- and zooplankton (Gladyshev *et al.*, 1993).

In summary, it appears that the concentrations of some highly toxic, particle-reactive contaminants with lipophilic properties, notably TBT, are high enough in some contaminated coastal waters, particularly semi-enclosed basins and harbours, probably to cause measurable toxic effects on some microlayer organisms. These predicted effects remain to be demonstrated by *in situ* studies. Moreover, their environmental significance is largely unknown and should be assessed in the context of concomitant ecological hazards associated with the same contaminants occurring in the underlying water column and sedimentary environments. A special issue of *Marine Environmental Research* (Vol. 23, No. 4, 1987), provides a comprehensive account of studies to that date on the effects of microlayer contamination on resident biota. Whether or not toxic effects on neuston occur in offshore oceanic microlayers remains unknown due to the lack of any study of such areas.

Effects of greenhouse warming

The increasing concentration of atmospheric greenhouse gases is one of the most important global environmental problems. Models predict that the relative temperature increase will be greatest at high latitude and that the ocean surface may experience a decrease in salinity toward the poles (Wright *et al.*, 1986). One could expect the neustonic communities to expand toward the poles from the tropical and subtropical areas. However, changing wind and current patterns could have a significant influence on the distribution and survival of neuston through their influence on larval retention and transport. General circulation global climate models predict a severe restriction in the distribution of some commercial fish species. For example, in response to a doubling of atmospheric CO_2 and resultant ocean warming, the neustonic larval reproductive stages of some species currently ranging from California to Alaska might be restricted only to the very northern part of their current range (Sibley and Strickland, 1985).

Global change and the microlayer

Particulate organic matter, as well as a variety of biogenic chemicals and gases, are produced by plankton in the water column and rise to the surface where they enter the microlayer with varying (mostly unknown) residence times. Such biologically mediated processes as dimethyl-sulphide (DMS) production and CO_2 transport can certainly be inhibited if UV-B radiation reaches sufficiently high levels.

In the absence of sufficient experimental or *in situ* measurements, scientists currently differ in their assessment of the potential influence of microneuston on air–sea exchange of CO_2. One view holds that the rate of respiratory breakdown of particulate organic carbon in the microlayer would not be great enough compared with air–sea gas exchange rates to affect the concentration of CO_2 within the microlayer.

On the other hand, although the microlayer represents about one part in 10^6 of the total euphotic zone, it may account for a greater fraction of organic matter respiration because of the affinity of surface-active materials for the sea-surface interface. Significant respiratory CO_2 production can occur in the surface microlayer with the potential for enhanced CO_2 diffusion into the atmosphere (Garabetian, 1990). Measurements of the vertical upward flux of particulate carbon are lacking, but such fluxes could be substantial. For example, assuming a significant proportion (e.g., 1%) of the particulate carbon produced in the water column reaches the surface and undergoes respiratory breakdown in the micro-layer itself, CO_2 partial pressure in the diffusion sub-layer may increase sufficiently to retard CO_2 input from the atmosphere. This 'flux capping' hypothesis warrants further investigation.

Conclusions and recommendations

The sea surface is a unique chemical and biological environment. The quantities and types of anthropogenic chemicals entering the earth's atmosphere and depositing on the ocean surface continue to grow. In addition, global increases in UV-B radiation and ocean temperature have the potential for negative impacts on the neustonic organisms living in the sea surface. However, time it is not possible to predict local, let alone global, effects of these changes with any accuracy, given our current state of knowledge.

An important characteristic of the sea-surface microlayer is the patchi-ness, but also the coincidence, of high concentrations of both organisms and contaminants, resulting in potential exposures and subsequent hazards for neustonic organisms. To characterize better the temporal and spatial distribution of both organisms and contaminants, improved sea-surface sampling programmes, including remote sensing instrumentation, need to be developed to enable real-time chemical and biological measurements.

Virtually all the studies to date of the chemical and biological characteristics of the sea-surface microlayer have been undertaken in coastal environments, where concentrations of contaminants in both the water column and the microlayer are expected to be particularly high. It appears that highly toxic, particle-reactive contaminants with lipophilic properties, notably TBT and some toxic organics, probably do cause measurable toxic effects on microlayer organisms in some contaminated coastal waters, particularly semi-enclosed basins and harbours. The environmental significance of these microlayer effects (which remain to be demonstrated by *in situ* studies) is largely unknown and should be assessed in the context of concomitant ecological hazards associated with the same contaminants occurring in the underlying water column and sediment environments.

Whether or not toxic effects on neuston occur in offshore oceanic microlayers remains unknown due to the lack of studies in such areas. To evaluate the global impact of microlayer enrichment, reliable measurements and process-oriented studies of the bioavailability of, and exposure to, contaminants in the microlayer in the open ocean are needed, including the southern hemisphere. Chemical studies would require state-of-the-art contamination-free sampling and analytical capabilities that are presently only available to a few groups. As part of any study in open ocean regions, rigorous inter-laboratory calibration exercises on sampling and on surface chemical and biological analyses must be conducted. This type of effort should also be undertaken for coastal studies.

Studies of the occurrence of natural radionuclides such as ^{210}Po/^{210}Pb in the sea-surface microlayer can help in understanding the fluxes of metals and other substances to and in the microlayer. Radionuclides are particularly useful as tracers because contamination problems during collection and analysis of samples are much less severe. Studies using stable isotopes would also be useful in understanding processes in the sea surface.

In the absence of sufficient experimental or *in situ* measurements, scientists currently differ in their assessment of the potential influence of microneuston on air–sea exchange of CO_2. However, 'flux capping', i.e., where respiratory breakdown in the microlayer leads to increased CO_2

partial pressure and retards CO_2 input from the atmosphere, warrants further investigation.

To assess the chemical and radiation hazard to neuston, a 'neuston watch' might be considered. This would consist of a series of selected geographic sites which would be monitored on a continuing basis at least annually. Information from these stations would represent a database for assessing long-term status and trends of the sea surface.

References

Ahlstrom, E. H. and Stevens E., (1975). Report on neuston (surface) collections made on an extended CalCOFI cruise during May 1972. *Calif. Coop. Fish. Investig.*, **28**, 1 July 1973 to 30 June 1975.

Albright, L. J. (1980). Photosynthetic activities of phytoneuston and phytoplankton. *Can. J. Microbiol.*, **26**, 389–92.

Alzieu, C., Sanjuan J., Michel, P., Borel, M. and Dreno, J. P. (1989). Monitoring and assessment of butyltins in Atlantic coastal waters. *Mar. Pollut. Bull.*, **20**, 22–6.

Bacon, M. P. and Elzerman, A. W. (1980). Enrichment of ^{210}Po and ^{210}Pb in the sea surface microlayer. *Nature*, **284**, 332–4.

Badie C., Peres, J. M. and Romano, J. C. (1987). Concentration de radionuclides artificiels dans des films organiques de surface en Mediterranee Nord-occidentale. *C. R. Acad. Sci. Paris*, **304** (III, 8), 177–80.

Baier, R. E. (1972). Organic films on natural waters: their retrieval, identification and modes of elimination. *J. Geophys. Res.*, **77**, 5062–75.

Baier, R. E., Goupil, D. W., Perlmutter, S. and King, R. (1974). Dominant chemical composition of sea surface films, natural slicks and foams. *J. Rech. Atmos.*, **8**, 571–600.

Behrenfeld, M., Hardy, J., Gucinski, H., Hanneman, A., Lee, H. and Wones, A. (1993a). Effects of ultraviolet-B radiation on primary production along latitudinal transects in the South Pacific Ocean. *Mar. Environ. Res.*, **35**, 349–63.

Behrenfeld, M. J., Chapman J. W., Hardy J. T. and Lee H., II (1993b). Is there a common response to ultraviolet-B radiation by marine phytoplankton? *Mar. Ecol. Prog. Ser.*, **102**, 59–68.

Bezdek, H. F. and Carlucci A. F. (1972). Surface concentrations of marine bacteria. *Limnol. Oceanogr.*, **17**, 566–9.

Blanchard, D. C. (1975). Bubble scavenging and the water-to-air transfer of organic material in the sea. In *Applied Chemistry at Protein Interfaces*, ed. R. E. Baier, *ACS Adv. Chem. Ser.*, **145**, 360–87.

Blanchard, D. C. (1983). The production, distribution, and bacterial enrichment of the sea-salt aerosol. In *Air–Sea Exchange of Gases and Particles*, ed. P. S. Liss and W. G. N. Slinn, pp. 407–54. Dordrecht: Reidel.

Blodgett, K. B. (1934). Monomolecular films of fatty acids on glass. *J. Am. Chem. Soc.*, **56**, 495.

Blumthaler, M. and Ambach, W. (1990). Indication of increasing solar ultraviolet-B radiation flux in alpine regions. *Science*, **248**, 206–8.

Booth, B. C., Lewin, J., and Lorensen, C. J. (1988). Spring and summer growth rates of subarctic Pacific phytoplankton assemblages determined from carbon uptake and cell volumes estimated using epifluorescence microscopy. *Mar. Biol.*, **98**, 287–98.

Brewer, G. D. (1981). Abundance and vertical distribution of fish eggs and larvae in the Southern California Bight: June and October 1978. *Rapp. P. V. Reun. Cons. Int. Explor. Mer.*, **178**, 165–7.

Brockmann, U. H., Kattner, G., Hentzschel, G., Wandschneider, K., Junge, D. H. and Huhnerfuss, H. (1976) Naturliche oberflaschenfilme in Seegebeit vor Sylt. *Mar. Biol.*, **36**, 135–46.

Brown, D. M. and Cheng, L. (1981). New net for sampling the ocean surface. *Mar. Ecol. Prog. Ser.*, **5**, 225–7.

Bruland, K. W. (1980). Oceanographic distributions of cadmium, zinc, nickel, and copper in the North Pacific. *Earth Planet. Sci. Lett.*, **47**, 176–98.

Bryan, G. W., Gibbs, P. E., Huggett, R. J., Curtis L. A., Bailey D. S. and Dauer, D. M. (1989). Effects of tributyltin pollution on the mud snail *Ilyanassa obsoleta* from the York River and Sarah Creek, Chesapeake Bay. *Mar. Poll. Bull.*, **20**, 458–62.

Budd, P. L. (1940). Deveopment of the eggs and early larvae of six California fishes. *State of California, Dept. of Nat. Res., Div. of Fish and Game, Bureau of Marine Fisheries. Fish Bull., No. 56.* 20 pp.

Calmet, D. and Fernandez, J. M. (1990). Cesium distribution in Northwestern Mediterranean seawater, suspended particles and sediments. *Continent. Shelf Res.*, **10**, 895–914.

Cameron, P. and Berg, J. (1992). Morphological and chromosomal aberrations during embryonic development in dab *Limanda limanda. Mar. Ecol. Prog. Ser.*, **91**, 163–9.

Carlson, D. J. (1982a). Surface microlayer phenolic enrichments indicate sea surface slicks. *Nature*, **296**, 426–9.

Carlson, D. J. (1982b). Phytoplankton in marine surface microlayers. *Can J. Microbiol.*, **28**, 1226–34.

Carlson, D. J. and Mayer, L. M. (1982). Enrichment of dissolved phenolic material in the surface microlayer of coastal waters. *Nature*, **286**, 482–3.

Carlson, D. J., Cantey, J. L. and Cullen, J. J. (1988). Description and results from a new surface microlaycr sampling device. *Deep-Sea Res.*, **35**, 1205–13.

Carlucci, A. F., Craven, D. B. and Henrichs, S. M. (1985). Surface-film microheterotrophs: amino acid metabolism and solar radiation effects on their activities. *Mar. Biol.*, **85**, 13–22.

Carpenter, E. J. and Romans, K. (1991). Major role of the cyanobacterium *Trichodesmium* in nutrient cycling in the North Atlantic Ocean. *Science*, **254**, 1356–8.

Champalbert, G. (1977). Variations locales de la repartition verticale et de l'abondance de l'hyponeuston en fonction des situations meteorologiques. *Cah. Biol. Mar.*, **18**, 243–55.

Cheng, L. (1975). Marine pleuston-animals at the sea–air interface. *Oceanogr. Mar. Biol. Ann. Rev.*, **13**, 181–212.

Cleary, J. J. (1992). Organotin in the marine surface microlayer and subsurface waters of southwest England: Relation to toxicity thresholds and the UK Enviromental Quality Standard. *Mar. Env. Res.*, **32**, 213–22.

Cleary, J. J. and Stebbing, A. R. D. (1987). Organotin in the surface microlayer and subsurface waters of southwest England. *Mar. Pollut. Bull.*, **18**, 238–46.

Cleary, J. J., McFadzen, I. R. B. and Peters, L. D. (1993). *Surface Microlayer Contamination and Toxicity in the North Sea and Plymouth Near-shore Waters.* ICES paper CM 1993/E:28. Copenhagen: International Council for the Exploration of the Sea.

Conover, R. J. (1982). Interrelations between microzooplankton and other plankton organisms. *Ann. Inst. Oceanogr. Paris*, **58**(S), 31–46.

Cosovic, B. and Vojvodic, V. (1982). Direct determination of surface-active substances in natural waters. *Mar. Chem.*, **28**, 183–98.

Cosovic, B. and Vojvodic, V. (1989). Adsorption behaviour of the hydrophobic fraction of organic matter in natural waters. *Mar. Chem.*, **28**, 183–98.

Cross, J. N., Hardy, J. T., Hose, J. E., Hershelman, G. P., Antrim, L. D., Gossett, R. W. and Crecelius, E. A. (1987). Contaminant concentrations and toxicity of sea-surface microlayer near Los Angeles, California. *Mar. Environ. Res.*, **23**, 307–23.

Dahlback, B., Hermansson, M., Kjelleberg, S. and Norkrans, B. (1981). The hydrophobicity of bacteria: an important factor in their initial adhesion at the air-water interface. *Arch. Microbiol.*, **128**, 267–70.

Davis, P. G. and Sieburth, J. McN. (1982). Differentiation of phototrophic and heterotrophic nanoplankton populations in marine waters by epifluorescence microscopy. *Ann. Inst. Oceanogr. Paris*, **58**(S), 249–60.

De Souza-Lima, Y. and Chretiennot-Dinet, M. J. (1984). Measurements of biomass and activity of neustonic microorganisms. *Estuar. Coast. Shelf Sci.*, **19**, 167–80.

Dethelfson, V., Cameron, V. and von Westernhagen, H. (1985). Üntersuchungen über die Haufigkeit von Missbildungen in Fischembryonen der südlichen Nordsee. *Inf. Fischwirtsch*, **32**, 22–7.

Donard, O. F., Rapsomanikis, S. and Weber, J. H. (1986). Speciation of inorganic tin and alkyl tin compounds by atomic absorption spectrometry using electrothermal quartz furnace after hydride generation. *Anal. Chem.*, **54**, 772–7.

Eakins, J. D. and Lally, A. E. (1984). The transfer to land of actinide-bearing sediments from the Irish Sea by spray. *Sci. Total Environ.*, **35**, 23–32.

Fasching, J. L., Courant, R. A., Duce, R. A. and Piotrowicz, S. R. (1974). A new surface microlayer sampler based on the bubble microtome. *J. Rech. Atmos.*, **8**, 650–2.

Gallagher, J. L. (1975). The significance of the surface film in salt marsh plankton metabolism. *Limnol. Oceanogr.*, **20**, 120–3.

Garabetian, F. (1990). CO_2 production at the sea–air interface. An approach by the study of respiratory processes in surface microlayer. *Int. Rev. Ges. Hydrobiol.*, **75**, 219–29.

Garabetian, F., Romano, J-C., Paul, R. and Sigoillot, J-C. (1993). Organic matter composition and pollutant enrichment of sea surface microlayer inside and outside slicks. *Mar. Environ. Res.*, **35**, 323–39.

Gardiner, W. (1992). Shoreline deposition of contaminated surface film and its effect on intertidal organisms. MS Thesis. Huxley College of Environmental Studies, Western Washington University.

Garrett, W. D. (1965). Collection of slick-forming materials from the sea surface. *Limnol. Oceanogr.*, **10**, 602–5.

Gladyshev, M. I. (1986). Neuston of inland waters (a review). *Hydrobiol. J.*, **22**(5), 1–7.

Gladyshev, M. I., Kalachova, G. S. and Sushchik, N. N. (1993). Free fatty acids of surface film of water in the Sydinsky Bay of the Krasnoyarsk reservoir. *Int. Rev. Ges. Hydrobiol.*, **78**, 575–87.

Goldberg, E. D. (1986). Tributyltins: an environmental dilemma. *Environment.*, **28** (8), 17–21.

Green, A. E. S., Cross, K. R. and Smith, L. A. (1980). Improved analytic characterization of ultraviolet skylight. *Photochem. Photobiol.*, **31**, 59–65.

Hall, L. W. Jr., Lenkevich, M. J., Scott Hall, W., Pinkney, A. E. and Bushong, S. J. (1987). Evaluation of butyltin compounds in Maryland waters of Chesapeake Bay. *Mar. Pollut. Bull.*, **18**, 78–83.

Hardy, J. T. (1971). Ecology of phytoneuston in a temperate marine lagoon. PhD Dissertation. Dept. of Botany, University of Washington, Seattle, 160 pp.

Hardy, J. T. (1973). Phytoneuston ecology of a temperate marine lagoon. *Limnol. Oceanogr.*, **18**, 525–33.

Hardy, J. T. (1982). The sea-surface microlayer: biology, chemistry and anthropogenic enrichment. *Prog. Oceanogr.*, **11**, 307–28.

Hardy, J. T. and Apts, C. W. (1984). The sea surface microlayer: phytoneuston productivity and effects of atmospheric particulate matter. *Mar. Biol.*, **82**, 293–300.

Hardy, J. T. and Apts, C. W. (1989). Phytosynthetic carbon reduction: high rates in the sea-surface microlayer. *Mar. Biol.*, **101**, 411–17.

Hardy, J. T. and Cleary, J. (1992). Surface microlayer contamination and toxicity in the German Bight. *Mar. Ecol. Prog. Ser.*, **91**, 203–10.

Hardy, J. T., and Gucinski, H. (1989). Stratospheric ozone depletion: implications for the marine environment. *Oceanography*, **2**(2), 18–21.

Hardy, J. T. and Valett, M. K. (1981). Natural and microcosm phytoneuston communities of Sequim Bay, Washington. *Estuar. Coast. Shelf Sci.*, **12**, 3–12.

Hardy, J. T., Apts, C. W., Crecelius, E. A. and Bloom, N. S. (1985). Sea-surface microlayer metals enrichments in an urban and rural bay. *Estuar. Coast. Shelf Sci.*, **20**, 299–312.

Hardy, J. T., Kiesser, S. L., Antrim, L. D., Stubin, A. I., Kocan, R. and Strand, J. A. (1987a). The sea-surface microlayer of Puget Sound. Part I. Toxic effects on fish eggs and larvae. *Mar. Environ. Res.* **23**, 227–49.

Hardy, J. T., Crecelius, E. A., Antrim, L. D., Broadhurst, V. L., Apts, C. W. Gurtisen, J. M. and Fortman, T. J. (1987b). The sea-surface microlayer of Puget Sound. Part 2. Concentrations of contaminants and relation to toxicity. *Mar. Environ. Res.*, **23**, 251–71.

Hardy, J. T., Coley, J. A., Antrim, L. D. and Kiesser, S. L. (1988). A hydrophobic large-volume sampler for collecting aquatic surface microlayers: characterization and comparison to the glass plate method. *Can. J. Fish. Aquat. Sci.*, **45**, 822–6.

Hardy, J. T., Crecelius, E. A., Antrim, L. D., Kiesser, S. L., Broadhurst, V. L., Boehm, P. D. and Steinhauer, W. G. (1990). Aquatic surface contamination in Chesapeake Bay. *Mar. Chem.*, **28**, 333–51.

Harvey, G. W. (1966). Microlayer collection from the sea surface: a new method and initial results. *Limnol. Oceanogr.*, **11**, 608–14.

Harvey, G. W. and Burzell, L. A. (1972). A simple microlayer method for small samples. *Limnol. Oceanogr.*, **11**, 798–801.

Hattori, H., Yuki, K., Zaitsev, Y. P. and Motoda, S. (1983). A preliminary observation on the neuston in Surgua Bay. *Umi-Mer,* **21**, 11–20.

Heyraud, M. and Cherry, R. D. (1985). Correlation of [210]Po and ^{210}Pb enrichments in the sea-surface microlayer with neuston biomass. *Continental Shelf Res.*, **13**, 283–93.

Hoffman, G. L., Duce, R. A., Walsh, P. R., Hoffman, E. J., Ray, B. J. and Fasching, J. L. (1974). Residence time of some particulate trace metals in the oceanic surface microlayer: significance of atmospheric deposition. *J. Rech. Atmos.*, **8**, 745–59.

Holdway, P. and Maddock, L. (1983). Neustonic distributions. *Mar. Biol.*, **77**, 207–93.

Hunter, K. A. (1980). Processes affecting particulate trace metals in the sea-surface microlayer. *Mar. Chem.*, **9**, 49–70.

Hunter, K. A. and Liss, P. S. (1977). The input of organic material to the oceans: air–sea interactions and the organic chemical composition of the sea surface. *Mar. Chem.*, **5**, 361–79.

Hunter, K. A. and Liss, P. S. (1981a). Organic sea surface films. In *Marine Organic Chemistry*, ed. E. K. Duursma and R. Dawson, pp. 259–98. Amsterdam: Elsevier.

Hunter, K. A. and Liss, P. S. (1981b). Polarographic measurement of surface-active material in natural waters. *Water Res.*, **15**, 203–15.

Hunter, K. A. and Liss, P. S. (1981c). Principles and problems of modelling cation enrichment at natural air–water interfaces. In *Atmospheric Pollutants in Natural Waters*, ed. S. J. Eisenreich, pp. 99–127. Ann Arbor: Ann Arbor Press.

International Atomic Energy Agency (1985). Sediment K_ds and concentration factors for radionuclides in the marine environment. *IAEA Technical Report Series No 247*, Vienna, 73 pp.

Jacoby, C. A. (1982). Behavioral responses of the larvae *Cancer magister Dana* to light, pressure, and gravity. *Mar. Behav. Physiol.*, **8**, 267–83.

Johnson, P. W., Xu, H-S. and Sieburth, J. (1982). The utilization of chroococcoid cyanobacteria by marine protozooplankters but not by calanoid copepods. *Ann. Inst. Oceanogr., Paris.*, **58**(S), 297–308.

Karbe, L. (1992). Toxicity of surface microlayer, subsurface water and sediment elutriates from the German Bight: summary and conclusions. *Mar. Ecol. Prog. Ser.*, **91**, 197–201.

Kendall, A. W. Jr and Clark, J. (1982). Ichthyoplankton off Washington, Oregon, and Northern California, April–May 1980. NWAFC Processed Report 82–11. Seattle, WA: Northwest and Alaska Fisheries Center, NMFS, NOAA.

Kjelleberg, S. and Hakansson, N. (1977). Distribution of lipolytic, proteolytic, and amylolytic marine bacteria between the lipid film and the subsurface water. *Mar. Biol.*, **39**, 103–19.

Laughlin, R. B. Jr, Guard H. E., and Coleman, W. M. III (1986). Tributyltin in seawater: speciation and octanol–water partition coefficient. *Environ. Sci. Technol.*, **20**, 201–4.

Longwell, A. C. and Hughes, J. B. (1980). Cytologic, cytogenic and developmental state of Atlantic mackerel eggs from sea surface waters of the New York Bight, and prospects for biological effects monitoring with ichthyoplankton. *Rapp. P.V. Reun. Cons. Int. Explor. Mer.* **179**, 275–91.

Lubin, D., Frederick, J., Booth, C., Lucas, T. and Neuschuler, D. (1989). Measurement of enhanced springtime ultraviolet radiation at Palmer Station, Antarctica. *Geophys. Res. Lett.*, **16**, 783–5.

Maguire, R. J. and Tkacz, R. J. (1987). Concentration of tributyltin in the surface microlayer of natural waters. *Water Pollut. Res. J. Can.*, **22**, 227–33.

Marty, J. C., Zutic, V., Precali, R., Saliot, A., Cosovic, B., Smodlaka, N. and Cauwet, G. (1988). Organic matter characterization in the northern Adriatic Sea with special reference to the sea surface microlayer. *Mar. Chem.*, **25**, 243–63.

Nestrova, D. A. (1980). Phytoneuston in the western Black Sea. *Gidrobiol. Z.*, **16**, 26–31.

Norris, R. E. (1965). Neustonic marine Craspedomonodales (choanoflagellates) from Washington and California. *J. Protozool.*, **12**, 589–602.

Ofstad, E. B., Lunde, G. and Drangsholt, H. (1979). Chlorinated organic compounds in the fatty surface film on water. *Int. J. Environ. Anal. Chem.*, **6**, 119–31.

Oslo and Paris Commissions (OSPAR) (1993). Report of the workshop *Assessment Criteria for Chemical Data of the Joint Monitoring Program*, Scheveningen, Netherlands, November 15–17, 36 pp.

Patterson, C. C. and Settle, D. M. (1976). The reduction of order of magnitude errors in lead analyses of biological materials and natural waters by evaluating and controlling the extent and sources of industrial lead contamination introduced during sample collection and analysis. In *Accuracy in Trace Analysis: Sampling, Sample Handling, and Analysis*, ed. P. LaFleur, pp. 321–51. Washington, DC: National Bureau of Standards Special Publication 422.

Provenzano, A. J. Jr, McConaugha, J. R., Philips, K. B., Johnson, D. F. and Clark, J. (1983). Vertical distribution of first stage larvae of the blue crab *Callinectes sapidus* at the mouth of the Chesapeake Bay. *Estuar. Coast. Shelf. Sci.*, **16**, 489–99.

Rapaport, R. A. and Eisenreich, S. J. (1988). Historical inputs of high molecular weight chlorinated hydrocarbons to eastern North America. *Environ. Sci. Technol.*, **22**, 931–41.

Regan, J. D., Carrier, W. L., Gucinski, H., Olla, B. L., Yoshida, H., Fujimura, R. K. and Wicklund, R. I. (1992). DNA as a solar dosimeter in the ocean. *Photochem. Photobiol.*, **56**, 35–42.

Safronov, S. G. (1981). On neuston in Kamchatka waters of the Sea of Okhotsk. *Biol. Morva.*, **4**, 73–4.

Sauer, T. C., Durrell, G. S., Brown, J. S., Redford, D. and Boehm, P. D. (1989). Concentrations of chlorinated pesticides and polychlorinated biphenyls in microlayer and seawater samples collected in open ocean waters of the US east coast and in the Gulf of Mexico. *Mar. Chem.*, **27**, 235–7.

Savenko, V. S. (1990). Determination of the salinity of the thin surface microlayer. *Oceanology*, **30**(3), 289–92.

Shenker, J. M. (1988). Oceanographic associations of neustonic larval and juvenile fishes and Dungeness crab megalopae off Oregon. *Fish Bull. US*, **86**, 299–317.

Sherr, E. B., Sherr, B. F., Fallon, R. D. and Newell, S. Y. (1986). Small aloricatc ciliates as a major component of the marine heterotrophic nanoplankton. *Limnol. Oceanogr.*, **31**, 177–83.

Sibley, T. H. and Strickland, R. M. (1985). Fisheries: some relationships to climate change and marine environmental factors. In *Characterization of*

Information Requirements for Studies of CO₂ Effects, ed. M. R. White, pp. 95–143. Washington, DC: DOE/ER–0236, US Department of Energy.

Sieburth, J. (1971). Distribution and activity of oceanic bacteria. *Deep-Sea Res.*, **18**, 1111–21.

Sieburth, J., Willis, P. J., Johnson, K. M., Burney, C. M., Lavoie, D. M., Hinga, K. R., Caron, D. A., French, F. W., Johnson, P. W. and Davis, P. G. (1976). Dissolved organic matter and heterotrophic microneuston in the surface microlayers of the North Atlantic. *Science*, **194**, 1415–18.

Smyth, P. O. (1980). Callinectes (Decapoda: Portunidae) larvae in the Middle Atlantic Bight, 1975–1977. *Fish. Bull. US*, **78**, 251–65.

Stebbing, A. R. D., Soria, S., Burt, G. R. and Cleary, J. J. (1990). Water quality assays in two Bermudan harbors using the ciliate *Euplotes vannus* in relation to tributyltin distribution. *J. Exp. Mar. Ecol.*, **138**, 159–66.

Stolarski, R., Bojkov, R., Bishop, L., Zerefos, C., Staehelin, C. and Zawodny, J. (1992). Measured trends in stratospheric ozone. *Science*, **256**, 342–9.

Thain, J. (1992). Use of the oyster *Crassostrea gigas* embryo bioassay on water and sediment elutriate samples from the German Bight. *Mar. Ecol. Prog. Ser.*, **91**, 211–13.

Tully, O. and O'Ceidigh, P. (1986). The ecology of *Idotea* spp. (Isopoda) and *Gammarus locusta* on surface driftwood in Galway Bay (west of Ireland). *J. Mar. Biol. Assoc. UK*, **66**, 931–42.

Williams, P. M., Carlucci, A. F., Henrichs, S. M., Van Fleet, E. S., Horrigan, S. G., Reid, F. M. H. and Robertson, K. J. (1986). Chemical and microbiological studies of sea-surface films in the southern Gulf of California and off the west coast of Baja California. *Mar. Chem.*, **19**, 17–98.

Williams, T. D. (1992). Survival and development of copepod larvae *Tisbe battagliai* in surface microlayer, water and sediment elutriates from the German Bight. *Mar. Ecol. Prog. Ser.*, **91**, 221–8.

Wright, D. G., Hendey, R. M., Loder, J. W. and Dobson, F. W. (1986). Oceanic changes associated with global increases in atmospheric carbon dioxide: preliminary report for the Atlantic coast of Canada. *Can. Tech. Rep. Fish. Aquat. Sci.*, **1426**, 78 pp.

Zaitsev, Y. P. (1971). *Marine Neustonology* (translated from Russian). Springfield, VA: National Marine Fisheries Service, NOAA and National Science Foundation, National Technical Information Service. 207 pp.

Zaitsev, Yu. P. (1992). Recent changes in the trophic structure of the Black Sea. *Fish. Oceanogr.*, **1**, 180–9.

3

Report Group 3 – Photochemistry in the sea-surface microlayer

J. M. C. PLANE (*Chair*), N. V. BLOUGH (*Rapporteur*),
M. G. EHRHARDT, K. WATERS, R. G. ZEPP and R. G. ZIKA

Introduction

Due to the enrichment of chemicals and biota within the sea-surface microlayer, there is the widely held presumption that the surface microlayer could act as a highly efficient and selective microreactor, effectively concentrating and transforming materials brought to the interface from the atmosphere and oceans by physical processes. Rapid photochemical, chemical, and biological reactions within the microlayer could produce a variety of interesting feedbacks. For example, photochemical reactions might destroy (or produce) surface-active species, thereby altering surface wave damping and gas exchange rates. Elevated levels of highly reactive intermediates produced within this zone could present a 'reaction barrier' to the transport of some chemicals and trace gases across the air–sea interface, thus affecting their flux to the atmosphere or ocean. Further, reactions occurring within the microlayer potentially could enhance (or deplete) the surface concentrations of certain gases relative to those of bulk seawater, chemically modify compounds during their transport across the interface, alter the redox state, speciation and biological availability of trace metals deposited by the atmosphere to the interface, as well as influence the types and distributions of microlayer materials introduced to the atmosphere by bubble injection and to the deep ocean by particle settling.

Although these processes are very intriguing and potentially of great importance, in many instances evidence supporting the existence of them is lacking. Possible increases in ground-level UV-B radiation resulting from the depletion of stratospheric ozone may have important consequences for the microlayer and the processes it mediates, but the current absence of information does not allow a straightforward assessment of these possibilities.

One case where the presence of surface-layer photochemistry has been demonstrated is the oxidation of hydrocarbons associated with petroleum contamination. While hydrocarbons usually are minor components of natural surface films, in spill situations they may easily exceed their solubilities in water (in the low $10^{-6}\,g\,l^{-1}$ range for low-molecular-weight aromatic hydrocarbons, decreasing with increasing molecular weight). In this case they form a separate phase on the sea surface and mix with components of the natural surface microlayer. Fossil hydrocarbons may become principal constituents of surface films, not only in an actual oil spill, but also in marine areas exposed to chronic low-level petroleum contamination.

Recent advances in bulk-water photochemistry combined with new work on the composition and physical properties of the microlayer can provide fresh insights to the photochemical processes that may occur within this regime. To this end, in this section we will first review the photochemical properties of the major bulk-water chromophores and, where possible, will include pertinent information from studies of the microlayer. Data presented within this review will then be employed to acquire rough estimates of the rates of photochemical reactions within the microlayer. These estimates will be used to explore the possible impacts of microlayer photochemical reactions on a variety of processes occurring at the air–sea boundary.

Photochemistry in the upper ocean

A fundamental tenet of photochemistry is that light must first be absorbed for a photochemical reaction to be initiated. Thus, knowledge of the absorption spectra of chromophores residing within the microlayer is essential. Unfortunately, almost nothing is known about the optical absorption (and emission) properties of these species. Carlson and Mayer (1980) and Carlson (1982a, 1983) have shown that material absorbing at 280 nm is consistently enriched in the microlayer relative to bulk waters, and they concluded that dissolved phenolic material was the predominant light-absorbing species. This material, referred to here as coloured or chromophoric dissolved organic matter (CDOM), is a chemically complex and ill-defined mixture of anionic organic oligoelectrolytes known to contain phenolic moieties (Aiken *et al.*, 1985) and to exhibit surface-active properties (Hayase and Tsubota, 1986). In addition to chromophores associated with organic matter, a variety of other light-absorbing substances may concentrate in the sea-surface layer. These include nitrate, nitrite, transition metal complexes, keto acids, riboflavin, pteridines, algal pigments, cyanocobalamine, thiamin, biotin, and aromatic ketones. Unfortunately, not nearly enough is known about the identity and concentrations of these substances in the surface layer to perform a rigorous assessment of their photochemistry. Furthermore, until this information is acquired, possible differences between the optical properties of the microlayer and the underlying waters cannot be defined. Differences of this sort could arise from the selective partitioning of the more hydrophobic components of bulkwater CDOM into the microlayer, potentially leading to photochemical reactions within the microlayer that differ both qualitatively and quantitatively from those of bulk waters.

Major photochemical reactions

The pathways by which photochemical processes occur in solution are of two types. First, if the absorption spectrum of the substrate overlaps with the spectrum of incoming radiation, the substrate may undergo photoreaction directly. Second, solar energy may be absorbed by other substances in the solution – known as sensitizers – that then react with the substrate of primary concern, resulting in an indirect phototransformation. These reactions are mediated by transient species that are rapidly consumed by subsequent reactions. For these mechanisms, the rate of reaction is determined by the quantity and type of substrate, sensitizer, and incoming radiation. Surface gradients involving any of these factors will lead to altered rates of photoprocesses near the air–sea interface.

The penetration of solar radiation into the upper water column depends strongly on the wavelength and degree of water colour. As shown in Table 3.1, UV-B radiation is attenuated much more rapidly than visible radiation, although in clear open ocean water UV-B still penetrates to below 10 m (Smith and Baker, 1979). Most of the light-absorbing substances listed above only absorb light in the UV ($\lambda < 400$ nm) region. However, some of the unidentified chromophores in CDOM absorb light throughout the visible ($\lambda < 700$ nm) region; even though the absorption is weak, this can generate significant photochemistry below 10 m in clear water (Plane *et al.*, 1987). Another point to note in Table 3.1 is that there is effectively no attenuation of light passing through the microlayer itself.

Major photochemical intermediates in the sea include singlet oxygen, $^1O_2$; superoxide/hydroperoxide, O_2^-/HO_2; hydrogen peroxide, H_2O_2; and peroxy radicals, RO_2. $^1O_2$ is formed primarily through energy

Report Group 3

Table 3.1 *Penetration of UV-B and visible radiation in seawater – the fraction of the radiation that penetrates from the surface to a specified depth in the water column*

	Water depth (m)				
	0.01	1	10	50	100
UV-B radiation ($\lambda = 310$ nm)					
Clearest open ocean water	1.00	0.86	0.22	6×10^{-4}	—
Moderately productive water	0.99	0.40	1×10^{-4}	—	—
Visible radiation ($\lambda = 500$ nm)					
Clearest open ocean water	1.00	0.98	0.78	0.29	0.08
Moderately productive water	1.00	0.90	0.33	4×10^{-3}	2×10^{-5}

transfer from the excited triplet states of CDOM to ground state dioxygen, $^3O_2$, and wavelengths in the UV-A (315–400 nm) and UV-B (280–315 nm) regions are most effective in its formation (Zepp *et al.*, 1985). Quantum yields (the fraction or percentage of absorbed photons which give rise to products) range from 1% to 3% and generally decrease with increasing wavelength (Haag *et al.*, 1984). Because the decay of $^1O_2$ by solvent relaxation is very rapid, its loss is dominated by this pathway; the levels of $^1O_2$-reactive constituents in natural waters are generally too low to affect its steady-state concentration significantly.

The primary source of O_2^- is believed to be CDOM, although the precise reactions producing this species remain unclear. Some workers have suggested that O_2^- is formed by direct electron transfer from the excited triplet states of CDOM to O_2. However, reduction of O_2 by radicals or radical ions produced by intramolecular electron transfer reactions, H-atom abstractions and/or homolytic bond cleavages, is equally, if not more, plausible (Blough and Zepp, 1995). In the absence of O_2^--reactive compounds, the loss of O_2^- is dominated by dismutation to form H_2O_2. However, transition metal complexes having one-electron reduction potentials falling between the O_2/O_2^- and O_2^-/H_2O_2 couples can catalyze dismutation. Other potentially important sinks include reaction with NO, NO_2, phenoxy and peroxy radicals.

The primary source of H_2O_2 in the ocean is the dismutation of O_2^-, while bacteria or small phytoplankton provide the principal sink (Lean *et al.*, 1992). Hydroxyl radical, OH, is produced primarily through other mechanisms such as the photolysis of nitrite or nitrate. OH reacts with Br^- to produce Br_2^-) (Zafiriou *et al.*, 1987), which in turn reacts with carbonate

(True and Zafiriou, 1987), producing carbonate radicals. The carbonate radical may self-terminate in competition with its oxidation of organic substrates. The fate of the other product of nitrite photolysis, NO, is likely to be dominated by reaction with O_2^- (Blough and Zafiriou, 1985). This reaction yields peroxynitrite, which is also a strong oxidant and produces strong oxidants during its decomposition. The photolysis of NO_3^- also generates OH as well as NO_2. Subsequent reactions of NO_2 can produce NO_2^- and NO_3^-, thereby coupling the dynamic cycles of NO_2^- and NO_3^- photolysis. Although the oxidation of reduced metals such Fe(II) and Cu(I) by H_2O_2 can also produce OH (Moffett and Zika, 1987; Millero and Sotolongo, 1989; Zepp *et al.*, 1992), this process will be limited to those waters where significant levels of reduced metals are formed. The photolysis of H_2O_2 also is not a significant source of OH despite the high quantum yield for this reaction (0.98), due to the poor overlap between its absorption spectrum and the ground-level solar spectrum.

Peroxy radicals are thought to be formed by O_2 addition to primary (carbon-centered) radicals (R) produced photochemically from CDOM and other sensitizers via intramolecular H-atom abstractions, electron transfer reactions, and homolytic bond cleavages. These species can also be generated from secondary radicals produced by H-atom abstraction or addition reactions of RO_2 with DOM, or one-electron oxidations of DOM by CO_3^- or Br_2^-. Earlier work by Mill *et al.* (1980) and Faust and Hoigné (1987) provided evidence for the photochemical production of this class of radicals, but could not supply information on the specific RO_2 species that were formed. Indirect evidence for the formation of specific RO_2 species has been acquired by employing stable nitroxide radicals to trap the carbon-centered radicals, the immediate precursors to the RO_2 (Blough, 1988; Kieber and Blough, 1990a,b). Using a highly sensitive fluorescence detection scheme (Blough and Simpson, 1988) combined with high performance liquid chromatography, a number of low-molecular-weight radicals have been detected in seawater, principally the acetyl and methyl radicals. Evidence for the formation of radical centres on high-molecular-weight CDOM has also been obtained, although their yield appears to be significantly less than that of the low-molecular-weight species.

Quantum yields for the formation of the major species, the acetyl and methyl radicals, exhibit a different wavelength dependence and are 10- and 100-fold lower, respectively, than those for nitroxide reduction. These results imply that O_2^- production dominates RO_2 formation in surface seawaters and that different chromophores within the CDOM are responsible for the formation of these radicals. Estimated midday rates for formation of the acetylperoxy radical in natural waters range from 10^{-13}–10^{-11} M s^{-1} as compared with 10^{-11}–10^{-8} M s^{-1} for O_2^- (Blough and Zepp, 1994).

While the acetylperoxy radical may react with superoxide to form peracetic acid, the fate of the methylperoxy radical as well as the high-molecular-weight radical centres is less clear. Possible pathways include termination reactions to form non-radical and non-peroxidic products, and H-atom abstractions or addition reactions to form organic peroxides and secondary radicals.

Photoalteration and photoproducts of CDOM

That aquatic dissolved organic carbon (DOC) is photoreactive has been established by several recent studies (Kouassi *et al.*, 1990; Amador *et al.*, 1989; Kieber *et al.*, 1989; Mopper *et al.*, 1991; Miller and Zepp, 1992; Valentine and Zepp, 1993). Evidence for such photoreactivity derives in part from observed changes in the electronic absorption spectrum of the organic matter as well as in the fluorescence intensity and spectrum. The sunlight-induced decrease in absorbance is not accompanied by a corresponding change in DOC, although conversion to various UV-transparent products does occur (see below). Other studies have shown that sunlight fades the fluorescence intensity of humic substances more rapidly than the absorbance (Kouassi *et al.*, 1990).

Miller and Zepp (1992) have found that these spectral changes produce greater bleaching of the ultraviolet portion of the CDOM spectra. Thus, photodegradation of CDOM results in deeper penetration of solar UV (as compared with visible) radiation into the sea and freshwaters. This effect alone may have important effects on the carbon cycle by reducing photosynthesis and enhancing the photodegradation of CDOM.

The photoinduced fading of CDOM is accompanied by the formation of a variety of organic and inorganic compounds. The photoproducts of the biologically refractory CDOM in natural waters are low-molecular-weight compounds that are biologically labile, e.g., formaldehyde, acetaldehyde, and the α-keto acid, glyoxylate (Kieber *et al.*, 1989). Photoproduction rates of these compounds are greatest for inland and coastal waters and smallest for open ocean waters. UV-B radiation is primarily responsible for the photodegradaton of marine DOC. Francko (1990) has reviewed the rather sparse literature on effects of solar radiation on the bioavailablity of CDOM.

Carbon cycling drives the flux of other elements through ecosystems so that any change in the former has significant effects on the latter. For example, the photodegradation of CDOM is likely to result in the release of inorganic nitrogen. Photoreactions involving CDOM also have been shown to increase the biological availability of phosphorus (Francko, 1990).

In addition, trace gases such as carbon dioxide, carbon monoxide and carbonyl sulphide are formed. Carbon dioxide is the major gaseous product (Miller and Zepp, 1992). Photooxidation of CDOM is believed to be an important source of carbon monoxide in seawater, while its loss has been ascribed primarily to microbial processes (Conrad *et al.*, 1982). As a result of these two processes, carbon monoxide emissions from the sea follow a diurnal pattern, with maximum near-surface ocean concentrations greatly exceeding saturation during daylight. Although the sea is thought to be a net source of carbon monoxide, great uncertainty exists regarding the strength of this source. For example, a widely cited value for marine emissions is 10 to 40 Tg y^{-1} (Logan *et al*, 1981), but Conrad and co-workers (1982) have suggested that the oceanic source strength could range up to 180 Tg y^{-1} based on a study in the Atlantic Ocean.

Carbonyl sulphide is the most concentrated sulphur-containing gas in the troposphere (Bates *et al.*, 1992; Khalil and Rasmussen, 1984). COS is considered to be the major source of sulphate aerosols in the stratosphere during periods of quiescent volcanic activity (Crutzen, 1976; Servant, 1986; Hofmann, 1990). As such, COS affects stratospheric temperature and chemistry, including chemical processes that affect the ozone layer. These studies indicate that COS is formed primarily by photooxidation processes in the upper layers of the ocean (Ferek and Andreae, 1984), with the highest fluxes in coastal/shelf waters. Studies by Zepp and Andreae (1990) indicate that the predominant pathway for photochemical formation of COS involves photosensitized oxidation of organo-sulphur compounds by the CDOM in seawater. Wavelength studies of COS formation in coastal seawater samples have confirmed that COS is predominantly formed by the action of UV-B radiation.

Hydrocarbon degradation

With the exception of condensed aromatics, most hydrocarbons occurring in seawater, whether fossil or recently biosynthesized, are transparent to solar UV radiation at sea level and should therefore be inactive photochemically. However, surface seawater contains a variety of substances which act as photosensitizers. They include components of the dissolved organic matter commonly referred to as marine humic material (Hoigné *et al.*, 1989), and anthropogenic substances such as polycyclic aromatic ketones (Ehrhardt *et al.*, 1982; Ehrhardt and Petrick, 1993). Their first excited singlet states are generated by absorption of solar UV light quanta, and these are then converted with high intersystem crossing yields to the first excited triplet states. These are sufficiently long-lived for energy transfer to, or reaction with, non-absorbing target molecules. In bulk seawater, the most important quencher of excited triplet states of sensitizers is dissolved molecular oxygen, which is converted to $^1O_2$. Analyses of photooxidation products both in field samples and in model experiments conducted under controlled conditions have shown that the generation of carbon-centered radicals by abstraction of a hydrogen atom is the first step in complex reaction sequences leading to the oxidation and decomposition of alkanes, phenylalkanes, and cycloalkanes (Ehrhardt and Petrick, 1984, 1985a, 1985b; Ehrhardt and Weber, 1991). $^1O_2$ is not energetic enough for hydrogen abstraction, although it is capable of adding to the double bonds of olefins, thereby mediating photooxidation reactions of this class of hydrocarbons (Kearns, 1971).

Elevated concentrations of triplet sensitizers and hydrocarbons in the sea-surface microlayer apparently lead to reactions among these species which result in hydrogen abstraction. The carbon-centered radicals thus generated react with molecular oxygen to form peroxy radicals. By a variety of mechanisms the peroxy radicals can be reduced to oxyl radicals, which experimental evidence strongly suggests are key intermediates in the photooxidation reactions of hydrocarbons. Abstraction of a hydrogen atom leads to the formation of alcohols; ketones are formed by reaction with molecular oxygen. It is still an open question whether cleavage of carbon-carbon bonds leading to lower-molecular-weight carbonyl compounds and terminal alkenes takes place by the photodecomposition of intermediate ketones, or whether oxyl radicals trigger the fragmentation reactions.

Alicyclic hydrocarbons form relatively stable hydroperoxides (Ehrhardt and Weber, 1995) which may account for the increased toxicity of sunlight-illuminated oil films (Pengerud *et al.*, 1984). On the other hand, many photooxidation products of hydrocarbons are decomposed more rapidly by microorganisms than are the parent compounds (Rontani *et al.*, 1987). This effect is especially pronounced in the case of alicyclic hydrocarbons (Trudgill, 1979). There is reason to believe that more complex functionalized molecules are photodecomposed following similar reaction pathways.

Hydrocarbons are also produced by near-surface photochemically mediated reactions. Alkenes are the most reactive of these compounds and are formed in the mixed layer via photosynthetic and photochemical processes. Isoprene is produced as a by-product of phytoplankton photosynthesis (Bonsang *et al.*, 1992). Isoprene is highly reactive with ozone and various radical species in the atmosphere. It should exhibit similar reactivity in seawater, particularly in the vicinity of the microlayer where reactions with enriched levels of oxidants may limit its transport to the atmosphere. Recent evidence (Ratte *et al.*, 1993) suggests that low-molecular-weight alkenes (i.e., C_2–C_4) are produced photochemically from CDOM in the surface ocean. Like isoprene, these compounds are susceptible to oxidation in the near-surface layer, but because they are produced photochemically from DOM, they may also exhibit a production gradient from enriched CDOM in the microlayer. The hydrophobic nature of the alkenes, and particularly the higher-molecular-weight compounds like isoprene, may assist in their incorporation and oxidation within the microlayer.

Probable photochemical processes in the microlayer

Assuming that microlayer CDOM exhibits absorption spectra and quantum yields similar to those of the underlying bulk waters, currently available data can be employed within a simple model to estimate photochemical formation rates and fluxes of species in the microlayer. In this model, the formation rates, $F_M(\lambda)$, and fluxes, $Y_M(\lambda)$, are given by

$$F_M(\lambda) = E_0(\lambda) \cdot \Phi(\lambda) \cdot a(\lambda)_{i,M} \tag{3.1}$$

$$Y_M(\lambda) = E_0(\lambda) \cdot \Phi(\lambda) \cdot a(\lambda)_{i,M} \cdot z_M \tag{3.2}$$

where $E_0(\lambda)$ is the downwelling irradiance at the surface, $\Phi(\lambda)$ is the photochemical quantum yield, $a(\lambda)$ is the absorption coefficient of the species undergoing reaction and z_M is the microlayer depth. Integration over wavelength provides the total photochemical formation rate (F_M) and flux (Y_M) in the microlayer:

$$F_M = \int E_0(\lambda) \cdot \Phi(\lambda) \cdot a(\lambda)_{i,M} \, d\lambda \tag{3.3}$$

$$Y_M = F_M \cdot z_M \tag{3.4}$$

For CDOM photoreactions, Y_M can be compared with the total water column flux $(z \rightarrow \infty)$,

$$Y = \int E_0 (\lambda) \cdot \Phi(\lambda) \, d\lambda \qquad (3.5)$$

assuming that light absorption by CDOM dominates, as it often does in coastal/shelf waters over the 300–400 nm wavelength range (Vodacek et al., 1994).

Equations 3.1–3.4 indicate that F_M and Y_M are directly proportional to the absorption coefficients of the material in the microlayer. Thus, for a two-fold enrichment in absorption, F_M and Y_M will be twice as large as those obtained for an equivalent depth interval just beneath the micro-layer. However, the dilution of material by sampling devices that collect over depths greater than that of the microlayer can produce significant underestimates of F_M and slight overestimates of Y_M. For example, if the true microlayer depth was 1 µm, the use of a 50 µm sampler would produce a 15-fold underestimate of F_M and a 3.3-fold overestimate of Y_M for an observed absorption enrichment of 1.4 relative to bulk water. These dilution effects can be calculated from the relation

$$a(\lambda)_M = a(\lambda)_B \cdot [1 + z_s/z_M(E-1)] \qquad (3.6)$$

where $a(\lambda)_M$ and $a(\lambda)_B$ are the microlayer and bulk absorption co-efficients, respectively, E is the apparent enrichment factor for samples collected from depth z_s, and z_M is the true microlayer depth.

In principle, the elevated levels of highly reactive species produced photochemically within the microlayer could act as a barrier to the exchange of certain compounds across the interface. However, ignoring atmospheric deposition, the 1.5- to 2-fold enrichments of light-absorbing material observed by Carlson (1982a) would produce, at most, a 1.5- to 2-fold increase in the steady-state levels of reactive species in the micro-layer (Equation 3.1). Thus, the levels of trace constituents passing through the microlayer would be reduced by only two-fold over that for an equivalent depth interval just under the microlayer. If z_M is signifi-cantly smaller than z_s (Equation 3.6), this loss would be reduced even further; while a 50-fold reduction in z_M for an observed enrichment of 1.4 yields a 15-fold increase in production rate, the residence time (τ) for a species passing through this smaller layer is decreased 2500-fold due to the squared length dependence in the diffusion equation,

$$\tau = z_M^2/ 2D \qquad (3.7)$$

where D is the molecular diffusion coefficient. The significantly larger fluxes of reactive species generated in the near-surface bulk waters will act as the principal determinant of the surface concentrations of these compounds. This 'barrier' will be important only if the material in the microlayer is significantly more photoreactive than that of the bulk, or if the microlayer material produces a different ensemble of reactive intermediates that exhibit selectivity for certain compounds.

In situ *trace gas production*

CO, CO_2, and COS (in coastal regions) will all be formed through the photochemical degradation of CDOM in the microlayer. However, given the low enrichments for CDOM, photoproduction rates are not likely to be more than two-fold over bulk, and thus will make a negligible contribution to the total water column flux.

Atmospheric inputs of trace species

As the microlayer forms the interface between the ocean and the atmosphere, it is instructive to compare the atmospheric deposition of reactive species such as H_2O_2 and peroxy radicals with their *in situ* photochemical production in the microlayer. The depositional fluxes of a number of atmospheric species have been modelled by Thompson and Zafiriou (1983). *In situ* production can be estimated by assuming a typical microlayer thickness of 50 μm, and a two-fold enrichment of organic chromophores in the microlayer over the bulk-water column. This analysis indicates that atmospheric deposition of H_2O_2 and CH_3OO will far outweigh their *in situ* photochemical production within the microlayer (Table 3.2), unless the microlayer contains material substantially more photoreactive than the bulk-water CDOM on which these calculations are based. Wet deposition of H_2O_2 from fog and rain may constitute a significant input of this species to the microlayer over brief time periods, but water column production generally exceeds this input. As shown in Table 3.2, the photochemical flux of O_2^- within the microlayer is similar to the atmospheric flux of HO_2. O_2^- is likely to be important in the microlayer due to its efficient reaction with NO, phenoxy radicals, peroxy radicals, transition metal ions and possibly chlorinated hydrocarbons. Although the water column fluxes of O_2^- and H_2O_2 are usually far larger than the sum of their atmospheric and microlayer fluxes (Table 3.2), the effectively higher inputs of these very reactive species within this boundary due to atmospheric deposition and *in situ* photochemistry should lead

Table 3.2 *Estimated formation rates and fluxes for species generated photochemically within the microlayer, within the water column, and through dry deposition from the atmosphere[a]*

Species	F_M ($cm^{-3} s^{-1}$)	Y_M ($cm^{-2} s^{-1}$)[b]	Y ($cm^{-2} s^{-1}$)	Air–sea flux[c] ($cm^{-2} s^{-1}$)
O_2^-/HO_2	7.6×10^{10}	3.8×10^8	3.8×10^{12}	1.0×10^8
H_2O_2	1.8×10^{10}	8.9×10^7	7.6×10^{11}	1.4×10^{10}
$CH_3\bullet(CH_3OO\bullet)$	1.3×10^8	6.6×10^5	5.1×10^9	1.1×10^8
$CH_3C(O)\bullet(CH_3C(O)OO\bullet)$	6.3×10^8	3.2×10^6	1.9×10^{10}	—
CO	2.0×10^9	9.8×10^6	9.2×10^{10}	—

[a]24 hour averages. [b]Assumes a 50 μm microlayer. [c]From Thompson and Zafiriou (1983).

to a more rapid oxidative turnover of materials within this 'microreactor' and potentially to reactions not observed in bulk waters. The largest depositional flux of an atmospheric oxidant is that of ozone, a compound known to play a key role in the release of volatile iodine from the sea surface (Garland and Curtis, 1981; Thompson and Zafiriou, 1983).

The recent suggestion that a lack of available Fe limits primary production in certain open ocean waters (Martin and Fitzwater, 1988) has intensified interest in the transport and photochemical reactions of Fe species in both seawater (Wells *et al.*, 1991; Wells and Mayer, 1991) and atmospheric aerosols (Duce, 1986; Behra and Sigg, 1990; Zhuang *et al.*, 1990, 1992; Zhuang and Duce, 1992; Zuo and Hoigné, 1992; Zhu *et al.*, 1993; Erel *et al.*, 1993). At the oxygen concentrations and pH of surface seawaters, very little biologically available Fe(III) is expected to exist due to the high thermodynamic stability of the colloidal iron (hydr) oxides. At lower pH, where Fe(II) is more slowly oxidized by O_2 and H_2O_2 (Moffet and Zika, 1987; Millero and Sotolongo, 1989), evidence for the photo-reductive dissolution of colloidal iron oxides by CDOM has been obtained by several groups (Waite and Morel, 1984; Sulzberger *et al.*, 1989). While photoreduction still occurs at the higher pH of seawater, the Fe(II) that is produced appears to be oxidized more rapidly than its detachment from the oxide surface, so that little if any escapes to solution (see, however, O'Sullivan *et al.*, 1991, and King *et al.*, 1991). Wells *et al.* (1991) have found that the chemical availability of iron as determined by chelation with 8-hydroxyquinoline (oxine) is strongly correlated with the growth rates of marine phytoplankton and that sunlight increases this availability. These workers have suggested that CDOM-driven cycles of light-induced reduction and rapid reoxidation lead to the formation of amorphous Fe(III) precipitates that do serve as a source of available Fe (Wells and Mayer, 1991; Wells *et al.*, 1991; Waite, 1990).

Clear evidence for the photochemical generation of significant levels of Fe(II) in atmospheric aqueous phases (at lower pH) has been acquired by a number of groups (Faust, 1994; Sulzberger *et al.*, 1994; Waite and Szymczak, 1994). The fate of this Fe after deposition to the sea surface is not clear. However, higher levels of CDOM within the microlayer could act to maintain or increase the biological availability of this Fe through complexation and photochemical reduction at this interface. Manganese oxides undergo light-induced reductive dissolution in seawaters (Sunda *et al.*, 1983; Sunda and Huntsman, 1988). The product, Mn(II), is kinetically stable to oxidation in the absence of certain bacteria that are known to be photoinhibited in the upper ocean. These two effects of sunlight combine to produce a surface maximum in soluble Mn, unlike most transition metals which are depleted in surface waters due to biological removal processes. A more detailed discussion of transition metal photochemistry and its role in environmental processes is provided in Blough and Zepp (1994).

In situ *halogen chemistry*

The steady-state concentrations of reactive halogen species in the surface microlayer should be high relative to their concentrations in the bulk water, being maintained by atmospheric deposition of oxidants and photoreactions generating OH. The principal source of these species is through the deposition of O_3, which will react with I^- to such intermediates as I_2, HOI and possibly I_2^- (Garland and Curtis, 1981). Another photochemical source of reactive iodine species is homolytic fission of iodo-organic compounds. Br_2^- is another reactive halogen species that is formed through photoreactions in the microlayer.

The primary reactions of these halogen species should be oxidation of organic substrates, but substitution reactions of aromatic and unsaturated compounds are also possible (Waite *et al.*, 1988). Halogenation of organic compounds can also take place via chloro- and bromo-peroxidases, which employ hydrogen peroxide as the oxidant (Zepp and Ritmiller, 1995 and references therein).

Conversion of DOC

The characteristics of the sea-surface microlayer combine to produce an environment that is unique with respect to transformation of dissolved (DOM) and particulate organic matter (POM). Hence, DOM within the interface may be converted to POM by aggregation or polymerization, or by compression associated with wave motions or bubble collapse, or by adsorption to other oceanic or atmospherically derived particles in the interface (Carlson, 1993). Collected microlayer samples can be strongly UV-absorbing relative to the bulk water (Carlson, 1982a,b), implying that photochemical processes may be important in transformations between and within the DOM and POM fraction. Harvey *et al.* (1983), for example, showed that lipids could undergo photoinduced oxidation and cross-linking to produce polymers having properties similar to marine humus. The air–sea interface forces the aggregation of photosensitizers and redox species. Reactive substrates and the rapid replenishment of these species should make the microlayer a dynamic photochemically active boundary layer between the ocean and atmosphere. This should lead not only to particle formation, but also fragmentation, solubilization, mineralization and structural modification of the POM and DOM fractions that reside in this layer.

The consequences of the photochemically active boundary layer on the DOM and POM are numerous. One of the important consequences involves particulate formation and subsequent sinking. The process could potentially remove deleterious anthropogenic compounds or toxic metals to the sediments. Processes of molecular fragmentation or mineralization could process toxic and refractory materials into more readily biodegradable components. Modification of hydrophobic compounds (e.g., hydrocarbons and pigments) into surfactants via introduction of polar functional groups could be important in terms of producing, maintaining and modifying the nature of the organic sea-surface layer.

Hydrocarbon degradation

There is some evidence to indicate that sunlight illumination increases the toxicity of surface films of petroleum (Pengerud *et al.*, 1984). The enhanced toxicity has been attributed to stable hydroperoxides (Larson *et al.*, 1992), which cycloalkanes form quite readily (Erhardt and Weber, 1995). Preliminary tests using luminescent marine bacteria as indicator organisms suggest that the photooxidation of phenyl alkanes results in somewhat lower toxicity, although the LC_{50} doses remained in the same order of magnitude (Burns and Ehrhardt, unpubl.). Effects of enhanced toxicity may be offset by easier microbial degradability of other hydrocarbon photooxidation products, such as alcohols and ketones.

Global change issues

Changes in ultraviolet radiation

The depletion of stratospheric ozone is causing an increase in the levels of UV-B radiation reaching the surface, particularly at high latitudes. This change will affect photochemical processes with action spectra that are significant in the UV-B spectral region, relative to longer wavelengths (see Figure 13.16). Such processes include the photolysis of ketones, halogenated hydrocarbons, phenols, and nitrate ions. A further consideration is that some processes will probably show a nonlinear rate enhancement with increasing UV-B. For instance, the O_2^- radical is produced photochemically but also undergoes significant removal by reaction with other species mediated by photochemistry, such as NO, NO_2 and transition metals.

Increased UV-B can enhance or reduce the production of atmospherically important trace gases in the sea through changes in biological and photochemical processes at the sea surface (SCOPE, 1993). These changes may trigger significant feedback mechanisms. For example, the increased evasion of CO reduces the hydroxyl radical concentration in the troposphere. This permits greater levels of CH_4 to reach the stratosphere, leading to increased stratospheric H_2O concentrations and a higher likelihood of stratospheric ice cloud formation and chlorine-catalysed O_3 depletion. Marine emissions of CH_3Cl and CH_3Br also affect stratospheric O_3, whereas bromoform and other brominated gases have been implicated in tropospheric O_3 removal in polar regions. Increased evasion rates of isoprene and other non-methane hydrocarbons increase tropospheric O_3 in regions with elevated concentrations of nitrogen oxides. COS provides a major source of sulphate aerosols in the lower stratosphere (see page 77). These aerosols provide nuclei for polar stratospheric clouds and may also be involved directly in heterogeneous chemistry at low temperatures, enhancing stratospheric ozone depletion.

Dimethyl sulphide (DMS) is a sulphur-containing substance that is produced in the ocean and is known to be oxidized in the troposphere to species such as methane sulphonic acid and sulphate. This oxidation process could increase the condensation nuclei available for cloud formation in the marine troposphere (Bates *et al.*, 1992). An increase in cloud cover would decrease the UV-B levels reaching the sea surface and also affect the tropospheric heat budget. UV-B radiation potentially can affect fluxes of DMS through effects on phytoplankton that produce its precursor, dimethylsulphoniopropionate, on bacteria that consume DMS (Kiene and Bates, 1990), and on the photooxidation of DMS (Brimblecombe and Shooter, 1986). Interactions of these processes have major effects on the sea-to-air flux of DMS: presently, only about 3% of the DMS that is produced in the upper ocean actually is transferred to the atmosphere (Kiene and Bates, 1990). Thus, future changes in solar UV-B could potentially have major effects on release of this gas from the sea surface.

Radiation balance

The ocean provides a large heat reservoir. Changes in the heat content of the ocean have the potential for global changes in climate. Therefore, microlayer processes could affect the transmission of radiation between the ocean and the atmosphere. The ocean, like any other medium, will heat or cool until the input energy matches the output energy. For the ocean the input energy includes the short-wave (solar) radiation from 0.3 to 4.0 µm, and the output energy includes the long-wave blackbody radiation peaked near 10 µm. The long-wave emission is controlled by the emissivity of the medium and the temperature. The short-wave input energy is controlled by many factors, including the reflectivity of the surface. This is determined by the real part of the index of refraction and the incident angle of the radiation.

The long-wave radiation could be affected by the microlayer if this has a different emissivity than water in the spectral region near 10 µm. While it is likely that the emissivity of an organic film is different from that of water, the microlayer is not thick enough to absorb enough of the long-wave blackbody radiation originating within the water to have any significant effect on the total long-wave emission.

The index of refraction of an organic microlayer is likely to be higher than that of water in the short-wave region of the spectrum. This would increase the reflectance and cause a decrease in the heat transport into the ocean. Note that the organic film must be sufficiently thick, of the order of the wavelength of interest, for Fresnel's equations of reflectance to apply.

If we assume an index of refraction for an organic microfilm of 1.5 (water = 1.34) and examine the reflectance for the case of a normally incident solar beam, the reflectance could increase from 2.1% to 4.0% in the presence of the organic film. For heat budget calculations, this could represent a significant decrease in short-wave energy input. The actual loss of radiation will be a function of the solar zenith angle and the capillary wave spectrum, which is itself governed by the nature of the microlayer.

Conclusions and recommendations

Available evidence suggests that photochemical processes in the micro-layer may not differ substantially from those of the near-surface bulk waters. Assuming the presence of a 50 μm thick microlayer that is enriched in organic chromophores by two-fold over the bulk, it is concluded that the fluxes of reactive intermediates appear to be too small to affect significantly the transfer of reactive species across the air–sea interface. Further, the flux resulting from photochemical production of gases from within the microlayer is insignificant with respect to the total flux from the water column. It should be noted, moreover, that these processes do not appear to be limited by the transfer of photoreactive material from the underlying bulk waters.

There are several contexts in which microlayer photochemical processes may prove to be important. These are likely to include: 1. photochemical reactions within hydrocarbon slicks, 2. the fate of certain atmospheric species following deposition, and 3. changes in physical properties of the microlayer due to phototransformation of its constituents.

It should be noted that most of our conclusions are not based on direct experimental evidence. This situation has arisen in part through the lack of suitable techniques for studying photochemical processes within the microlayer on the required spatial and temporal scales. These limitations may be redressed by the application of new chemical and spectroscopic techniques.

Recommendations for further work in this area include (i) determination of the spectral characteristics of the microlayer in a range of environments; (ii) acquisition of action spectra for photochemical reactions of individual compounds that are enriched in the microlayer; (iii) determination of the effects of the photochemical alteration of the microlayer on the air–sea exchange of atmospherically important trace gases; (iv) evaluation of the impact of the deposition of O_3 and other tropospheric oxidants on the microlayer; (v) determination of the influence of the microlayer on the sea-surface albedo and its susceptibility to photochemical alteration; and (vi) evaluation of the rates of photo-induced particle formation and removal and the role of these processes in the scavenging of toxic compounds from the upper ocean.

References

Aiken, G. R., McKnight, D. M., Wershaw, R. L. and MacCarthy, P. (eds.) (1985). *Humic Substances in Soil, Sediment and Water: Geochemistry, Isolation and Characterization*. New York: Wiley Interscience, 691 pp.

Amador, J. A., Alexander, M. and Zika, R. G. (1989). Sequential photochemical and microbial degradation of organic molecules bound to humic acid. *Appl. Environ. Microbiol.*, **55**, 2843.

Bates, T. S., Lamb, B. K., Guenther, A., Dignon, J. and Stoiber, R. E. (1992). Sulphur emissions to the atmosphere from natural sources. *J. Atmos. Chem.*, **14**, 315–37.

Behra, S. and Sigg, L. (1990). Evidence for the redox cycling of iron in atmospheric water droplets. *Nature*, **344**, 419–21.

Blough, N. V. (1988). Electron paramagnetic resonance measurements of photochemical radical production in humic substances. 1. Effects of O_2 and charge on radical scavenging by nitroxides. *Environ. Sci. Technol.*, **22**, 77–82.

Blough, N. V. and Simpson, D. J. (1988). Chemically-mediated fluorescence yield switching in nitroxide–fluorophore adducts: optical sensors of radical/ redox reactions. *J. Am. Chem. Soc.*, **110**, 1915–17.

Blough, N. V. and Zafiriou, O. C. (1985). Reactions of superoxide with nitric oxide to form peroxonitrate in alkaline aqueous solution. *Inorg. Chem.*, **24**, 3502–4.

Blough, N. V. and Zepp, R. G. (1995). Reactive oxygen species in natural waters. In *Active Oxygen in Chemistry*, ed. C. S. Foote *et al.*, pp. 280–333. London: Chapman and Hall.

Bonsang, B., Polle, C. and Lambert, G. (1992). Evidence for marine production of isoprene. *Geophys. Res. Lett.*, **19**, 1129–32.

Brimblecombe, P. and Shooter, D. (1986). Photooxidation of dimethylsulphide in aqueous solution. *Mar. Chem.*, **19**, 343–53.

Carlson, D. J. (1982a). Surface microlayer phenolic enrichments indicate sea surface slicks. *Nature*, **296**, 426–9.

Carlson, D. J. (1982b). A field evaluation of plate and screen microlayer sampling techniques. *Mar. Chem.*, **11**, 189–208.

Carlson, D. (1983). Dissolved organic materials in surface microlayers: temporal and spatial variability and relation to sea state. *Limnol. Oceanogr.*, **28**, 415–31.

Carlson, D. J. (1993). The early diagenesis of organic matter: reaction at the air– sea interface. In *Organic Geochemistry*, ed. M. H. Engel and S. A. Macko, pp. 255–68. New York: Plenum.

Carlson, D. J. and Mayer, L. M. (1980). Enrichment of dissolved phenolic organic material in the surface microlayer of coastal waters. *Nature*, **286**, 482–3.

Conrad, R., Seiler, W., Bunse, G. and Giehl, H. (1982). Carbon monoxide in seawater (Atlantic Ocean). *J. Geophys. Res.*, **87**, 8852–93.

Crutzen, P. J. (1976). The possible importance of COS for the sulphate layer of the stratosphere. *Geophys. Res. Lett.*, **3**, 73–6.

Duce, R. A. (1986). The impact of atmospheric nitrogen, phosphorus, and iron species on marine biological productivity. In *The Role of Air–Sea Exchange in Geochemical Cycling*, ed. P. Buat-Menard, pp. 497–529. Norwell, MA: D. Reidel.

Ehrhardt, M. G. and Petrick, G. (1984). On the sensitized photo-oxidation of alkylbenzenes in seawater. *Mar. Chem.*, **15**, 47–85.

Ehrhardt, M. G. and Petrick, G. (1985a). The sensitized photo-oxidation of n-pentadecane as a model for abiotic decomposition of aliphatic hydro-carbons in seawater. *Mar. Chem.*, **16**, 227–38.

Ehrhardt, M. G. and Petrick, G. (1985b). The generation of γ-diketones by sensitized photo-oxidation of n-alkanes under simulated environmental con-ditions. *Proc. Am. Chem. Soc.*, **189**, 35.

Ehrhardt, M. G. and Petrick, G. (1993). On the composition of dissolved and particle associated fossil fuel residues in Mediterranean surface water. *Mar. Chem.*, **42**, 57–70.

Ehrhardt, M. G. and Weber, R. R. (1991). Formation of low molecular weight carbonyl compounds by sensitized photochemical decomposition of aliphatic hydrocarbons in seawater. *Fresnius J. Anal. Chem.*, **339**, 772–6.

Ehrhardt, M. G. and Weber, R. R. (1995). Sensitized photo-oxidation of methyl-cyclohexane as a thin film on seawater by irradiation with natural sunlight. *Fresnius J. Anal. Chem.*, **352**, 357–63.

Ehrhardt, M. G., Bouchertall, F. and Hopf, H-P. (1982). Aromatic ketones con-centrated from Baltic Sea water. *Mar. Chem.*, **11**, 449–61.

Erel, Y., Pehkonen, S. O. and Hoffman, M. R. (1993). Redox chemistry of iron in fog and stratus clouds. *J. Geophys. Res.*, **98**, 18 423–34.

Faust, B. C. (1994). A review of the photochemical redox reactions of Fe(III) species in atmospheric, oceanic and surface waters: influences on geo-chemical cycles and oxidant formation. In *Aquatic and Surface Photo-chemistry*, ed G. R. Helz, R. G. Zepp and D. G. Crosby, pp. 3–37. Ann Arbor, MI: Lewis Publishers.

Faust, B. C. and Hoigné, J. (1987). Sensitized photooxidation of phenols by fulvic acid and in natural waters. *Environ. Sci. Technol.*, **21**, 957–64.

Ferek, R. J. and Andreae, M. O. (1984). Photochemical production of carbonyl sulphide in marine surface waters. *Nature*, **307**, 148–50.

Francko, D. A. (1990). Alteration of bioavailability and toxicity by photo-transformation of organic acids. In *Organic Acids in Aquatic Ecosystems*, ed. E. M. Perdue and E. T. Gjessing, pp. 167–77. Chichester: John Wiley and Sons.

Garland, J. A. and Curtis, H. (1981). Emissions of iodine from the sea surface in the presence of ozone. *J. Geophys. Res.*, **86**, 3183–6.

Haag, W. R., Hoigné, J., Gassman, E. and Braun, A. M. (1984). Singlet oxygen in surface waters. Part II. Quantum yields of its production by some natural humic materials as a function of wavelength. *Chemosphere*, **13**, 641–50.

Harvey, G. R., Boran, D. A., Chesal, L. A. and Tokar, J. M. (1983). The structure of marine humic acids. *Mar. Chem.*, **12**, 119–32.

Hayase, K. and Tsubota, H. (1986). Monolayer properties of sedimentary humic acid at the air-water interface. *J. Colloid. Intr. Sci.*, **114**, 220–6.

Hofmann, D. J. (1990). Increase in the stratospheric background sulfuric acid aerosol mass in the past ten years. *Science*, **248**, 996–1000.

Hoigné, J., Faust, B. C., Haag, W. R., Scully, F. E. Jr and Zepp, R. G. (1989). Aquatic humic substances as sources and sinks of photochemically produced transient reactants. *A CS Adv. Chem. Ser.*, **219**, 363–81.

Kearns, D. R. (1971). Physical and chemical properties of singlet molecular oxygen. *Chem. Rev.*, **71**, 395–427.

Khalil, M. A. K. and Rasmussen, R. A. (1984). Global sources, lifetimes and mass balances of carbonyl sulphide (OCS) and carbon disulphide (CS_2) in the Earth's atmosphere. *Atmos. Environ.*, **18**, 1805–13.

Kieber, D. J. and Blough, N. V. (1990a). Fluorescence detection of carbon-centered radicals in aqueous solution. *Free Rad. Res. Comm.*, **10**, 109–17.

Kieber, D. J. and Blough, N. V. (1990b). Determination of carbon-centered radicals in aqueous solution by liquid chromatography with fluorescence detection. *Anal. Chem.*, **62**, 2275–83.

Kieber, D. J., McDaniel, J. and Mopper, K. (1989). Photochemical source of biological substrates in sea water: implications for carbon cycling. *Nature*, **341**, 637–9.

Kiene, R. and Bates, T. S. (1990). Biological removal of dimethyl sulphide from seawater. *Nature*, **345**, 702–5.

King, D. W., Lin, J. and Kester, D. R. (1991). Determination of Fe(II) in seawater at nanomolar concentrations. *Anal. Chim. Acta.*, **247**, 125–32.

Kouassi, M., Zika, R. G. and Plane, J. M. C. (1990). Photochemical modelling of marine humus fluorescence in the ocean, *Neth. J. Sea Res.* **27**, 33–41.

Larson, R. A., Bott, T. L., Hunt, L. L. and Rogenmuser, K. (1992). Photo-oxidation products of a fuel oil and their antimicrobial activity. *Environ. Sci. Technol.*, **13**, 965–9.

Lean, D. R. S., Cooper, W. J. and Pick, F. R. (1992). Hydrogen peroxide (H_2O_2) formation and decay in lakewaters. *Preprint extended abstracts, Division of Environmental Chemistry, A CS National Meeting, San Francisco*, Vol. 32, pp. 80–3.

Logan J. A., Prather, J. M., Wofsy, S. C. and McElroy, M. B. (1981). Tropospheric chemistry: a global perspective. *J. Geophys. Res.*, **96**, 7210–54.

Martin, J. H. and Fitzwater, S. E. (1988). Iron deficiency limits phytoplankton growth in the northeast Pacific subarctic. *Nature*, **331**, 341–3.

Mill, T., Hendry, D. G. and Richardson H. (1980). Free-radical oxidants in natural waters. *Science*, **207**, 886–7.

Miller, W. L. and Zepp, R. G. (1992). Photochemical carbon cycling in aquatic environments: formation of atmospheric carbon dioxide and carbon monoxide. *Preprint extended abstracts, Division of Environmental Chemistry, A CS National Meeting, San Francisco*, Vol. 32, pp. 158–60.

Millero, F. J. and Sotolongo, S. (1989). The oxidation of Fe(II) with H_2O_2 in seawater. *Geochim. Cosmochim. Acta*, **53**, 1867–73.

Moffet, J. W. and Zika, R. G. (1987). Reaction kinetics of hydrogen peroxide with copper and iron in seawater. *Environ. Sci. Technol.*, **21**, 804–10.

Mopper, K., Zhou, X., Kieber, R. J. *et al.* (1991). Photochemical degradation of dissolved organic carbon and its impact on the oceanic carbon cycle. *Nature*, **353**, 60–2.

O'Sullivan, D., Hanson, A. K., Miller, W. L. and Kester, D. R. (1991). Measurement of Fe(II) in surface water of the equatorial Pacific. *Limnol. Oceanogr.*, **36**, 1727–41.

Pengerud, H., Thingstad, F., Tjessem, K. and Aaberg, A. (1984). Photo-induced toxicity of North Sea crude oils toward bacterial activity. *Environ. Sci. Technol.*, **15**, 142–6.

Plane, J. M. C., Zika, R. G., Zepp, R. G. and Burns, L. A. (1987). Modeling applied to natural waters. In *Photochemistry of Environmental Aquatic Systems*, ed. R. G. Zika and W. J. Cooper, pp. 250–67. *ACS Adv Chem. Series*, **327**.

Ratte, M., Plass-Dülmer, C., Koppsmann, R. and Rudolph, J. (1993). Production mechanism of C_2–C_4 hydrocarbons in seawater. *Global Biogeochem. Cycles*, **7**, 369–78.

Rontani, J. F., Bonin, P. and Giusti, G. (1987). Mechanistic study of interactions between photo-oxidation and biodegradation of n-nonylbenzene in seawater. *Mar. Chem.*, **22**, 1–12.

SCOPE (1993). *Effects of Increased Ultraviolet Radiation on Global Ecosystems.* Paris, France: Scientific Committee on Problems of the Environment (SCOPE).

Servant, J. (1986). The burden of the sulphate layer of the stratosphere during volcanic 'quiescent' periods. *Tellus*, **38B**, 74–9.

Smith, R. C. and Baker, K. S. (1979). Penetration of UV-B and biologically effective dose-rates in natural waters. *Photochem. Photobiol.*, **29**, 311–23.

Sulzburger, B., Suter, D., Siffert, C. *et al.* (1989). Dissolution of Fe(III) (hydr) oxides in natural waters: laboratory assessment of the kinetics controlled by surface chemistry. *Mar. Chem.*, **28**, 127–44.

Sulzburger, B., Laubscher, H. and Karametaxas, G. (1994). Photoredox reactions at the surface of iron (III) (hydr)oxides. In *Aquatic and Surface Photochemistry*, G. R. Helz, R. G. Zepp and D. G. Crosby, pp. 53–73. Ann Arbor, MI: Lewis Publishers.

Sunda, W. G. and Huntsman, S. A. (1988). Effect of sunlight on redox cycles of manganese in the southwestern Sargasso Sea. *Deep-Sea Res.*, **35**, 1297–317.

Sunda, W. G., Huntsman, S. A. and Harvey, G. R. (1983). Photoreduction of manganese oxides in seawater and its geochemical and biological implications. *Nature*, **301**, 234–6.

Thompson, A. M. and Zafiriou, O. C. (1983). Air–sea fluxes of transient atmospheric species, *J. Geophys. Res.*, **88**, 6696–708.

Trudgill, P. W. (1979). Microbial degradation of alicyclic hydrocarbons. In *Developments in Biodegradation of Hydrocarbons-1*, ed. R. J. Watkinson, pp. 47–84. London: Applied Science Publishers.

True, M. B. and Zafiriou, O. C. (1987). Reaction of Br_2^- produced by flash photolysis of sea water with components of the dissolved carbonate system. In *Photochemistry of Environmental Aquatic Systems*, ed. R. G. Zika and W. J. Cooper, pp. 106–15. *ACS Adv. Chem. Ser.*, **327**.

Valentine, R. L. and Zepp, R. G. (1993). Formation of carbon monoxide from the photodegradation of terrestrially dissolved organic carbon in natural waters. *Environ. Sci. Technol.*, **27**, 409–12.

Vodacek, A., Green, S. A. and Blough, N. V. (1994). An experimental model of the solar-stimulated fluorescence of chromophoric dissolved organic matter. *Limnol. Oceanogr.*, **39**, 1–11.

Waite, T. D. (1990). Photoprocesses involving colloidal iron and manganese oxides in aquatic environments. In *Effects of Solar Ultraviolet Radiation on Biogeochemical Dynamics in Aquatic Environments*, ed. N. V. Blough and R. G. Zepp, pp. 97–101. Woods Hole Oceanographic Institution Technical Report WHOI-90-9.

Waite, T. D. and Morel, F. M. M. (1984). Photoreductive dissolution of colloidal iron oxides in natural waters. *Environ. Sci. Technol.*, **18**, 860–8.

Waite, T. D. and Szymczak, R. (1994). Photoredox transformations of iron and manganese in marine systems: review of recent field investigations. In *Aquatic and Surface Photochemistry*, ed. G. R. Helz, R. G. Zepp and D. G. Crosby, pp. 39–52. Ann Arbor, MI: Lewis Publishers.

Waite, T. D., Sawyer, D. T. and Zafiriou O. C. (1988). Oceanic chemical reactive chemical transients. *Appl. Geochem.*, **3**, 9–17.

Wells, M. L., and Mayer, L. M. (1991). The photoconversion of colloidal iron oxyhydroxides in seawater. *Deep-Sea Res.*, **38**, 1379–95.

Wells, M. L., Mayer, L. M., Donard, O. F. X., de Souza Sierra, M. M. and Ackelson, S. G. (1991). The photoloysis of colloidal iron in the oceans. *Nature*, **353**, 248–50.

Zafiriou, O. C., True, M. B. and Hayon, E. (1987). Consequences of OH radical reaction in sea water: Formation and decay of Br_2^- ion radical. In *Photochemistry of Environmental Aquatic Systems*, ed. R. G. Zika and W. J. Cooper, pp. 89–105. *ACS Adv. Chem. Ser.*, **327**.

Zepp, R. G. and Andreae, M. O. (1990). Photosensitized formation of carbonyl sulphide in seawater. In *Effects of Solar Ultraviolet Radiation on Biogeochemical Dynamics in Aquatic Environments*, ed. N. V. Blough and R. G. Zepp, p. 180. Woods Hole Oceanographic Technical Report WHOI–90–09.

Zepp, R. G. and Ritmiller, L. F. (1995). Photoreactions providing sinks and sources of halocarbons in aquatic environments. In *Aquatic Chemistry: Interfacial and Interspecies Processes*, eds C. P. Huang, C. R. O'Melia, and J. J. Morgan, Chapter 13, pp. 253–78. *ACS Adv. Chem. Ser.*, **244**.

Zepp, R. G., Schlotzhauer, P. and Sink, M. R. (1985). Photosensitized transformations involving electronic energy transfer in natural waters: Role of humic substances. *Environ. Sci. Technol.*, **19**, 74–81.

Zepp, R. G., Faust, B. C. and Hoigné, J. (1992). Hydroxyl radical formation in aqueous reactions (pH 3–8) of iron(II) with hydrogen peroxide: the photo-Fenton reaction. *Environ. Sci. Technol.*, **26**, 313–19.

Zhu, X. R., Prospero, J. M., Savoie, D. L., Millero, F. J., Zika, R. G. and Saltzman, E. S. (1993). Photoreduction of iron(III) in mineral aerosol solutions. *J. Geophys. Res.*, **98**, 9039–46.

Zhuang, G. and Duce, R. A. (1992). The chemistry of iron in marine aerosols. *Global Biogeoeochem. Cycles*, **6**, 161–73.

Zhuang, G., Duce, R. A. and Kester, D. R. (1990). The dissolution of atmospheric iron in surface seawater of the open ocean. *J. Geophys. Res.*, **95**, 16207–16.

Zhuang, G., Yi, J., Duce, R. A. and Brown, P. R. (1992). Link between iron and sulphur cycles suggested by detection of Fe(II) in remote marine aerosols. *Nature*, **355**, 537–9.

Zuo, Y. and Hoigné, J. (1992). Formation of hydrogen peroxide and depletion of oxalic acid in atmospheric water by photolysis of iron(III)-oxalato compounds. *Environ. Sci. Technol.*, **26**, 1014–22.

4

Transport processes in the sea-surface microlayer

LUTZ HASSE

Abstract

The gas flux of weakly soluble gases through the air–sea interface is controlled by the transport mechanism in the aqueous diffusive boundary layer. The combination of molecular and turbulent transport and of secondary motions near the interface determines the exchange rate. This layer is difficult to access experimentally, so a combination of observation and physical interpretation is necessary.

Typical modes of fluid motion at the interface and their potential to further gas exchange are reviewed: organized motions, like cell and helicoidal rolls or Langmuir circulation on one hand, waves and wind induced shear flow on the other. Special attention is given to wave dissipation in the form of wave breaking. The secondary flow and irregular motions of breaking waves, as well as possible rolling motions of smaller waves on the slopes of larger waves, are seen as enhancing gas transfer through surface renewal.

Observations of surface streaming obtained by dying the sea surface are discussed in terms of the above-mentioned models of surface renewal. A set of observations by Gemmrich is used to assess the effectiveness of secondary motions, as found at a given wind speed at sea, to enhance gas transfer. It is found for the natural mix of wind speeds (e.g. for the North Atlantic Ocean) that enhancement of gas transfer, compared with undisturbed boundary-layer flow, occurs in about 20–25% of cases.

The above-mentioned observations are taken under conditions of typical wind distribution without, however, the influence of rain. Current concepts of how rain influences the sea surface are reported, but these are insufficient to determine the effect of rain on gas exchange.

Introduction

This section deals with transport processes in the microlayer. The basic concept of the microlayer envisages a bulk layer of the upper ocean that is fully mixed by turbulence, and a layer of vanishing turbulence at the interface, where molecular motion dominates. For the transport of admixtures perpendicular to the interface, the way that molecular and turbulent transports combine is of interest. It is assumed that the reader is familar with the two major concepts that have been developed as a semi-theoretical tool to describe gas transfer through the aqueous side of the air–sea interface, namely the boundary-layer model and the surface-renewal model.

Both models take it as established that in the bulk water at some distance from the interface, say of the order of a few centimetres and deeper, the flow is fully turbulent and the water well mixed. At the interface, turbulence is damped and molecular forces dominate. The boundary-layer model describes the mode that we expect to prevail in a steady state, when only molecular and turbulent motions exist, or when the time constant of the surface renewal is extremely long. Surface renewal could provide a more efficient transport of admixtures, when continuously or intermittently bulk water is brought to the surface by some processes that have yet to be identified.

With respect to air–sea gas exchange, it seems important to decide whether the boundary model or the surface-renewal model is applicable. This is due to the fact that because of the slow molecular diffusion of gases, gas exchange experiments are often not very conclusive. Hence, we are inclined to extrapolate from heat exchange to gas exchange. The extrapolation involves the $-\frac{2}{3}$ power of the Schmidt number in the boundary-layer model, and the $-\frac{1}{2}$ power in the surface-renewal model (Deacon, 1977; Holmén and Liss, 1984).

In the context of the present book, we will restrict the discussion to processes that are likely to be influential in the transport of gases and materials. We need not consider larger scale primary motions of the fluid, where the particles keep their position relative to each other. However, secondary motions, where the particles are rearranged, are important. Since we are considering transport processes in the microlayer, we are interested in a rather small scale: thicknesses of the order of fractions of a millimetre. Hence, conditions that are inhomogeneous on scales larger than a few metres may be considered homogeneous for flows at the interface.

Micrometeorologists are used to thinking of air–sea fluxes in terms of friction velocity and a stability parameter, such as the Monin–Obukhov length. We should note that in problems where molecular properties (like molecular diffusion or surface tension) play an important role, the mean temperature of the water is a relevant variable. With respect to buoyancy driven motions, it is appropriate to recall that the density of seawater is a nonlinear function of salinity and temperature. Heat loss at the sea surface may therefore lead to qualitatively similar, but quantitatively different, motions for different temperature regimes.

There is not much direct information on transport processes in the microlayer. This lack is mainly due to wave motion: the superposition of waves of different frequencies and different directions produces a rather random wave pattern, which makes it practically impossible to position an instrument such that it can measure inside the microlayer.

Since at present we cannot measure directly in the oceanic microlayer, most of our knowledge about transport processes in the microlayer stems from indirect evidence from a variety of observations and measurements of different degrees of resolution. These may be roughly categorized as follows:

(i) Bulk measurements, for example, the skin temperature deviation as a function of air–sea heat flux and wind speed. The interpretation is based on model concepts.

(ii) Visual inspection, for example of surface streaming, which needs model concepts to draw conclusions on the transport processes. The interpretation is less ambiguous due to the constraints imposed by continuity.

(iii) Measurements of flow patterns or temperature profiles in the microlayer itself.

Before inferences are drawn from indirect evidence, it is useful to recall some of the basic physics involved in microlayer processes.

Basic concepts

Equations of motion and boundary conditions

The forces that determine fluid motions (liquid and gaseous phase) are described by the Navier–Stokes equations. Quite different motions can be described by the same equations. They range from the global circulations of the oceans and the atmosphere to flow in a cloud chamber or wind–water tunnel, to the flow out of a tea pot, and to the lubrication of the axle of a bicycle. Both laminar and turbulent flows are described by the Navier–Stokes equations. The character of the flow, as described by solution of the Navier–Stokes equations, will depend on the boundary conditions. Only when the geometric boundaries are similar, and only when at corresponding geometric places the boundary conditions (appropriately scaled) are equal, will the solutions (properly scaled) be equal.

The importance of the boundary conditions needs to be appreciated. This is especially so in the present context, where, for lack of sufficient data from the oceanic microlayer, inferences are drawn from other experiments. The flow of a fluid at a solid wall, where we have a no-slip condition, is different from the flow at a liquid interface, where we have a free surface condition, even when in both cases the geometric boundaries are identical (for example infinite width and depth, flat, level interface).

- Consider the flow in a convection tank. By differential heating and cooling at the bottom and top of the tank, we force a mean transport. This is likely to take the form of a closed circulation that is confined or influenced by the side walls of the tank. The actual structure of the circulation will depend *inter alia* on the horizontal dimensions of the tank relative to the vertical scale.

- Consider a wind-water flume. The mean flow and the waves feel the friction from the bottom and from the side walls. The flow is faster in the centre of the cross-section, and typically sucks in water from the sides, such that a pair of longitudinal helicoidal rolls result, which grow with fetch.

- Consider the condition for generation of turbulence. In the commonly encountered flows of a boundary layer, we find a shear flow induced by some driving force, and shaped by friction at the boundary. The interaction between mean shear flow and turbulence provides the kinetic energy of turbulence. Most velocity profiles at a boundary are strongly curved, such that the maximum turbulence generation is near to the boundary. This is quite different from an experiment in, for example, a tank where the turbulence is generated by movements of a grid (Asher and Pankow, 1991). In the latter case, the turbulent kinetic energy decays on its way from the grid towards the interface. Account needs to be taken of the different distributions of turbulent kinetic energy when turbulent transports are important.

In my judgement, most laboratory experiments are biased towards the surface-renewal model, not necessarily because this is the natural mode in the open sea, but because the inevitable geometric confinements in a laboratory experiment force secondary motions (Hasse, 1990). These secondary motions, when rapid enough, produce surface renewal. Thus, we cannot directly transpose the results of laboratory experiments to the open sea.

The motions in the microlayer obey the Navier–Stokes equations. In our case, we need the solution of the Navier–Stokes equations for a free surface with surface tension. The general solution is not known, but we do know a few typical modes of motion. For ease of discussion, we may divide these into energetically weak and strong processes, which pose different conditions at the boundary and, consequently, different patterns of transport.

Hasse and Liss (1980) have argued that the proper boundary condition governing turbulent motion near the interface is given by two-dimensional continuity in the interface proper. This was based on the assertion that the kinetic energy of turbulence is too low to form locally new surface in the presence of surface tension. The two-dimensional continuity condition leads to the same theory of decay of turbulence on approach to the interface as at a solid wall (e.g., Monin and Yaglom, 1965). This is the basis of the boundary model (Figure 4.1). The cases where surface tension essentially governs the flow at the interface may be seen as weak processes. We can also conceive organized motions, which do no net work against surface tension, say in the form of a small-scale closed circulation (secondary motions).

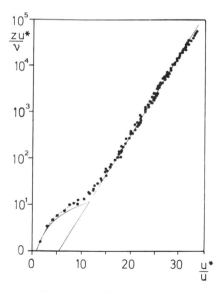

Figure 4.1. Velocity profile from fluid motions at a solid wall, shown on a logarithmic scale. Here, u is velocity, u^* friction velocity, z height above the surface, and v kinematic viscosity. The straight line corresponds to the logarithmic velocity law, while the linear velocity profile at the wall is distorted due to the logarithmic scale. The curve shows the profile according to the boundary-layer model (Reichardt, 1951), assuming constant momentum flux through the layer. Adapted from Schlichting (1982).

At sea, wave breaking is a widespread phenomenon. In this case, the wave field has collected energy for some time, so that there is enough energy to disrupt the surface. This is an energetically strong process where the notion of a microlayer is perhaps misleading.

Secondary motions: polygonal cells and helicoidal rolls

Typical flow patterns of boundary-layer circulations are quasi-hexagonal cells and roll vortices or Langmuir circulations.

Cells: In laboratory experiments, regular, convectively driven cells, called Benard cells, have been found to exist in viscous fluids. In the atmosphere, cells of considerable size have been observed in unstable situations, where they are made visible by cloud structures. In the laboratory, the cells organize themselves in a honeycomb configuration with the cells having a hexagonal form. In the atmosphere, the cells are less perfectly organized and deviate from a pure hexagonal form (Figure 4.2). With more vigorous atmospheric convection, the walls of the cells are formed by rising motions, while sinking motion is in the open centre (open cells). The reverse pattern, with rising motion in the centre (closed cells), is predominantly found with less vigorous convection. It is interesting to note that in any given situation the cells appear to be of about the same size.

It appears that at the sea surface small-scale cells exist only under calm or near calm conditions. With increasing wind velocity, cells are replaced by roll vortices. This is somewhat in contrast to the atmosphere, where

Figure 4.2. Sketch of polygonal cells, as are found with more vigorous convection.

intensive deep convection in the form of polygonal cells is frequently observed in cold air outbreaks at the rear of mid-latitude depressions.

Roll vortices are found as helicoidal motions orientated along the mean flow direction. In the atmosphere, roll vortices become visible as cloud streets that look well organized and have considerable horizontal extent. The buoyantly driven part of the system usually covers less area than the resulting return flow. Roll vortices are similar to Langmuir circulations that are found in the ocean (for a review see Pollard, 1977).

A variety of reasons has been advanced for the existence of roll vortices in the atmosphere, notably buoyancy and inflexion-point instability. In the latter case, the energy is taken from the kinetic energy of the shear flow (Etling and Brown, 1993). Cloud streets over the ocean are typically formed in unstable conditions of the boundary layer. Different explanations have also been proposed for the Langmuir circulations in the oceans. For the smaller scales, I assume that static instability due to cooling at the surface could lead to sinking motions (Figure 4.3). These, together with the rising return flow, must be organized in such a way that in the presence of wind-induced shear helicoidal motions result.

It is interesting to see that in the Craik and Leibovich (1976) theory of Langmuir circulations, the energy of the secondary motion is taken from the mean shear flow. This theory is applicable to unstable and stable situations (where heat is pumped against buoyancy). The description given is for distances of 5–300 m between slicks. Smith (1980) has given a graphic summary (Figure 4.4) of the theory that, in simple terms, explains how, from an initial horizontal disturbance of a Stokes drift, a pair of helicoidal rolls is induced. This mechanism would give one explanation for the smaller-scale convergences in the form of lines that have been observed by dyeing the sea surface.

For the roll vortices at the sea surface there is some indication that the downwind motion is fastest in the lines of convergence. This is reasonable since the action of the wind will accelerate the surface water, while the upwelling water has less momentum.

Cloud streets in the marine boundary layer and Langmuir circulations of the mixed layer appear rather uniformly well organized. The stripes of small-scale surface streaming look less well organized, pointing to a more transient process and random excitation.

The analogy between rolls in the atmospheric boundary layer and small-scale surface streaming in seawater is not complete. In the atmosphere, cloud streets can occur over a solid surface, that is, with a 'no slip' lower boundary condition. However, in the case of rolls at the sea

L. Hasse

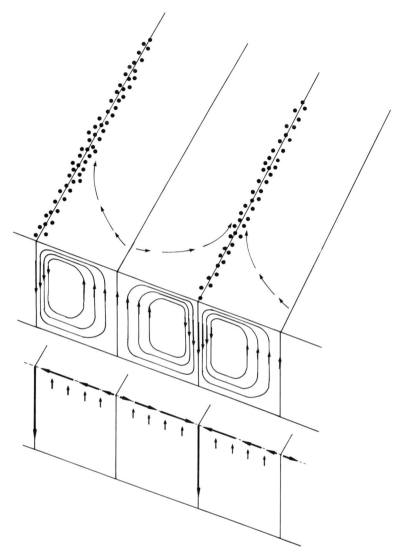

Figure 4.3. Cross-sections through roll vortices. Upper – conventional perception of a closed circulation with helicoidal motions, where convergence lines become visible through admixtures (indicated by full dots). Lower – alternative scheme of flow that produces a similar appearance of convergence lines. This pattern has a broader area of divergence in the surface.

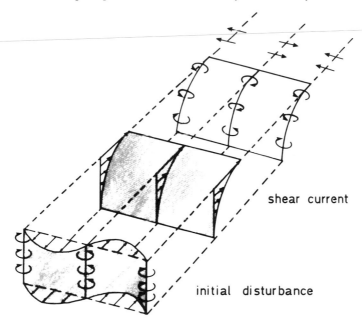

shear current

initial disturbance

Figure 4.4. Graphical representation of conditions for generation of Langmuir circulations (adapted from Smith, 1980). In the original scheme, the shear flow is thought to be Stokes or Ekman drift with a depth scale of several metres. The concept also works with the small-scale shear at the interface – any eddy that penetrates into the shear layer will be bent over to form a roll vortex. This is a special case of the tilting term of the vorticity equation that describes the vorticity balance. With a random excitation, this could explain some observed line structures of surface streaming.

surface, the fluid in the interface may move (the two-dimensional continuity condition). A closed circulation would do no net work against surface tension when the fluid at the surface moves.

In the atmospheric boundary layer and in the mixed layer of the ocean, roll vortices are confined in their vertical extent by the stronger static stability of an inversion or the thermocline respectively. In case of small-scale surface streaming a kind of closed circulation (Figure 4.3) is inferred, but there is no suggestion as to the depth reached. Convection, where the denser water sinks to some unknown depth (large compared to the observed horizontal scale), and with a return flow of almost negligible upwards velocity, could also explain the observed surface pattern.

The types of boundary-layer motion discussed so far are by no means exclusive. In fluid flows at a solid wall, eddies are observed to develop at the wall that detach and move into the interior of the fluid. These eddies may take the form of horseshoe eddies, which develop in the shear layer at the wall with a caterpillar motion. This type of motion is mainly outside the viscous sublayer (the caterpillar belt rests at the surface). So far, we do not have any indication that eddies detach from the sea surface with a pattern that includes motion in the surface proper.

Secondary motions: surface drift, waves, and wave breaking

Waves are ubiquitous on the oceans. They form an essential part of air–sea interaction since they participate in the momentum transfer between atmosphere and oceans. The total momentum transfer may be considered as partitioned into momentum transfer to waves (some type of form drag, acting at the sea surface mainly as wave-coherent pressure fluctuations) and viscous drag at the air–water interface. The viscous drag is the result of the movements of air molecules that are necessarily associated with transfer of momentum. This mechanism acts independently of whether there are waves. The momentum taken up by the waves is distributed over a layer depth of order $\lambda/2\pi$, where λ is the wavelength. The momentum given to the sea surface accelerates the liquid at the interface. The resulting drift current is strongest in a viscous sublayer of thickness d proportional to v/u^*, where v is kinematic viscosity and u^* is friction velocity. The shear flow at the surface is part of the oceanic Ekman drift. While the shear flow itself is horizontal (or, rather, parallel) to the surface and will not directly contribute to transports perpendicular to the surface, it will contribute to the generation of turbulence and will thus be important. It has been shown by Banner and Phillips (1974) that the surface drift layer produced by wind stress becomes important in the breaking of smaller waves.

Surface waves are described by solutions of the Navier–Stokes equation, assuming small perturbations of a mean state. This linearization provides a good description of non-breaking ocean surface waves, even for finite amplitudes. This corresponds to the mathematical formulation of a wave field by superposition of monochromatic, sinusoidal waves of different wave numbers or frequencies – the Fourier components. Wave spectra are a tool suitable for describing instabilities, wave–wave interactions, and molecular dissipation, but do not lend themselves to the description of sharp edges. A step function, for example, is described by a series of higher harmonics.

Real surface gravity waves of finite amplitude have peaked crests and rounded troughs. A mathematical model of them is the Gerstner wave, which consists of a circular motion and a translation, where the amplitude of the circular motion decays with depth. This is a finite amplitude solution of the inviscid equations of motion. There may be other forms of finite amplitude surface gravity waves. The actual form and its mathematical formulation are not directly important, in the present context. Rather, the form of the waves at the crest determines the balance of forces that leads to wave breaking.

We need not consider ordinary wave motions. As long as waves remain linear, their motions are ordered motions that do not directly contribute to transport in the microlayer. The orbital wave motions, on average, result in a mass transport velocity, called the Stokes drift. This is a second order term of the solutions. It stems from the decay of the orbital wave motions with depth, where at each depth the forward motions are slightly larger than the return speed (see e.g. Phillips, 1977; Dobson, 1986). The original Stokes theory was for an inviscid fluid. For a viscous fluid, slightly higher values result, especially near the surface (Figure 4.5). The Stokes drift decays exponentially with distance from the interface, over a layer depth that is determined by the orbital wave motions. It can be seen as the mass transport resulting from the momentum given by the atmosphere to the waves. The horizontal transport by Stokes drift is probably unimportant for gas exchange.

In a real wave field of gravity waves, the superposition of waves of different frequencies and directions locally results in strong accelerations at the wave crest, whence the waves become unstable and break. It is

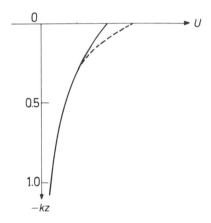

Figure 4.5. Stokes drift. The dashed line shows the velocity profile U for a viscous flow. The depth is scaled with wave number k. The drawing is for a monochromatic wave. In a real sea, contributions from smaller and larger waves would be added according to the wave spectrum, leading to a more peaked profile (from Longuet-Higgins, 1969).

L. Hasse

spilling   plunging

Figure 4.6. Wave breaking. The sketch shows spilling (left) and plunging breaking (right). Plunging is often seen in waves shoaling in on a gently sloping beach, but also occurs on the ocean. The less spectacular form of an intermediate stage of plunging (centre) may well start the spilling process.

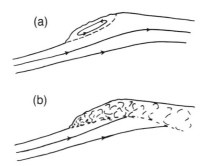

Figure 4.7. Spilling breaker. (a) The traditional view with a surface 'roller'; (b) viewed as the source of a turbulent region (from Banner and Peregrine, 1993). The pattern is seen in a frame of reference moving with the crest of the wave. The wave face is towards the left. The wave crest moves faster than the water particles, hence the relative fluid motion is mainly to the right, as indicated by arrows. We would expect turbulence also behind the 'roller' in part (a). The scheme is obtained from visualization of flow in laboratory studies with forced breaking.
Surface renewal is expected at the forward face of the spilling wave.

perhaps fair to say that the empirical finding, that waves can be well described by linear theory, can be seen as a result of dissipation of wave energy. In a given wind field, energy is continuously fed to the waves; at equilibrium, the same amount is dissipated. A considerable part of the dissipation is thought to occur through wave breaking. Wave breaking may take a variety of forms that range from the production of trains of capillary waves at the leeward face of steep waves to the vigorous overturning of steeper gravity waves (Figure 4.6) that produces white caps and foam. It appears that the secondary motions of the breaking process are irregular. These motions are fast and vigorous enough to produce surface renewal.

It has been pointed out by Banner and Phillips (1974) that waves with glassy crests also break, but without the concomitant air entrainment. This process is called microscale breaking. A scheme for the flow pattern (Figure 4.7) has been given by Banner and Peregrine (1993). It shows an

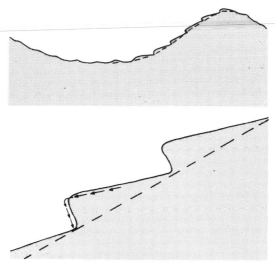

Figure 4.8. Upper – Steepening of wavelets on the down slope of larger waves. Lower – Hypothetical motion of a spilling wavelet on a slope.

area of divergent flow at the crest of a breaking wave, followed by an area of convergence or irregular motion. The resulting pattern may be taken as a secondary circulation superposed on the ordinary wave motion. The area of this secondary motion appears to be small, of an order of 10% of total wavelength.

Observations show that often the larger waves have at their forward face a dense structure of wavelets with steep leading edges, which may also occur without air entrainment. The spatial scale of these irregular steps is of an order of 20 cm. In fact, waves smaller than the dominant waves are always seen in a wave field as soon as the waves have outgrown the initial stage of ripples. I believe that these smaller waves at the down slope of larger waves are deformed to have steep leading edges. This is likely to be part of the dissipation of wave energy. The phenomenon is interesting also with regard to transport processes. I have not seen a description or theory of the motion in such a field of wavelets, so have to speculate what the characteristic motions may be. My conception of the motions of such a 'wavelet' is given in Figure 4.8. If translations resulting from the orbital motion of the larger wave are subtracted, the secondary motion of the wavelet looks similar to the caterpillar motion, where the lower portion of the caterpillar belt is at rest, while the upper part moves at twice the mean speed of the wavelet. For lack of a more detailed inventory of flow through these wavelets, we can only conclude that there necessarily is

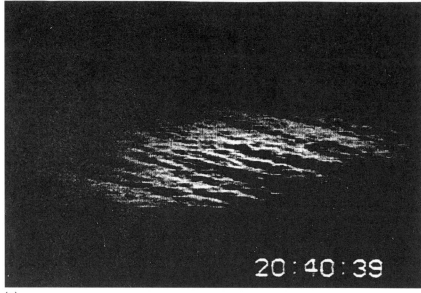

(a)

(b)

Figure 4.9. Surface streaming. Samples of observed patterns reproduced from video recordings. (a) Lines – distance between lines is of the order of 30 cm. (b) Lines, possibly produced from wave breaking. Note that the visible white structures are from cork powder, not from bubbles. (c) Cells – diameter of the open cells is of the order of 1 cm. From Gemmrich and Hasse (1992).

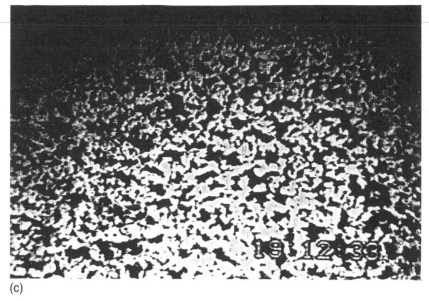

(c)

surface renewal and that the motions are rapid enough to contribute to gas exchange.

We may note that the descriptive term 'wavelet' is somewhat misleading. We are used to thinking of waves as a linear process, while the motions of the wavelets at the forward face of larger waves look rather nonlinear. Wavelets with a steep leading edge would require a series of Fourier components for their empirical description. Hence, it is difficult to discover a nonlinear motion, like down slope wavelets, in wave spectra.

Surface streaming

Observations of surface streaming

One way to study the motions in the microlayer is to dye the surface. Woodcock (1941) reported the results of experiments he conducted during clear nights in shallow fresh and salt waters, where he dyed the surface with lycopodium spores. He observed surface water movements that formed lines and streaks fast enough for the eye not to be able to follow the details of the motions. Gemmrich has conducted a series of similar experiments under a variety of conditions on open water (Figure 4.9). He summarizes his results in descriptive terms (Gemmrich and Hasse, 1992):

Table 4.1 *Secondary motions of the sea surface that may be relevant for transport in the microlayer*

Type	Horizontal scale	Environmental conditions	Effective for vertical transport	Coverage (%)	Note
Polygonal cells	Few cm	Calm	Yes	3	Estimated to cover the total surface at 40% of time with wind strength Bft 0 and 1.
Lines, helicoidal rolls	c. 5 cm	Low wind	Yes	8	Estimated to occur in 30% of cases with Bft 2 and 3. (At higher wind speeds, lines are assumed to indicate microscale breaking.)
Marginal fingers	c. 10 cm	Low wind	Weak	1	Estimated to occur in 15% of cases with Bft 2 and 3, but movements include only part of the marked patch.
Accelerated stripes	c. 10 cm	Light wind, stable	No	—	Accelerated stripes, taken to be ineffective for surface renewal.
Wind rows, Langmuir circulations	5–300 m	Low to moderate winds	No	—	Langmuir circulation or 'wind rows' of a scale of 5 m to 300 m are considered ineffective in microlayer transport, but are instrumental in the transport of heat from the interface into the interior of the mixed layer.
Microscale breaking	0.1 λ	Wind above 2 or 3 m s^{-1}	Yes	3	Microscale breaking, estimated to cover only 1/10 of wavelength. However, it appears that not all waves show microbreaking. Rather, every 3rd or 4th wave appears to influence the marked patch, hence only about 3% of the area is active.
Breaking with white caps	0.1 λ	Moderate to high winds	Yes	0.5–2.5	Wave breaking with white caps. White cap coverage has been studied extensively, hence a direct estimate is available from a white cap atlas by Spillane *et al.* (1986). For the North Atlantic Ocean and the natural mix of wind speeds a range of 0.5% to 2.5% white cap area is given.
Down-slope wavelets	0.1–0.5 m	Moderate to high winds	Yes	1–5	Estimated to cover a slightly larger area than white caps, say, by a factor of two.

- *Cells* were only observed under near calm conditions, which are rather rare on the open sea.
- *Lines or streaks* were the most frequent pattern. These were interpreted either as an indication of roll vortices or as the result of wave breaking. Lines appeared at wind speeds of 2 or 3 m s^{-1} or higher.
- *Accelerated stripes*. Less frequently, Gemmrich found that in the marked patch narrow stripes developed in which the tracer moved much faster than in the spaces in between. I classify these motions as not being effective in surface renewal, since the motion seems to be in the downwind direction, along the stripes, giving no reason for upwelling.
- *Marginal fingers*. Some observations showed that at the upwind and downwind sides of the originally marked patch, the homogeneous field was broken up into stripes or fingers. This gives the impression that, under the force of the wind, a thin surface film moves faster than the bulk of the water below. At present, it is unclear whether this represents some motion originating on the aqueous side, or accelerated motions forced by some structure in the air. Accelerated stripes and marginal fingers were found mainly under stable conditions of the atmospheric surface layer.

Except for the few observations of cell-like patterns, there appear to be few patterns below about 3 m s^{-1} that give reason to suspect surface renewal. Above *c.* 3 m s^{-1}, lines appear more frequently (Figure 4.10).

Assessment

For the calculation of gas exchange rates through the microlayer it is necessary to decide what model is applicable. More specifically, we want to know how relevant the different processes are that may lead to surface renewal, e.g. wave areal coverage at a given time. Occurrence of most of the processes depends on wind speed. Hence, we estimate the percentage of area for a given Beaufort (Bft) strength and take a weighted average, considering the frequency distribution of Bft forces. For a rough estimate, the average of Bft forces at the North Atlantic Ocean has been used. The calculation is based on Table 4.1, where the relevant processes are listed. Some of the estimates are taken from the observations of Gemmrich. Certainly, this is not a large sample, but for the present a rough estimate must suffice.

The present observations from the open sea are more in favour of the

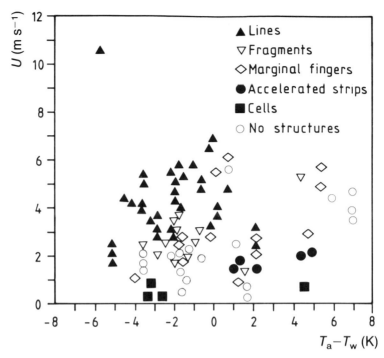

Figure 4.10. Observations of secondary flow patterns versus air–sea temperature difference and wind speed. From Gemmrich and Hasse (1992).

boundary model than for surface renewal. The latter would be the appropriate model, on average, for an area of order 20%, while over most of the area the boundary model is applicable.

Influence of rain

It has been a common belief among seafarers that rain reduces the roughness of the sea, a belief that has existed long before its first mention in the scientific literature. It appears that primarily the smaller-scale waves are damped. How rain affects the transport processes in the microlayer is less evident.

Rain has several interrelated effects at the surface layer of the ocean, some of which are discussed below:

- Change of salinity and hence density stratification.
- Momentum transfer to the sea surface by rain.
- Production of small waves by impinging drops.

- Production of splash drops.
- Generation of turbulence.
- Damping of waves.

Salinity

Fresh water from rain dilutes the salinity in the mixed layer of the ocean. Ostapoff *et al.* (1973) have given an example, where heavy rainfall from enhanced convection in the ITCZ led to a decrease in salinity by 0.14‰ in a layer of 11 m thickness, corresponding to 43 mm of rain. The dilution rendered the top of the mixed layer stable, inhibiting exchange of the deeper parts of the mixed layer with the atmosphere. Their observation is an integrated effect for a time of 18 hours, and includes vertical mixing and perhaps advection. The effect on the microlayer is less evident. For the latter, it is more interesting to know how deep rain drops penetrate initially.

Katsaros and Buettner (1969) have measured the penetration of fresh-water drops of 3 mm and $\leq$ 1.2 mm diameter into salt water in a laboratory tank. Depths of 7 mm and 4 mm respectively were affected, as indicated by the change in salinity. Waves were absent during their experiment. We may conclude from their findings that at least typical rain drops have enough energy to penetrate the surface. This necessarily leads to some turbulence in the microlayer. Rain spectra typically also contain smaller drops, that probably form a shallow layer of freshwater atop the saline water, accompanied by a strong stabilizing effect. It is unclear to me what the net effect on gas transfer may be.

Momentum transfer

The second primary effect of rain drops at a water surface is the exchange of momentum. The situation of an isolated drop falling vertically on a water surface at rest has already been studied by Reynolds. Circular waves are generated that progress outwards from the point of impact. A realistic theory of momentum transport in a wind field needs to consider the angle of incidence of rain impinging on propagating waves. A calculation has been given by Le Méhauté and Khangaonkar (1990). Their paper deals with the momentum transfer of rain to waves. It does not answer the question of what transport processes are induced in the microlayer. We lack the information as to what depth the rain drops penetrate in the first instance (before mixing), and how they distribute their momentum on the way to this depth.

Some thought has been given as to how rain affects the total air–sea exchange of momentum. The original idea was that rain has a higher density than air and transfers more momentum. Caldwell and Elliott (1972) have refined the idea, considering that rain modifies the wind profile while falling through the boundary layer and exchanging momentum with the surrounding air. The effect on the wind profile is hardly noticeable. Personally, I find the change of the wind profile less important. The driving force in the planetary boundary layer is the geostrophic wind. It does not really matter for the sea surface whether some of the momentum, generated by the ageostrophic components in the boundary layer and given to the sea surface as shear stress, is slightly modified by rain. The main fact is that rain carries momentum taken from higher layers, say cloud height, to the surface. A rough calculation shows that with heavy rain the momentum transferred to the sea surface by rain is of the same order of magnitude as the typical shear stress. Averaged annually, momentum transferred by rain is two orders of magnitude smaller than the wind-induced shear stress.

Rain-induced turbulence and wave damping

Drops penetrating through the interface proper necessarily disturb the ordered wave motions. It is difficult to envisage the incorporation of the rain drops in the water as a linear process. Hence, it is reasonable to assume that rain drops interfere with wave motion and shear currents and produce turbulence. Presently, no detailed concept is available that describes the amount and vertical distribution of raindrop-produced turbulence in the interfacial layers.

The observed calming of the seas by rain and the concomitant damping of small-scale waves is most often ascribed to the turbulence produced by rain (e.g. Nystuen, 1990). The observation of circular waves resulting from the impact of a single drop on still water has also led to the suggestion that small-scale waves are damped by wave–wave interaction with raindrop-produced waves (e.g. Tsimplis, 1992). Additionally, there is the possibility that a thin layer of freshwater above the saltier water below will interact with the smaller-scale waves and redistribute wave energy.

Heat flux

It will also be noted that rain falling on the sea surface typically has a slightly lower temperature than the sea surface (if rain drops are warmer than the air in the surface layer they will be cooled by evaporation to some temperature close to the wet bulb temperature). Hence, due to the lower temperature, the kinematic viscosity is slightly increased. The main effect of the different temperature, however, is the heat flow associated with the fresh water flux (a sample calculation is given by Katsaros, 1976).

It is evident from the preceding discussion that experimental investigations are required for a better understanding of turbulence generation and mixing in water – possibly with shear flow.

Experimental studies of the microlayer: miscellaneous techniques

Early investigations

An early study of the microlayer was conducted by Woodcock (1941), who investigated the motions in shallow fresh and salt waters during clear nights using dye. He found rapid small-scale motions that he termed surface streaming.

Most of the earlier work dealt with the cool skin of the ocean (for a compilation see Hasse, 1963). There are the pioneering measurements by Bruch (1940) with a floating thermocouple, by Woodcock and Stommel (1947) with a fine mercury in glass thermometer, and by Ball (1954) with an infrared radiation thermometer. Also, measurements of temperature and humidity profiles have been used to estimate the surface film temperature (Hasse, 1963). The temperature of a cool or warm film at the sea surface was found to be in good agreement with the heat exchange at the interface, when explained by the boundary model using the empirical coefficient from flow at a solid surface (Hasse, 1971). It is worthwhile to note that, for several reasons (see Katsaros and Businger, 1973), it is rather difficult to make infrared measurements of the sea-surface temperature with the necessary accuracy of 0.1 K or better, an accuracy of roughly 1:1000. A review of measurements and concepts of the aqueous sublayer at an air–water interface was given by Katsaros (1980). Her review comprises results from *in situ* measurements at the sea surface and from laboratory experiments. A more recent compilation of dynamical processes of transfer at the sea surface is given in Thorpe (1995).

Turbulence generation at the boundary

Kim *et al.* (1971) studied fluid flow at a solid wall using hydrogen bubbles, dye, and hot wire measurements. Hydrogen bubbles were produced at vertical and horizontal wires. They showed that turbulence generation near the wall occurs intermittently as a bursting process, and describe the motions that are associated with this process. However, inspection of their figures shows that the observed flow patterns are outside the viscous sublayer. We cannot infer from their observations whether flow patterns in the microlayer are induced by the bursts and how these would look. Personally, I find it justified to infer that a bursting process near the wall, outside the microlayer, would also induce motions in this layer. However, we do not have evidence of how these motions decay on approach to the wall. Unfortunately, we are not permitted to assume that the flow at the free fluid surface would be similar, since the boundary conditions are different.

Radon evasion technique

A geochemical method that allows determination of the gas exchange rate directly is the radium/radon method. Seawater contains a certain amount of radium that decays with a known decay rate to radon. Assuming that there is negligible exchange through the thermocline, the amount of radon in the mixed layer is a variable that depends on the source strength (radium) and the sink strength, the gas exchange rate through the air–sea interface. Measurements by this method have been made with considerable experimental effort in the ocean (Peng *et al.*, 1979). Unfortunately, it was not possible to relate the gas exchange rate to external parameters (Hasse and Liss, 1980).

Velocity profiles near the interface

McLeish and Putland (1975) measured the velocity profile in the top 2 mm of the water in a wind–water tunnel. Mean profiles were derived from clouds of microscopic hydrogen bubbles that were practically neutrally buoyant. Results from three experiments are reproduced. The profiles could be fitted by an exponential curve, though some small deviations remain visible that could be induced by random fluctuations. The profiles have a resolution of 0.05 mm and look rather good and consistent (Figure 4.11). However, the small deviations from the fitted curves do not lend themselves to a simple interpretation or conceptual model.

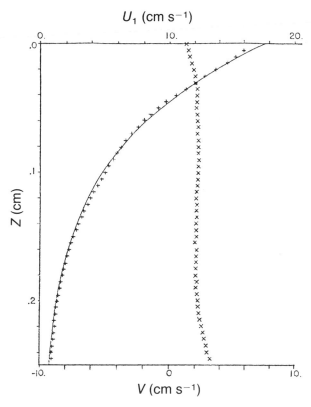

Figure 4.11. Velocity profile measured in a wind–water tunnel, '+' downwind and '×' crosswind. The fitted line is an exponential curve. From McLeish and Putland (1975).

Temperature profiles at the interface

For measurements of temperature profiles in the microlayer (Figure 4.12) a probe that pierces through the interface has been developed. In order to avoid difficulties with the signal at the interface proper, a probe that rises from below through the microlayer is required. Its upward velocity needs to be larger than the wave-induced vertical velocities of the interface. Such a sonde has been built with a fast temperature sensor and has been demonstrated to be feasible to operate (Mammen and von Bosse, 1990). In principle, a fast temperature sonde could help to measure the temperature profile and thus distinguish between the surface renewal and the boundary-layer model. However, the temperature profiles according to the two hypotheses are not that much different, and in open, natural

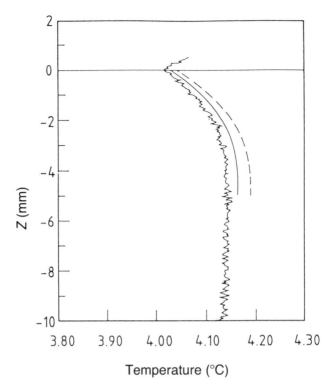

Figure 4.12. Temperature profile measured in Kieler Förde at the pier of the Institut für Meereskunde. Solid line: fitted polynomial. Dashed line: function after time-constant correction (lines are shifted by 0.025 K). From Mammen and von Bosse (1990).

waters there is some inevitable scatter in individual temperature profiles. The temperature profiles obtained so far have not been sufficient to distinguish between the two theories.

Forced heat flow technique

A forced heat flow technique has recently been used in the laboratory to probe the microlayer. The method operates a strong infrared light source (1000 W) in a pulsed mode to heat the water surface by absorption. From the known heat flux into the water and the measured surface temperature deviation, the effective conductivity of the microlayer can be calculated (Jähne et al., 1989). This method has also been advocated as a tool to determine gas exchange rates. The latter application, however, implies the choice of a model to extrapolate to the lower molecular diffusivity of gases.

Conclusion

On the open sea, the microlayer is almost inaccessible to direct physical measurement. Some sophisticated laboratory measurements are available, as well as concepts developed from physical arguments. It is necessary that such interferences be tested on the open sea, in a natural sea state. This holds for the three groups of processes that involve motions of the microlayer, namely surface streaming, wave breaking and the influence of rain. Potential for research is seen in optical probing of microlayer movements, e.g. with multiple wavelengths, which has not yet been fully explored; but see Korenowski (Chapter 15, this volume) for a fuller description.

References

Agee, E. M., Chen, T. S. and Dowell, K. E. (1973). A review of mesoscale cellular convection. *Bull. Am. Meteorol. Soc.*, **54**, 1004–12.

Asher, W. E. and Pankow, J. F. (1991). Prediction of gas/water mass transport coefficients by a surface-renewal model. *Environ. Sci. Technol.*, **25**, 1294–300.

Ball, F. K. (1954). Sea surface temperatures. *Aust. J. Phys.*, **7**, 649–52.

Banner, M. L. and Peregrine, D. H. (1993). Wave breaking in deep water. *Ann. Rev. Fluid Mech.*, **25**, 373–97.

Banner, M. L. and Phillips, O. M. (1974). On the incipient breaking of small scale waves. *J. Fluid Mech.*, **65**, 647–56.

Bruch, H. (1940). Die vertikale Verteilung von Windgeschwindigkeit und Temperatur in den untersten Metern über der Meeresoberfläche. *Veröffentlichungen des Instituts für Meereskunde an der Universität Berlin*, Neue Folge A 38, 66 pp.

Caldwell, D. R. and Elliott, W. P. (1972). The effect of rainfall on the wind in the surface layer. *Boundary-Layer Meteorol.*, **3**, 146–51.

Craik, A. D. D. and Leibovich, S. (1976). A rational model for Langmuir circulation. *J. Fluid Mech.*, **73**, 401–26.

Deacon, E. L. (1977). Gas transfer to and across an air–water interface. *Tellus*, **29**, 363–74.

Dobson, F. (1986). Introductory physical oceanography. In *Introductory Physics of the Atmosphere and Ocean*, ed. L. Hasse and F. Dobson, pp. 53–119. Dordrecht: Reidel.

Etling, D. and Brown, R. A. (1993). Roll vortices in the planetary boundary layer, a review. *Boundary-Layer Meteorol.*, **65**, 217–48.

Gemmrich, J. and Hasse, L. (1992). Small-scale surface streaming under natural conditions as effective in air–sea gas exchange. *Tellus*, **44B**, 150–9.

Hasse, L. (1963). On the cooling of the sea surface by evaporation and heat exchange. *Tellus*, **15**, 363–6.

Hasse, L. (1971). The sea surface temperature deviation and the heat flow at the sea–air interface. *Boundary-Layer Meteorol.*, **1**, 368–79.

Hasse, L. (1990). On the mechanism of gas exchange at the air–sea interface. *Tellus*, **42B**, 250–3.

Hasse, L. and Dobson, F. (1986). *Introductory Physics of Atmosphere and Ocean*. Dordrecht: D. Reidel. 126 pp.

Hasse, L. and Liss, P. S. (1980). Gas exchange across the air–sea interface. *Tellus*, **32**, 470–81.

Holmén, K. and Liss, P. S. (1984). Models for air–water gas transfer: an experimental investigation. *Tellus*, **36B**, 92–100.

Jähne, B., Libner, P., Fischer, R., Billen, T. and Plate, E. J. (1989). Investigating the transfer processes across the free aqueous viscous boundary layer by the controlled flux method. *Tellus*, **41B**, 177–95.

Katsaros, K. B. (1976). Effects of precipitation on the eddy exchange in a wind driven sea. *Dynam. Atmos. Oceans*, **1**, 99–126.

Katsaros, K. B. (1980). The aqueous thermal boundary layer. *Boundary-Layer Meteorol.*, **18**, 107–27.

Katsaros, K. B. and Buettner, K. J. K. (1969). Influence of rainfall on temperature and salinity of the ocean surface. *J. Appl. Meteorol.*, **8**, 15–18.

Katsaros, K. B. and Businger, J. A. (1973). Comments on the determination of the total heat flux from the sea with a two-wavelength radiometer system as developed by McAlister. *J. Geophys. Res.*, **78**, 1964–70.

Kim, H. T., Kline, S. J. and Reynolds, W. C. (1971). The production of turbulence near a smooth wall in a turbulent boundary layer. *Fluid Mech.*, **50**, 133–60.

Le Méhauté B. and Khangaonkar, T. (1990). Dynamic interaction of intense rain with water waves. *J. Phys. Oceanogr.*, **20**, 1805–12

Longuet-Higgins, M. S. (1969). A nonlinear mechanism for the generation of sea waves. *Proc. R. Soc.*, **A311**, 371–89.

Mammen, T. C. and von Bosse N. (1990). STEP – a temperature profiler for measuring the oceanic thermal boundary layer at the ocean–air interface. *J. Atmos. Oceanic Technol.*, **7**, 312–22.

McLeish, W. and Putland, G. E. (1975). Measurement of wind-driven flow profiles in the top millimeter of water. *J. Phys. Oceanogr.*, **5**, 516–18.

Monin, A. S. and Yaglom, A. M. (1965). *Statistical Fluid Mechanics* (transl. from Russian), Vol. I, 1971, p. 769, Vol. II, 1975, p. 874. Cambridge, MA: MIT Press.

Nystuen, J. A. (1990). A note on the attenuation of surface gravity waves by rainfall. *J. Geophys. Res.*, **95**, 18 353–5.

Ostapoff, F., Tarbeyev Y. and Worthem S., (1973). Heat flux and precipitation estimates from oceanographic observations. *Science*, **180** (4089), 960–2.

Peng, T. H., Broecker, W. S., Mathieu, G. G., Li, Y-H. and Bainbridge, A. E. (1979). Radon evasion rates in the Atlantic and Pacific Oceans as determined during the GEOSECS Program. *J. Geophys. Res.*, **84, C5**, 2471–86.

Phillips, O. M. (1977). *The Dynamics of the Upper Ocean*, 2nd edn, Cambridge: Cambridge University Press. 336 pp.

Pollard, R. T. (1977). Observations and theories of Langmuir circulations and their role in near surface mixing. In *A Voyage to Discovery*, ed. M. Angel, pp. 235–51. Supplement to *Deep Sea Res.*

Reichardt, H. (1951). Vollständige Darstellung der turbulenten Geschwind-
igkeitsverteilung in glatten Leitungen. *Z. Angewandte Math. Mech.*, **31**,
208–219.

Schlichting, H. (1982). *Grenzschicht-Theorie*, 8th edn., p. 843. Karlsruhe:
Braun. (English edn: *Boundary-Layer Theory*, 7th edn. 1979, New York:
McGraw-Hill.)

Smith, J. A. (1980). Waves, currents and Langmuir circulation. PhD
Dissertation, Institute Oceanography, Dalhousie University, Halifax.
242 pp.

Spillane, M. C., Monahan, E. C., Bowyer P. A., Doyle D. M. and Stabeno,
D. J. (1986). Whitecaps and global fluxes. In *Oceanic Whitecaps*, ed.
E. C. Monahan and G. MacNiocaill, pp. 209–18. Dordrecht: D. Reidel.

Thorpe, S. A. (1995). Dynamical processes of transfer at the sea surface.
Progr. Oceanogr., **35**, 315–52.

Tsimplis, M. N. (1992). The effect of rain in calming the sea. *J. Phys.
Oceanogr.*, **22**, 404–12.

Woodcock, A. H. (1941). Surface cooling and streaming in shallow fresh and
salt waters. *J. Mar. Res.*, **4**, 153–61.

Woodcock A. H. and Stommel, H. (1947). Temperatures observed near the
surface of a freshwater pond at night. *J. Meteorol.*, **4**, 102–3.

5

The role of organic films in air–sea gas exchange

NELSON M. FREW

Abstract

Various parameterizations of gas exchange with wind speed at the ocean surface are poorly constrained by field measurements using natural and artificial tracers. One of the factors leading to uncertainty for *in situ* estimates of the gas transfer velocity is the presence of organic films at the air–sea interface. Such films are derived from bulk seawater dissolved organic matter, from terrestrial sources (natural and anthropogenic) and from petroleum seeps and spills. The ubiquitous background of degraded natural biopolymeric and geopolymeric materials in the sea potentially contributes to surface accumulations of organic matter even in very oligotrophic waters. Specific inputs during phytoplankton blooms and from neuston in the microlayer also contribute to the enrichment of surface-active matter at the interface.

Organic films can affect air–sea gas exchange through both static and dynamic mechanisms. The static effect arises from the presence of additional mass transfer resistance due to the physical barrier provided by the film. This effect is not considered to be important at the sea surface, since it requires the presence of condensed, solid type surfactant films that are easily dispersed under typical oceanic conditions of wind and waves. Much more significant is the hydrodynamic effect of a film that arises from the viscoelastic property of a surfactant-influenced interface. The presence of non-zero viscoelasticity modifies the surface boundary conditions affecting hydrodynamic processes which govern gas exchange. The effects include the reduction of the length and velocity scales of the near-surface turbulence, inhibition of wave growth and enhancement of wave energy dissipation (wave damping).

Results from laboratory investigations of gas exchange at surfactant-influenced air–water interfaces in various flow systems are reviewed. These results provide evidence of first-order effects of surfactant films in reducing the mass transfer coefficient. Gas transfer is significantly reduced by even slight surface accumulations of organic materials. This effect would be most important for coastal waters where biological productivity is high. However, even at the low surface film pressures typical of the ocean surface, the viscoelastic modulus is estimated to be sufficient to cause gas exchange reductions. Evidence from wind–wave tank studies show strong inverse correlations between gas exchange rate and various measures of seawater organic matter and surfactant concentrations. Thus, significant spatial and temporal variations in gas transfer are expected for different geographical and biological regimes for a given set of physical forcing conditions, e.g. wind stress and fetch. In view of this, it is suggested that a unique relation between wind speed and gas transfer velocity is unlikely to exist for the oceans and other natural water bodies containing dissolved organic matter.

A high degree of correlation is demonstrated between gas transfer velocity and mean square wave slope. This correlation is remarkably robust for air–water gas transfer under a variety of surface conditions, including variable surfactant concentrations and types. Thus, mean square wave slope may provide a very useful parameterization of gas transfer that integrates the surfactant film effect. Such a parameterization could be applied over large spatial scales using satellite-based microwave sensors.

Introduction

The microlayer and air–sea processes

This chapter focuses on organic surface films in the marine microlayer and their role in modulating air–sea gas exchange. The marine microlayer, the thin interfacial boundary between the atmosphere and the ocean, plays a critical role in air–sea interactions, including heat, mass and momentum exchange. The physical mechanisms governing these exchange processes are of interest in the modelling of surface roughness, gas fluxes, thermal structure and mixed layer depths, ocean circulation and seasonal cycles of biological productivity. An understanding of these processes also is essential to more encompassing coupled models of atmosphere–ocean interactions and global climate. In addition, the microlayer has potentially important roles in photochemical and biologically mediated transformations of organic matter that adsorbs at the sea surface. The microlayer also may play an important role in reproductive cycles of marine species and exacerbate the impact of anthropogenic pollutants on the marine food chain.

Air–sea gas exchange

Accurate estimates of air–sea gas fluxes are essential for understanding the global cycles of carbon dioxide, methane, nitrous oxide, dimethyl-suphide and other trace gases that affect the earth's radiation budget. Gas fluxes can be determined from a knowledge of the ambient air–sea gas concentration gradients and the gas transfer velocity, k_w. (Here the subscript w denotes that the gas exchange rate is controlled by the transfer resistance on the water side of the air–water interface, as is true for sparingly soluble gases.) Major campaigns to measure atmospheric and oceanic gas concentrations in conjunction with large international field programmes such as the Joint Global Ocean Flux Study (JGOFS) and the World Ocean Circulation Experiment (WOCE) will significantly augment the existing database on regional variations in gas concentration gradients. However, a weak link is the current understanding of the gas transfer process itself, and the lack of a robust parameterization of the gas transfer velocity, k_w.

Various methods have been used to estimate k_w from *in situ* measurements. Steady-state distributions of natural ^{14}C have been used to estimate a global long-term average (Craig, 1957; Broecker and Peng, 1974; Broecker, Peng and Engh, 1980a), while bomb-^{14}C distributions have yielded estimates of average rates over large ocean areas (Münnich & Roether, 1967; Broecker, Peng and Takahashi, 1980b; Broecker and Peng, 1984; Broecker *et al.*, 1985). Shorter-term estimates have been given by the ^{222}Rn deficit method (Broecker & Peng, 1971, 1974; Peng, Takahashi and Broecker, 1974; Peng *et al.*, 1979; Kromer and Roether, 1983), as well as from studies using deliberate tracers, such as sulphur hexafluoride (Wanninkhof, Ledwell and Broecker, 1985; Upstill-Goddard *et al.*, 1990; Watson, Upstill-Goddard and Liss, 1991; Wanninkhof and Bliven, 1991; Wanninkhof *et al.*, 1993). Tracer techniques are limited to relatively long time scales (days to weeks). Thus, while they provide long-term estimates of average fluxes, tracer methods are not well suited for constraining parameterizations of k_w, which is governed by a number of variables that change over short time scales (minutes to hours). The eddy correlation technique (Wesley *et al.*, 1982; Smith and Jones, 1985, 1986; Smith *et al.*, 1991) potentially provides instantaneous flux measurements, but generally has yielded much higher fluxes than tracer techniques and remains controversial (Broecker *et al.*, 1986).

Existing field estimates of k_w have been summarized by Wanninkhof and colleagues (Wanninkhof *et al.*, 1985; Wanninkhof, 1992) and by Liss and Merlivat (1986) and generally show strong nonlinear correlations with wind-speed, u. The transfer velocity is thus frequently parameterized in terms of wind speed (Smethie *et al.*, 1985; Liss and Merlivat, 1986; Tans, Fung & Takahashi, 1990; Wanninkhof, 1992), which is easily measured from surface ships and buoys, estimated from satellite-borne scatterometers (Etcheto & Merlivat, 1988) or computed from general atmospheric circulation models (Erickson, 1989). Unfortunately, such

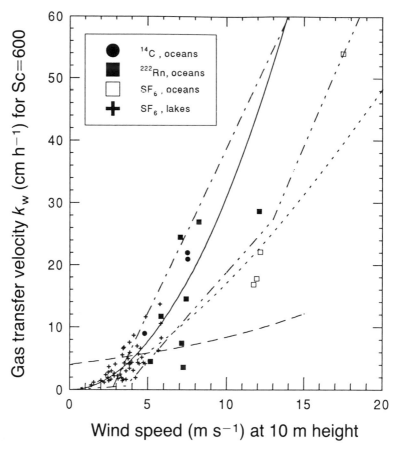

Figure 5.1. A summary of empirical gas transfer–wind-speed relationships and gas transfer results over the oceans and freshwater lakes at various wind speeds. All data are corrected to Sc=600. The k_w–u relationships are those of Wanninkhof (1992) for short-term winds (solid line), Smethie *et al.* (1985) (dash–dot line), Liss and Merlivat (1986) (dash–double dot line), Deacon (1980) (long dash line), and Hartman and Hammond (1985) (short dash line). Adapted from Wanninkhof (1992).

u–k_w relationships are not tightly constrained by the field data for oceans and lakes (Figure 5.1). Recently, Wanninkhof (1992) has considered various factors leading to bias in u–k_w relationships, including short-term variability in wind-speed combined with the nonlinearity of the u–k_w relation, variable fetch and chemical enhancement effects for reactive species such as CO_2. Wanninkhof concluded that the variance in laboratory and field data is not fully explained by these factors alone.

Transfer velocity is more directly related to the friction velocity, $u_w{}^*$

(where $u_w^* = (\tau/\varrho_w)^{1/2}$, where τ is the wind stress, ϱ_w is the liquid phase density, and the subscript w denotes water as the liquid phase) (Ledwell, 1984; Jähne *et al.*, 1987). In the absence of direct measurements, τ must be estimated from u using an empirical drag coefficient. The neutral buoyancy condition implicit in this approach is frequently not fulfilled. The drag coefficient, and therefore u_w^*, vary with boundary-layer stability and also may be influenced by the long wave field (Smith *et al.*, 1992; Geernaert *et al.*, 1993). Erickson (1993) has proposed an alternate model for k_w based on the thermal stability of the air–seawater interface. That model yields a global area-weighted k_w in good agreement with the mean determined from ^{14}C data, but shows poor agreement with results obtained using the Liss–Merlivat relation at low to moderate wind-speeds. Other factors including wind fetch, sea-surface microlayers and variable surface ocean bubble populations were cited as additional sources of uncertainty not addressed in the stability model.

Laboratory studies of gas exchange in wind–wave tanks allow more precise control of wind stress and other conditions and have led to a better understanding of the factors controlling gas exchange, especially the role of waves. Wind–wave tank results have been summarized by Jähne *et al.* (1987). However, it is difficult to compare laboratory studies of k_w with field studies because the length scales and wave fields vary among different wave tanks and are not necessarily comparable to real ocean surface conditions.

The role of biogenous organic matter

One of the unconstrained factors leading to the large uncertainty in u–k_w relations is the phenomenon of surface films. Here, the term 'surface film' is used to mean any surface excess of surface-active organic material at the air–water interface and is not restricted to condensed films or 'slicks', which visibly damp small gravity and capillary waves. That surface films affect air–water gas exchange has been recognized previously by various investigators, either through the inadvertent presence of 'nuisance' films in their experimental systems or through the deliberate introduction of concentrated surfactant films to suppress small-scale waves (Broecker, Peterman and Siems, 1978; Jähne, Münnich and Siegenthaler, 1979; Jähne *et al.*, 1984, 1985, 1987). Other early work focused primarily on the use of insoluble lipid films to retard evaporation in water storage areas (Jarvis, Timmons and Zisman, 1962; La Mer and Healy, 1965; Garrett, 1971). From an oceanographic point of view, the traditional treatment of gas exchange primarily as a physics problem has led to a certain neglect of the role of ocean chemistry and biology in modulating the exchange process, particularly biological activity that results in production of dissolved organic matter (DOM). This neglect has been reinforced by emphasis on static rather than dynamic effects of films and the misconception that gas exchange is significantly affected only in the presence of concentrated insoluble surfactant films or 'slicks', a surface condition not prevalent in the open ocean (Liss, 1983).

The potential effect of surface-active organic matter on gas exchange at the sea surface has only recently been considered in detail (Goldman, Dennett and Frew, 1988; Frew *et al.*, 1990; Frew, Goldman and Bock, unpublished data). Goldman *et al.* (1988) were the first to show that gas transfer was likely to vary with biological productivity. Their laboratory measurements of oxygen evasion rates for different natural waters in a small stirred system showed reductions in gas transfer velocities ranging from 5–15% for oligotrophic waters to 40–60% for inshore waters. Similar observations applied to exudates from batch cultures of various species of marine phytoplankton (Frew *et al.*, 1990). More recent measurements in experimental wind–wave systems have shown considerably larger effects. (Frew *et al.*, unpublished data). Gas exchange across a very clean water interface was shown to vary smoothly in proportion to the square of the linear speed of the rotor used to generate wind stress (Figure 5.2). In the presence of low concentrations of a water-soluble surfactant, however, the u–k_w relation acquired the bilinear character (Figure 5.2) frequently observed for natural waters and contaminated laboratory systems (Liss and Merlivat, 1986; Jähne *et al.*, 1987; Wanninkhof, 1992). Strong inverse correlations of gas transfer velocity with various measures of organic matter (Frew *et al.*, unpublished data) and with wave field characteristics (Bock and Frew, unpublished data) have been observed for seawater collected along productivity gradients. In Figure 5.3, gas transfer velocities measured in a laboratory wind–wave tank are shown as a function of time for seawater from a series of stations along a transect from Narragansett, RI to Bermuda. Several observations can be made concerning these data: 1. all of the samples yielded transfer velocities that were significantly lower than that measured for clean distilled water under identical conditions; 2. transfer velocities measured for coastal water samples were as much as five times lower than those measured for oligotrophic water samples; and 3. although the wind stress was constant, the transfer velocities varied with time, in this case decreasing monotonically as surface-active materials accumulated at the

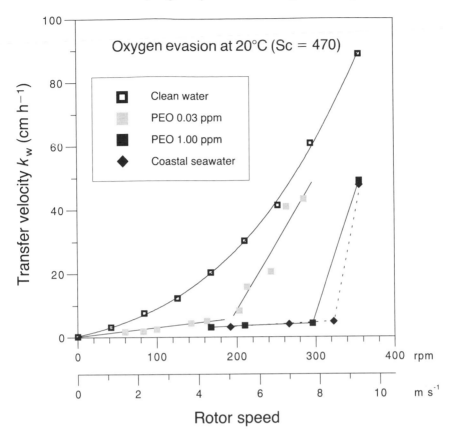

Figure 5.2. Variation of gas transfer velocities (k_w) with rotor speed for a clean distilled water–air interface and for interfaces with polyethyleneoxide (PEO) films in dynamic equilibrium with bulk PEO concentrations of 0.03 ppm and 1 ppm. Also shown are values obtained with natural adsorbed films from Vineyard Sound seawater equilibrated at the indicated rotor speeds. Transfer velocities are for O_2 evasion measured in an annular wind–wave channel with a channel width and depth of 10 cm. Indicated rotor speeds are mean (mid-channel) linear velocity for rotor blades at 5 cm height above mean water surface. From unpublished data of Frew, Goldman and Bock.

interface. In Figure 5.4, significant correlations between gas transfer velocities (at $t=0$) and surfactant concentration, chlorophyll fluorescence, dissolved organic carbon (DOC), and coloured dissolved organic matter (CDOM) fluorescence are illustrated for the Narragansett–Bermuda samples and for other samples from Monterey Bay. Gas transfer velocities are thus likely to vary spatially and temporally due to changing organic matter concentrations, in addition to variation of other

N. M. Frew

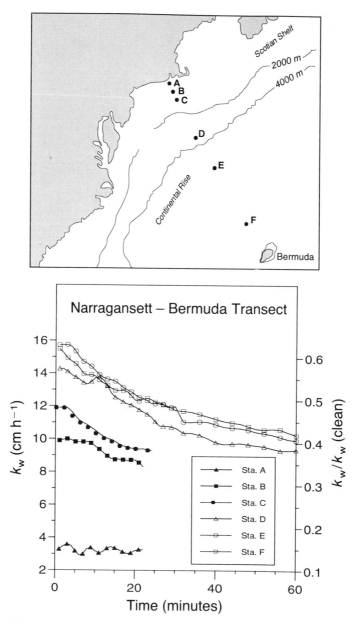

Figure 5.3. Chart of sampling stations for a Narragansett–Bermuda transect (upper panel) and time series measurements of k_w for samples from transect stations A–F (lower panel). Experimental conditions as in Figure 5.2. All measurements made at a uniform wind stress applied at a mean linear rotor speed of 5 m s^{-1}. Scale at right indicates transfer velocities relative to those obtained for clean distilled water. From unpublished data of Frew, Goldman and Bock.

governing parameters. It should be noted that the correlations shown in Figure 5.4 pertain to bulk-water concentrations of organic matter. While these correlations are significant, one would not necessarily expect k_w to be tightly coupled to bulk-water dissolved organic matter (DOM) concentrations, because the pertinent physical mechanisms involve surface phenomena and depend on the actual condition of the interface with respect to adsorbed surfactants. The latter is affected not only by diffusion from the bulk and adsorption/desorption kinetics, but by other dynamic processes such as current shear and upwelling that deliver organic matter to the interface (Moum, Carlson and Cowles, 1990) at varying rates.

It is evident from these laboratory results that surfactant films exert a first-order effect on gas exchange at low to moderate wind speeds, and that no unique relation between wind speed and transfer velocity is likely to exist for natural waters containing organic matter. The probable influence of surfactants in higher wind speed regimes, in the presence of breaking waves and bubbles, remains to be explored. This chapter will consider briefly the chemical nature of marine surfactants and their surface physical properties, and will then move to a consideration of the mechanisms by which surfactants may influence the gas exchange process. While the discussion will emphasize gas exchange across the marine microlayer, the concepts apply to all natural waters, including rivers, lakes and reservoirs, as well as artificial systems such as water treatment facilities and reactor vessels, where surface-active organic matter is present.

Surface-active matter in the marine microlayer

Sources and chemical nature

Sea-surface films are derived from bulk seawater dissolved organic matter, from terrestrial sources (natural and anthropogenic) delivered by atmospheric transport and runoff, and petroleum seeps and spills. The general background levels of degraded biopolymeric and geopolymeric materials in the sea potentially contribute to surface accumulations of organic matter even in oligotrophic waters. Inputs of fresh biogenous materials from localized elevations in productivity further contribute to the surfactant pool. Surfactants are concentrated at the air–sea interface by a number of physical processes including diffusion, turbulent mixing, bubble and particle transport, larger scale convergent circulations driven by wind, tidal forces and internal waves, and direct atmospheric deposi-

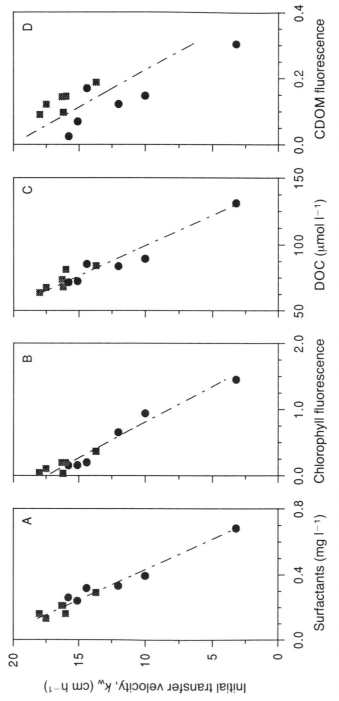

Figure 5.4. Correlations of k_w with (A) surfactant concentration, (B) *in situ* chlorophyll fluorescence, (C) dissolved organic carbon (DOC) and (D) coloured dissolved organic matter (CDOM) fluorescence at 450 nm, for seawater samples collected in Monterey Bay (■) and along a Gulf Stream transect from Narragansett to Bermuda (●). Experimental conditions as in Figure 5.3. Transfer velocities vary with surface film conditions, which are time-dependent due to diffusion and adsorption of surfactants at the interface; therefore, for comparative purposes, these correlation plots utilize initial k_w values obtained for freshly formed surfaces. From unpublished data of Frew, Goldman and Bock; DOC data from E. T. Peltzer, pers. commun.

tion. The composition of sea-surface films remains largely undefined even though significant enrichments of numerous specific compound classes have been observed (Hunter and Liss, 1981; Williams, 1986; Williams *et al.*, 1986). Of the DOM in surface films, as with bulk seawater DOM, less than half of the organic matter can be characterized structurally; the remainder is assumed to be macromolecular in nature and to encompass a very wide range of molecular weight and solubility in seawater.

The single largest source of surfactants is production by autochthonous marine organisms, principally phytoplankton, which exude natural surfactants as metabolic by-products (Zutic *et al.*, 1981). Excretion products of many types of marine phytoplankton include complex β-glucans and heteropolysaccharides of high molecular weight (Allan, Lewin and Johnson, 1972; Mykelstad, 1974; Smestad, Haug and Myklestad, 1975; Percival, Rahman and Weigel, 1980; Mykelstad, Djurhuus and Mohus, 1982; Ramus, 1982) and are frequently found in surface waters during blooms (Sakugawa and Handa, 1985a,b). Most macromolecular materials exhibit some degree of surface activity (Leenheer, 1985). Although quite soluble, some of these polysaccharides are likely to be conjugated to hydrophobic moieties sufficient to make them at least weakly surface active. Proteins and lipids are also present in phytoplankton exudates but are generally less abundant than carbohydrates (Hellebust, 1974). However, their contribution to the surface physical properties of microlayer films may be disproportionately large despite their low concentration levels (Van Vleet and Williams, 1983).

In addition, surface-active materials are contributed by the breakdown of dead organisms. The intermediate and end products of subsequent chemical and microbiological transformations are likely to include a variety of compounds whose molecular structures contain hydrophobic and hydrophilic moieties. Zutic *et al.* (1981) suggested that a large proportion of the surfactants produced in plankton cultures are secondary products of excretion rather than high-molecular-weight metabolic end products. These materials may result from rapid degradation by bacteria and extracellular enzymes. Another possible production pathway is condensation of relatively low-molecular-weight exudates to form surface-active macromolecular structures (Nissenbaum, 1974). Phytoplankton exudates are thought to be components of humic materials (Rashid, 1985; Gillam & Wilson, 1985), which are known to be surface-active. Further discussion of the chemistry and biology of sea-surface microlayer material and its properties, as well as how it is sampled, can be found in Hunter (Chapter 9) and Hardy (Chapter 11), in this volume.

Surface physical properties

It is useful to review several aspects of the surface physical properties of microlayer films before looking at their impact on gas exchange.

Surface tension, surface pressure and static elasticity

Due to the high Gibbs surface free energy (G) of the clean air–water interface, a truly organic-free water surface is likely to be found only in experimental systems where extraordinary measures are taken to exclude organic matter (Scott, 1972, 1975). In natural waters, ubiquitous surface-active compounds adsorb at the air–water interface, lowering G and, therefore, the surface tension, γ, which is the incremental change in free energy per unit change in surface area, A (at constant temperature, T and pressure, P):

$$\gamma = (G/A)_{T,P} \tag{5.1}$$

In considering the effects of adsorbed films, it is useful to define the terms surface pressure, π (units, mN m^{-1}), which is the surface tension of the film-influenced interface relative to that of a clean interface:

$$\pi = \gamma_{\text{clean}} - \gamma_{\text{film}} \tag{5.2}$$

and the dynamic surface viscoelastic modulus, $\acute{\varepsilon}$, the change in surface pressure with surface area element, A_e:

$$\acute{\varepsilon} = - \, d\pi/d\ln A_e \tag{5.3}$$

The viscoelastic modulus (units, mN m^{-1}) describes the resistance of the surface to a change in area. For insoluble films, the surfactant molecules are localized as monolayers in the interface. If the film-covered surface behaves in a purely elastic manner (that is, with no surface shear viscosity or dilational viscosity), then $\acute{\varepsilon} = \varepsilon_0$, the Gibbs or equilibrium elasticity. For such films, surface pressure responds instantaneously to changes in surface area and ε_0 can be evaluated from plots of π measured statically as a function of surface area; these plots are known as 'π–A isotherms'.

The properties of insoluble monolayers of pure compounds have been well studied over many decades (for example, see Gaines, 1966). Depending on the nature of the lipophilic and hydrophilic moieties of the surfactant molecule, the π–A isotherms for monolayers can be classified into three broad types: 1. solid or condensed, 2. liquid, and 3. gaseous or expanded (Figure 5.5A), reflecting the two-dimensional physical state of the film molecules. Compounds with small hydrophilic head groups and long linear hydrocarbon tails are gaseous at very low surface pressures

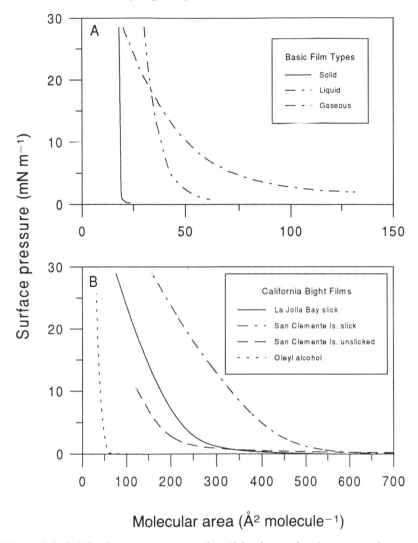

Figure 5.5. (A) Surface pressure–area (π–A) isotherms for three general monolayer film types: solid-like films represented by stearyl alcohol, liquid-like films
represented by oleyl alcohol, and gas-like films represented by polyethylene
glycol (300) monolaurate. Adapted from Barger (1985). (B) Isotherms for
representative sea-surface microlayer films from the California Bight, scaled
according to apparent mean molecular area using a mean molecular weight of 4.5
kdalton determined by size exclusion chromatography. The π–A isotherm for
oleyl alcohol is shown for comparison. Adapted from Frew & Nelson (1992a).

but transition to a condensed state by assuming a close-packed surface orientation when compressed. Compounds with large or multiple hydrophilic groups or with unsaturated or branched hydrophobic moieties that inhibit a close-packed orientation, form expanded films over a wide range of compression. The inherent elasticity of an air–water interface influenced by the different film types varies greatly.

The surface physical properties of sea-surface films, particularly their π-A characteristics and static (Gibbs) elasticities, have been studied extensively by several investigators (Jarvis *et al.*, 1967; Garrett, 1965; Barger, Daniel and Garrett, 1974; Barger, 1985; and Barger and Means, 1985; Frew and Nelson, 1992a,b; Bock and Frew, 1993). The π-A isotherms of these films generally indicate expanded liquid or gaseous film types (Figure 5.5B). The isotherms have been interpreted as being characteristic of surfactants having more extended, highly oxygenated polar head groups (Barger & Means, 1985) than classic insoluble lipids that form solid or condensed films. Analogous to pressure–volume relationships for gases, π-A isotherms for pure insoluble surface films at low pressures can be described using two-dimensional equations-of-state (Barger & Means, 1985; Frew & Nelson, 1992b). Virial equation-of-state fits of π-A isotherms for sea-surface films suggest average molecular weights of a few kilodaltons.

Gibbs elasticities have been determined for a large number of films collected at coastal and open ocean sites. Maximum ε_0 values in the range of 20–40 mN m^{-1} have been reported for surface pressures ranging from 10–15 mN m^{-1} (Barger and Means, 1985; Frew and Nelson, 1992a,b; Bock and Frew, 1993); at much lower surface pressures, in the range 0.5–1 mN m^{-1}, ε_0 values ranging from 2 to 8 mN m^{-1} have been reported, still sufficient to produce significant hydrodynamic effects such as damping of capillary waves.

Natural films as mixtures and their dynamic surface response

It is important to emphasize that natural films are derived from complex mixtures of compounds present in seawater. Isolation of sea-surface film materials using solid phase extraction (Frew & Nelson, 1992a) has allowed comparisons of film surface physical properties with chemical properties. Scaling of film π-A isotherms according to chemical attributes, as a means of normalizing them to a specific area, reveals considerable variability among films from different slicks and films derived from sub-surface samples (Frew & Nelson, 1992b). The surface response of these mixed natural films is not purely elastic. This is apparent even from quasi-static measurements of microlayer film π–A isotherms. It is commonly observed that the isotherms are not reversible, exhibiting a pronounced hysteresis upon repeated compression and dilation. These characteristics have been attributed to varying proportions of components with markedly different π–A characteristics (Frew and Nelson, 1992b; Bock and Frew, 1993).

The surface properties of individual film fractions can be shown to vary systematically with polarity. This can be demonstrated by chromatographic fractionation of microlayer organics and measurement of the surface properties of the individual fractions (Frew and Nelson, 1992a). For example, Figure 5.6 illustrates the results of a simple separation of surfactants from a California Bight microlayer sample by normal phase thin layer chromatography (TLC) and measurement of the π–A isotherms in a film balance using seawater as the subphase. The crude fractions (still complex mixtures) have π–A isotherms ranging from condensed to gaseous types. Components with higher mobility on the silica TLC plate (that is increasingly hydrophobic components) exhibit higher ε_0 values at a given surface pressure than less mobile (more polar) components. The unfractionated sample exhibits intermediate ε_0 values. As a crude approximation, the Gibbs elasticity of the whole film is represented by a weighted sum (based on carbon content) of the ε_0 values determined for the TLC fractions.

This same π–A data can be interpreted in terms of a scaling law surface solution theory for polymers (Villanove and Rondelez, 1980; Kim and Chung, 1988; Gaines, 1991),

$$\pi = C_s{}^m = C_s{}^{2\alpha/(2\alpha-1)} \tag{5.4}$$

where C_s is the surface concentration (mass/unit area), m is the slope of the log π – log C_s plot and α is the characteristic scaling law exponent relating the radius of gyration of the isolated polymer chain to the polymer molecular weight. The exponent can be used as an indicator of the quality of the air–water interface as a solvent for a surface-active polymer. Theoretical values of α are 0.50 for a subphase (for example, H_2O) which acts as a poor solvent and 0.77 for a subphase acting as a good solvent. The scaling law plots for the data of Figure 5.6 are presented in Figure 5.7; these plots are approximately linear over a surface pressure range of 0.3–7 mN m^{-1}. The derived scaling exponent (α) ranges from 0.56 for hydrophobic material that migrated to the top half of the TLC plate, to a value of 0.71 for the polar materials remaining at the spotting origin. Though only crudely separated by TLC, these component

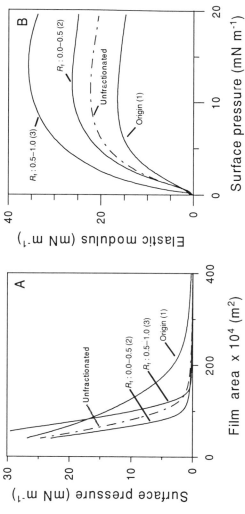

Figure 5.6. Comparison of the surface pressure–area isotherms (A) and quasi-static elasticity–surface pressure plots (B) for normal phase thin layer chromatography (TLC) fractions of a California Bight microlayer sample with those for the unfractionated organic material (dot–dash line) from the same sample. Each film contained an equivalent amount of carbon (15 μg). The fractions are (1) most polar material, remaining at the spotting origin, no migration with mobile phase; (2) intermediate polarity, lower half of TLC plate, Rf = 0.0–0.5, and (3) least polar material, upper half of TLC plate, Rf = 0.5–0.10. Rf is the ratio of distances travelled by the analyte and solvent, respectively, on the TLC plate. From Frew & Nelson, unpublished data.

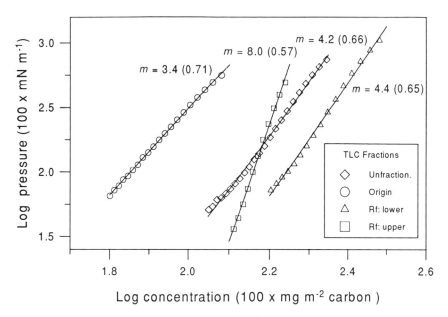

Figure 5.7. Scaling law plots (Equation 5.4) of the surface pressure–surface concentration data of Figure 5.6 according to polymer surface solution theory (Villanove and Rondelez, 1980). Scaling exponents (α) shown in parentheses, where slope $m = 2\alpha/(2\alpha - 1)$. Theoretical values are $\alpha = 0.50$ for a subphase acting as a poor solvent and $\alpha = 0.77$ for a subphase that is a good solvent for polymeric surfactants. From Frew and Nelson, unpublished data.

fractions have α values which nearly span the expected theoretical range. Such data suggest the presence of compounds of widely different lipophilic–hydrophilic balance and equilibrium spreading pressures.

Due to the complex mix of components, the composition and molecular arrangement of microlayer films can vary in response to physical forcing. Such films are capable of undergoing relaxation processes on a variety of timescales that lead to a surface physical response which is not purely elastic (van den Tempel and Lucassen-Reynders, 1983; Lucassen-Reynders, 1986). Instead, the film-influenced water surface will exhibit a complex stress response to compressional and dilational straining (Bock and Frew, 1993). For relatively insoluble films, these relaxation processes include very fast molecular reorientations and conformational changes, and less rapid losses from the film due to desorption, micelle formation or collapse to multilayer solid phases. Since these relaxation processes result in energy losses, they can be represented formally as a dilational viscosity. The surface viscoelastic modulus, $\acute{\varepsilon}$, then becomes a complex, frequency-dependent quantity (Levich, 1962; Hansen and Mann, 1964; Lucassen-Reynders and Lucassen, 1969; Bock and Mann, 1989; Fernandez et al., 1992):

$$\acute{\varepsilon} = \varepsilon_d + i\,\omega\,\eta_d \tag{5.5}$$

where ε_d and η_d are the surface dilational elasticity and the surface dilational viscosity, respectively, and ω is the radian frequency ($\omega = 2\pi f$) of the straining process. In general, both the elastic and viscous terms in Equation 5.5 are dependent on frequency. Shear viscosity effects are usually small (Lucassen-Reynders, 1986) and therefore neglected in this formulation; $\acute{\varepsilon}$ is also referred to as the surface dilational modulus.

In the case of surfactants with appreciable solubility in the bulk phase and therefore capable of diffusional exchange between the interface and the bulk phase, an additional component of dilational viscosity is introduced (although some water-soluble polymeric surfactants form monolayers which behave like insoluble films). For dilute solutions of water-soluble surfactants, the equilibrium surface tension is related to the surface concentration (Γ, moles/unit area) and bulk concentration (C, moles/unit volume) via the Gibbs equation:

$$\Gamma = -\frac{1}{RT}\left(\frac{d\gamma}{d\ln C}\right) \tag{5.6}$$

During compression and dilatation of the water surface, for example by waves or eddy turbulence, the system strives toward the equilibrium, Γ, but with a time lag dependent on the rate constant for diffusional exchange relative to the frequency of straining. This can be expressed as (Cini et al., 1987):

$$\varepsilon_d = \varepsilon_0\,\frac{1 + (\omega_D/2\omega)^{1/2}}{1 + \omega_D/\omega + (2\omega_D/\omega)^{1/2}}$$

$$\tag{5.7}$$

$$\varepsilon_i = \varepsilon_0\,\frac{(\omega_D/2\omega)^{1/2}}{1 + \omega_D/\omega + (2\omega_D/\omega)^{1/2}}$$

where ω_D, the diffusional frequency, represents the effect of diffusional exchange and is defined in terms of the diffusion coefficient of the surfactant, D_s ($cm^2\,s^{-1}$), the bulk concentration, C (mole cm^{-3}), and the surface concentration, Γ (mole/cm^2) (Cini et al., 1987):

$$\omega_D = D_s \left(\frac{dC}{d\Gamma} \right)^2 \tag{5.8}$$

Diffusional exchange lowers the magnitude of $\acute{\epsilon}$ by levelling out surface tension gradients (Lucassen-Reynders, 1986; Dysthe and Rabin, 1986).

The surface dilational modulus is a critical parameter which modulates gas transport mechanisms, including wave-induced turbulence and eddy turbulence derived from sub-surface interior mixing. However, the functional relation between k_w and $\acute{\epsilon}$ is not known. In the laboratory, $\acute{\epsilon}$ can be derived from measurements of distance damping coefficients using artificially generated wave packets containing a range of frequency components (Bock and Mann, 1989; Fernandez *et al.*, 1992). Unfortunately, it is difficult to measure $\acute{\epsilon}$ experimentally either *in situ* or in wind–wave tanks in the presence of wind shear, and few literature reports are available. Alpers and Hühnerfuss (1989) have used the ratio of equilibrium wave spectra from slicked and unslicked ocean surfaces to estimate the viscoelastic modulus. Using a similar approach, Cini, Lombardini and Hühnerfuss (1983) and Lombardini *et al.* (1989) computed wave spectra and dilational moduli using sun glint imagery and point measurements of wave height from a Goubau line system. In general, however, the range of viscoelasticities prevalent at the ocean surface in both coastal and oligotrophic regimes is not well known and the determination of $\acute{\epsilon}$ of the ocean surface is currently an active area of research.

Surfactants and mechanisms of gas exchange

Static versus dynamic effects of surfactant films

In an earlier review, Liss (1983) categorized film effects into static effects related to the intrinsic transfer resistance of monolayers and dynamic (viscoelastic) effects which fundamentally alter hydrodynamical flow. The direct effect of a surface film is the contribution of an additional liquid phase resistance to mass transfer due to the presence of a static physical barrier. In non-flow systems and for gases for which mass transfer is controlled by resistance in the gaseous boundary layer (Liss, 1983), considerable mass transfer resistances have been demonstrated for insoluble surfactant films of the condensed type, for example, with 1-hexadecanol (Plevan and Quinn, 1966; Sada and Himmelblau, 1967; Burnette and Himmelblau, 1970; Springer and Pigford, 1970). Numerous studies have demonstrated reductions in mass transfer of H_2O, for example, using condensed films of insoluble surfactants (Jarvis *et al.*,

1962; La Mer, 1962; Frenkiel, 1965; La Mer and Healy, 1965; Garrett, 1971; Martinelli, 1979).

This barrier effect is highly dependent on film type and requires a rigid, close-packed lattice characteristic of condensed, insoluble monolayers. Springer and Pigford (1970) estimated apparent diffusion coefficients for gas molecules diffusing through a condensed monolayer to be of order 10^{-9} cm^2 s^{-1}, indicating that such films act as solid barriers. Soluble surfactants as well as insoluble surfactants that form expanded type films offer no appreciable resistance (Jarvis *et al.*, 1962). Springer and Pigford (1970), by comparing mass transfer resistances for stagnant and stirred cells with and without films, were able to separate static effects and dynamic effects for insoluble and soluble surfactant films. They showed clearly that films of soluble surfactants were too permeable to offer significant liquid phase resistance.

In the case of gases for which the exchange is controlled by the liquid phase transfer resistance, no significant static effect has been observed for films of either insoluble or soluble surfactants (Petermann, 1976; Martinelli, 1979).

Studies of gas transfer across water surfaces contaminated by petroleum oil films have also been reported (Liss and Martinelli, 1978; Martinelli, 1979), although such films are not generally surfactant monolayers, but rather liquid phases that float on the water surface, introducing an additional interface (Martinelli, 1979). Observed reductions in gas exchange across oil films are consistent with the above results in that the barrier effect is more pronounced for exchange controlled by the gas phase transfer resistance than for liquid phase control. However, much thicker oil films (thickness of order 10^{-2}–10^{-4} cm) are required for significant effects as compared with monolayer surfactant films (thickness of order 10^{-6}–10^{-7} cm).

As summarized by Liss (1983), it is unlikely that static effects for a range of surfactant film types are important under typical conditions of turbulence found at the sea surface, since the close-packed configuration would not be maintained in the presence of wind and breaking waves.

In contrast to non-flow systems, gas exchange observations in various types of flow systems, including stirred cells (Downing and Truesdale, 1955; Goodridge and Robb, 1965; Metzger, 1968; Springer and Pigford, 1970; Lee, Tsao and Wankat, 1980; Asher and Pankow, 1986, 1989, 1991a,b; Goldman *et al.*, 1988; Frew *et al.*, 1990), falling films (Cullen and Davidson, 1956), channel flows (Mancy and Okun, 1965; Moo-Young and Shoda, 1973), and wind–wave tanks (Broecker *et al.*, 1978; Jähne *et al.*, 1979, 1984, 1985, 1987; Frew *et al.*, unpublished data), have indicated pronounced hydrodynamic effects due to films. These arise from the viscoelastic nature of a film-influenced interface, represented by έ. The condition of non-zero viscoelasticity may affect the rate of gas transfer by changing the near-surface turbulence, the characteristics of the wave field, or both.

Virtually no field data exist that evaluate the impact of films on gas exchange *in situ*, with the exception of a report by Brockmann *et al.* (1982), who found reductions in gas transfer across an artificial slick of oleyl alcohol. In the following sections, evidence from different experimental laboratory systems will be reviewed. As a basis for discussing these, it is necessary to introduce a brief discussion of basic gas transfer models. More detailed consideration of these models can be found in the references cited.

Physical models of air–water gas transfer

Various models have been proposed to describe the exchange of gases across a liquid–gas interface for turbulent systems in the absence or presence of wind shear (Higbie, 1935; Danckwerts, 1951; Fortescue and Pearson, 1967; Lamont and Scott, 1970; Münnich and Flothmann, 1975; Ledwell, 1984; Coantic, 1986; Back and McCready, 1988; Brumley and Jirka, 1988).

Models based on turbulence considerations have been reviewed by Brumley and Jirka (1988). These include both conceptual and explicit hydrodynamic models. The turbulence models are basically variations of the surface-renewal model first proposed by Higbie (1935) and later extended by Danckwerts (1951). The mass transfer coefficient, k_w, can be expressed generally as:

$$k_w \approx \left(\frac{V}{L} D \right)^{1/2} \qquad (5.9)$$

where V and L are the characteristic velocity and length scales describing the turbulent liquid motions very near the interface, induced either by mechanical shear in the interior of the liquid or by wind shear at the surface. The models differ among themselves in the choice of the appropriate length and velocity scales.

In the surface-renewal model, turbulent eddies cause the liquid surface to be randomly replaced by elements of the bulk fluid. During residence at the surface, these fluid elements tend to equilibrate with the gas phase by molecular diffusion according to the concentration gradient. The residence time (Higbie, 1935) of the fluid element in the diffusive sublayer, t_* (units, s), controls the rate of exchange such that

$$k_w = 2 \, (D / \pi t_*)^{1/2} \qquad\qquad (5.10)$$

Rather than assume that t_* is constant, Danckwerts (1951) suggested a statistical distribution of surface element residence times and calculated the fraction s (units, s^{-1}) of the surface renewed in unit time. The gas transfer velocity then becomes:

$$k_w = (D \, s)^{1/2} \qquad\qquad (5.11)$$

The fact that the surface-renewal rate, s, cannot easily be related to measurable turbulence parameters has led to hydrodynamically explicit models (Fortescue and Pearson, 1967; Lamont and Scott, 1970) in which k_w remains proportional to $D^{1/2}$ and the turbulence is parameterized using the the integral turbulent length scale and turbulent kinetic energy, assuming large and small eddies respectively. Brumley and Jirka (1988) took a different approach. Instead of choosing a single length scale, they assumed that eddies of all scales approach and interact with the surface, causing a fluctuating surface divergence over the time and length scales of the eddy. In their model, stagnation type flow characterized by the surface divergence effectively pumps gas to and from deeper regions where rotational mixing occurs. Brumley and Jirka (1988) pointed out that, in their model, the transfer process would be sensitive to even slight surface contamination that would reduce surface divergence fluctuations.

Turbulent boundary layer models (Deacon, 1977; Ledwell, 1984) represent the other major type of model frequently used to describe gas transfer. In contrast to renewal models, these models assume continuity of the surface, that is, that the velocity fluctuations normal to the interface become zero at the interface. The model of Ledwell (1984) allows surface divergence such that new liquid molecules can appear at the surface. In Deacon's model for a smooth rigid wall, however, surface divergence is not allowed. Turbulent energy is not sufficient to overcome surface tension and create new surface (Hasse and Liss, 1980). Momentum transfer and mass transfer are linked using the universal velocity profile over a smooth rigid surface.

Assuming continuity of stress at the interface, the gas transfer velocity can be shown to be proportional to the water-side friction velocity, $u_w{}^*$, and the ratio of the transfer coefficients for mass (molecular diffusivity, D) and momentum (kinematic viscosity, v) as:

$$k_w = \frac{1}{\beta} \left(\frac{D}{v} \right)^{n} u_w^* \qquad (5.12)$$

where the Schmidt number exponent, n, is equal to 0.67 for a surface acting as a smooth rigid wall (Deacon, 1977) and is equal to 0.5 for a free surface (Ledwell, 1984). The parameter β is the dimensionless resistance to momentum transfer across the water-side viscous boundary layer (Jähne et al., 1987); β is constant for the rigid wall case and, in the case of the free wavy surface, is dependent on the wave field.

Thus, for a free surface in the absence of surface contamination, both the surface renewal and boundary-layer models predict a $D^{1/2}$ dependence for k_w and are difficult to distinguish experimentally. Hasse (1990) has cautioned against extrapolating wind–wave tank results that appear to favour surface-renewal models to gas exchange at the ocean surface, in view of secondary flows that force additional renewal in laboratory tanks.

Experimental studies of films in stirred systems

Due to the importance of surfactants in industrial processes, much of what is known about the effect of surfactants on gas exchange in turbulent systems has come from chemical engineering studies in various flow systems. An important contribution was that of Davies (Davies, 1966, 1972; Davies and Rideal, 1963), who used the surface renewal concept to interpret experimental observations of surfactant effects on mass transfer in stirred systems. Davies argued that surface renewal by turbulent eddies approaching a film-influenced interface is opposed by an additional tangential stress that is due to the film pressure, π, shown conceptually in Figure 5.8. As a result, the thicknesses of the viscous and diffusive sublayers effectively increase, reducing the rate of mass transfer. Davies' theoretical model gives the gas transfer velocity of a film-influenced interface as:

$$k_w[\text{film}] = \frac{k_w[\text{clean}]}{\left[1 + 0.25(v_0\gamma_{eq}v\pi)^{-1}\left(\frac{\pi\eta B}{\pi+\eta B}\right)^2\right]^{1/2} + 0.5(v_0\gamma_{eq}v\pi)^{-1/2}\left(\frac{\pi\eta B}{\pi+\eta B}\right)} \qquad (5.13)$$

where k_w [clean] is given by $D^{1/2}\varrho^{1/2}v_0^{3/2}\gamma^{-1/2}$, and v_0 is the characteristic velocity of an eddy, v is the kinematic viscosity, η is the bulk viscosity, B is taken as $100v_0$ and γ_{eq} is the equivalent surface tension, which includes both surface tension and a gravity term accounting for surface deformation by the eddy. The factor $(\pi\eta B/\pi+\eta B)$ is the combined resistance to flow in the surface plane and the viscous sublayer, due to the film. The

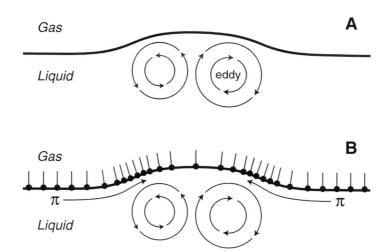

Figure 5.8. Schematic representation of surface renewal by a turbulent eddy, adapted from Davies (1972). Approach of eddy to (A) clean surface and (B) to surface with a surfactant film present. In the case of the film-influenced interface, the eddy motions in the plane of the surface are opposed by an additional force due to the film surface pressure, π.

reduction in k_w is dependent on the turbulence level and the degree of contamination of the surface. For $\pi = 0$, Equation 5.13 reduces to k_w [film] $= k_w$ [clean].

The reduction in k_w due to a film as predicted by Equation 5.13 can be compared with experimental data (Figure 5.9) for CO_2 absorption in a stirred cell with a plane surface (Davies, 1972). The theoretical curves underpredict the reduction in k_w at low π (Figure 5.9A). This is due to the use of π as the resisting force in the derivation of Equation 5.13 rather than $\acute{\varepsilon}$, which is generally larger than π except for the case of ideal gaseous films (Figure 5.9B). Using the experimental data (Figure 5.9B) for ε_0 as a function of π to approximate $\acute{\varepsilon}$, one can calculate the enhanced reduction in k_w [film] when ε_0 is substituted for π in Equation 5.13; this gives profiles of $R = k_w$[film]/k_w[clean] as a function of π which represent the experimental CO_2 data more closely (Figure 5.9A, dashed lines). Similar profiles for R versus π have been reported for oxygen evasion at film-influenced interfaces in a stirred system for natural water samples (Goldman et al., 1988) and for surface-active phytoplankton exudates (Frew et al., 1990). These authors found reductions in R that reached 50–60% even at high levels of turbulence and transfer velocities; these were shown to be correlated with surface physical parameters and dissolved

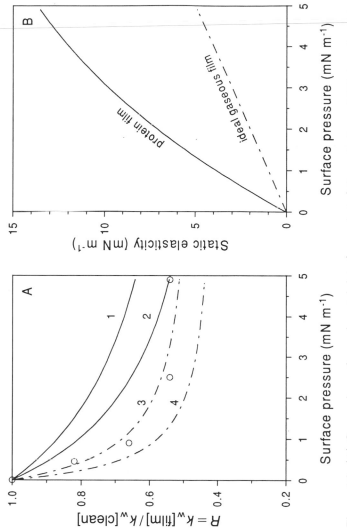

Figure 5.9. (A) Comparison of experimental data (open circles) for reduction of CO_2 invasion rate into water in a stirred cell (plane surface) with theoretical reduction predicted by Equation 5.13 for v_0 = 5.2 cm s^{-1}, γ_{eq} = 72 mN m^{-1} (curve 1) and v_0 = 6.6 cm s^{-1}, γ_{eq} = 144 mN m^{-1} (curve 2) for a range of surface pressures in the presence of a protein film. Curves 3 and 4 were derived from Equation 5.13 substituting ε_0 data for π. Adapted from Davies (1972). (B) Comparison of ε_0 versus π for a protein film (solid line) and an ideal gaseous film (dashed line).

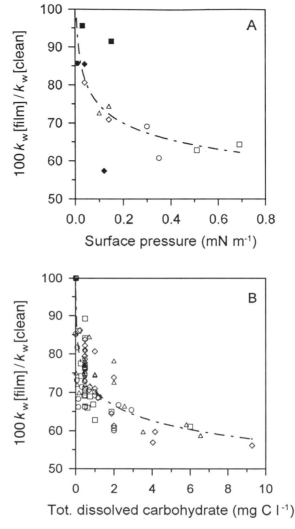

Figure 5.10. Relative changes in O_2 evasion rates ($100k_w$[film]/k_w[clean]) measured in a small stirred system as a function of (A) surface pressure and (B) total dissolved carbohydrate for exudates of marine phytoplankton species. Symbols: ▽, *Isochrysis galbana*; ○, *Phaeodactylum tricornutum*; ◇, *Chaetoceros curvisetus*; □, *Olisthodiscus luteus*; ●, *Chlorella stigmaphatora*; ■, *Nitzschia frustulum*; ◆, *Porphyridium* sp. Adapted from Frew *et al.* (1990).

organic matter concentrations (Figure 5.10). Their results showed that even at the very low surface pressures typical of the ocean surface ($\pi < 1$ mN m^{-1}), reductions in gas exchange would be expected due to surfactants. The sensitivity of k_w to low levels of surface contamination is a consequence of the strong positive deviation of the π-$\acute{\varepsilon}$ relation compared with ideal gaseous films (Figure 5.9B). The reductions in k_w predicted by Equation 5.13 for the film-influenced interface at $\pi < 1$ mN m^{-1} (about a factor of two) are very consistent with the reductions in k_w illustrated in Figure 5.3 for seawater samples relative to clean water under conditions of wind-driven turbulence. This may be taken as strong evidence that gas transfer at the turbulent boundary layer in the oceans will also be strongly diminished by the presence of surfactant-enriched films.

The Danckwerts distribution function for surface age has been shown to be approximately valid experimentally for stirred systems in the presence of films at low π (Springer and Pigford, 1970; Lee et al., 1980). This is consistent with surface visualization studies (Davies, 1972) which have shown that the frequency of surface renewal is not affected at low π, but that the mean area involved with each clearance decreases with increasing π. Above a critical surface pressure, however, renewal becomes increasingly less frequent. Lee et al., (1980) compared fluctuations in oxygen concentration profiles in the presence and absence of surfactant films. They showed that the hydrodynamic effect of a film is due to an increase in the characteristic length scale, L, and a decrease in the velocity scale, V, of eddies approaching the film-covered interface. The suppression of low-energy eddies (small and slow) results in a redistribution of eddy sizes toward larger average L. Equation 5.13 predicts an asymptotic limit in R as π increases. This occurs when the frequency of surface area renewal becomes vanishingly small; the surface is immobilized and takes on the characteristics of a solid wall, for which k_w is proportional to $D^{2/3}$, as in the boundary-layer model of Deacon (1977).

The hydrodynamic effect is dependent on surfactant type. While for a given surfactant, R is a smooth function of surface concentration and pressure (Moo-Young and Shoda, 1973; Lee et al., 1980; Goldman et al., 1988), this is not the case when data for different surfactant types are compared. Although both the length and velocity scales of eddies are altered by a film, Lee et al. (1980) observed that k_w, V, and L are affected by differing degrees at a given π for different molecular structures. Macromolecular protein surfactants were more effective in reducing k_w than extremely soluble ionic surfactants of low molecular weight. Springer and Pigford (1970) showed that under turbulent conditions, soluble surfactants were more effective than insoluble surfactants in suppressing k_w and suggested that the more rapid reestablishment of soluble films after disruption by turbulence was responsible. They pointed out that film recovery times were dependent on such factors as molecular weight, solubility, adsorption rate, and bulk and surface diffusivities. According to Meijboom and Vogtlander (1974), the recovery time of a film is proportional to the elasticity and inversely proportional to the adsorption coefficient of the surfactant. These effects would generally be implicit in the use of the surface dilational modulus $\acute{e}$ rather than π in Equation 5.13.

While useful conceptually, it is difficult to apply the Davies model directly to experimental systems due to the lack of explicit descriptions of turbulence scales. Asher and Pankow (1986, 1989, 1991a,b) tested the hydrodynamically explicit surface-renewal models of Fortescue and Pearson (1967) and Lamont and Scott (1970) using a grid-stirred system to provide well-defined turbulence levels in the presence of varying degrees of surface contamination. They found that, above a critical turbulence level, k_w was linear with the scaling parameters of both the large eddy and turbulence dissipation models, consistent with surface renewal (Figure 5.11). However, both the critical turbulence levels and the slopes of the linear regressions were found to be dependent on the amount of contamination. Increased surface contamination raised the critical turbulence level and decreased the slope of the k_w relation. Below the critical turbulence level, the rate of increase of k_w with turbulence was much lower and k_w was not well represented by surface-renewal models.

Measurement of concentration fluctuation depths (Asher and Pankow, 1989, 1991a) confirmed that, for clean or nearly clean interfaces, surface concentration fluctuations occur mainly within the concentration boundary layer, but that the mean concentration fluctuation depth increases with decreasing turbulence and that the fraction of concentration fluctuations within the boundary layer also decreases. For surfaces covered by coherent insoluble films, concentration fluctuations in the diffusive sublayer were not observed at any of the measured turbulence levels, that is, surface renewal was non-existent.

The experimental evidence for film effects in turbulent systems presented in this section supports the general conceptual model of Davies (1966) in which surfactants reduce the near-surface turbulence. The introduction of finite viscoelasticity to the air–water interface provides an additional tangential stress that opposes surface renewal by turbulent eddies; this additional resistive force is effective even at extremely low surface pressures.

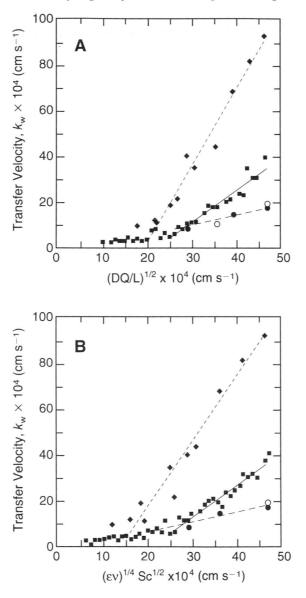

Figure 5.11. Measurements of k_w for surfactant-influenced water surfaces in a grid-stirred system, plotted according to the surface renewal models of (A) Fortescue and Pearson (1967), where D is diffusivity, L is the turbulence length scale and Q is a measure of the turbulence kinetic energy, and (B) Lamont and Scott, where ε is the turbulent eddy dissipation rate, ν is kinematic viscosity, and $Sc = \nu D^{-1}$. Symbols: ◆, rayon/vacuumed surface; ■, lens paper cleaned surface; ●, 1-octadecanol monolayer; and ○, uncleaned surface. Redrawn from Asher and Pankow (1986).

Experimental studies of films in wind–wave systems

Influence of waves on gas exchange

Numerous studies of gas transfer have been carried out in wind–wave facilities of varying scales (Kanwisher, 1963; Broecker et al., 1978; Cohen, Cocchio and Mackay, 1978; Liss et al., 1981; Mackay and Yeun, 1983; Merlivat and Memery, 1983; Jähne et al., 1985, 1987; Wanninkhof and Bliven, 1991; Frew et al., unpublished data). In these studies, it invariably has been observed that, below a critical wind velocity, gas exchange increases only very slowly with an increase in wind speed. Beyond the critical wind velocity and coincident with the onset of capillary-small gravity waves, the gas exchange rate abruptly increases. It is evident that the gas exchange process involves fundamentally different hydrodynamic regimes above and below this critical wind velocity. The observations in wind–wave tanks are thus strikingly similar to those of Asher and Pankow (1986) for stirred systems in the presence of surface-active contaminants. The functional dependence of k_w on the Schmidt number ($Sc = \nu D^{-1}$) has been tested experimentally using dual tracer exchange experiments (Jähne et al., 1984). That study confirmed the transition in Schmidt number exponent from $n \approx 0.7$ at low $u_w{}^*$, to $n \approx 0.5$ above the critical $u_w{}^*$. Within the low $u_w{}^*$ regime, k_w is approximately linear with $u_w{}^*$. Above the critical wind velocity, the dependence appears to be greater than linear. The dimensionless transfer velocity $k_w{}^+$ (defined as $k_w/u_w{}^*$) increases with $u_w{}^*$ and then tends toward a constant value. Thus, by comparison with the boundary-layer model, there appears to be an additional enhancement due to waves.

The mechanisms by which waves enhance gas–liquid mass transfer are not clearly understood. The boundary layer model (Equation 5.12) does not account for the influence of small-scale waves other than to modify the magnitude of the drag coefficient. A model by Witting (1971) proposed that wave motions introduce periodic surface divergence without turbulence interactions; however, this model can support only small enhancements in k_w for the range of wave slopes typical of wave tanks and the sea surface. Other recent models have focused on nonlinear interactions between small-scale waves and the surface shear current as a source of additional turbulence and enhanced gas transfer (McCready and Hanratty, 1984; Coantic, 1986; Back and McCready, 1988). Hanratty (1991) has argued that for irregular wavy surfaces, random velocity fluctuations normal to the liquid surface, as distinct from turbulence, might directly enhance gas transfer. Csanady (1990) has proposed that intense surface divergences associated with breaking wavelets might also play an important role; however, whether microscale breaking occurs at low friction velocities is unclear. What does seem apparent from the various proposed models, is that parameters affecting the characteristics of the wave field will be extremely important in controlling gas transfer rates.

Gas exchange in the presence of films

It is now understood that the slope break in the $u–k_w$ relation observed even in relatively clean experimental systems is the result of residual surfactant films. These adventitious films are extremely difficult to eliminate due to low levels of dissolved organics that are present even in extensively purified water or due to contamination from surrounding air and from wave tank construction materials. Recent experimental work in a small annular wind–wave channel designed specifically to study surfactant effects (Frew, Goldman and Bock, unpublished data) demonstrated that no slope break occurs in very clean systems. In that study, k_w varied smoothly as u increased, going roughly as $k_w \propto$ (rotor speed)^2 (see Figure 5.2). Capillary ripples were observed for $u_w{}^* < 0.05$ cm s^{-1}. This is consistent with earlier work by Scott (1972), who found no evidence for a critical wind velocity for the initiation of wind waves on assiduously cleaned surfaces using a completely different tank geometry.

Another important result of that study was that, for a surfactant-free system, k_w was linear with $u_w{}^*$ (that is, $k_w{}^+$ was constant) over the entire range of $u_w{}^*$ measured ($0.03 < u_w{}^* > 1.2$ cm s^{-1}). Thus, another possible interpretation of the data of Jähne *et al.* (1987), showing a greater than linear dependence of k_w on $u_w{}^*$, is that surfactant films continue to suppress the exchange rate even above the critical $u_w{}^*$. This is supported by the data of Figure 5.2 that show that even above the critical wind velocity, k_w values for the surfactant-influenced interface still fall well below the clean water $u–k_w$ relation. The slopes of the linear portions of the $u–k_w$ plot and the wind velocity at which the slope break occurs both vary with surfactant concentration. There is some evidence that these also depend on the chemical structure of the surfactant. Similar effects are seen in the $k_w{}^+$ data of Jähne *et al.* (1987) collected at constant wind speed but for varying degrees of adventitious contamination and therefore wave patterns (Figure 5.12A). The dimensionless transfer velocity $k_w{}^+$ is nearly independent of $u_w{}^*$, varying by a factor of five, while the friction velocity decreases only slightly due to modification of the drag coefficient by the films. It seems reasonable to assume that the data trend

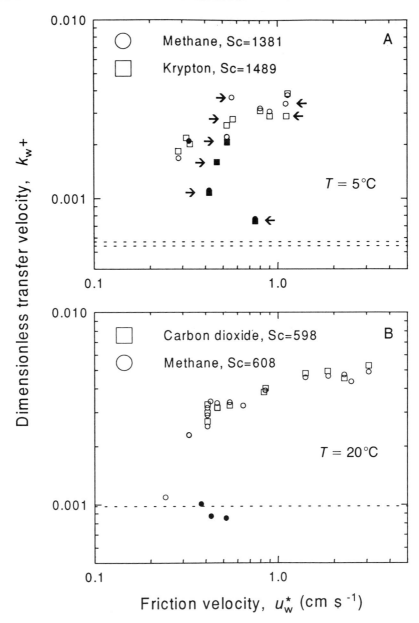

Figure 5.12. Dimensionless transfer velocities ($k_w^+ = k_w/u_w^*$) plotted against u_w^* for various tracers in circular wind–wave tanks, for varying levels of surface contamination and wave patterns. Data collected at a common wind speed, but different wave patterns and contamination levels are indicated by arrows ($\rightarrow$, $u=3.4$ m s^{-1}; $\leftarrow$, $u=5.0$ m s^{-1}). Dashed horizontal lines represent the prediction of the smooth-wall model of Deacon (1977). Adapted from Jähne *et al.* (1987).

toward the solid wall model with increasing contamination, although contamination levels were not reported in that study. Similar data of Jähne *et al.* (1987) show that dimensionless transfer velocities tend to attain a constant value only at high friction velocities at surfactant-influenced interfaces and, at lower u_w^*, generally fall below that constant level in k_w/u_w^*–u_w^* plots (Figure 5.12B).

The small annular wind–wave system used in the studies of Frew *et al.* (unpublished data) allowed continuous measurements of transfer velocity and wave slope on a short time scale (response time of system, 60–90 s). This facilitated monitoring of the transfer velocity with time-dependent changes in surface elasticity. An example of time-dependent decreases in k_w has already been given in Figure 5.3. In the present context, it is useful to note that all of the data in Figure 5.3 represent k_w, (at nearly constant wind stress) for wavy surfaces with the exception of the *Station A* sample. The steady decrease in k_w over the interval sampled occurred in the presence of a uniform but evolving field of small-scale waves; for the *Station A* sample, waves were absent, representing transfer in the smooth wall regime. A second example is given in Figure 5.13, in which time series data are presented for oxygen evasion from distilled water in the presence of surface-active exudates from the marine chrysophyte *Olisthodiscus luteus* (Frew *et al.*, unpublished data). After establishment of a stable wave field, the exudate was introduced to the tank as a sub-surface injection at $t=0$, to give a DOC concentration of 125 μM. An extremely rapid decrease in k_w occurred upon addition, followed by a slower asymptotic decrease to a limiting value which is identical to that in Figure 5.3 (*Station A*) for the same wind speed. The initial rapid drop in k_w suggests that a finite surface excess and consequent visco-elasticity were very quickly established, possibly by relatively soluble surfactants present in the exudate. Adsorption of less soluble surfactants eventually increased the viscoelastic modulus sufficiently to immobilize the surface.

In these and similar experiments, reductions in k_w of nearly an order of magnitude were observed at constant wind speed, with decreases in u_w^* of 15% or less. Clearly, such data suggest that in the presence of films, Equation 5.12 is not a sufficient parameterization of k_w.

Influence of surfactants on the wave field

Despite the lack of knowledge of the explicit mechanisms by which waves promote gas exchange, it seems clear that surfactant effects in modifying the wave field will play an important role. If the exchange mechanism

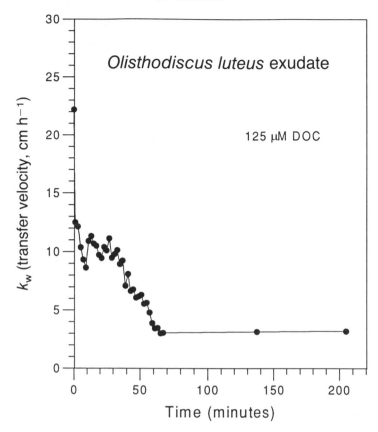

Figure 5.13. Time series measurements of k_w during the evolution of surface films from surface-active exudates of the marine chrysophyte, *Olisthodiscus luteus*, injected into distilled water at time $t=0$, to give a DOC concentration of 125 μmol l^{-1}. The surface viscoelasticy is increasing with time as surfactants adsorb at the interface. The transfer velocity decreases to an asymptotic limit corresponding to mass transfer at a smooth rigid wall, $k_w \approx D^{2/3}$. Measurements are for O_2 evasion in small circular tank at 20 °C and a rotor speed of 5 m s^{-1}. From unpublished data of Frew, Goldman and Bock.

involves conversion of wave energy to near-surface turbulence, then the presence of interfacial viscoelasticity will have effects similar to those discussed for stirred systems. More direct effects related to wave slope would also depend on the viscoelastic modulus. Surfactant effects on the wave field include inhibition of wave growth, enhancement of wave energy dissipation (damping), and reduced exchange of energy between waves of different frequencies via nonlinear mixing.

Of these, the damping effects of monolayer films have been investigated in the most detail (Levich, 1962; Lucassen, 1968; Lucassen-Reynders and Lucassen, 1969; Cini and Lombardini, 1978; Cini *et al.*, 1987; Alpers and Hühnerfuss, 1989; Bock and Mann, 1989). The damping effect of a film arises from the creation of transient surface tension gradients during compression and dilation of the surface by transverse (gravity and capillary) waves as they propagate. The resulting longitudinal or Marangoni waves (Lucassen, 1968) are highly damped due to the surface elasticity, which acts as the restoring force in a tangential direction. The coexistence of transverse and longitudinal waves modifies the hydrodynamic boundary conditions governing wave propagation. The solutions to the equations of motion for a film-damped interface have been presented by a number of investigators. A rigorous derivation of a dispersion relation incorporating viscoelastic effects has been formulated (Lucassen, 1968; Bock and Mann, 1989), which allows computation of the wave damping coefficient as a function of frequency. The damping effect is usually expressed as the ratio of damping coefficients for the film-influenced and clean interface. An approximate formulation (Cini *et al.*, 1987) is given by

$$y_\omega = \frac{1 \pm 2\theta + 2\theta^2 - X + Y(X + \theta)}{1 \pm 2\theta + 2\theta^2 - 2X + 2X^2} \tag{5.14}$$

where $X = \varepsilon_o \varkappa^2/(2\nu\varrho\omega^3)^{1/2}$, $Y = \varepsilon_o \varkappa/4\nu\varrho\omega$, $\theta = (\omega_D/2\omega)^{1/2}$. The wave number $\varkappa$ assumes the Kelvin dispersion relation, $\omega^2 = (\gamma\varkappa^3/\varrho) + g\varkappa$. The damping ratio is generally found to have a maximum in the frequency range of 3–30 Hz. However, since the magnitude of the damping ratio and the frequency at which the ratio is a maximum are dependent on the value of the surface dilational modulus, different surfactant types yield different damping curves as a function of frequency. Figure 5.14 shows the theoretical damping curves for different values of ε_o and the diffusional frequency, ω_D (Cini and Lombardini, 1978). The magnitude of the damping maximum is seen to decrease with decreasing ε_o and increasing ω_D, while the frequency of the maximum increases with decreasing ε_o. These considerations suggest that if the mechanisms enhancing gas exchange at a wavy surface are sensitive to a particular wave number range, for example capillary-small gravity waves, then surfactant films will have a significant impact via the damping mechanism, and that there may also be second-order effects dependent on surfactant characteristics, including solubility.

The variation of k_w with wave field characteristics has been the subject of several studies (Jähne *et al.*, 1984, 1987; Frew, Goldman and Bock,

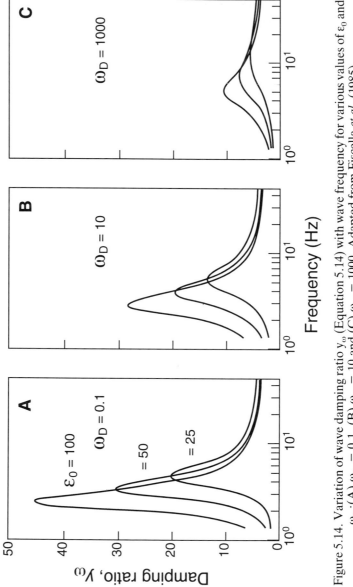

Figure 5.14. Variation of wave damping ratio y_ω (Equation 5.14) with wave frequency for various values of ε_0 and ω_D: (A) $\omega_D = 0.1$, (B) $\omega_D = 10$ and (C) $\omega_D = 1000$. Adapted from Fiscella *et al.* (1985).

unpublished data). A strong correlation between k_w and mean square wave slope was demonstrated for wave facilities of different scales (Jähne *et al.*, 1987). That study suggested that through nonlinear interactions, waves of all scales contribute to a cascade of energy which eventually becomes turbulence, enhancing k_w. The relationship between surfactants, wave slope and k_w has been shown clearly in experimental work (Bock and Frew, unpublished data) which followed the wave field during the transition from a clean interface to a film-influenced interface. In that work, the gas transfer velocity closely tracked mean square slope as surface-active materials adsorbed at the interface and gradually increased the surface elastic modulus (Figures 5.15 to 5.17). In contrast to the finding of Jähne *et al.* (1987), the transfer velocity was more linearly correlated with higher frequency waves (that is, capillary ripples). Conceptually, this would be consistent with the known damping effects of surfactants (see Figure 5.14), since low initial elasticity values would first affect capillary waves. Only when the elasticity reached relatively high values would the longer waves begin to be damped. This hypothesis needs further quantitative confirmation in a large-scale wind–wave facility in the presence of longer waves. The correlation between k_w and mean square slope (Figure 5.18) is surprisingly robust for a variety of surfactant types over a wide range of wind-speeds (Bock and Frew, unpublished data). Thus, a parameterization of k_w in terms of mean square slope appears to be quite promising since it would integrate the effects due to surfactant films.

Concluding remarks

This paper has selectively reviewed a large body of experimental evidence obtained for various laboratory flow systems that indicates that surface-active compounds have first-order effects on mass transfer at the air–water interface. The effects are hydrodynamic in nature and are a consequence of the finite viscoelasticity of the film-influenced interface. Can these observations be extrapolated to large-scale systems like the oceans, and what is the magnitude of these effects on oceanic gas exchange? Several points can be made to attempt an answer to these questions. First, the phenomenon of reduction of gas exchange in the presence of films is remarkably consistent despite the fact that quite disparate experimental systems, including those with mechanically driven bulk turbulence and with wind-driven turbulence and various geometries, have been used to study the problem. In a broad sense, these

N. M. Frew

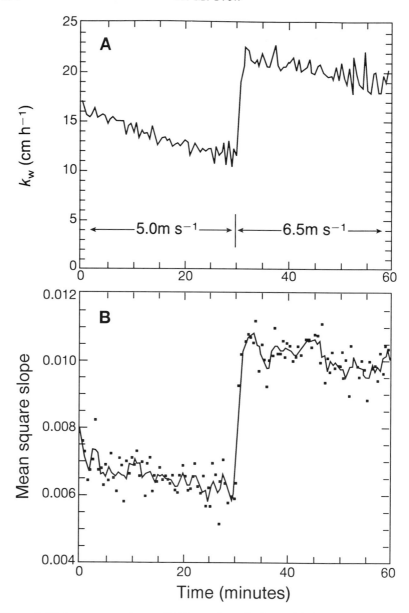

Figure 5.15. Intercomparisons of (A) transfer velocity and (B) mean square wave slope as functions of time in a coastal seawater sample, as measured for O_2 evasion at 20 °C in the annular wind–wave tank. At $t=30$ min., the rotor speed was increased from 5.0 m s^{-1} to 6.5 m s^{-1}. The gradual decreases in the values of the two parameters during periods of constant rotor speed were the result of accumulation of surfactants at the air–seawater interface. From unpublished data of Frew and Bock.

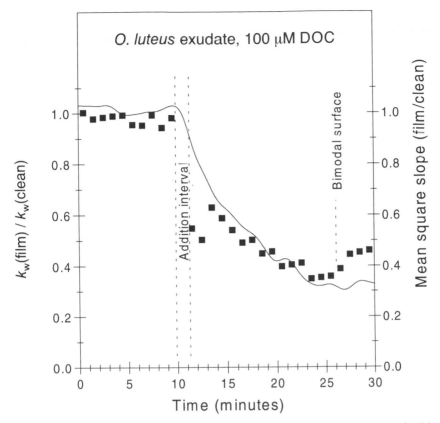

Figure, 5.16. Comparison of time profiles of k_w (■) and mean square slope (solid line) for a time series experiment similar to that in Figure 5.13. Both parameters have been normalized to clean surface values for comparison. Exudate was added at $t = 10$ min. At $t = 26$ min. surface roughness became bimodal with development of a small slicked area. From unpublished data of Frew, Goldman and Bock.

systems are not unlike the oceans in that they all have in common the turbulent boundary layer. It is the near-surface turbulence that governs the exchange process, including oceanic gas exchange, rather than bulk circulations and secondary flows, which must vary in detail from one experimental system to another. Second, the general uniformity of experimental results in wave tanks of different scales (Jähne *et al.*, 1987) and the fact that gas transfer velocities measured in wind–wave tanks are similar to field measurements (within a factor of two) when compared for the same range of $u_w{}^*$, suggests that waves have a dominant role in determining the near-surface turbulence. Jähne *et al.* (1987) have pointed out that, compared with the energy transferred by wind to the surface

N. M. Frew

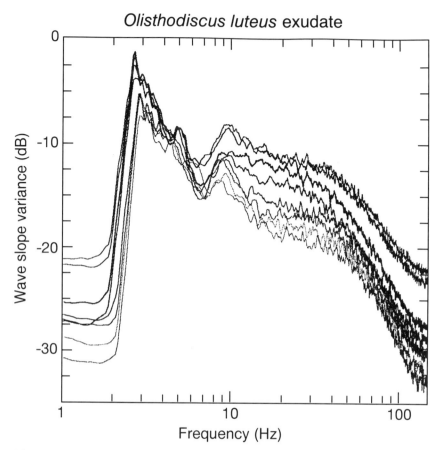

Figure 5.17. Wave slope spectra corresponding to 7.5 minute intervals beginning at $t = 0$ min. for the *Olisthodiscus luteus* exudate experiment in Figure 5.16. The slope variance decreases steadily with time after addition of the exudate solution at $t = 10$ min. From unpublished data of Frew, Goldman and Bock.

shear current, the energy transferred to and stored by waves is much larger, and that only a small fraction of this energy needs to be transformed into near-surface turbulence to have a large impact on gas transfer. If waves are indeed the principal agents promoting near-surface turbulence and gas transfer at the air–sea interface, then the hydrodynamic effects of surfactants must come into play. The problem then revolves around the abundance and distribution of sea-surface films, or more germanely, the temporal and spatial variation of surface viscoelasticity in the oceans.

The global distribution of biogenous sea-surface films is largely

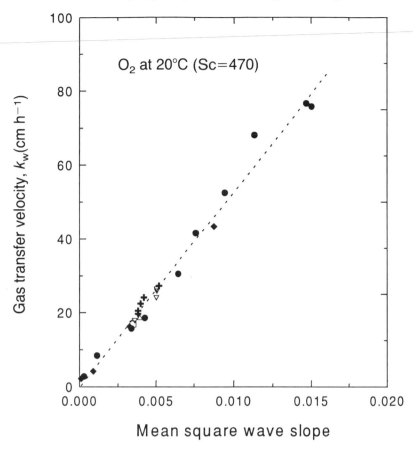

Figure 5.18. Correlation of gas transfer velocity with mean square slope over all frequencies for a range of rotor speeds (1–9.3 m s⁻¹) and surface conditions. Symbols: +, •; Clean water, ▼; polyethylene oxide, ◆, isopropanol; △, sodium dodecyl sulphate, SDS, 500 ppm; □, SDS, 1000 ppm. From unpublished data of Frew and Bock.

unknown. Large-scale features in sea-surface roughness, presumed to be due to these films, have been observed in the Mediterranean, in the eastern coastal North Atlantic and in large areas of the Indian and Pacific Oceans during space shuttle missions using optical (sunglint) and synthetic aperture radar imagery, suggesting that these are not rare events (Kaltenbach *et al.*, 1984; Scully-Powers, 1986). Garrett, (1986) has considered the potential occurrence and distribution of surface slicks, using climatological winds and primary productivity estimates to predict areas of likely slick fomation. Large areas of the open ocean are subject to low to moderate winds for a significant portion of the year and also experience periodically enhanced production of organic matter, conditions conducive to, but not guaranteeing, the formation of sea slicks. Using similar considerations, Asher (Chapter 8, this volume) has estimated the potential impact of surface films on the global air–sea CO_2 exchange budget and shown that the impact for various hypothetical slick coverages would be significant.

It is generally observed that coherent films (slicks) are stable and persistent only up to wind speeds of 5–6 m s^{-1} (Garrett, 1986; Romano and Marquet, 1991); beyond this limit, they are eroded by breaking waves and further dispersed by the wind. Thus, although coherent slicks would have the most dramatic effect in reducing gas exchange by totally inhibiting surface renewal, the effect of soluble surfactants is likely to be more important quantitatively. Surface enrichments of soluble surfactants in dynamic equilibrium with the bulk water can be maintained at higher wind speeds and are much more readily re-established after the continuity of the surface has been broken by plunging waves. Such effects would be most important for coastal microlayers where inputs from biological productivity and terrestrial sources are highest. However, even very oligotrophic waters have dissolved organic carbon concentrations of 60–80 µmol 1^{-1} and are capable of supplying the microlayer with soluble surfactants. As shown earlier, surface pressures need not be large for significant viscoelastic effects.

The lack of *in situ* data on the range of viscoelasticity typical of various ocean regimes and particularly under different forcing conditions limits further prediction. The enrichments of surface-active materials commonly found in the marine microlayer (Hunter and Liss, 1981; Carlson, 1983; Williams *et al.*, 1986) strongly suggest that the hydrodynamic boundary conditions that govern gas exchange in laboratory experimental systems will be similar at the ocean surface and that gas exchange therefore will be subject to similar alterations. It seems probable that much of the scatter seen in field measurements of k_w, particularly at low wind speeds, is due to surfactants. A more definitive answer to the questions posed above must await further research aimed at *in situ* measurements and quantification of surfactant effects. An improved understanding will require advances in techniques to measure *in situ* gas transfer rates and governing environmental parameters on appropriately short time scales. Simultaneous measurements of transfer velocity, wind stress, wave slope, surfactant concentration and viscoelasticity are needed. Such data could be used to construct more accurate and robust parameterizations of gas transfer rates that would be applicable over large spatial scales using satellite-based sensors. The role of surfactants in other surface processes influencing gas exchange, including wave breaking and control of bubble size distributions, bubble hydrodynamics and stability, and bubble-induced turbulence, also needs to be explored.

Acknowledgments

The preparation of this review and portions of the research described herein were supported by grants from the Office of Naval Research (N00014-89-J-1080, N00014-94-1-0400), the National Science Foundation (OCE-9000146, OCE-9301334) and the National Aeronautics and Space Administration (NAGW-2431). The author gratefully acknowledges the interaction, collaboration, and contributions of his colleagues, E. J. Bock, J. C. Goldman, T. Hara, R. K. Nelson, N. V. Blough and E. T. Peltzer in various aspects of the experimental work reviewed here. This is contribution No. 8875 of the Woods Hole Oceanographic Institution.

References

Allan, G. G., Lewin, J. and Johnson, P. G. (1972). Marine polymers. IV. Diatom polysaccharides. *Bot. Mar*, **25**, 102–8.

Alpers, W. and Hühnerfuss, H. (1989). The damping of ocean waves by surface films: a new look at an old problem. *J. Geophys. Res.*, **94** (C2), 6251–65.

Asher, W. E. and Pankow, J. F. (1986). The interaction of mechanically generated turbulence and interfacial films with a liquid phase controlled gas/liquid transport process. *Tellus*, **38B**, 305–18.

Asher, W. E. and Pankow, J. F. (1989). Direct observation of concentration fluctuations close to a gas/liquid interface. *Chem. Engng. Sci.*, **44**, 1451–5.

Asher, W. E. and Pankow, J. F. (1991a). The effect of surface films on concentration fluctuations close to a gas/liquid interface. In *Air–water Mass Transfer*, ed. S. C. Wilhelms and J. S. Gulliver pp. 68–79. New York: American Society of Civil Engineers.

Asher, W. E. and Pankow, J. F. (1991b). Prediction of gas/water mass transport coefficients by a surface-renewal model. *Environ. Sci. Technol.*, **25** (7), 1294–300.

Back, D. D. and McCready, M. J. (1988). Effect of small-wavelength waves on gas-transfer across the ocean surface. *J. Geophys. Res.*, **93**, 5143–52.

Barger, W. R. (1985) Naturally-occurring surface-active material concentrated at the air–water interface. PhD thesis, University of Maryland.

Barger, W. R. and Means, J. C. (1985). Clues to the structure of marine organic material from the study of physical properties of surface films. In

164   *N. M. Frew*

Marine and Estuarine Geochemistry, ed. A. C. Sigleo and A. Hattori (pp. 47–67). Chelsea, MI: Lewis.

Barger, W. R. Daniel, W. H. and Garrett, W. D. (1974). Surface chemical properties of banded sea slicks. *Deep-Sea Res.*, **21**, 83–9.

Bock, E. J. and Frew, N. M. (1993). Static and dynamic response of natural multicomponent oceanic surface films to compression and dilation: laboratory and field observations. *J. Geophys. Res.*, **98** (C8), 14599–617.

Bock, E. J. and Mann, J. A. Jr (1989). On ripple dynamics. II. A corrected dispersion relation for surface waves in the presence of surface elasticity. *J. Colloid Interface Sci.*, **129**, 501–5.

Brockmann, U. H., Hühnerfuss, H., Kattner, G., Broecker, H. C. and Hentzchel, G. (1982). Artificial surface films in the sea area near Sylt. *Limnol. Oceanogr.*, **27**, 1050–8.

Broecker, W. S. and Peng, T-H. (1971). The vertical distribution of radon in the BOMEX area. *Earth Planet. Sci. Lett.*, **11**, 99–108.

Broecker, W. S. and Peng, T-H. (1974). Gas exchange rates between air and sea. *Tellus*, **24**, 21–35.

Broecker, W. S. and Peng, T-H. (1984). Gas exchange measurements in natural systems. In *Gas Transfer at Water Surfaces*, ed. W. Brutsaert and G. H. Jirka (pp. 479–93). Hingham, MA: D. Reidel.

Broecker, H. C., Petermann, J. and Siems, W. (1978). The influence of wind on CO_2 exchange in a wind–wave tunnel, including the effects of monolayers. *J. Mar. Res.*, **36**, 595–610.

Broecker, W. S., Peng, T-H. and Engh, R. (1980a). Modeling of the carbon system. *Radiocarbon*, **22**, 565–98.

Broecker, W. S., Peng, T-H. and Takahashi, T. (1980b). A strategy for the use of bomb-produced radiocarbon as a tracer for the transport of fossil fuel CO_2 into deep-sea source regions. *Earth Planet. Sci. Lett.*, **49**, 463–8.

Broecker, W. S., Peng, T-H., Ostlund, G. and Stuiver, M. (1985). The distribution of bomb radiocarbon in the ocean. *J. Geophys. Res.*, **90**, 6953–70.

Broecker, W. S., Ledwell, J. R., Takahashi, T., Weiss, L. M. R., Memery, L., Peng, T-H., Jähne, B. and Münnich, K. O. (1986). Isotopic versus micrometeorologic ocean CO_2 fluxes: a serious conflict. *J. Geophys. Res.*, **91**, 10 517–27.

Brumley, B. H. and Jirka, G. H. (1988). Air–water transfer of slightly soluble gases: turbulence, interfacial processes and conceptual models. *Physico-chem. Hydrodynam.*, **10**, 295–319.

Burnette, J. C. and Himmelblau, D. M. (1970). The effect of surface active agents on interphase mass transfer. *Am. Inst. Chem. Engng. J.*, **16**, 185.

Carlson, D. J. (1983). Dissolved organic material in surface microlayers: temporal and spatial variability and relation to sea state. *Limnol. Oceanogr.*, **28**, 415–31.

Cini, R. and Lombardini, P. P. (1978). Damping effect of monolayers on surface wave motion in a liquid. *J. Colloid Interface Sci.*, **65**, 387–9.

Cini, R., Lombardini, P. P. and Hühnerfuss, H. (1983). Remote sensing of marine slicks utilizing their influence of wave spectra. *Int. J. Remote Sensing*, **4**, 101–10.

Cini, R., Lombardini, P. P., Manfredi, C. and Cini, E. (1987). Ripple damping due to monomolecular films. *J. Colloid Interface Sci.*, **119**(1), 74–80.

Coantic, M. (1986). A model of gas transfer across air–water interfaces with capillary waves. *J. Geophys. Res.*, **91**, 3925–43.

Cohen, Y., Cocchio, W. and Mackay, D. (1978). Laboratory study of liquid-phase volatilization rates. *Environ. Sci. Technol.*, **12**, 553–8.

Craig, H. (1957). The natural distribution of radiocarbon and the exchange time of carbon dioxide between the atmosphere and the sea. *Tellus*, **9**, 1–17.

Csanady, G. T. (1990). The role of breaking wavelets in air–sea gas-transfer. *J. Geophys. Res.*, **95**, 749–59.

Cullen, E. J. and Davidson, J. E. (1956). The effect of surface active agents on the rate of absorption of carbon dioxide by water. *Chem. Engng. Sci.*, **6**, 49.

Danckwerts, P. V. (1951). Significance of liquid-film coefficients in gas absorption. *Industr. Engng. Chem. Res.*, **43**, 1460–7.

Davies, J. T. (1966). The effects of surface films in damping eddies at a free surface of a turbulent liquid. *Proc. R. Soc. London*, **A290**, 515–26.

Davies, J. T. (1972). *Turbulence Phenomena*. New York: Academic Press.

Davies, J. T. and Rideal E. K. (1963). *Interfacial Phenomena*. New York: Academic Press.

Deacon, E. L. (1977). Gas transfer to and across an air–water interface. *Tellus*, **29**, 363–74.

Deacon, E. L. (1980). Sea–air transfer: the wind speed dependence. *Boundary-Layer Meteorol.*, **21**, 31–7.

Downing, A. L. and Truesdale, G. A. (1955). Some factors affecting the rate of solution of oxygen in water. *J. Appl. Chem.*, **5**, 570–81.

Dysthe, K. and Rabin, Y. (1986). Damping of short waves by insoluble surface films. In *Role of Surfactant Films on the Interfacial Properties of the Sea Surface*, ed. F. Herr and J. Williams C-11–86, pp. 187–213. London: Office of Naval Research.

Erickson, D. J. III (1989). Variations in the global air–sea transfer velocity field of CO_2. *Global Biogeochem. Cycles*, **3**, 37–41.

Erickson, D. J. III (1993). A stability dependent theory for air–sea gas exchange. *J. Geophys. Res.*, **98**(C5), 8471–88.

Etcheto, J. and Merlivat, L. (1988). Satellite determination of the carbon dioxide exchange coefficient at the ocean–atmosphere interface: a first step. *J. Geophys. Res.*, **93**, 15669–78.

Fernandez, D. M., Vesecky, J. F., Napolitano, D. J., Khuri-Yakub, B. T. and Mann, J. A. (1992). Computation of ripple wave parameters: a comparison of methods. *J. Geophys. Res.*, **97**(C4), 5207–13.

Fiscella, B., Lombardini, P. P., Triverno, P., Pavese, P. and Cini, R. (1985). Measurements of the damping effect of a spreading film on wind-excited sea ripples using a two-frequency radar. *Il Nuovo Cimento C*, **8**, 175–83.

Fortescue, G. F. and Pearson, J. R. A. (1967). On gas absorption into a turbulent liquid. *Chem. Engng. Sci.*, **22**, 1163–76.

Frenkiel, J. (1965). *Evaporation Reduction*. Arid Zone Research Series XXVII. Paris: UNESCO, pp. 22–6.

Frew, N. M. and Nelson, R. K. (1992a). Isolation of marine microlayer film surfactants for ex situ study of their surface physical and chemical properties. *J. Geophys. Res.*, **97**(C4), 5281–90.

Frew, N. M. and Nelson, R. K. (1992b). Scaling of marine microlayer film surface pressure-area isotherms using chemical attributes. *J. Geophys. Res.*, **97**(C4), 5291–300.

Frew, N. M., Goldman, J. C., Dennett, M. R. and Johnson, A. S. (1990). Impact of phytoplankton-generated surfactants on air–sea gas exchange. *J. Geophys. Res.*, **95**, 3337–52.

Gaines, G. L. Jr (1966). *Insoluble Monolayers at Liquid–Gas Interfaces*. New York: Wiley-Interscience.

Gaines, G. L. Jr (1991). Monolayers of polymers. *Langmuir*, 7, 834–9.

Garrett, W. D. (1965). Collection of slick-forming materials from the sea surface. *Limnol. Oceanogr.*, **10**, 602–4.

Garrett, W. D. (1971). Retardation of water drop evaporation with monomolecular surface films. *J. Atmos. Sci.* **28**, 816–19.

Garrett, W. D. (1986). Physicochemical effects of organic films at the sea surface and their role in the interpretation of remotely sensed imagery. In *Role of Surfactant Films on the Interfacial Properties of the Sea-surface*, Rep. C-11–86, pp. 1–17. Arlington, VA.: Office of Naval Research.

Geernaert, G. L., Hansen, F., Courtney, M. and Herbers, T. (1993). Directional attributes of the ocean surface wind stress vector. *J. Geophys. Res.*, **98**, 16 571–82.

Gillam, A. H. and Wilson, M. A. (1985). Pyrolysis-GC-MS and NMR studies of dissolved seawater humic substances and isolates of a marine diatom. *Org. Geochem.*, **8**, 15–25.

Goldman, J. C., Dennett, M. R. and Frew, N. M. (1988). Surfactant effects on air–sea gas exchange under turbulent conditions. *Deep-Sea Res.* **35**(12), 1953–70.

Goodridge, F. and Robb, I. D. (1965). Mechanisms of interfacial resistance in gas absorption. *Ind. Engng. Chem. Fundament.*, **4**, 49.

Hanratty, T. J. (1991). Effect of gas flow on physical absorption. In *Air–Water Mass Transfer*, ed. S. C. Wilhelms and J. S. Gulliver, pp. 10–33. New York: American Society of Civil Engineers.

Hansen, R. S. and Mann, J. A. Jr (1964). Propagation characteristics of capillary ripples. I. The theory of velocity dispersion and amplitude attenuation of plane capillary waves on viscoelastic films. *J. Appl. Phys.*, **35**, 152–61.

Hartman, B. and Hammand, D. E. (1985). Gas exchange in San Francisco Bay. *Hydrobiologica*, **129**, 59–68.

Hasse, L. (1990). On the mechanism of gas exchange at the air–sea interface. *Tellus*, **42B**, 250–3.

Hasse, L. and Liss, P. S. (1980). Gas exchange across the air–sea interface. *Tellus*, **32**, 470–81.

Hellebust, J. A. (1974). Extracellular products. In *Algal Physiology and Biochemistry*, ed. W. D. P. Stewart, pp. 838–63. Berkeley: University of California Press.

Higbie, R. (1935). The rate of absorption of a pure gas into a still liquid during short periods of exposure. *Am. Inst. Chem. Engng.*, **35**, 365–89.

Hunter, K. A. and Liss, P. S. (1981). Organic sea surface films. In *Marine Organic Chemistry*, ed. E. K. Duursma and R. Dawson, pp. 259–98. New York: Elsevier.

Jähne, B., Münnich, K. O. and Siegenthaler, U. (1979). Measurements of gas exchange and momentum transfer in a circular wind–water tunnel. *Tellus*, **31**, 321–9.

Jähne, B., Huber, W., Dutzi, A., Wais, T. and Ilmberger, J. (1984). Wind/wave tunnel experiment on the Schmidt number and wave field dependence of air/water gas exchange. In *Gas Transfer at Water Surfaces*, ed. W. Brutsaert and G. H. Jirka, pp. 303–9. Hingham, MA: D. Reidel.

Jähne, B., Wais, T., Memery, L., Caulliez, G., Merlivat, L., Münnich, K. O. and Coantic, M. (1985). He and Rn gas exchange experiments in the large wind–wave facility of IMST. *J. Geophys. Res.*, **90**, 11 898–997.

Jähne, B., Münnich, K. O., Bösinger, R., Dutzi, A., Huber, W. and Libner, P. (1987). On the parameters influencing air–water gas exchange. *J. Geophys. Res.*, **92**, 1937–49.

Jarvis, N. L., Timmons, C. O. and Zisman, W. A. (1962). The effect of mono-molecular films on the surface temperature of water. In *Retardation of Evaporation by Monolayers: Transport Processes*, ed. V. K. La Mer, pp. 41–58. New York: Academic Press.

Jarvis, N. L., Garrett, W. D., Schieman, M. A. and Timmons, C. O. (1967). Surface chemical characterization of surface-active material in seawater. *Limnol. Oceanogr.*, **12**, 88–96.

Kaltenbach, J. L. (1984). The view from the Shuttle-Orbiter: observing the oceans from manned space flights. *Proc. Int. Soc. Optical Engng. (SPIE)*, **489**, *Ocean Optics*, VII, 203–7.

Kaltenbach, J. L., Helfert, M. R. and Wells, G. L. (1984). The view from the Shuttle-Orbiter: observing the oceans from manned space flights. *Proc. Int. Soc. Opt. Engng. (SPIE)*, **489**, *Ocean Optics*, VIII, 203–7.

Kanwisher, J. (1963). On the exchange of gases between the atmosphere and the sea. *Deep-Sea Res.*, **10**, 195–207.

Kim, M. W. and Chung, T. C. (1988). Two dimensional properties of surfactant-like polymer monolayers. *J. Colloid Interface Sci.*, **124**, 365–70.

Kromer, B. and Roether, W. (1983). Field measurements of air–sea gas exchange by the radon deficit method during JASIN (1978) and FGGE (1979). *Meteorol. Forschung.*, **A/B 24**, 55–75.

La Mer, V. K. (1962). Preface. In *Retardation of Evaporation by Monolayers: Transport Processes*, ed. V. K. La Mer, pp. xiii–xiv. New York: Academic Press.

La Mer, V. K. and Healy, T. W. (1965). Evaporation of water: its retardation by monolayers. *Science*, **148**, 36–42.

Lamont, J. C. and Scott, D. S. (1970). An eddy cell model of mass transfer into the surface of a turbulent liquid. *Am. Inst. Chem. Engng. J.*, **16**, 513–19.

Ledwell, J. R. (1984). The variation of the gas transfer coefficient with mole-cular diffusivity. In *Gas Transfer at Water Surfaces*, ed. W. Brutsaert & G. H. Jirka, pp. 293–302. Hingham, MA: D. Reidel.

Lee, Y. H., Tsao, G. T. and Wankat, P. C. (1980). Hydrodynamic effect of surfactants on gas–liquid oxygen transfer. *Am. Inst. Chem. Engng. J.*, **26**, 1008–12.

Leenheer, J. A. (1985). Fractionation techniques for aquatic humic substances. In *Humic Substances in Soil, Sediment and Water*, ed. G. R. Aiken, D. M. McKnight, R. L. Wershaw and P. MacCarthy, pp. 409–29. New York: John Wiley & Sons.

Levich, V. G. (1962). *Physicochemical Hydrodynamics*. Englewood Cliffs, NJ: Prentice Hall.

Liss, P. S. (1983). Gas transfer, experiments and geochemical implications. In *Air–Sea Exchange of Gases and Particles*, ed. P. S. Liss and G. N. Slinn, pp. 241–98. Dordrecht: D. Reidel.

Liss, P. S. and Martinelli, F. N. (1978). The effect of oil films on the transfer of oxygen and water vapour across an air–water interface. *Thalass. Jugoslav.*, **14**, 215–20.

Liss, P. S. and Merlivat, L. (1986). Air–sea gas exchange rates: introduction and synthesis. In *The Role of Air–Sea Exchange in Geochemical Cycling*, ed. P. Buat–Menard, pp. 113–29. Hingham, MA: D. Reidel.

Liss, P. S., Balls, P. W., Martinelli, F. N. and Coantic, M. (1981). The effect of evaporation and condensation on gas transfer across an air–water interface. *Oceanol. Acta*, **4**, 129–38.

Lombardini, P. P., Fiscella, B., Trivero, P., Cappa, C. and Garrett, W. D. (1989). Modulation of the spectra of short gravity waves by sea surface films: slick detection and characterization with a microwave probe. *J. Atmos. Oceanic Technol.*, **6**, 882–90.

Lucassen, J. (1968). Longitudinal capillary waves. *Trans. Faraday Soc.*, **64**, 2221–30.

Lucassen-Reynders, E. H. (1986). Dynamics properties of film-covered surfaces. In *Role of Surfactant Films on the Interfacial Properties of the Sea-Surface*, ed. F. Herr and J. Williams, pp. 175–86. London: Office of Naval Research.

Lucassen-Reynders, E. H. and Lucassen, J. (1969). Properties of capillary waves. *Adv. Colloid Interface Sci.*, **2**, 347–95.

Mackay, D. and Yeun, A. T. K. (1983). Mass transfer coefficient correlations for volatilization of organic solutes from water. *Environ Sci. Technol.*, **17**, 211–17.

Mancy, K. H. and Okun, D. A. (1965). The effects of surface active agents on aeration. *J. Water Pollut. Control Fed.*, **37**, 212–227.

Martinelli, F. N. (1979) The effect of surface films on gas exchange across the air–sea interface. Doctoral dissertation, University of East Anglia.

McCready, M. J. and Hanratty, T. J. (1984). A comparison of turbulent mass transfer at gas-liquid and solid-liquid interfaces. In *Gas Transfer at Water Surfaces*, ed. W. Brutsaert and G. H. Jirka, pp. 283–92. Hingham, MA: D. Reidel.

Meijboom, F. W. and Vogtlander, J. G. (1974). The influence of surface active agents on the mass transfer for gas bubbles in liquid – I. *Chem. Engng. Sci.*, **29**, 857.

Merlivat, L. and Memery, L. (1983). Gas exchange across an air–water interface: experimental results and modeling of bubble contribution to transfer. *J. Geophys. Res.*, **88**, 707–24.

Metzger, I. (1968). Surface effects in gas absorption. *Environ. Sci. Technol.*, **2**, 784–6.

Moo-Young, M. and Shoda, M. (1973). Gas absorption rates at the free surface of a flowing water stream: effect of surfactant and of surface baffles. *Industr. Engng. Process. Design and Devel.*, **12**, 410.

Moum, J. N., Carlson, D. J. and Cowles, T. J. (1990). Sea slicks and surface strain. *Deep-Sea Res.*, **37**, 767–75.

Münnich, K. O. and Flothmann, D. (1975). Gas exchange in relation to other air/sea interaction phenomena. In *Air/Sea Interaction Phenomena*, ed. J. M. Prospero, pp. 310–20. Washington, DC: The National Research Council, National Academy of Sciences.

Münnich, K. O. and Roether, W. (1967). Transfer of bomb carbon-14 and tritium from the atmosphere to the ocean: internal mixing of the ocean on the basis of tritium and carbon-14 profiles. In *Radioactive Dating and Methods of Low-level Counting*, pp. 93–104. Vienna: IAEA-ICSU Symposium.

Mykelstad, S. (1974). Production of carbohydrates by marine planktonic diatoms. I. Comparison of nine different species in culture. *J. Exp. Mar. Biol. Ecol.*, **15**, 261–74.

Mykelstad, S., Djurhuus, R. and Mohus, A. (1982). Determination of exo-(B-1,3)-D-glucanase activity in some planktonic diatoms. *J. Exp. Mar. Biol. Ecol.*, **56**, 205–11.

Nissenbaum, A. (1974). The organic geochemistry of marine and terrestrial humic substances. In *Advances in Organic Geochemistry 1973*, ed. F. Bierner and B. Tissot, pp. 39–52. Paris: Editions Technip.

Peng, T-H., Takahashi, T. and Broecker, W. S. (1974). Surface radon measurements in the North Pacific Ocean station PAPA. *J. Geophys. Res.*, **79**, 1772–80.

Peng, T-H., Broecker, W. S., Mathieu, G. G., Li, Y. H. and Bainbridge, A. E. (1979). Radon evasion rates in the Atlantic and Pacific oceans as determined during the GEOSECS program. *J. Geophys. Res.*, **84**, 2471–86.

Percival, E., Rahman, M. A. and Weigel, H. (1980). Chemistry of the poly-saccharides of the diatom *Coscinodiscus nobilis*. *Phytochemistry*, **19**, 809–11.

Petermann, J. (1976) Der einfluss der oberflachenspannung wassriger systeme auf die kinetjik des gasaustausches. Doctoral dissertation, Universität Hamburg.

Plevan, R. E. and Quinn, J. A. (1966). Effect of monomolecular films on the rate of gas absorption into a quiescent liquid. *Am. Inst. Chem. Engng. J.*, **12**, 894.

Ramus, J. (1982). The production of extracellular polysaccharides by the unicellular red alga *Porphyridium aerugineum*. *J. Phycol.*, **8**, 97–111.

Rashid, M. A. (1985). *Geochemistry of Marine Humic Compounds*. New York: Springer-Verlag.

Romano, J-C. & Marquet, R. (1991). Occurrence frequencies of sea-surface slicks at long and short time-scales in relation to wind speed. *Estuarine Coastal Shelf Sci.*, **33**, 445–8.

Sada, E. and Himmelblau, D. M. (1967). Transport of gases through insoluble monolayers. *Am. Inst. Chem. Engng. J.*, **13**, 860.

Sakugawa, H. and Handa, N. (1985a). Isolation and chemical characterization of dissolved and particulate polysaccharides in Mikawa Bay. *Geochim. Cosmochim. Acta*, **49**, 1185–93.

Sakugawa, H. and Handa, N. (1985b). Chemical studies on dissolved carbo-hydrates in the water samples collected from the North Pacific and Bering Sea. *Oceanol. Acta*, **8**, 185–96.

Scott, J. C. (1972). The influence of surface-active contamination on the initiation of wind waves. *J. Fluid Mech.*, **56**, 591–606.

Scott, J. C. (1975). The preparation of water for surface-clean fluid mechanics. *J. Fluid Mech.*, **69**, 339–351.

Scully-Powers, P. (1986). Navy Oceanographer shuttle observations STS 41-G: mission report. *Technical Document 7611*, San Diego: Navy Underway Systems Center.

Smestad, B., Haug, A. and Myklestad, S. (1975). Structural studies of the extracellular polysaccharide produced by the diatom *Chaetoceros curvisetus Cleve*. *Acta Chem. Scand.*, **29**, 337–40.

Smethie, W. M., Takahashi, T. T., Chipman, D. W. and Ledwell, J. R. (1985). Gas exchange and CO_2 flux in the tropical Atlantic Ocean determined from ^{222}Rn and pCO_2 measurements. *J. Geophys. Res.*, **90**, 7005–22.

Smith, S. D. and Jones, E. P. (1985). Evidence for wind-pumping of air–sea gas exchange based on direct measurements of CO_2 fluxes. *J. Geophys. Res.*, **90**, 869–75.

Smith, S. D. and Jones, E. P. (1986). Isotopic and micrometeorological ocean CO_2 fluxes: different time and space scales. *J. Geophys. Res.*, **91**, 10 529–32.

Smith, S. D., Anderson, R. J., Jones, E. P., Desjardins, R. L., Moore, R. M., Hertzman, O. and Johnson, B. D. (1991). A new measurement of CO_2 eddy flux in the nearshore atmospheric surface-layer. *J. Geophys. Res.*, **96**, 8881–7.

Smith, S. D., Anderson, R. J., Oost, W. A., Kraan, C., Maat, N., DeCosmo, J., Katsaros, K. B., Bumke, K., Hasse, L. and Chadwick, H. M. (1992). Sea surface wind stress and drag coefficients: the HEXOS results. *Boundary-Layer Meteorol.*, **60**, 109–142.

Springer, T. G. and Pigford, R. L. (1970). Influence of surface turbulence and surfactants on gas transport through liquid interfaces. *Industr. Engng. Chem. Fundament.*, **9**, 458–65.

Tans, P. P., Fung, I. Y. and Takahashi, T. (1990). Observational constraints on the global atmospheric CO_2 budget. *Science*, **247**, 1431–8.

Upstill-Goddard, R. C., Watson, A. J., Liss, P. S. and Liddicoat, M. I. (1990). Gas transfer in lakes measured with SF_6. *Tellus*, **42B**, 364–77.

van den Tempel, M. and Lucassen-Reynders, E. H. (1983). Relaxation processes at fluid interfaces. *Adv. Colloid Interface Sci.*, **18**, 281–301.

Van Vleet, E. S. and Williams, P. M. (1983). Surface potential and film pressure measurements in seawater systems. *Limnol. Oceanogr.*, **28**, 401–14.

Villanove, R. and Rondelez, F. (1980). Scaling description of two-dimensional chain conformations in polymer monolayers. *Phys. Rev. Lett.*, **45**, 1502–5.

Wanninkhof, R. (1992). Relationship between wind speed and gas exchange over the ocean. *J. Geophys. Res.*, **97**(C5), 7373–82.

Wanninkhof, R. and Bliven, L. (1991). Relationship between gas exchange, wind speed and radar backscatter in a large wind–wave tank. *J. Geophys. Res.*, **96**, 2785–96.

Wanninkhof, R., Ledwell, J. R. and Broecker, W. S. (1985). Gas exchange–wind speed relationship measured with sulfur hexafluoride on a lake. *Science*, **227**, 1224–6.

Wanninkhof, R., Asher, W., Weppernig, R., Chen, H., Schlosser, P., Langdon, C. and Sambrotto, R. (1993). Gas transfer experiment on Georges Bank using two volatile deliberate tracers. *J. Geophys. Res.*, **98**(C11), 20 237–48.

Watson, A., Upstill-Goddard, R. and Liss, P. S. (1991). Air–sea exchange in rough and stormy seas, measured by a dual tracer technique. *Nature*, **349**, 145–7.

Wesley, M. L., Cook, D. R., Hart, R. L. and Williams, R. M. (1982). Air–sea exchange of CO_2 and evidence for enhanced upwelling fluxes. *J. Geophys. Res.*, **87**, 8827–32.

Williams, P. M. (1986). The chemical composition of the sea-surface microlayer and its relation to the occurrence and formation of natural sea-surface films. In *Role of Surfactant Films on the Interfacial Properties of the Sea Surface*, ed. F. Herr and J. Williams, Report C-11-86, pp. 79–110. London: Office of Naval Research.

Williams, P. M., Carlucci, A. F., Henrichs, S. M., Van Vleet, E. S., Horrigan, S. G., Reid, F. M. H. and Robertson, K. J. (1986). Chemical and microbiological studies of sea-surface films in the southern Gulf of California and off the west coast of Baja California. *Mar. Chem.*, **19**, 17–98.

Witting, J. (1971). Effects of plane progressive irrotational waves on thermal boundary layers. *J. Fluid Mech.*, **50**, 321–34.

Zutic, V. B., Cosovic, B., Marcenko, E. and Bihari, N. (1981). Surfactant production by marine phytoplankton. *Mar. Chem.*, **10**, 505–20.

Symbolic notation

A	surface area
C	surfactant concentration, bulk phase
C_s	surfactant concentration, surface (mass/unit area)
D	molecular diffusivity
D_s	molecular diffusivity, surfactant
g	gravitational acceleration constant
G	Gibbs surface free energy
k_w	gas transfer velocity, water-side resistance dominant
k_w^+	gas transfer velocity, dimensionless
L	turbulence length scale
n	Schmidt number exponent
P	pressure
R	gas law constant
s	surface-renewal rate
Sc	Schmidt number
t	time
t_*	residence time of fluid element in diffusive sublayer
T	temperature
u	wind-speed
u_w^*	friction velocity, water-side
V	turbulence velocity scale
v_o	characteristic velocity of an eddy
$y\omega$	damping ratio, frequency-dependent
α	scaling law exponent
β	momentum transfer resistance, dimensionless
γ	surface tension
γ_{eq}	equivalent surface tension
Γ	surfactant concentration, surface (mol/unit area)
η	viscosity, bulk fluid
$\varkappa$	wave number
ν	kinematic viscosity
π	surface pressure
ϱ	density
τ	wind stress
ω	frequency, radian
ω_D	frequency, diffusional
ε_o	Gibbs elasticity
$\acute{\varepsilon}$	modulus of surface viscoelasticity

6

Bubbles and their role in gas exchange

DAVID K. WOOLF

Abstract

Bubbles can be generated at the sea surface by many mechanisms, but the main source is by the entrainment of air in breaking waves. Bubbles will scavenge material from the surrounding water, thus contributing to the cycling of dissolved and particulate organic material. When they burst at the sea surface these bubbles generate a sea salt aerosol contaminated with material scavenged from the sea-surface microlayer and below. Gases will be exchanged between a bubble and the surrounding water while it is submerged. In addition, the breaking waves and surfacing bubble plumes disrupt the surface microlayer, and this may enhance transfer of gases directly through the sea surface.

The net transfer of a gas between a bubble and the surrounding water, from entrainment until the bubble bursts or fully dissolves, contributes to the total transfer of that gas between atmosphere and ocean. This bubble-mediated transfer has some special properties that set it apart from direct air–sea transfer of poorly soluble gases. Bubble-mediated air–sea transfer velocities depend on the solubility in addition to the molecular diffusivity of the gas in seawater. Bubble-mediated transfer is not proportional to air–sea concentration difference, but is biased toward injection and the forcing of supersaturation. The entrainment of air by breaking waves increases rapidly in intensity with higher wind speeds. Field measurements of whitecap coverage and the sub-surface distribution of bubbles provide an empirical basis for modelling of the oceanic effects of bubble-mediated transfer. The forcing of supersaturation is associated with bubble clouds several metres deep which consist mainly of very small bubbles. Significant contributions to the air–sea transfer velocity of a gas are associated with the flushing of the near-surface by generally larger bubbles.

The equilibrium supersaturations of very poorly soluble gases such as oxygen and nitrogen can be a few per cent, but are much smaller for more soluble gases such as carbon dioxide. The contribution of bubble-mediated transfer to the air–sea transfer velocity is significant for most poorly soluble gases, but again decreases with increasing solubility. Bubble-mediated gas transfer increases in proportion to whitecap coverage, implying high supersaturations and high transfer velocities in storm conditions. By including a plausible contribution of bubble-mediated transfer, a parameterization of total transfer velocity which is compatible with isotopic determinations of the global exchange of carbon dioxide can be constructed. It should not be inferred, however, that a reliable parameterization of air–sea gas transfer has been achieved. Moreover, there is still very little understanding of air sea transfer processes, especially at high wind speeds.

Introduction

Bubbles in the upper ocean are involved in many physical processes of geochemical significance. The production of bubbles by four mechanisms – whitecaps, rain, snow and supersaturation – was considered by Blanchard and Woodcock (1957). It is almost certain that the whitecapping associated with breaking waves is the overwhelmingly dominant mechanism. Air is entrained from the atmosphere in breaking waves and forms bubbles. Some of these bubbles are advected to several metres depth and their clouds can be organized by Langmuir circulation (Thorpe, 1982; Thorpe and Hall, 1983).

Bubbles contribute to air–sea gas transfer and the transport of organic material while submerged and produce aerosol when they burst at the sea surface. Bubbles scavenge material from the upper ocean while submerged and from the surface microlayer on bursting (Sutcliffe, Baylor and Menzel, 1963; MacIntyre, 1974; Blanchard, 1975; Scott, 1975; Wallace and Duce, 1978; Weisel *et al.*, 1984). Material can be brought to the sea surface on surfacing bubbles, and can be entrained into the mixed layer by breaking waves or removed on spray drops. Much of the marine aerosol is derived from the film drops and jet drops generated by bursting bubbles (Blanchard, 1963, 1983; Woolf, Bowyer and Monahan, 1987). Material on the sea surface can alter the surface behaviour of bubbles, so that there is a complicated interaction of whitecaps and microlayer (Woolf and Monahan, 1988).

The possibility that the air–sea transfer of gases might be enhanced by bubble clouds was recognized by Kanwisher (1963). Later, it was found that gas transfer at very high wind stresses in laboratory wind–wave facilities appeared to depend on solubility, and this could most easily be explained if part of the transfer passed through bubbles (Merlivat and Memery, 1983; Broecker and Siems, 1984). Bubble-mediated transfer is also distinguished by an exceptional behaviour in near-saturation conditions (Memery and Merlivat, 1985). Strong evidence that bubble-mediated transfer is important at sea has been presented recently by Wallace and Wirick (1992) and by Farmer, McNeil and Johnson (1993).

Bubble-mediated transfer of gases is the main subject of this paper. In the next section, the various mechanisms of air–sea gas transfer, including bubble-mediated transfer, are briefly considered. The special properties of bubble-mediated transfer, the oceanic distribution of bubbles and estimates of the resulting gas transfer are described in the following three sections. All of the geochemical effects of bubbles (e.g., scavenging, aerosol production) depend on the size of the bubble and in most cases its path in the upper ocean. Therefore, much of the description of bubble distributions and fluxes, and modelling of their behaviour, is relevant to these other branches. In the two concluding sections the implications of the predicted properties of bubble-mediated gas exchange for our understanding of air–sea gas exchange as a whole are discussed and summarized.

Mechanisms of air–sea gas transfer

The basic principles of air–sea gas exchange have been explained by Liss (1983). On the assumption that gas transfer between the atmosphere and ocean is an example of simple diffusional transfer between two reservoirs, one would predict that the net flux, F (mass per unit time, per unit area), will be proportional to the air–sea concentration difference (by convention measured in the liquid phase), ΔC. Thus, the basic flux equation can be written as,

$$F = K_T \, \Delta C$$

where K_T, which has the dimensions of velocity, is termed the air–sea transfer velocity or piston velocity. The transfer velocity depends on the efficiency of transfer through the interfacial layers. For poorly soluble gases (CO_2, dimethyl sulphide (DMS), O_2, N_2, inert gases . . .) the limiting step is transfer across the marine microlayer. For these gases, K_T will be the sum of contributions from various mechanisms of transfer across the sea-surface microlayer.

Further discussion is restricted to the rough water conditions (moderate and high wind speeds) when wave breaking is possible. The two basic categories of gas transfer are shown schematically in Figure 6.1. Gas can be transferred directly across the sea surface or between a bubble and the water surrounding it. It is convenient to treat these two types of transfer separately and write,

$$K_T = K_o + K_b$$

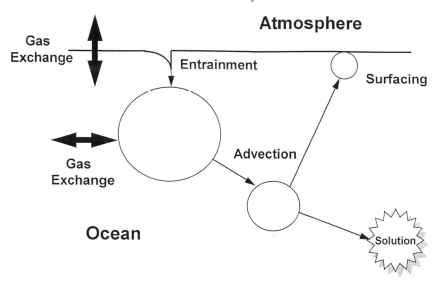

Figure 6.1. A schematic of bubble evolution after entrainment into the upper ocean.

where K_o is the contribution of **direct** transfer and K_b the contribution of **bubble-mediated** transfer. The transfer of gas between atmosphere and ocean *via* a bubble is a three-stage process. Part of the atmosphere is encapsulated in a bubble. Gas diffuses between the bubble and the surrounding water while the bubble is submerged. Eventually, either the bubble will dissolve entirely (in which case that fragment of the atmosphere has been injected into the ocean) or the bubble will surface and burst, releasing its contents to the atmosphere. The net transfer of gas is the difference between the initial and final contents of the bubble, and can be in either direction. Note that the sea-surface microlayer has been bypassed and instead transfer is limited by the microlayer around the bubble and also the finite capacity of the bubble.

In order to understand the context of bubble-mediated transfer, the properties of direct transfer must first be considered. Direct transfer across the sea surface by a combination of molecular and turbulent diffusion depends on the molecular diffusion constant, D, of the dissolved gas and the wind and wave field (see e.g., Jähne *et al.*, 1987). It is widely agreed that in rough water conditions the transfer velocity of a gas will be proportional to the square root of its diffusion constant or,

$$K_o \propto Sc^{-0.5}$$

where Sc = v/D (Sc is the Schmidt number, v is the kinematic viscosity of the seawater).

A number of different mechanisms and models have been proposed for the effect of the wind and wave field on transfer velocities. While the mechanism is not widely agreed, most models predict transfer proportional to the friction velocity of the wind in the 'rough water regime'(Jähne *et al.*, 1987, present evidence that using mean square slope instead removes most of the variation with facility geometry found in laboratory studies). Csanady (1990) has argued that the main turbulent mechanism is rollers on breaking wavelets. It is to be expected that longer breaking waves and whitecapping have some effect on the transfer velocity. Banner, Jones and Trinder (1989) have shown that wavelets can be suppressed by large breaking waves, and Csanady has inferred from this that the rise of transfer velocity with wind speed will be weaker than linear at high wind speeds. Soloviev and Schluessel (1994) describe a surface-renewal model of heat and gas transfer with a transition from a breaking wavelet regime to a regime of long breaking waves at high wind speeds. Their conclusions reinforce Csanady's inference. Mechanisms associated with whitecapping which may enhance gas transfer must be weighed against wavelet suppression. Exceptionally high levels of near-surface turbulence are associated with breaking waves (Agrawal *et al.*, 1992; Osborn *et al.*, 1992) and there should be a related increase in surface transfer (Kitaigorodskii, 1984). Rising bubbles generate turbulence and Monahan and Spillane (1984) have hypothesized that as a result whitecaps are a low-resistance vent through the sea surface. The main problem is that it remains largely a matter of conjecture as to what the main mechanisms of direct air–sea transfer are. The reason is that there are few practical methods of studying these processes in the field, while the applicability of laboratory studies is always in doubt (Jähne *et al.*, 1987).

An important experimental approach which is reasonably practical in the field is to study the transfer of heat directly through the sea surface. Direct transfer of heat and gas through the marine microlayer are very similar (Jähne *et al.*, 1987), though the diffusion constant of heat is much greater than that of any gas. There are two special benefits of using heat as a proxy tracer for understanding air–sea gas transfer. Firstly, these methods select direct transfer and exclude bubble-mediated transfer, making it especially powerful in combination with a method (e.g., any chemical budget technique) which measures total transfer. Secondly, it is possible to investigate transfer processes at high spatial and temporal resolution. For instance, Jessup (1996) has used an infrared scanner to investigate breaking waves in detail. Jähne *et al.* (1989) have developed a technique using a controlled irradiation of the surface. It is also possible to infer transfer velocities from parameterizations of the thermal skin effect (e.g., Grassl, 1976; Schluessel *et al.*, 1990). At a given heat flux, the temperature difference across the marine microlayer will be inversely proportional to the transfer velocity of heat, which itself is proportional to the transfer velocity of a gas.

Soloviev and Schluessel (1994) have shown that measurements of the skin effect imply a suppression of transfer (a weaker than linear increase) rather than an enhancement by breaking waves at high wind speeds. This result implies that extraordinarily high values of direct transfer will not be encountered at high wind speeds. This inference should, however, be treated with caution. One major problem is that in the original studies, each retrieved surface temperature was an average over a fairly large radiometer footprint. If within this footprint there was a small fractional area (e.g., the location of a whitecap) where the microlayer was highly disrupted, this would have only a small effect on the calculated temperature (and inferred transfer velocity), even though this small area may contribute very substantially to the transfer of both heat and gas. It has been widely opined that the thermal skin effect will 'vanish' at high wind speeds as a result of the disruption of the microlayer by breaking waves (Ewing and McAlister, 1960); but the corollary that the transfer of poorly soluble gases must then be massively enhanced has been largely disregarded. Jessup (1996) has found that the surface disruption by breaking waves is more limited than earlier supposed, but that it is still considerable. The experimental investigation of this phenomenon is complicated by the likelihood of the breaking waves and bubbles altering the thermal emissivity of the sea surface. Further studies of the disruption of the microlayer and alterations of emissivity by breaking waves are required. In the meantime, it is difficult to assess the importance of whitecapping as a mechanism of *direct* air–sea gas transfer.

It is likely that bubble-mediated transfer will be very rapid at high wind speeds (Woolf, 1993). Theoretical arguments lead to the deduction that energy dissipation in breaking waves will be proportional to the cube of the friction velocity of the wind (Phillips, 1985; Thorpe, 1993; Melville, 1994). It also seems likely that whitecapping will be at least approximately proportional to this energy dissipation. (Lamarre and Melville, 1991, have shown that a large fraction of the energy is expended in entraining air.) Unfortunately, the dependence of energy dissipation and whitecapping on the sea state is not fully understood. For this reason, it is difficult to extrapolate levels of wave breaking (or associated gas transfer) from the laboratory or lake to the open sea using data on friction velocity and wave characteristics. A reasonable alternative is to use measurements of whitecap coverage. The variation of oceanic whitecap coverage with wind speed, atmospheric stability, water temperature and fetch has been investigated statistically (Monahan and O'Muircheartaigh, 1980, 1986; Monahan and Monahan, 1986). The results generally support the theory that whitecap coverage will be approximately proportional to the friction velocity cubed (Wu, 1979), but the effects of fetch and wind duration are not fully understood. Thus, bubble-mediated transfer is very sensitive to wind speed and will be most significant in storm conditions. Precise details of the environmental dependence of bubble injection remain obscure. In the absence of measurements of the sub-surface bubbles, it is better to assume bubble injection will be proportional to whitecap coverage rather than assume a simple wind speed dependence. Margins of error will be much reduced if sub-surface bubble concentration or whitecap coverage are measured, otherwise a statistical dependence of whitecap coverage on the measured environmental parameters may be used.

Properties of bubble-mediated transfer

In this section, the behaviour of individual bubbles and the basic properties of bubble-mediated transfer are described. Much of this and the following two sections is a summary of theory and results described in two previous publications (Woolf and Thorpe, 1991; Woolf, 1993).

The transfer of gases via a bubble depends both on the history of that bubble (especially its depth versus time) and the characteristics of turbulent transfer between the bubble and the water at each instant. The history of a bubble depends on upper water motions (advection), the buoyancy of the bubble (terminal rise speed) and its growth or shrinking (hydrostatic pressure and net exchange of all gases). Any model incorporating all these processes is fairly complicated. Instantaneous transfer between a bubble and the surrounding water is simply an example of diffusive transfer between two reservoirs, and can be written,

$$J = 4\pi r^2 j \left(C_w - S p_b \right)$$

The flux, J (mass per unit time) depends on the surface area ($4\pi r^2$), the **individual bubble transfer velocity** of the gas (j) and its partial pressure in the bubble (p_b), the solubility of the gas (S) and its concentration in the bulk of the water (C_w). Some characteristics of a rising bubble have been solved analytically (see e.g., Levich, 1962; Batchelor, 1967), but accurate solutions for large and non-spherical bubbles are more elusive. Clift *et al.* (1978) have reviewed much of the theoretical and experimental work on this problem and present various formulae for terminal rise speed and *j*. One important factor is the concentration of surface-active material on the bubble. Surface-active material affects both the rise speed and *j* by transforming the surface of the bubble from being essentially fluid for a 'clean' bubble to essentially rigid for a 'dirty' bubble. There is a link here between the cycling of surface-active material and bubble-mediated gas exchange. A complete treatment of bubble evolution would include the transfer of particles between bubbles and the surrounding water (Thorpe, Bowyer and Woolf, 1992) and the dynamical response of the bubbles. Only a stop-gap solution is described here, assuming that bubbles advected fairly deep will usually be dirty (Woolf and Thorpe, 1991), while large bubbles flushed briefly through the surface waters will be clean (Woolf, 1993). Boundary-layer theories predict that for a clean bubble:

$$j \propto D^{1/2}$$

and for a dirty bubble:

$$j \propto D^{2/3}$$

Another limitation of existing models is that the formulae for *j* apply to bubbles rising at terminal velocity in a quiescent liquid while bubble plumes are turbulent. This 'external' turbulence will be insufficient in mature bubble clouds to alter the dynamics of these bubbles (Thorpe, 1982), but the situation in the vicinity of an active breaking wave may be very different. An adequate description of transfer coefficients within a dense bubble plume is one of the most important, but also most difficult, requirements of a complete model of air–sea gas transfer.

The most distinctive properties of bubble-mediated transfer can be traced to the fact that the partial pressure of a gas in a bubble, p_b, will usually differ from its partial pressure in the atmosphere, p_a. One cause of this difference is the hydrostatic pressure on a submerged bubble, which raises the partial pressure of each gas in a bubble. As a result even when a gas is slightly supersaturated (with respect to the atmosphere) there will be a net transfer to the water from many bubbles. If the traditional air–sea gas exchange equation is used, K_b behaves anomalously near saturation (see for example, Memery & Merlivat, 1985). In effect, the diffusive transfer between air and sea occurs with part of the atmospheric gas in direct contact with the sea (that in the bubbles) raised to a higher pressure. The net flux, F, can be defined,

$$F = (K_o + K_b) [C_w - Sp_a (1 + \Delta)]$$

where K_b is independent of saturation and the additional term, Δ, is the 'average' fractional extra pressure on the gas in contact with the sea; Δ is also the **equilibrium supersaturation** at which there will be no net air–sea transfer of the gas. Differences between the partial pressures in the atmosphere and the bubbles also arise as a result of the changing composition of the bubble as various gases diffuse between a bubble and the surrounding water at different rates. The partial pressures of the most soluble gases vary most quickly in a bubble. A diffusive flux between neighbouring reservoirs drives these two reservoirs toward equilibrium. In the case of a submerged bubble, there can be time for the more soluble gases to approach equilibrium, reducing the net flux of that gas thereafter. The finite size of the bubbles therefore introduces a solubility dependence to both K_b and Δ. Both K_b and Δ will generally be smaller for more soluble gases. There are a few interesting limits for the net transfer of a gas within its lifetime. If a bubble completely dissolves then its contribution to Δ will be inversely proportional to the solubility of the gas. (In total, Δ will be less sensitive to solubility than this as bubbles that only partially dissolve, or even grow, will contribute significantly to Δ.) If a gas reaches equilibrium then the contribution of that bubble to K_b will be inversely proportional to the solubility of the gas and independent of j. If a bubble surfaces before the partial pressure of a gas can change significantly, then this contribution to K_b will be independent of solubility and proportional to j. Opposite limits may apply to the same gas in different bubbles, and to different gases in the same bubble. At the very extremes of solubility – and assuming a well-bound distribution of bubbles of which a total volume, V, is flushed through the surface per unit area, per unit time (thus V has dimensions of velocity) – there should be the following limits:

$$(\beta \gg 1) \qquad\qquad K_b = V/\beta$$

where β is the Ostwald solubility.

$$(\beta \ll 1) \qquad K_b \text{ is independent of } \beta \text{ but a function of D}$$

If there is a common relationship between j and D for all the bubbles, then the limit ($\beta \ll 1$) can be further defined. For example, if $j \propto D^{1/2}$ then,

$$(\beta << 1) \qquad\qquad\qquad K_b \propto D^{1/2}$$

The value of the equilibrium saturation, Δ, of each gas will also depend on the saturation level of nitrogen and oxygen (Woolf & Thorpe, 1991). This arises from a general alteration in the dynamics of a bubble as it shrinks or grows in response to the net flux of air between the bubble and the water.

The introduction of an additional term, Δ, to the air–sea gas equation avoids transfer velocities that depend on the saturation level. There are a number of experiments in which the results have been fitted to the traditional equation, though there was almost certainly significant bubble-mediated transfer. In the case of pure evasion ($P_a \approx 0$),

$$F = K_T C_w$$

$$F = K_e C_w$$

and the calculated transfer velocity, K_e, and the true transfer velocity (as defined in the modified air–sea gas equation), K_T, are identical. Thus, there is no real difficulty here in interpreting the results of the radon deficit method or the dual tracer method. (The much greater difficulty associated with the complicated dependence of K_b on solubility and the diffusion constant will be discussed later.) In the case of pure invasion ($C_w \approx 0$):

$$F = -K_T S p_b (1 + \Delta)$$

$$F = -K_i S p_b$$

and the calculated transfer velocity, K_i, will slightly exceed K_T:

$$K_i = K_T (1 + \Delta)$$

Note that one can rewrite the modified flux equation as

$$F = K_e C_w - K_i S p_a$$

which is similar to the form adopted by Keeling (1993). The results of many laboratory experiments approximate either measurements of K_e (for an evasion experiment, $C_w >> S p_a$) or K_i, but errors will be large if data from near-saturation are used. It would be better if the results were immediately fitted to the modified gas flux equation, thereby determining both K_T and Δ (though uncertainties in chemical analysis may be too great to give a meaningful estimate of Δ). The greatest difficulty with laboratory results is that of extrapolating sensibly to the ocean surface.

Considerable variability in transfer velocities at the same friction velocity, but in different wind–wave tanks, has been reported (Jähne *et al.*, 1987; Wanninkhof and Bliven, 1991). It is only reasonable to extrapolate contributions by whitecapping if there is a measurement of whitecap coverage of sub-surface bubble concentration in the simulation. The value of Δ in a laboratory experiment will not usually be realistic as all vertical mixing processes will not be closely simulated.

The oceanic distribution of bubbles

The magnitude of the bubble-mediated contribution to the air–sea transfer velocity, K_b, and the equilibrium supersaturation of each gas, Δ, depend on the number and size of bubbles produced and details of their dispersion. The exact dependencies of K_b and Δ on the solubility and diffusion constant of the gas also depend on the distribution of the bubbles. The present-day knowledge of the distribution of bubbles in the upper ocean and its response to environmental factors is reviewed here. Fairly simple descriptions of the flux and dispersion of bubbles are derived from considerations of the few measurements. These models together with the mass transfer equations of bubbles described in the last section can be used to estimate K_b and Δ for various gases as a function of environmental factors (presently just whitecap coverage or wind speed). It is important to note the scarcity of the existing measurements and the corresponding limitations of the derived models. Models of bubble-mediated gas transfer can only be as good as the bubble measurements. Since these measurements are scarce and incomplete, quantitative model predictions must necessarily be tentative.

Measuring the size and concentration of bubbles in the upper ocean is technically difficult and there are still only a few published results. There are a number of reports of the mean concentration and size distribution from half a metre to a few metres depth determined by photographic methods. Data from three studies (Kolavayev, 1976; Johnson and Cooke, 1979; Walsh and Mulhearn, 1987) at a similar wind speed and depth are collected in Figure 6.2. The concentration of 100 μm radius bubbles was almost identical (which may be coincidence), but there are significant differences in the size distribution. The maximum in each of these distributions and the concentration of the smaller bubbles are almost certainly artefacts of the method. The concentrations of larger bubbles are relatively reliable and suggest both some degree of consistency in concentration and size distribution, but also enough variability to

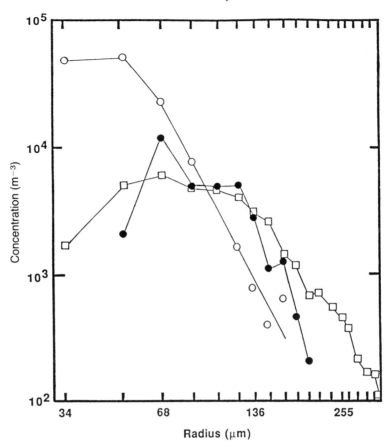

Figure 6.2. The mean concentration of bubbles at 1.5–1.8 metres depth and at a wind speed of 11–13m s^{-1}. Concentrations in consecutive ranges of 17 μm radius width are plotted on a logarithmic scale. Based on a figure of Walsh and Mulhearn (1987). Filled circles: Walsh and Mulhearn (1987), 1.7 metres depth. Open circles: Johnson and Cooke (1979), 1.8 metres depth. Open squares: Kolavayev (1976), 1.5 metres depth.

warrant further investigations of the factors determining bubble concentration and size distribution. A few effects have already been modelled by Thorpe, Bowyer and Woolf (1992). It is common to fit a distribution of the form

$$dN/dr \propto r^{-b} \exp(-z/L)$$

to experimental data, where dN/dr is the concentration of bubbles per unit volume, per increment radius of radius, r, at a depth z. The size distribution is described by a single, dimensionless **slope parameter** '*b*', and the depth dependence by an attenuation depth, *L*. Woolf (1988) estimated for each data set in Figure 6.2 that for the larger bubbles $b = 3.2$ (Kolavayev), 4 (Walsh), or 4.4 (Johnson). The slope parameter seems to increase slightly with depth (Woolf, 1988) but a single value, $b = 4$, is a reasonable description for depths greater than 0.5 m. Vagle and Farmer (1992) have measured the concentration of bubbles from 8 μm to 130 μm in radius at a large range of depths by acoustical backscatter. Their results show that a slope, $b = 4$, is a reasonable fit for radii as small as 30 μm, but there is a maximum in dN/dr typically at 25 μm radius. Vagle and Farmer also reported that the slope, *b*, increased considerably toward the base of bubble clouds. Statistics on large bubbles are very poor, and it is difficult to deduce more than that there are very few of these bubbles at a metre or more in depth; it would be rash to assume the slope, *b*, is unchanged for radii greater than 150 μm.

The attenuation with depth of bubble concentrations also varies significantly about a typical attenuation depth, $L = 1$ m. A few studies of bubble clouds (Thorpe, 1986; Woolf, 1995), all with the same acoustical instrument, show a large variation in attenuation depth (Table 6.1). The available database is too limited to infer with confidence the main factors determining the attenuation, but there are some indications. It is probable that clouds are generally deeper in convective conditions. The lowest attenuation depths were measured in the spring in the North East Atlantic, early in the spring bloom, shortly after a shallow seasonal thermocline is established. The results, $L(U)$, of a simple linear regression of *L* on the wind speed, *U*, for each study is also included in Table 6.1. The correlation between depth and wind speed is generally weak but positive. This positive relationship is unsurprising given the role of the wind in upper ocean mixing.

There are fewer measurements of bubbles in the upper ocean within half a metre of the sea surface. It is difficult to operate an *in situ* device very close to the surface, but Medwin and Breitz (1989) did succeed in measuring bubbles 0.25 m beneath breaking waves with an acoustical resonator. Blanchard and Woodcock (1957) measured bubbles near the sea surface but only several seconds after a wave had broken. Upward-looking acoustical instruments can measure bubble distributions remotely (Thorpe, 1986; Crawford and Farmer, 1987; Vagle and Farmer, 1992) but are confused by 'shadowing' and 'multiple scattering' in the densest bubble plumes. The results reported by Medwin and Breitz imply a very distinct change in the size distribution of bubbles as the surface is

Table 6.1 *Attenuation depths, L, of bubble clouds measured with the acoustical instrument, ARIES*

Deployment	Temperature range (°C)	Number in sample	L ($\pm$ SD) (m)	$L(U)$ (m)
NE Atlantic April–May 1990	12.2–12.6	22	0.62 ($\pm$0.14)	$0.22 + 0.051U$
DB2 Sept.–Oct. 1984	14.7–17.3	23	1.57 ($\pm$0.45)	$0.92 + 0.057U$
DB2 Dec. 1984–Feb. 1985	10.7–11.6	18	0.85 ($\pm$0.14)	$0.48 + 0.035U$
Irish Sea	6.1–6.6	4	1.08 ($\pm$0.34)	$-0.33 + 0.131U$

From Thorpe (1986); Woolf (1995)

approached; for radii less than 60 μm b is still 4, but for radii of 60 to 240 μm, $b = 2.5$.

It is difficult to predict the flux of bubbles through the sea surface from oceanic measurements of bubble concentration (Woolf, 1990; Bowyer, 1992). Larger, more buoyant bubbles tend to surface quickly and are unlikely to be advected to a metre depth. The spectrum of bubbles near a breaking wave will change rapidly in the first seconds. Most of the easier sampling strategies are likely to be biased against measuring large bubbles as a result. It is necessary either to develop a good model of the relationship between the surface flux and the concentration far beneath the surface, or to find the flux by another method.

The production rate of bubbles per increment radius, dF/dr, can be parameterized as

$$\mathrm{d}F/\mathrm{d}r \propto r^{-c}$$

with a flux slope parameter, c. While most measurements have been of the spectrum some time after a wave breaks (or a time-averaged spectrum) there have been a few attempts to measure the spectrum immediately after the wave breaks or the total flux of surfacing bubbles. Note that for $c < 3$ the volume flux, V, increases indefinitely as larger bubbles are considered unless there is some cut-off. Monahan and Zietlow (1969) measured bubble statistics for a laboratory simulation of a whitecap, produced by two colliding waves, and found a slope parameter of 2.5. The statistics were accumulated from photographs of bubbles on the water surface from 2–14 seconds after the breaking wave. The effects on the measured spectrum of the time at which bubbles surface and their surface residence time are unclear, but this does not appear to be a reliable measurement of the flux parameter, c. Cipriano and Blanchard (1981) used a small weir as a simulation of breaking waves. In this set-up they were able to measure bubble fluxes quite comprehensively. Their measurements of bubble concentration at the centre of the surfacing plume and at two separations from the centre are shown in Figure 6.3 along with results from two field studies (Blanchard and Woodcock, 1957; Johnson and Cooke, 1979) and another laboratory study (Baldy and Bourguel, 1985). As discussed above, the slope is fairly consistent away from a breaking plume, but is probably different close to the plume. From this standpoint, the simulation used by Cipriano and Blanchard looks fairly realistic, but this is not certain. Baldy and Bourguel (1987) also found a decrease in the slope parameter, b, very near the surface in wind–wave tank experiments. The flux of bubbles measured by Cipriano and Blanchard cannot be described by a single value of the parameter c; c increases with increasing radius. This does have the desirable result that the volume flux peaks towards the middle of the measured size range, and the measurements are in fact a sufficiently complete description of the size distribution to enable estimations of the associated gas transfer. Bowyer (1992) has made an indirect measurement of the initial distribution of the bubbles produced by a pouring event; the distribution can be fitted by a value of c between 1 and 1.5. The distribution is significantly different than that measured by Cipriano and Blanchard, but is sufficiently similar to confirm the usefulness of simulations.

The flux of bubbles measured by Cipriano and Blanchard (1981) appears to be the best available for estimating oceanic fluxes. A scaling argument can be used to predict oceanic fluxes at a photographically measured whitecap coverage (W or W_B) from the fluxes and whitecap area reported by Cipriano and Blanchard (Monahan, 1986; Woolf, 1993). With a different measurement protocol using video, the measured whitecap coverage, W_A, is much lower (Monahan and Woolf, 1989). There is not yet an adequate description of the dispersion of the initial influx of bubbles into the upper ocean. Descriptions of bubble-mediated gas exchange (Woolf and Thorpe, 1991; Woolf, 1993) are necessarily piecemeal. The dispersion of bubbles in the upper ocean was modelled by Woolf and Thorpe (1991) only after an initial phase in which it was assumed some of the bubbles would reach a depth of 0.5 m. As a result, the flux of bubbles (slope parameter of 4) that quite successfully mimicked the distributions measured by Johnson and Cooke (1979) is

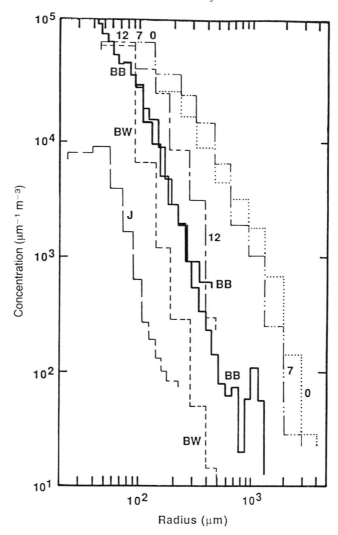

Figure 6.3. The concentration of bubbles per increment radius plotted against radius. 0, Cipriano and Blanchard (1981) – at the plume centre; 7, Cipriano and Blanchard – 7 cm from the plume centre; 12, Cipriano and Blanchard – 12 cm from the plume centre. BB, Baldy and Bourguel (1985) – below the wave troughs, with a wave maker; BW, Blanchard and Woodcock (1957); J, Johnson and Cooke (1979). $U = 11$–13 m s^{-1}, 0.7 metres depth.

probably very different from the true surface flux (Woolf, 1993). The fluxes and concentrations of bubbles near breaking waves and their relationship to whitecap coverage and other remotely sensed variables require careful and detailed study.

Estimates of bubble-mediated gas transfer

Models of the flux and dispersion of bubbles in the upper ocean can be combined with equations describing gas transfer between bubbles and the water surrounding them to calculate estimates of bubble-mediated transfer between the atmosphere and oceans. The equilibrium supersaturation, Δ, and the bubble-mediated transfer velocity, K_b, are dealt with separately. One model describes the surface flux but includes only the simplest description of bubble motion (Woolf, 1993). Another model describes the dispersion of bubbles after the injection phase (Woolf and Thorpe, 1991).

The supersaturating effect of a bubble increases linearly with its depth. Modelling of the dispersion of bubbles shows that either turbulent diffusion (Thorpe, 1984) or advection in Langmuir cells (Woolf and Thorpe, 1991) can be strong enough to disperse very many small bubbles ($< 150\,\mu m$ radius) to a few metres depth, but only a tiny fraction of larger, more buoyant bubbles will reach one metre in depth. Woolf and Thorpe found that as a result most of the supersaturating effect of bubbles was associated with the small bubbles. This is fortuitous since, as was explained in the last section, there is very little information on the concentration of large bubbles at depths of a metre or so. By fitting the flux and dispersion parameters to bubble data reported by Johnson and Cooke (see also Figures 6.2 and 6.3), Woolf and Thorpe reached an estimate of gas fluxes in near-saturation conditions. There was some uncertainty due to the scarcity of data on very small bubbles, but this can now be resolved thanks to new data on these bubbles (Vagle and Farmer, 1992). The best estimates of Δ lie somewhere between those calculated for a 30 μm radius cut-off of the input and those for a 20 μm cut-off (typically 25% greater). The equilibrium supersaturation is largely the result of a balance between the direct flux (approximately linear in wind speed or friction velocity) and the bubble-mediated flux (which has an approximately cubic relationship to wind speed). As a result, Δ is expected to increase approximately with the square of wind speed,

$$\Delta \cong (U/U_i)^2 \%$$

where U_i is the wind speed at which the supersaturation of a particular gas equals 1% (nitrogen, $7.2\,m\,s^{-1}$; oxygen, $9\,m\,s^{-1}$; argon, $9.6\,m\,s^{-1}$; carbon dioxide, 49 m s^{-1} – Woolf and Thorpe, 1991 (30 μm cut-off)). The equilibrium supersaturation of carbon dioxide will be very small, but those of less soluble gases are more significant. Very large differences between Δ in individual instances and these predicted values can be expected, largely as a result of the variability of whitecapping at a single wind speed and the variability of the attenuation depth of bubble clouds. Variations in attenuation depth alone may account for supersaturations varying by up to a factor of ten. The measurements reported by Wallace and Wirick (1992) imply particularly high equilibrium supersaturations. Since vertical mixing generally increases with wind speed, the dependence of Δ on wind speed may often be stronger than quadratic. Woolf and Thorpe (1991) did not take account of the influence of near-surface bubbles on supersaturation. Keeling (1993) has shown that a flux of large bubbles penetrating to 0.25 m can force significant supersaturations. These values of supersaturation are again greatest for the most poorly soluble gases, but are considerably less than those attributed to deep bubbles by Woolf and Thorpe. The relative contribution of shallow bubbles is greater for more soluble gases, but the supersaturation of carbon dioxide is still expected to be quite small ($\approx 0.1\%$). As wind speeds increase and the supersaturations of very poorly soluble gases reach a few per cent, the role of the shallow bubbles and consequently the wind speed dependence of Δ will change. Very high supersaturations can only be forced by deep bubbles or small bubbles that entirely dissolve, and will be opposed by shallow bubbles. Note for instance that if the supersaturation is 5%, there should be a net flux of gas from the water column into bubbles in the upper half metre. Supersaturations may approach a plateau where the flux through deep bubbles is balanced by the flux through shallow bubbles. This makes the large supersaturations reported by Wallace and Wirick (1992) all the more remarkable. Large supersaturations ($>5\%$) can only be forced by bubbles if clouds are deep as well as intense.

The bubble-mediated transfer velocity, K_b, is associated largely with a shallow flushing of bubbles, with the exception of gases of extremely low solubility for which the few long-lived bubbles will be important. A considerable constraint on the transfer velocity is given by the inequality $K_b < V/\beta$ (described earlier). On the basis of the air flux, V, measured by Cipriano and Blanchard (1981),

$$V = 6.25\, W_B \text{ litres (m}^2\text{ s)}^{-1}$$

This is a much greater air flux than predicted by Keeling (in part as a result of confusion as to whether the whitecap area reported by Cipriano and Blanchard corresponded to the 'traditional' definition of whitecap coverage (W or W_B), or W_A, a definition which was introduced later).

The size spectrum of bubbles measured by Cipriano and Blanchard is also used. In order to calculate transfer velocities it is also necessary to define the lifetime of the bubbles and the mass transfer coefficients for bubble-to-water transfer, j. Bubbles may be advected anything up to half-a-metre depth in the main plume. The average penetration is likely to be much less than this, particularly in spilling breakers. Gas transfer is calculated on the assumption of a lifetime equal to the time for the bubble to rise just 0.1 m. Mass transfer coefficients are particularly problematic for the crowded and turbulent plume. Results using formulae for clean bubbles rising in quiescent water ($j \propto D^{1/2}$) are presented here. Figure 6.4 is a contour plot of K_b as a function of the molecular diffusion constant, D, and solubility, β, of the gas at a whitecap coverage (W_B) of 1%. The dependence of K_b on D and β is complicated (a fit of the form $K_b \propto \beta^{-m}D^n$ is unsatisfactory). The following formulation can be fitted fairly well to model results:

$$K_b = \frac{V}{\beta}\left[\, 1+(e\beta D^n)^{-1/f}\,\right]^{-f}$$

This formula has the correct asymptotes ($K_b \cong V/\beta$ at large β, $K_b \propto D^n$ at small β). For clean bubbles, $n = 0.5$, and there is a good fit to model data with parameters $\{e = 1.4 \times 10^4 \; s^{1/2} \; m^{-1}; f = 1.2\}$.

The model results depend critically on a number of assumptions about bubble flux, penetration depth and transfer coefficients. As described above, the flux of bubbles is inferred only from a laboratory experiment, while the penetration has minimal basis. Transfer coefficients are a particular problem, especially as the effect of turbulence is not known. The use of transfer coefficients for clean bubbles in quiescent water is highly arbitrary and yet critical to the result. For instance, Woolf (1993) predicted a mean transfer velocity of 8.5 cm h^{-1} for carbon dioxide using formulae for clean bubbles, but a mean of only 2.6 cm h^{-1} using formulae for dirty bubbles. In addition, for dirty bubbles $n = 2/3$. The simple model also takes no consideration of bubbles which are mixed to a considerable depth. The result of this deficiency is that parameters e and f and the transfer velocity of very poorly soluble gases are all likely to be under-estimated.

D. K. Woolf

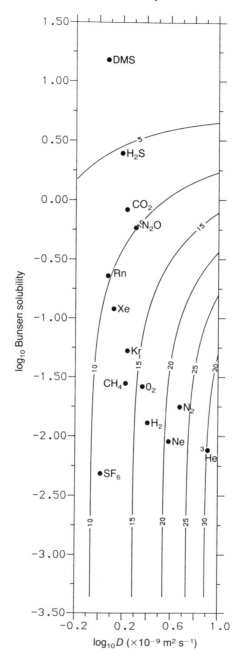

Figure 6.4. Bubble-mediated transfer velocities, K_b, at a whitecap coverage of 1% as a function of solubility and diffusion constant, D Contours of K_b in cm h^{-1} and in increments of 5cm h^{-1}.

Discussion

Bubble-mediated transfer is an important mechanism of air–sea gas exchange. Its special properties must be considered when constructing an overall model of air–sea gas exchange.

Difficulties are created by the fact that the contribution of bubbles to the air–sea transfer velocity of gases, K_b, has a complicated dependence on the solubility, β, and molecular diffusion constant, D, of the gas. It is common to infer the transfer velocity of a gas of interest from an estimate of the transfer velocity of another gas under the assumption of a simple dependence on D or the Schmidt number (e.g., $K_T \propto Sc^{-1/2}$). The dual tracer method (Watson, Upstill-Goddard and Liss, 1991) further requires such an assumption before the transfer velocity of either tracer (helium-3 or sulphur hexafluoride) can be calculated. The model results (Figure 6.4) imply that there is no significant deviation from $K_T \propto Sc^{-1/2}$ between ^{3}He and SF_6, and thus the transfer velocity of these gases can reasonably be calculated under this assumption. This conclusion must be very tentative as long-lived bubbles (excluded from the simple model) are likely to contribute significantly to the transfer velocity of these gases. The model also predicts that the bubble-mediated transfer of carbon dioxide will be only 60% of the value calculated for it on the assumption that $K_T \propto Sc^{-1/2}$. In general, if the solubility dependence is ignored the transfer velocity of a gas is liable to be overestimated from the transfer velocity of a less soluble gas. Wanninkhof *et al.* (1993) have estimated transfer velocities from a dual tracer experiment assuming $K_T \propto Sc^{-n}$, where values of n were adopted from a laboratory simulation (Asher *et al.*, 1992). Asher *et al.* found that values of n significantly less than 0.5 were appropriate for describing the relative transfer of ^{3}He and SF_6. The details of the direct and bubble-mediated transfer underlying their results are as yet unclear. Wanninkhof *et al.* estimate using a 'variable n' relatively large transfer velocities, but their estimates for gases with a Schmidt number of 600 ignore any solubility dependence, and thus are likely to be too high for carbon dioxide. The transfer of gases by 'dirty' bubbles ($j \propto D^{2/3}$) may also increase the effective value of n. On balance, it is more likely that the transfer velocity of carbon dioxide will be overestimated rather than underestimated by the dual tracer method assuming $n = 0.5$. The dependence of transfer velocity on β and D needs further attention. One possible approach is to extend the dual tracer method with a number of additional tracers of various β and D.

The bubble-mediated transfer of gases is distinguished by an asymmetry between invasion and evasion. In order to avoid transfer velocities that depend on saturation, one can write

$$F = (K_o + K_b) [C_w - \mathrm{Sp_a} (1 + \Delta)]$$

The value of Δ does not affect the evasion rate of a gas, but is exceptionally important when considering gases near saturation. Time series of the natural concentration of gases in near saturation (for example: nitrogen, oxygen and inert gases) offer an opportunity to estimate both K_T and Δ. Farmer et al. (1993) have already estimated transfer velocities from measurements of dissolved nitrogen and oxygen. (They have found a transfer velocity in storm conditions approximately a factor of three greater than that predicted by Liss and Merlivat (1986) and that measured by Watson et al. (1991) in the sole determination using the dual tracer method at a similar wind speed.) Given only a single pair of values of flux and saturation, K_T and Δ cannot both be determined, and one or other must be estimated by another means. Given a time series of gas concentrations, wave breaking and bubble cloud statistics and other environmental measurements, there is a good possibility of establishing the dependence of K_o, K_b and Δ on environmental factors.

The bubble-mediated transfer of gases will be very sensitive to wind speed and approximately proportional to whitecap coverage. Monahan and O'Muircheartaigh (1980) calculated the statistically optimal estimate of fractional whitecap coverage from wind speed alone,

$$W_B = 3.84 \times 10^{-6} U^{3.41}$$

This strong wind-speed dependence implies that most bubble-mediated transfer will occur in fairly high winds in spite of the relative rarity of those winds. Detailed statistics can be calculated starting from an estimate of the probability of occurrence of wind speeds. Statistics of individual ship observations of wind speeds in the North Atlantic (Kent et al, 1991) are shown in Figure 6.5a. The frequency-weighted whitecapping, $f\mathrm{d}U$, occurring when the wind-speed is in the range $U \rightarrow U + \mathrm{d}U$, is related to the probability of those wind speeds, $p(U)\,\mathrm{d}U$, by

$$f\,\mathrm{d}U = W_B(U)\,p(U)\,\mathrm{d}U$$

An estimate of the probability density function of whitecap coverage based on the North Atlantic winds is shown in Figure 6.5b. Since bubble-mediated transfer is approximately proportional to whitecap coverage, this plot reveals the range of conditions requiring study. Most bubble-mediated transfer is associated with high wind speeds ($U > 10$ m s^{-1}). A

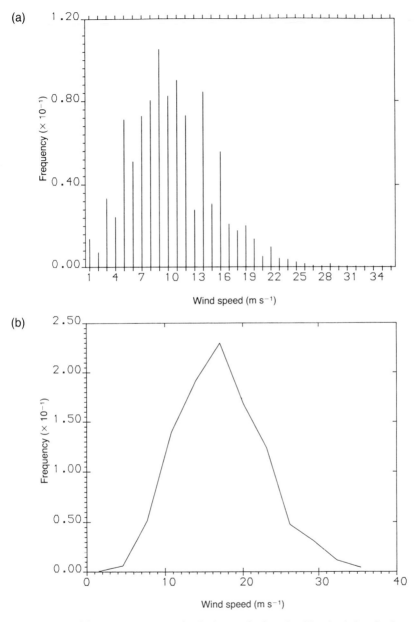

Figure 6.5. (a) A histogram of wind speeds for the North Atlantic from a voluntary ship observing program (VSOP-NA; Kent *et al.*, 1991). Frequency of observation of wind-speeds in 36 equal and contiguous ranges of wind speed from 0–72 knots (0–37 m s^{-1}). (b) Probability density function of whitecap coverage (and bubble-mediated transfer) in the North Atlantic estimated from the VSOP-NA data set.

significant fraction of the total transfer is associated with the rare occasions when wind speeds are greater than 20 m s^{-1}. Estimates of gas exchange at very high wind speed are largely based on extrapolation from less challenging conditions, and must be considered fairly speculative. Accurate estimates of mean (global, regional, seasonal . . .) transfer velocities will require accurate estimates of both the probability of storms and transfer velocities in storm conditions.

Quantitative estimates of bubble-mediated transfer are unlikely to be accurate as yet, but it is at least interesting to investigate how the model predictions compare with existing models and field measurements of the total transfer velocity. Direct and bubble-mediated transfer occur in parallel. As discussed earlier, the relationship of the direct transfer velocity, K_o, to environmental factors is not fully understood, but at least in part is expected to be related linearly to the friction velocity of the wind. More generally, K_T may be described by

$$K_T = K_u(\text{Sc}, u^* \text{ or mean square slope}) + K_w(\text{Sc}, \beta)W$$

where part of the transfer, K_u, is related to the friction velocity or surface roughness, while an additional contribution (generally both direct and bubble-mediated), K_w, W is related to whitecapping. This is a hybrid of some models of direct transfer (e.g., Deacon, 1977), which suggest transfer velocities proportional to the friction velocity, and whitecap models (e.g., Monahan & Spillane, 1984) which assume transfer velocities are linearly dependent on whitecap coverage. If it is assumed the bubble-mediated transfer alone is directly related to whitecapping and is described by the model results

$$K_b = K_w W$$

$$K_b = 2450W/\{\beta[1 + (14\beta\text{Sc}^{-0.5})^{-1/1.2}]^{1.2}\} \text{ cm h}^{-1}$$

in general and

$$K_b = 850W \text{ cm h}^{-1}$$

for CO_2 at 20 °C. Similarly, if the influence of whitecapping on direct transfer is neglected,

$$K_o = K_u$$

Jähne *et al.* (1987) reported the following expression for rough water conditions (a fluid surface boundary condition) in a circular windwave tank with no whitecapping:

$$K_o = 1.57 \times 10^{-4} u^* \, (600/Sc)^{1/2}$$

though they also show that mean square slope is a more reliable parameter of transfer than friction velocity. Jähne *et al.* did not propose that this expression would hold universally, but it is adopted here as a provisional formula for direct transfer in the field. A relationship of K_o in terms of wind speed alone can be inferred by combining this expression with formulae for the neutral drag coefficient (Large & Pond, 1981), C_{DN}.

In summary, for carbon dioxide at 20 °C, the following parameterization is proposed:

$$K_T = 1.57 \times 10^{-4} u^* \, (600/Sc)^{1/2} + 850W \text{ cm h}^{-1}$$

or more approximately in terms of wind speed alone:

$$K_T = 1.57 \times 10^{-4} \, U \, (600 C_{DN}/Sc)^{1/2} + 3.264 \times 10^{-3} \, U^{3.41} \text{ cm h}^{-1}$$

Note that this is a 'clean ocean' model in which both the sea surface and the surface of the bubbles are assumed to be essentially fluid. It is likely that surface organic material will often significantly reduce both the direct and bubble-mediated transfer. This model can be compared with two well-established models. The model of Liss and Merlivat (1986) is well supported by dual tracer results from lakes and the southern North Sea (Watson *et al.*, 1991), but is at odds with isotopic determinations of the global mean transfer velocity (Broecker *et al.*, 1986). The main virtue of the relationship favoured by Tans, Fung and Takahashi (1990) is it 'gives the right answer' for the global mean, but it is also supported by the 'variable *n*' interpretation of a dual tracer experiment on Georges Bank (Wanninkhof *et al.*, 1993). All three $K_T(U)$ relationships are plotted in Figure 6.6 along with the dual tracer results (assuming $n = 0.5$) from the North Sea and George Bank. (As discussed earlier in this section, comparison of tracer results is greatly complicated by the dependence of transfer velocities on solubility in addition to diffusivity.)

Accurate calculation of mean transfer velocities requires careful treatment of the statistics of wind speed and other environmental parameters (e.g., Thomas *et al.*, 1988; Etcheto and Merlivat, 1988; Erickson, 1989, 1993). Approximate values of the global mean transfer velocity of CO_2 are calculated by taking a Rayleigh distribution of mean 7.4 m s^{-1} to represent global winds (Wanninkhof, 1992) and transfer velocities for CO_2 in seawater at 20 °C. The results are given in Table 6.2., along with corresponding values calculated for the North Atlantic winds, and the ratio from these two distributions. (The expressions for K_o and K_b are

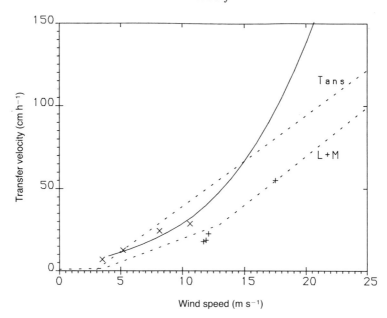

Figure 6.6. Air–sea transfer velocity plotted against wind speed. The new model presented here is plotted as a full curve. Two older models (Tans: Tans *et al.* (1990); L+M: Liss and Merlivat (1986)) are plotted as dashed curves. Results from dual tracer experiments from the North Sea (+) and from Georges Bank (×) are also shown.

very unlikely to be good predictions for wind speeds less than 4 m s^{-1}. The values in Table 6.2 were calculated by extrapolating to calm conditions, but wind speeds of less than 4 m s^{-1} only contributed a very small fraction to the total.)

The new model predicts a global mean transfer velocity compatible with isotopic determinations, and transfer velocities in good agreement with the Georges Bank results. The predicted values of transfer velocity are, however, much higher than those predicted by Liss and Merlivat and those apparent from the North Sea results. It would be worth investigating whether transfer velocities are lower in the southern North Sea than the oceans in general as a result of an atypical environment. In this context, the effect of surface-active materials on transfer velocities and the significance of sea state require further study. The high sensitivity (particularly of whitecap coverage and bubble-mediated transfer) to the distribution of wind speeds, as demonstrated in Table 6.2, highlights the importance of good wind-speed distributions and suggests that air–sea

Table 6.2 *Mean wind speed, whitecap coverage and transfer velocity of carbon dioxide associated with two wind speed distributions and the ratio of the two*

	Rayleigh	VSOP-NA	Ratio
Wind speed	7.4	10.2	1.38
Whitecap coverage (%)	0.83	2.28	2.75
K_b	7.04	19.36	2.75
K_o	14.85	21.63	1.46
K_T, this model	21.9	41.0	1.87
K_T, Liss and Merlivat (1986)	12.6	23.0	1.82
K_T, Tans *et al.* (1990)	24.8	40.5	1.63

transfer velocities will be extremely variable, regionally and seasonally. This variability is greatly underestimated by those relying solely on long-term averages of wind speed. A minor carbon dioxide–climate feedback through the sensitivity of storm frequency and strength to climate can also be inferred.

The new model predicts a relationship with wind speed reasonably close to those of Wanninkhof (1992) and Erickson (1993), but the principles and derivation are different from either of these. The emphasis here is not to reach the best empirical relationship on the basis of existing determinations of transfer velocity, but to build from knowledge of the underlying physical processes. As more is learnt about the physical mechanisms of air–sea gas transfer, the parameterization can be modified appropriately.

There is still only a little evidence from tracer experiments in the field of increased transfer velocities at high wind speeds, and we can not be certain yet that bubble-mediated transfer is significant to these enhancements. Hopefully, further experiments will clarify the dependence of transfer on wind speed (and other variables) and the role of bubbles. Methods that help to separate the contribution of direct mechanisms and bubble-mediated mechanisms will be very valuable. Direct transfer may be studied by measuring the transfer of tracers, e.g., heat, through the surface microlayer. Estimates of bubble-mediated transfer will also benefit from improved measurement of the distribution of bubbles (particularly near the surface) in the upper ocean, and studies of the behaviour of the bubbles. The alteration of transfer coefficients by surface-active material is critical, since it affects not only the magnitude of transfer but also the dependence of transfer on solubility and molecular diffusion constant. This is closely related to the state of the surface microlayer. It is likely that when a wave breaks surface-active material on the sea surface is transferred directly to the surface of bubbles. Material will also be scavenged by a bubble from the upper ocean. Some material may slough off the bubble when its surface is saturated. Thus, bubble-mediated gas transfer is tied to the complicated interaction of bubbles and the surface microlayer. There is an obvious need for new techniques both to measure accurately air–sea transfer velocities and, in parallel, indicators of both direct and bubble-mediated mechanisms with attention to the influence of surface-active material.

Summary

1. Bubble-mediated transfer is an important mechanism of air–sea gas exchange at moderate and high wind speeds. Estimates of bubble-mediated transfer should preferably be based on (*in situ* or remote sensing) measurement of bubble clouds, wave breaking or whitecap coverage rather than wind speed. Whitecapping also disrupts the surface microlayer and possibly suppresses short surface waves, but the total effect on direct transfer is difficult to assess.

2. Bubble-mediated transfer is asymmetric with a net invasion of gas at saturation. The gross rate of bubble-mediated transfer and the supersaturating effect of bubbles can be described by a transfer velocity, K_b, and the equilibrium supersaturation, Δ, respectively. The net air–sea flux of a gas is described by

$$F = (K_o + K_b) [C_w - \mathrm{Sp}_a (1 + \Delta)]$$

3. K_b and Δ are both dependent on the solubility, β, and the molecular diffusion constant, D, of the gas.

4. The equilibrium supersaturation, Δ, will be approximately proportional to the square of wind speed. Δ may be a few per cent for poorly soluble gases such as oxygen, but will be relatively insignificant for more soluble gases, including carbon dioxide.

5. The bubble-mediated air–sea transfer velocity, K_b, will be approximately proportional to whitecap coverage. K_b decreases with increasing solubility but is significant for most gases including carbon dioxide.

6. It is possible to construct a model of transfer velocity, with $K_o \propto u^*$ and $K_b \propto W$, that is both physically reasonable and compatible with isotopic estimates of the mean transfer velocity of CO_2 (Broecker *et al.*, 1986). This model, however, does require greater transfer

velocities at moderate and fairly high wind speeds than most field measurements imply.

7. The study of bubble-mediated transfer is at an early stage. Several important properties have been established, but the accuracy of quantitative estimates is still likely to be poor. The 'limited success' described in 6. should not obscure the huge uncertainties in all quantitative estimates of gas transfer velocities (particularly at high wind speeds). Better measurement techniques and superior insight into air–sea transfer processes at high wind speeds will be required for any major progress.

References

Agrawal, Y. C., Terray, E. A., Donelan, M. A., Hwang, P. A., Williams, A. J., III, Drennan, W. M., Kahma, K. K. and Kitaigorodskii, S. A. (1992). Enhanced dissipation of kinetic energy beneath surface waves. *Nature*, **359**, 219–20.

Asher, W. E., Farley, P. J., Wanninkhof, R., Monahan, E. C. and Bates, T. S. (1992). Laboratory and field measurements concerning the correlation of fractional area whitecap coverage with air/sea gas transport. In *Precipitation Scavenging and Atmosphere–Surface Exchange, Volume 2 – The Semonin Volume: Atmosphere–Surface Exchange Processes*, ed. S. E. Schwartz and W. G. N. Slinn, pp. 815–28. Washington DC: Hemisphere.

Baldy, S. and Bourguel, M. (1985). Measurements of bubbles in a stationary field of breaking waves by a laser-based single-particle scattering technique. *J. Geophys. Res.*, **90**, 1037–47.

Baldy, S. and Bourguel, M. (1987). Bubbles between the wave trough and wave crest levels. *J. Geophys. Res.*, **92**, 2919–29.

Banner, M. L., Jones, I. S. F. and Trinder, J. C. (1989). Wavenumber spectra of short gravity waves. *J. Fluid Mech.*, **198**, 321–44.

Batchelor, G. K. (1967). *An Introduction to Fluid Dynamics*. Cambridge: Cambridge University Press. 615 pp.

Blanchard, D. C. (1963). The electrification of the atmosphere by particles from bubbles in the sea. *Progr. Oceanogr.*, **1**, 71–202.

Blanchard, D. C. (1975). Bubble scavenging and the water-to-air transfer of organic material in the sea. In *Applied Chemistry at Protein Interfaces* (*Advances in Chemistry Series, 145*), ed. R. Baier, pp. 360–87. Washington, DC: American Chemical Society.

Blanchard, D. C. (1983). The production, distribution and bacterial enrichment of the sea-salt aerosol. In *Air–Sea Exchange of Gases and Particles*, ed. P. S. Liss and W. G. N. Slinn, pp. 407–54. Dordrecht, Holland: Kluwer Academic Publishers.

Blanchard, D. C. and Woodcock, A. H. (1957). Bubble formation and modification in the sea and its meteorological significance. *Tellus*, **9**, 145–58.

Bowyer, P. A. (1992). The rise of bubbles in a glass tube and the spectrum of bubbles produced by a splash. *J. Mar. Res.*, **50**, 521–43.

202   *D. K. Woolf*

Broecker, H. Ch. and Siems, W. (1984). The role of bubbles for gas transfer from water to air at higher windspeeds: experiments in the wind–wave facility in Hamburg. In *Gas Transfer at Water Surfaces*, ed. W. Brutsaert and G. H. Jirka, pp. 229–36. Dordrecht, Holland: Kluwer Academic Publishers.

Broecker, W. S., Ledwell, J. R., Takahashi, T., Weiss, R., Merlivat, L., Memery, L., Peng, T-H., Jähne, B. and Munnich, K. O. (1986). Isotopic versus micrometeorological ocean CO_2 fluxes: a serious conflict. *J. Geophys. Res.*, **91**, 10 517–27.

Cipriano, R. J. and Blanchard, D. C. (1981). Bubble and aerosol spectra produced by a laboratory 'breaking wave'. *J. Geophys. Res.*, **86**, 8085–92.

Clift, R., Grace, J. R. and Weber, M. E. (1978). *Bubbles, Drops and Particles*. New York: Academic Press. 380 pp.

Crawford, G. B. and Farmer, D. M. (1987). On the spatial distribution of bubbles. *J. Geophys. Res.*, **92**, 8231–43.

Csanady, G. T. (1990). The role of breaking wavelets in air–sea gas transfer. *J. Geophys. Res.*, **95**, 749–59.

Deacon, E. L. (1977). Gas transfer to and across an air–water interface. *Tellus*, **29**, 363–74.

Erickson, D. J. III. (1989). Variations in the global air–sea transfer velocity field of CO_2. *Global Biogeochem. Cycles*, **3**, 37–41.

Erickson, D. J. III. (1993). A stability dependent theory for air–sea gas exchange. *J. Geophys. Res.*, **98**, 8471–88.

Etcheto, J. and Merlivat, L. (1988). Satellite determination of the carbon dioxide exchange coefficient at the ocean–atmosphere interface: a first step. *J. Geophys. Res.*, **93**, 15 669–78.

Ewing, G. and McAlister, E. D. (1960). On the thermal boundary-layer of the ocean. *Science*, **131**, 1374–6.

Farmer, D. M., McNeil, C. L. and Johnson, B. D. (1993). Evidence for the importance of bubbles in increasing air–sea gas flux. *Nature*, **361**, 620–3.

Grassl. H. (1976). The dependence of the measured cool skin of the ocean on wind stress and total heat flux. *Boundary-Layer Meteorol.*, **10**, 465–74.

Jähne, B., Munnich, K. O., Bosinger, R., Dutzi, A., Huber, W. and Libner, P. (1987). On the parameters influencing air–water gas exchange. *J. Geophys. Res.*, **92**, 1937–49.

Jähne, B., Libner, P., Fischer, R., Billen, T. and Plate, E. J. (1989). Investigating the transfer processes across the free aqueous viscous boundary layer by the controlled flux method. *Tellus*, **41B**, 177–95.

Jessup, A. T. (1996) The infrared signature of breaking waves. In *The Air–Sea Interface*, ed. M. A. Donelan, W. H. Hui and W. J. Plant. Toronto, Canada: University of Toronto Press (in press).

Johnson, B. D. and Cooke, R. C. (1979). Bubble populations and spectra in coastal waters: a photographic approach. *J. Geophys. Res.*, **84**, 3761–6.

Kanwisher, J. (1963). On the exchange of gases between the atmosphere and the sea. *Deep-Sea Res.*, **10**, 195–207.

Keeling, R. F. (1993). On the role of large bubbles in air–sea gas exchange and supersaturation in the ocean. *J. Mar. Res.*, **51**, 237–71.

Kent, E. C., Truscott, B. S., Hopkins, J. S. and Taylor, P. K. (1991). The accuracy of ships' meteorological observations: results of the VSOP-NA. *Marine Meteorology and Related Oceanographic Activities, Report No. 26*, 86 pp. Geneva: WMO.

Kitaigorodskii, S. A. (1984). On the fluid dynamical theory of turbulent gas transfer across an air–sea interface in the presence of breaking wind waves. *J.
Phys. Oceanogr.*, **14**, 960–72.

Kolavayev, P. A. (1976). Investigation of the concentration and statistical size
distribution of wind-produced bubbles in the near-surface ocean layer
(English translation). *Oceanology*, **15**, 659–61.

Lamarre, E. and Melville, W. K. (1991). Air entrainment and dissipation in
breaking waves. *Nature*, **351**, 469–72.

Large, W. S. and Pond, S. (1981). Open ocean momentum flux measurements in
moderate to strong winds. *J. Phys. Oceanogr.*, **11**, 324–36.

Levich, V. G. (1962). *Physico-chemical Hydrodynamics*. New York: Prentice-
Hall. 700 pp.

Liss, P. S. (1983). Gas transfer: experiments and geochemical implications. In
Air–Sea Exchange of Gases and Particles, ed. P. S. Liss and W. G. N. Slinn,
pp. 241–98. Dordrecht, Holland: Kluwer Academic Publishers.

Liss, P. S. and Merlivat, L. (1986). Air–sea gas exchange rates: introduction and
synthesis. In *The Role of Air–Sea Exchange in Geochemical Cycling*, ed.
P. Buat-Menard, pp. 113–27. Dordrecht, Holland: Kluwer Academic
Publishers.

MacIntyre, F. (1974). Chemical fractionation and sea-surface microlayer processes. In *The Sea (Marine Chemistry)*, Vol. 5, ed. E. D. Goldberg, pp. 245–
99. New York: John Wiley and Sons.

Medwin, H. and Breitz, N. D. (1989). Ambient and transient bubble spectral
densities in quiescent seas and under spilling breakers. *J. Geophys. Res.*, **94**,
12 751–9.

Melville, W. K. (1994). Energy dissipation by breaking waves. *J. Phys.
Oceanogr.*, **24**, 2041–9.

Memery, L. and Merlivat, L. (1985). Modelling of gas flux through bubbles at the
air–water interface. *Tellus*, **37B**, 272–85.

Merlivat, L. and Memery, L. (1983). Gas exchange across an air–water interface:
experimental results and modeling of bubble contribution to transfer. *J.
Geophys. Res.*, **88**, 707–24.

Monahan, E. C. (1986). The ocean as a source of atmospheric particles. In *The
Role of Air–Sea Exchange in Geochemical Cycling*, ed. P. Buat-Menard,
pp. 129–63. Dordrecht, Holland: Kluwer Academic Publishers.

Monahan, E. C. & Monahan, C. F. (1986). The influence of fetch on whitecap
coverage as deduced from the Alte Weser Light-station observer's log.
Abstract in *Oceanic Whitecaps and Their Role in Air–Sea Exchange
Processes*, ed. E. C. Monahan and G. MacNiocaill, p. 294. Dordrecht,
Holland: Kluwer Academic Publishers.

Monahan, E. C. and O'Muircheartaigh, I. G. (1980). Optimal power-law
description of oceanic whitecap coverage dependence on wind speed. *J.
Phys. Oceanogr.*, **10**, 2094–9.

Monahan, E. C. and O'Muircheartaigh, I. G. (1986). Whitecaps and the passive
remote sensing of the ocean surface. *Int. J. Remote Sensing*, **7**, 627–42.

Monahan, E. C. and Spillane, M. C. (1984). The role of oceanic whitecaps in air–
sea gas exchange. In *Gas Transfer at Water Surfaces*, ed. W. Brutsaert and G.
H. Jirka, pp. 495–503. Dordrecht, Holland: Kluwer Academic Publishers.

Monahan, E. C. and Woolf, D. K. (1989). Comments on 'Variations of whitecap
coverage with wind stress and water temperature'. *J. Phys. Oceanogr.*, **19**,
706–11.

Monahan, E. C. and Zietlow, C. R. (1969). Laboratory comparisons of fresh-water and salt-water whitecaps. *J. Geophys. Res.*, **74**, 6961–6.

Osborn, T., Farmer, D. M., Vagle, S., Thorpe, S. A. and Curé, M. (1992). Measurements of bubble plumes and turbulence from a submarine. *Atmosphere–Ocean*, **30**, 419–40.

Phillips, O. M. (1985). Spectral and statistical properties of the equilibrium range in wind-generated gravity waves. *J. Fluid Mech.*, **156**, 505–31.

Schluessel, P., Emery, W. J., Grassl, H. and Mammen, T. (1990). On the bulk-skin temperature difference and its impact on satellite remote sensing of sea surface temperature. *J. Geophys. Res.*, **95**, 13341–56.

Scott, J. C. (1975). The preparation of water for surface-clean fluid mechanics. *J. Fluid Mech.*, **69**, 339–51.

Soloviev, A. V. and Schluessel P., (1994). Parameterization of the cool skin of the ocean and of the air–ocean gas transfer on the basis of modelling surface renewal. *J. Phys. Oceanogr.*, **24**, 1339–46.

Sutcliffe, W. H. Jr., Baylor, E. R. and Menzel, D. W. (1963). Sea surface chemistry and Langmuir circulation. *Deep-Sea Res.*, **10**, 233–43.

Tans, P. P., Fung, I. Y. and Takahashi, T. (1990). Observational constraints on the global atmospheric CO_2 budget. *Science*, **247**, 1431–8.

Thomas, F., Perigaud, C., Merlivat, L. and Minster, J-F. (1988). World-scale monthly mapping of the CO_2 ocean atmosphere gas-transfer coefficient. *Phil. Trans. R. Soc., London*, **A325**, 71–83.

Thorpe, S. A. (1982). On the clouds of bubbles formed by breaking wind-waves in deep water, and their role in air–sea gas transfer. *Phil. Trans. R. Soc., London*, **A304**, 155–210.

Thorpe, S. A. (1984). A model of the turbulent diffusion of bubbles below the sea surface. *J. Phys. Oceanogr.*, **14**, 841–54.

Thorpe, S. A. (1986). Measurements with an automatically recording inverted echo sounder: ARIES and the bubble clouds. *J. Phys. Oceanogr.*, **16**, 1462–78.

Thorpe, S. A. (1993). Energy loss by breaking waves. *J. Phys. Oceanogr.*, **23**, 2498–502.

Thorpe, S. A. and Hall, A. J. (1983): The characteristics of breaking waves, bubble clouds, and near-surface currents observed using side-scan sonar. *Cont. Shelf Res.*, **1**, 353–84.

Thorpe, S. A., Bowyer, S. A., and Woolf, D. K. (1992). Some factors affecting the size distributions of oceanic bubbles. *J. Phys. Oceanogr.*, **22**, 382–9.

Vagle, S. and Farmer, D. M. (1992). The measurement of bubble-size distri-butions by acoustical backscatter. *J. Atmos. Oceanic Technol.*, **9**, 630–44.

Wallace, D. W. R. and Wirick, C. D. (1992). Large air–sea gas fluxes associated with breaking waves. *Nature*, **356**, 694–6.

Wallace, G.T. Jr and Duce, R. A. (1978). Transport of particulate organic matter by bubbles in marine waters. *Limnol. Oceanogr.*, **23**, 1155–67.

Walsh, A. L. and Mulhearn, P. J. (1987). Photographic measurements of bubble populations from breaking wind waves at sea. *J. Geophys. Res.*, **92**, 14 553–65.

Wanninkhof, R. (1992). Relationship between wind speed and gas exchange over the ocean. *J. Geophys. Res.*, **97**, 7373–82.

Wanninkhof, R. H. and Bliven, L. F. (1991). Relationship between gas exchange, wind speed and radar backscatter in a large wind–wave tank. *J. Geophys. Res.*, **96**, 2785–96.

Wanninkhof, R., Asher, W., Wepperning, R., Chen, H., Schlosser, P., Langdon, C. and Sambrotto, R. (1993). Gas transfer experiment on Georges Bank using two volatile deliberate tracers. *J. Geophys. Res.*, **98**, 20 237–48.

Watson, A. J., Upstill-Goddard, R. C. Liss, P. S. (1991). Air–sea gas exchange in rough and stormy seas measured by a dual-tracer technique. *Nature*, **349**, 145–7.

Weisel, C. P., Duce, R. A., Fasching, J. L. and Heaton, R. W. (1984). Estimates of the transport of trace metals from the ocean to the atmosphere. *J. Geophys. Res.*, **89**, 11 607–18.

Woolf, D. K. (1988). The role of oceanic whitecaps in geochemical transport in the upper ocean and the marine environment. PhD dissertation, 331 pp. Dublin: National University of Ireland.

Woolf, D. K. (1990). Comment on an article by Wu. *Tellus*, **42B**, 385–6.

Woolf, D. K. (1993). Bubbles and the air–sea transfer velocity of gases. *Atmosphere–Ocean*, **31**, 517–40.

Woolf, D. K. (1995). Vertical mixing in the upper ocean and air–sea gas transfer. In *Air–Water Gas Transfer*, ed. B. Jaehne and E. C. Monahan, pp. 59–67. Hanau, Germany: AEON.

Woolf, D. K. and Monahan, E. C. (1988). Laboratory investigations of the influence on marine aerosol production of the interaction of oceanic whitecaps and surface-active material. In *Aerosols and Climate*, ed. P. V. Hobbs and M. P. McCormick, pp. 1–8. Hampton, VA: A. Deepak.

Woolf, D. K. and Thorpe, S. A., (1991). Bubbles and the air–sea exchange of gases in near-saturation conditions. *J. Mar. Res.*, **49**, 435–66.

Woolf, D. K., Bowyer, P. A. and Monahan, E. C. (1987). Discriminating between the film drops and jet drops produced by a simulated whitecap. *J. Geophys. Res.*, **92**, 5142–50.

Wu, J. (1979). Oceanic whitecaps and sea state. *J. Phys. Oceanogr.*, **9**, 1064–8.

7

The physical chemistry of air–sea gas exchange

LEON F. PHILLIPS

Abstract

The thermodynamics and kinetics of air–sea exchange are discussed in terms of the underlying theory and on the basis of numerical calculations. The thermodynamic driving force for gas exchange is dependent on the air–sea temperature difference as well as on the partial pressure or concentration difference across the interface. The kinetics of the exchange process are strongly affected by the surface temperature of the water, as controlled by the fluxes of sensible and latent heat. The results of calculations for a model which incorporates a turbulent air layer are compared with the experimental data of Liss *et al.* (1981), Smith and Jones (1985) and Smith *et al.* (1991). This comparison clearly demonstrates the importance of coupling, in the sense of Onsager's irreversible thermodynamics, of the fluxes of sensible heat and matter across the interface. The calculations also suggest a possible new approach to the measurement of air–water exchange rates for trace gases such as carbon dioxide.

Introduction

The rate of air–sea exchange of carbon dioxide is a topic which has been the subject of some controversy in recent years. This exchange rate, which is of great practical importance in connection with global warming, is very difficult to determine experimentally, and the difficulty is compounded by the need to obtain values for both the long-term average of the rate on a global scale and the local exchange rate at a particular instant. Sophisticated indirect estimates of the globally and temporally averaged rate and directly measured values of the local rate at one specific time have sometimes appeared to be in conflict, and many values of the transfer rate which purport to derive from direct measurements in fact depend on a particular assumption about the dependence of the rate on the difference in CO_2 concentration between air and sea. In this kind of situation it is often worthwhile to approach the problem afresh, at the simplest, most fundamental level, and this we now do.

The occurrence of any physico-chemical process depends on the existence of two things, namely, a thermodynamic driving force for the process and a kinetic pathway by which the process can occur. Thermodynamics governs how far the process can go: kinetics governs how fast it goes. The thermodynamic driving force is commonly expressed in terms of a difference in chemical potential, Gibbs free energy, or entropy between the initial and final states of the process, which in the present instance correspond to CO_2 gas at a specified temperature and partial pressure in the air and dissolved CO_2 at a specified concentration and temperature in the ocean. The effectiveness of the kinetic pathway can be expressed in terms of a conductance for the air–sea interface, whose value depends on such variables as the temperature, thickness, area, degree of agitation, turbulence, and extent of contamination of the interfacial region. This physico-chemical approach to the problem differs from the more usual one in which gas transfer is considered solely in terms of hydrodynamic effects (see, for example, Coantic, 1986; Jähne et al., 1987). Its success can be measured by the fact that it accounts quantitatively for some experimental observations which are not readily accommodated by hydrodynamic theories, namely, gas transfer in the opposite direction to that which would be expected from the partial pressure difference.

Gas exchange at the surface of an ocean or lake has often been described by the equation

$$\text{Flux} = K_w\, \alpha(P_{air} - P_{sea}) \tag{7.1}$$

where Flux means the quantity of gas transferred from air to sea per unit area (measured normal to the flux vector) per second, α is the solubility of the gas in water at some temperature measured not too far from the surface, P_{air} is the partial pressure of the gas in air and P_{sea} is the partial pressure of gas that would be in equilibrium with the water at the existing concentration and temperature. The factor $\alpha(P_{air} - P_{sea})$, which is roughly equivalent to the difference in concentration of dissolved CO_2 between the surface and a mixed layer which is assumed to be present a short distance below the surface, provides the driving force for gas transfer and so should reflect the thermodynamics of the process. The transfer velocity, K_w, which is a function of wind speed, sea-surface temperature and the nature of the gas being transferred, embodies the kinetics of the process. Etcheto and Merlivat (1988) concluded that for

CO_2 the gas-transfer coefficient, i.e. the product of K_w and α, should be a function only of wind speed, since the effects of temperature on α and K_w (via the Schmidt number) largely cancel one another. Some of the problems encountered in arriving at this convenient parameterization of the gas-transfer rate are summarized by Liss *et al.* (1981).

Figure 7.1 demonstrates that, in quantitative terms, Equation 7.1 provides a rather poor description of the exchange process. At a wind speed of 8 m s^{-1} the range of the measured values of transfer velocity amounts to a factor of five and, as Upstill-Goddard *et al.* (1990) point out, the scatter of the points in Figure 7.1 greatly exceeds the limits of experimental error, a clear indication that one or more uncontrolled variables are present. Further, the use of Equation 7.1 to estimate air–sea fluxes of CO_2, on the basis of accepted values of K_w, leads to problems in balancing the earth's annual budget of anthropogenic CO_2 (the calculated air–sea fluxes are too small) and in accounting for the observed small inter-hemisphere concentration gradient of atmospheric CO_2 (Tans *et al.*, 1990). Etcheto *et al.* (1991) present most of the same data as is given in Figure 7.1, plotted together with some new wind-tunnel data and the more recent results of Watson *et al.* (1991) for the North Sea, as evidence in support of Liss and Merlivat's 1986 formulation. However, their plot emphasizes data obtained at high wind speeds, where the agreement possibly is better. From the same data they derive a globally averaged transfer rate of CO_2 (for an average wind speed near 8 m s^{-1}), which differs from the figure based on ^{14}C exchange by a factor of 1.6 and, to this author at least, their explanations of the discrepancy are not convincing. Also the apparently better agreement at high wind speeds may well be merely a reflection of the paucity of data at high wind speeds.

Various authors have tried different ways of getting around the difficulties arising from the use of Equation 7.1. Because the oceanic sink appeared too small, Tans *et al.* (1990) postulated the existence in the northern hemisphere of an additional, terrestrial sink for CO_2. Sarmiento and Sundquist (1992) balanced the CO_2 budget by including transport of CO_2 to the ocean by rivers. Robertson and Watson (1992) increased the long-term, time-averaged air–sea fluxes by adjusting the value of α in Equation 7.1 to the value appropriate for a 'cool skin' on the surface of the ocean. Broecker and Peng (1992) accounted for the small inter-hemisphere concentration gradient in terms of transport of CO_2 between hemispheres by the oceanic circulation. Wanninkhof (1992), in a general review of the relationship between wind speed and gas exchange, scaled K_w in proportion to the inverse square root of the Schmidt number, and

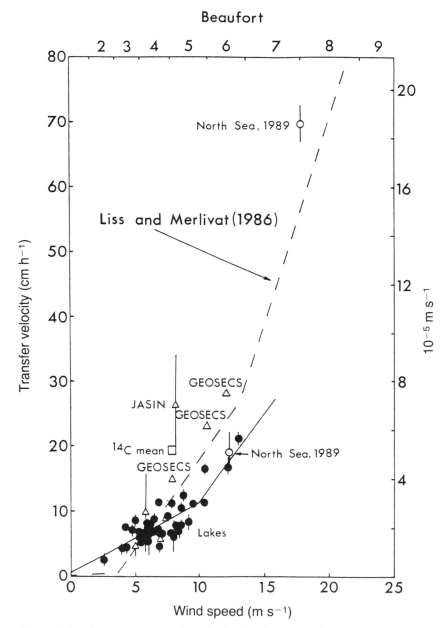

Figure 7.1. Air–water gas transfer velocity K_w plotted as a function of wind speed. All data refer to CO_2 at 20 °C. Different gases are normalized to CO_2 at a windspeed of 10 m s⁻¹. From an update by Turner (1990), based mainly on Upstill-Goddard *et al.* (1981), using data presented by Liss and Merlivat (1986).

suggested that other factors besides wind speed might affect the value of K_w including wind speed variability, the nature of the wave field, and chemical enhancement of CO_2 transfer at low wind speeds. Sarmiento *et al.* (1992) circumvented the whole problem by calculating the oceanic uptake of CO_2 by means of a perturbation approach in a three-dimensional global general ocean circulation model, rather than using data based on Equation 7.1, and concluded that the value of the air–sea exchange rate has very little influence on the cumulative oceanic uptake. Erickson (1993) extended the 'whitecap' model of Monahan and Spillane (1984) to include allowance for the varying thermal stability of the air–sea interface, thereby obtaining a transfer velocity which is a function of the temperature difference between the air and the sea surface as well as wind speed. However, the fundamental inadequacy of Equation 7.1 as an expression of the kinetics and thermodynamics of air–sea exchange has seldom been addressed.

Previous articles on the theory of air–sea exchange of CO_2 dealt with the thermodynamic driving force for the process (Phillips, 1991a,b) and, in a preliminary way, with the kinetics (Phillips, 1992). Subsequently, the steady-state model used in these calculations was extended to incorporate a turbulent air layer, enabling the results to be compared with experimental data from the literature (Phillips, 1994a). This provided an unequivocal demonstration of the correctness of the physico-chemical approach to the problem and suggested a possible new method of measuring air–sea exchange rates. More recent work has shown that the application of irreversible thermodynamics to gas–liquid exchange applies strictly to the stagnant air layer adjacent to the liquid surface (Phillips, 1994b), that the effect of cooling of the sea surface by long-wavelength radiation is generally negligible (Phillips, 1994c), and that the dual-tracer method of Watson *et al.* (1991) can be expected to work very well when a surface-renewal model applies and only slightly less well when a boundary-layer model applies (Phillips, 1995), provided there is no strong heating or cooling of the sea surface. The aim of the present chapter is to present a synthesis of the material in these papers, together with a discussion and refutation of some recent criticisms (Doney, 1994, 1995a,b).

Thermodynamics: the steady-state model

We aim to calculate the thermodynamic driving force for a steady-state flux of carbon dioxide between air and sea, a steady state being defined as one in which the values of relevant variables, such as the CO_2 concentrations and temperatures of air, sea and the air–sea interface, the relative humidity of the air, the conductances of the interface for heat and matter, and the fluxes of heat, water vapour and CO_2 are changing very slowly on the timescale of interest. The importance of the timescale in steady-state treatments is discussed elsewhere (Phillips, 1991b). The essential conditions for a steady-state theory to apply at a particular location are that any transients associated with exposure of the ocean surface to a fresh air mass have had time to die away, and that the time over which fluxes are averaged is of similar magnitude to the response time of the instrumentation that is being used to measure the fluxes. Smith *et al.* (1991) have observed a negative correlation between the measured CO_2 flux and the hourly rate of change of wind speed, an observation which may be related to the need for a finite time to elapse before a steady-state regime is established. Under such conditions the present theory would not be expected to apply. A steady state is not to be confused with a stationary state, in which the values of one or more fluxes are equal to zero. The concept of the steady state represents a mathematical approximation, one which works very well in practice provided the above conditions are satisfied. In the field of chemical kinetics, for example, much of the progress made since about 1920 has depended on the application of steady-state theory.

In the steady state the values of state functions in all parts of the system are constant on the timescale of the measurements, and the flux through every part of the system is the same, from the point where CO_2 partial pressure is measured in the turbulent air layer to the place where the bulk CO_2 concentration is measured in the liquid. The relative resistance of the different regions to gas transfer is then reflected in the magnitudes of the local concentration gradients and, because the available concentration range is limited by the difference between the bulk air and sea concentrations, this means that the relative resistance of a region is related to its effective thickness. The layer of liquid immediately below the surface has a very high resistance and its thickness is measured in microns; the turbulent air layer has a fairly low resistance and its thickness is measured in metres. The statement is sometimes made that, because of its high resistance, the surface layer of liquid controls the magnitude of the flux for a slightly soluble gas such as CO_2. This statement is misleading in relation to the steady state. Certainly, the concentration gradient is largest in the region of highest resistance, but it does not follow that gradients in the rest of the system are unimportant, nor that processes in other parts of the system have a negligible effect on the magnitude of the flux.

For a steady-state model the fluxes must be described in terms of Onsager's irreversible thermodynamics, of which an excellent introductory account is given in the book by Denbigh (1951). Since that book was written, the field of non-equilibrium thermodynamics has developed in a way which tends to make current research papers appear very complicated and forbidding and almost impossible to relate to experiment. Fortunately, a relatively simple treatment, based on Denbigh's discussion of thermal diffusion through a membrane, is sufficient for our present purpose. This treatment leads to an expression for the driving force for CO_2 exchange which differs from the driving-force factor of Equation 7.1 in that it includes the temperature difference as well as the partial pressure difference across the interface.

We wish to calculate the effective thermodynamic driving force for CO_2 transport between two regions adjacent to and on opposite sides of the interface. The regions are assumed to have temperatures T_a and T_s and CO_2 concentrations corresponding to partial pressures P_a and P_s. Since we are not considering the kinetics of transport, the structure of the interface, and in particular the detailed temperature profile through the interface, does not concern us here. The interface could be replaced by a solid membrane or an impervious stone wall without affecting the present discussion. For the present we shall also ignore the flux of water vapour, which proves to have a remarkable effect on the kinetics of the transfer process via its effect on the temperature profile through the interface but has no direct effect on the thermodynamic driving force for CO_2 transport. The results which we shall obtain for CO_2 are also applicable to water vapour and to any other gas that is involved in transport through the interface. We assume that the interface possesses an average temperature T_m and an effective thickness Δz, and that the liquid is in thermal and vapour-pressure equilibrium with an extremely thin film of gas immediately adjacent to the liquid surface. The regions on either side of the interface are assumed to be of infinite extent and of fixed temperature and composition. For a system with two coupled fluxes the basic equations of Onsager's theory are:

$$J_1 = L_{11}X_1 + L_{12}X_2 \tag{7.2a}$$

$$J_2 = L_{21}X_1 + L_{22}X_2 \tag{7.2b}$$

where J_1 and J_2 are the fluxes (quantities transferred per unit area in unit time) of heat and carbon dioxide, respectively, X_1 and X_2 are the thermodynamic driving forces responsible for these fluxes when they occur singly, the coefficients L_{11} and L_{22} are the conductances of the interface for heat and carbon dioxide, respectively (L_{11} and L_{22} can be related to an average thermal conductivity and an average molecular diffusion coefficient for the interface, respectively) and L_{12} and L_{21} are coupling coefficients, which express the dependence of the heat flux on the driving force for CO_2 transport and the dependence of the CO_2 flux on the driving force for heat flow, i.e. the temperature gradient, when the two fluxes occur simultaneously. This set of equations is readily expanded to deal with any number of coupled fluxes. The coefficients L_{ij} are constants for a specified system, independent of the driving forces X, and in general the coupling terms L_{mn} are non-zero, although in particular cases they may be negligible for all practical purposes. For the present system, for example, we can neglect the direct coupling coefficients L_{ij} between fluxes of different gases, including water vapour. Onsager showed (and many others since have shown by a variety of methods) that for the kind of system we are considering

$$L_{mn} = L_{nm} \qquad (7.3)$$

i.e. the matrix of coefficients L_{ij} is symmetrical.

Our first objective is to eliminate the coupling coefficient L_{21} from Equation 7.2b. The standard procedure for doing this is as follows. We consider the situation in which there is no temperature gradient, i.e. X_1 is zero. The flux of CO_2 given by Equation 7.2b is then simply $L_{22}X_2$, while the flux of heat from air to sea, given by Equation 7.2a as $L_{12}X_2$, is equal to the flux of CO_2 multiplied by Q, the heat of solution of CO_2 in seawater. Hence we obtain

$$L_{12}X_2 = Q\, L_{22}X_2 \qquad (7.4)$$

or, in view of Equation 7.3,

$$L_{21} = Q\, L_{22} \qquad (7.5)$$

and so

$$J_2 = L_{22}(Q\, X_1 + X_2) \qquad (7.6)$$

Next, we consider the driving forces, X_1 and X_2. In Onsager's version of the theory, which we shall use, the driving forces are defined so that the product of a flux J_m by the corresponding force X_m is equal to the rate of dissipation of energy in unit volume, expressed as the product of temperature and the rate of entropy production due to the irreversible process. (In the alternative treatment of de Groot and his school (de

Groot and Mazur, 1962) the product of flux and driving force is set equal to the rate of entropy production without the T factor. This does not affect the final results.) Here *unit volume* refers to unit volume of the interface. Fluxes occurring outside the interfacial zone are regarded as reversible, i.e. they lead to no net production of entropy. By rate of entropy production we mean the rate of increase of the entropy of the universe, since it is this increase which provides the driving force for the irreversible process. It is only in an isolated system that the local entropy production determines the direction of change corresponding to a spontaneous process. Reversible processes, defined as processes during which the system remains in equilibrium, contribute nothing to the entropy change of the universe and so do not contribute to the thermodynamic driving force. The statement that ΔS is zero for a reversible process can be regarded as a corollary of the Second Law of Thermodynamics (see, for example, Atkins, 1994: 123). Elsewhere (Phillips, 1994b) it is shown that the region of the interface we are considering is actually the stagnant air layer immediately above the liquid surface. At the bottom of this stagnant layer the gas and liquid are in equilibrium, i.e. gas–liquid exchange is reversible in the thermodynamic sense. At the top of the stagnant layer the gas composition and temperature merge with those of the turbulent air layer. In the steady state, transport of gas across the stagnant layer necessarily results in release of Q, the heat of solution (or condensation, in the case of water vapour), at the liquid surface.

For the heat flux Denbigh derives

$$X_1 = -T^{-1}dT/dz \tag{7.7}$$

which can be integrated from T_a to T_s over a layer of small but finite thickness Δz to yield

$$X_1 = -\ln(T_s/T_a)/\Delta z \tag{7.8}$$

For the flux of CO_2 between air at temperature T_a and CO_2 partial pressure P_a and seawater at temperature T_s with a CO_2 activity in solution corresponding to partial pressure P_s, we note that the entropy change of interest amounts to the entropy difference between an initial state with a small quantity of CO_2 on the air side of the interface and a final state in which this same quantity of CO_2 has been transferred to the sea-side of the interface, but nothing else has altered. Thus, the entropy production is simply the change in entropy of the CO_2 as it goes from the initial state to the final state. Because entropy is a state function, this entropy change is independent of the path by which the change of state takes place, so we may consider the transfer as occurring in two stages – first an irreversible gas-phase transition from state P_a, T_a to state P_s, T_s, and then a reversible process of solution at P_s, T_s. The process of solution, being reversible, leads to no net production of entropy, while the entropy change per mole of CO_2 during the irreversible part of the process amounts to

$$\delta S = C_p \, Ln(T_s/T_a) - R \, Ln(P_s/P_a) \tag{7.9}$$

where C_p is the constant-pressure heat capacity of CO_2 gas, R is the ideal gas constant, and Equation 7.9 is obtained by summing the two entropy changes for a mole of CO_2 going by a reversible path first from T_a to T_s at constant pressure and then from P_a to P_s at constant temperature. Hence, since the volume of unit area of interface is Δz we find

$$X_2 = T_m \, [C_p \, Ln(T_s/T_a) - R \, Ln(P_s/P_a)]/\Delta z \tag{7.10}.$$

When the results of Equations 7.8 and 7.10 are substituted into Equation 7.6, we obtain

$$J_2 = -L_{22} \, [Q^* \, Ln(T_s/T_a) + RT_m \, Ln(P_s/P_a)]/\Delta z \tag{7.11}$$

where

$$Q^* = Q - C_p T_m \tag{7.12}$$

is called the *heat of transport* by analogy with the similar quantity encountered in the treatment of thermal diffusion through a membrane. The factor by which the conductance L_{22} is multiplied in Equation 7.11 represents the effective driving force for air–sea transport of the gas. Elsewhere (Phillips, 1991b), it is shown that Equation 7.11 reduces to Equation 7.1 for the special case of isothermal transfer with a small fractional partial-pressure difference $(P_s - P_a)/P_a$, and that the ratio Q^*/RT_m, which measures the relative importance of the temperature and pressure terms in the thermodynamic driving force, is strongly dependent on the identity of the gas being transferred. For most common gases, Q^* is a positive quantity at ordinary temperatures. Notable exceptions are SF_6, H_2 and He.

The above derivation has been criticised by Doney (1994, 1995a), who gives an alternative derivation in which the term involving the heat capacity is absent from the thermodynamic driving force of Equation 7.10, and who argues that the term Q on the right-hand side of Equation 7.12 is not to be identified with the heat of solution or condensation, but is in reality much smaller. By these expedients he arrives at an expression for the gas flux which is entirely consistent with Equation 7.11.

The first part of Doney's argument depends on his derivation of the result

$$\frac{T\sigma}{A} = -\frac{\Delta T}{T}J_{q,2} - RT\sum_i \frac{\Delta Pi}{Pi}J_i \qquad (7.13)$$

which is equation (22) of his paper (Doney, 1994). The quantity on the left-hand side of Equation 7.13 is termed the 'dissipation function' for the system. It is the product of temperature and the total rate of entropy production per unit area of interface due to the combination of the conductive heat flux $J_{q,2}$ and the diffusive fluxes J_i of gases through the interface. Note that the factors $\Delta T/T$ and $\Delta P_i/P_i$ are limiting forms of the logarithmic factors of Equation 7.11 for small pressure and temperature differences ΔP and ΔT. The significance of this equation is that all of the thermodynamic forces for a particular system can be extracted from the system's dissipation function. The result (Equation 7.13) is easily shown to be incorrect on the basis of elementary thermodynamics.

Equation 7.13 is notable for the fact that the term containing the gas fluxes does not involve the temperature difference across the interface. Therefore, according to Equation 7.13, the entropy production due to the gas fluxes must be *independent* of the temperature difference across the interface. If this were true, the entropy produced by transferring a small quantity of gas through the interface would be the same, for a given final temperature T_f, for two different initial gas temperatures T_1 and T_2. Since entropy is a state function, i.e. the difference in entropy between two states is independent of the path taken between the states, the entropy change for the gas going directly from T_1 to T_f via the interface must be the same as the entropy change for the gas going first from T_1 to T_2 at constant pressure, prior to passage through the interface, and then from T_2 to T_f via the interface. If the entropy changes were indeed identical for the two direct passages through the interface, this would require that the entropy of the gas did not change on going from T_1 to T_2 at constant pressure prior to passage through the interface. Such a conclusion is clearly incorrect and in fact amounts to a violation of the Second Law of Thermodynamics. Therefore Equation 7.13 is wrong.

The entropy change for a gas going from T_1 to T_2 at constant pressure is $C_p\,Ln(T_2/T_1)$, or $C_p\Delta T/T$ for a small temperature difference ΔT. As Doney points out, this term *is* present in the dissipation function that corresponds to Equation 7.9 above, and it ensures that the present treatment does not violate the second law. The error in Equation 7.13, which Doney arrives at by two different routes, arises in part from his inclusion of the entropy of reversible solution or condensation of the gas in the total entropy production and in part from his treatment of the enthalpy transferred along with the gas as if it were an external heat change q. Note also that the argument in the preceding paragraph is independent of the model chosen to describe the system, i.e. of whether the gas and liquid regions are semi-infinite, as in the present treatment, or finite sub-systems in a larger but still finite system, as in Doney's treatment.

Doney's second argument, that the quantity Q in Equation 7.12 is not to be identified with the heat of solution or condensation, is based on the statement that 'the bulk of the heat released or absorbed during the phase transfer comes from the liquid rather than the gas phase'. This statement derives from consideration of a detailed molecular model of the liquid surface and is at best irrelevant. Equation 7.2a states that the heat flux in the steady state has two components, one being the ordinary conductive flux and the other associated with the gas flux. The heat flux associated with the gas flux is identical, apart from a small correction (Phillips, 1994b), with the latent heat flux, so an equivalent way of expressing Equation 7.2a is to say that the total heat flux measured at some point below the interface is equal to the sum of the sensible and latent heat fluxes as measured at some point above the interface. As far as Equation 7.2a is concerned, it can make no difference precisely where the release of latent heat occurs in relation to the extremely thin region over which the medium changes from gas to liquid, or even whether such a region exists at all. Thermodynamics does not concern itself with models at the molecular level.

The conductance L_{22} can be evaluated for the present system by requiring Equation 7.11 to reduce to Fick's first law of diffusion for an isothermal system with a thin layer of gas at temperature T_m immediately adjacent to the surface. The result is

$$L_{22} = D_m C_m / R T_m \qquad (7.14)$$

where D_m is the gas-phase diffusion coefficient at temperature T_m and C_m is the gas-phase concentration corresponding to partial pressure P_m. Here we must use the *gas-phase* values of concentration and diffusion coefficient, because the irreversible part of the gas exchange is the transfer of the CO_2 from P_a, T_a in the air some distance away from the surface to P_m, T_m immediately adjacent to the surface. As noted earlier, the gas is assumed to be in equilibrium with the liquid for an extremely thin film at temperature T_m adjacent to the surface, and the change from the gas to the liquid phase produces no increase in the total entropy of the universe when gas and liquid are in equilibrium. This relieves us from having to estimate diffusion coefficients and other transport parameters for the difficult interface region over which the properties of the medium change very rapidly from those of a liquid to those of a gas. When the result (Equation 7.14) is inserted into Equation 7.11 and the thickness Δz is allowed to tend to zero, we obtain

$$J_2 = -D_m C_m \left[(Q^*/RT_m) \, T_m'/T_m + C_m'/C_m \right] \tag{7.15}$$

where a prime indicates a derivative with respect to height z and a subscript m now indicates a value measured in the gas phase at a point immediately adjacent to the surface. Equation 7.15 is the most useful form for numerical calculations.

Kinetics: 1. A surface-renewal model

Unlike the thermodynamics of gas exchange, which is independent of any assumed model of the surface, the kinetics of gas exchange can only be discussed in relation to a particular model. The Higbie–Danckwerts surface-renewal model (see Danckwerts, 1970) provides a fairly realistic picture of the nature of a liquid surface, which is assumed to consist of a mosaic of elements of liquid of different ages, with fresh parcels of bulk liquid being brought to the surface from time to time and the distribution $p(t)$ of ages t at any instant being given by

$$p(t) = \text{constant} \times t^{-1} \exp(-f\,t) \tag{7.16}$$

where f, the fraction of surface renewed in unit time, can be regarded as the reciprocal of an effective renewal time τ. This model suffers both from being over-simplified and from a shortage of experimental values of the parameter f, but it does provide an adequate and agreeably simple framework for numerical calculations. The boundary-layer model, which assumes a smooth, continuous change from molecular diffusion to eddy diffusion with increasing depth, is less convenient for our present purpose.

The surface-renewal model has been criticised by Hasse (1990), in comparison with the boundary-layer model, on the grounds that it cannot predict the observed enhancement of gas-transfer rate which occurs with the onset of capillary waves, because of the very different timescales for capillary waves (0.1 s) and surface renewal, when the renewal time is taken to be about 10 s at a wind speed of 6 m s^{-1}. This criticism depends in part on an identification of the surface-renewal time with the time taken for a particular depth of cool skin on the water to re-establish itself after it has been disturbed. Ewing and McAlister (1960) found the time for re-establishment of a cool surface layer to range from 5 to 40 s under the conditions of their experiments, in which they measured the intensity of infrared radiation emitted by the top 0.1 mm of liquid at low wind speeds. In fact, the critical liquid layer for transfer of CO_2 is much thinner than 0.1 mm at moderate wind speeds, and the observation that the onset of capillary waves with a timescale of the order of 0.1 s corresponds to the onset of enhanced gas transfer is probably better interpreted as an indication that the effective renewal time is about 0.1 s under these conditions. Taking the timescale to be 10 s for a layer of thickness 0.1 mm, the corresponding thickness for a timescale of 0.1 s is 0.01 mm, which is not unreasonable in the light of the present calculations.

The idea that the surface of the ocean has a 'cool skin' is of long standing (see, for example, Paulson and Simpson, 1981, and references therein), but only a few measured temperature profiles through the interface, with vertical resolution adequate to show the fine structure of the boundary layer, have been published (Azizyan *et al.*, 1984; Volkov and Soloviev, 1986). The surface layer, whose temperature is controlled in part by evaporation or condensation of water, in part by its proximity to air at a temperature different from that of the ocean and in part by emission or absorption of long-wavelength radiation, might be expected to have a significant effect on the conductance of the interface for a gas such as CO_2, and numerical calculations using a surface-renewal model show that this is indeed the case. In addition to the bulk air and sea temperatures and partial pressures of CO_2, and the average renewal time, other important parameters in these calculations include the relative humidity of the air, which will be expressed in terms of the dew-point temperature (the temperature at which liquid water would be in equilibrium with the existing partial pressure of water vapour in the air), and the thickness Δz of the stagnant air layer, above which the air is assumed to be well-mixed by turbulence.

Some results of numerical calculations of average CO_2 fluxes as a function of surface-renewal time are given in Figure 7.2 for a very thin stagnant air layer (four mean free paths of H_2O in air: $c.\ 4 \times 10^{-5}$ cm) and in Figures 7.3 and 7.4 for relatively thick stagnant air layers (10^{-3} and 10^{-2} cm, respectively). The surface-renewal time and stagnant layer thickness are adjustable parameters in these calculations. As explained elsewhere (Phillips, 1992), the values of temperature and CO_2 concentra-

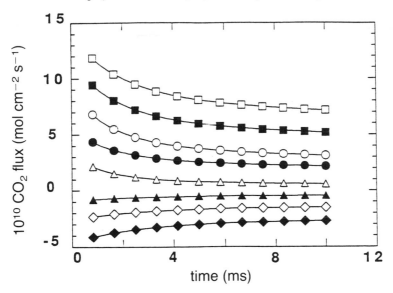

Figure 7.2. Calculated air–sea CO_2 flux as a function of surface-renewal time for fixed air and sea partial pressures of CO_2 (4.0 and 3.6×10^{-4} atm.) and bulk sea temperature (288 K), with a very thin stagnant air layer (4×10^{-5} cm). Values of air temperature and dew-point (K), reading from top to bottom: (278, 278), (288, 278), (298, 278), (288, 288), (298, 288), (278, 298), (288, 298), (298, 298).

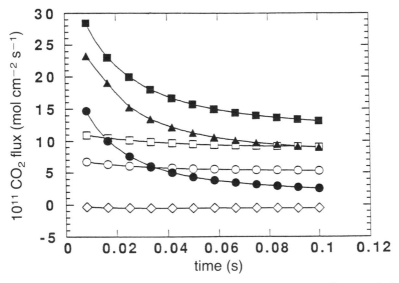

Figure 7.3. As in Figure 7.2, but for a thicker stagnant air layer (10^{-3} cm). Sea temperature 288 K. Values of air temperature and dew-point (K), reading from top to bottom: (298, 278), (298, 288), (298, 298), (278, 278), (278, 288), (278, 298).

L. F. Phillips

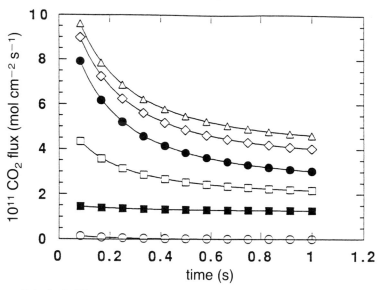

Figure 7.4. As in Figure 7.2, for a still thicker stagnant air layer (10^{-2} cm). Sea temperature 288 K. Values of air temperature and dew-point (K), reading from top to bottom: (298, 278), (298, 288), (298, 298), (288, 288), (278, 278), (278, 298).

tion at the liquid surface were adjusted at each integration step to conserve heat, including latent heat, and CO_2 fluxes.[†] Cooling of the surface by radiation was omitted, the effect of radiative cooling being subsumed in the similar effect produced by evaporation of water (Phillips, 1994c). Each of the curves in Figures 7.2 to 7.4 corresponds to a different combination of air temperature and dew-point, for the same value of bulk sea temperature (288 K) and the same air and sea partial pressures of CO_2 (4.0 and 3.6×10^{-4} atm., respectively). Coupling of heat and matter fluxes was included for the stagnant air layer adjacent to the interface, this layer being treated as a single volume element, and the ordinary equations for thermal conduction and for diffusion of a solute were used in other volume elements. Although the model is over-simplified and the range of dew-point temperatures somewhat greater than is likely to be encountered in practice, the qualitative conclusions to be drawn from these results are both relevant and interesting.

[†] Fortran listings of computer programs and typical input data files for calculations mentioned here can be obtained by e-mail from *phillips@chem.canterbury.ac.nz*

The results in Figure 7.2, for a very thin stagnant layer (4×10^{-5} cm) and short renewal times, are similar to those published previously (Phillips, 1992) and show the same reduction in air–sea flux of CO_2 when the sea surface is warmed by either conduction or condensation. The sign of the air–sea partial pressure difference is such that the CO_2 flux predicted by Equation 7.1 would be from air to sea. Nevertheless, for the three curves with high dew-point temperature (298 K), the flux is from sea to air. Also, although Equation 7.10 shows that the driving force for air–sea transport increases with increasing air temperature, this is outweighed by the effect of increasing air temperature on the temperature of the surface, so that the flux from air to sea decreases with increasing air temperature.

The results in Figure 7.3 are for a much thicker stagnant layer (10^{-3} cm) and longer renewal times, with the same sea temperature and CO_2 partial pressures as in Figure 7.2. For the three sets of data shown as filled points in Figure 7.3, the air temperature was 298 K. These points show similar behaviour to the results in Figure 7.2, with varying renewal time and dew-point temperature. The flux decreases markedly with increasing renewal time, for two reasons. One is that the liquid-phase concentration gradient of CO_2 at the surface rapidly decreases with time after a surface is freshly exposed to the air. The other is that the surface temperature increases with time as a result of conduction of heat from the air side of the stagnant layer. On moving from one curve to another among these three, the flux decreases with increasing dew-point because of the effect of evaporation or condensation on the temperature of the surface.

The three curves with unfilled points in Figure 7.3 were calculated for an air temperature of 278 K. For these curves, the effect of varying the dew-point is the same as for the other three, but the dependence of the flux on renewal time is much less because cooling of the surface by conduction offsets the effect of the decrease in CO_2 concentration gradient with time after the surface is exposed to air. For the lowest curve in Figure 7.3, the imbalance between cooling of the surface by evaporation and warming by conduction is such that the surface gradually warms, and the very small flux out of the sea actually increases with increasing renewal time. The effect of increasing air temperature in Figure 7.3 is in the opposite direction from that in Figure 7.2, because in Figure 7.2 the stagnant layer was thin enough for conduction of heat to the surface to be important, whereas in Figure 7.3 the surface temperature changes only slowly as a result of conduction, and the air temperature exerts its main effect via the thermodynamic driving force.

The results in Figure 7.4, for a still thicker stagnant layer (10^{-2} cm) and renewal times up to 1 s, show similar effects to those in Figure 7.3, but the effect of varying air temperature on the thermodynamic driving force is now much more important than its effect on the surface temperature, and the effect of varying dew-point temperature on the conductance of the surface is smaller because the flux of water vapour is much less with the thicker stagnant layer.

The pattern of temperature effects revealed in Figures 7.2 to 7.4 is quite complex. Nevertheless, it is clear that the flux of CO_2 is strongly affected by the surface temperature of the water. This is because the temperature dependence of the solubility of CO_2 in seawater is large and negative (Weiss, 1974) and, although the diffusion coefficient in the liquid does vary with temperature, the diffusion coefficient always remains a positive quantity. As mentioned before, the sea surface is in thermal equilibrium with the air on the sea side of the stagnant air layer and the partial pressures of water vapour and CO_2 on the sea side of the stagnant layer are also in equilibrium with the surface layer of water (see Figure 7.5). For a given partial pressure of CO_2 in the gas phase, if the temperature at the surface is high enough for the equilibrium concentration of CO_2 at the surface to be less than the bulk concentration of CO_2 in the liquid, the concentration gradient in the liquid will be in the right direction for the flux of CO_2 to be from sea to air, and the concentration gradient across the stagnant air layer will adjust itself as necessary to conserve flux. This is the origin of the negative fluxes found for the lowest curves in Figure 7.2.

The condition for a positive flux into the sea, namely, that the concentration of CO_2 at the surface should exceed the bulk concentration, can be written as

$$\alpha_{\text{surface}} \left(P_{\text{air}} + \delta \right) > \alpha_{\text{sea}} \, P_{\text{sea}} \qquad (7.17)$$

where α_{surface} and α_{sea} are the solubilities of CO_2 at the surface and bulk temperatures of the sea, respectively, and the CO_2 partial pressure difference δ across the stagnant air layer will usually be negligible in comparison with P_{air}. (See Figure 7.5; note that δ is negative in this figure.) When the opposite inequality holds, the direction of the CO_2 flux will be from sea to air, even if P_{air} exceeds P_{sea}. The strong dependence of the surface temperature on the flux of water vapour provides a kinetic (as opposed to thermodynamic) coupling between the fluxes of CO_2 and H_2O, a coupling which can cause CO_2 transport to occur in the direction opposite to the thermodynamic driving force for transport of CO_2 alone.

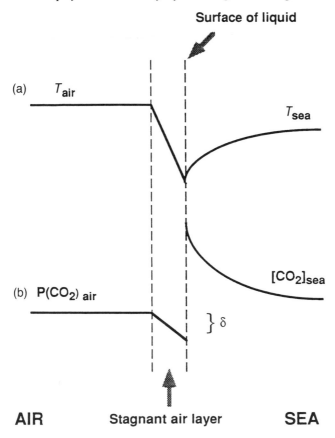

Figure 7.5. Diagrammatic profiles through the air–sea interface for the case of a surface layer which has been cooled by evaporation, with CO_2 flux into the sea. (a) The form of the temperature profile, with conductive heat flux towards the surface from both sides of the interface, as required to balance the latent heat flux. (b) The corresponding steady-state CO_2 partial pressure (air) and concentration (sea) profiles, showing the definition of the quantity δ, which appears in equation 7.17. When the flux is into the sea, δ is negative. The ratio of aqueous-phase concentration to gas-phase partial pressure of CO_2 at the interface is equal to the solubility α at the temperature of the liquid surface.

This situation is reminiscent of the well-known process of 'active transport' of ions such as K^+ through the membrane of a living cell, transport occurring in a direction opposite to that which would be predicted from the chemical potential gradient of the ion. Such processes are common in biology but are seldom encountered in non-living systems.

It is important to notice that the above conclusions, regarding the controlling influence of sea-surface temperature on the magnitude and direction of the CO_2 flux, are not dependent on the correctness of the surface-renewal model. The surface-renewal model merely provides a convenient framework for the calculations. The same conclusions would be reached for any model in which instantaneous fluxes of heat, water vapour and carbon dioxide were calculated, whether or not the distribution of times over which the fluxes were averaged happened to resemble the distribution in Equation 7.15.

According to the data of Weiss (1974), a temperature change from 10 °C to 8 °C produces a 7% increase in the solubility of CO_2 in seawater, a change which will often exceed the percentage difference between P_{air} and P_{sea}. Ewing and McAlister (1960) observed the first 0.1 mm of liquid to be cooler than the bulk by 0.6 °C at very low wind speeds. Schluessel *et al*, (1990) observed bulk-to-skin temperature differences in the North Atlantic Ocean ranging between $+1.0$ K and -1.0 K during a six-week period, but their measurements too did not yield actual temperature profiles. The surface-to-bulk temperature increases shown by the profiles of Azizyan *et al.*, (1984) are only of the order of 0.5 °C; however, most of their results were obtained at rather low wind speeds and the temperature changes occupy a much greater distance than in most of the present theoretical calculations. The average renewal times, τ, for the air and water layers near the interface, estimated from their profiles by means of the formula from diffusion theory

$$z^2 = 2D_x\tau \qquad\qquad (7.18)$$

where z is the rms distance over which the temperature change occurs and D_x is the thermal diffusivity (thermal conductivity divided by the product of density and specific heat), are both of the order of one second, corresponding to the longest timescale of the present calculations.

Calculated values of CO_2 flux for fixed values of renewal time are shown in Figures 7.6 to 7.8 as a function of air temperature for different dew-point temperatures, with the same stagnant-layer thicknesses as in Figures 7.2 to 7.4. Also shown are the results of similar calculations in which the coupling of heat and matter fluxes, via the factor Q^*, was set to zero. These graphs demonstrate the interaction between the two opposite effects of varying air temperature, namely, the effect arising from coupling of the heat and matter fluxes and the effect of changing the temperature of the liquid surface by conduction. The effect of increasing the dew-point temperature is always to decrease the flux from air to sea (or increase the flux from sea to air) but the direction of the effect of increasing the air temperature varies with the thickness of the stagnant air

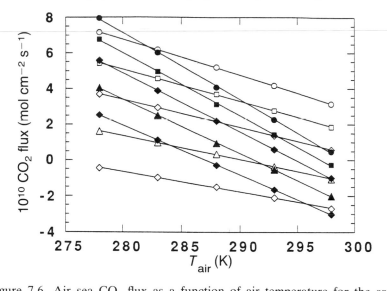

Figure 7.6. Air–sea CO_2 flux as a function of air temperature for the same conditions as in Figure 7.2 (renewal time 10^{-2} s, air and sea partial pressures of CO_2 4.0 and 3.6×10^{-4} atm., respectively, bulk sea temperature 288 K, and stagnant air layer thickness 4×10^{-5} cm). Filled points have Q^* set to zero. Dew-point temperatures, reading from top to bottom: 278, 283, 288, 293 and 298 K.

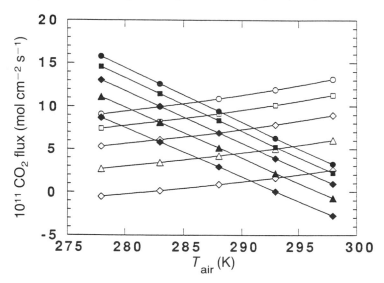

Figure 7.7. Air–sea CO_2 flux as a function of air temperature for a renewal time of 0.1 s, air and sea partial pressures of CO_2 equal to 4.0 and 3.6×10^{-4} atm., respectively, bulk sea temperature 288 K, and stagnant air layer thickness 10^{-3} cm, as in Figure 7.3. Filled points have Q^* set to zero. Dew-point temperatures, reading from top to bottom: 278, 283, 288, 293 and 298 K.

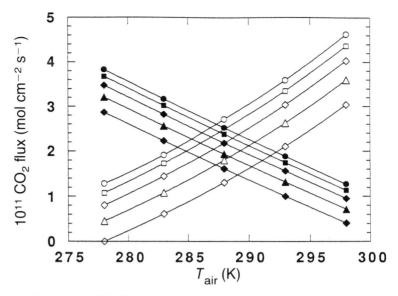

Figure 7.8. Air–sea CO_2 flux as a function of air temperature for a renewal time of 1.0 s, air and sea partial pressures of CO_2 equal to 4.0 and 3.6 × 10^{-4} atm., respectively, bulk sea temperature 288 K, and stagnant air layer thickness 10^{-2} cm, as in Figure 7.4. Filled points have Q^* set to zero. Dew-point temperatures, reading from top to bottom: 278, 283, 288, 293 and 298 K.

layer. When the stagnant air layer is thin enough for the surface to be heated or cooled significantly by conduction during a time of the order of the surface-renewal time, the effect of changing the surface temperature predominates and the flux decreases with increasing air temperature. When the stagnant layer is much thicker, the effect of coupling predominates and the flux increases with increasing air temperature. When the coupling factor Q^* is set to zero (actually, setting Q^* to zero turns off both the coupling of fluxes in the Onsager sense and the effect of temperature on the chemical potential difference that is responsible for ordinary diffusion) only the interaction between the air and surface temperatures remains.

Despite the complications arising from variations in the dew-point of the air and in the air–sea temperature difference, the present calculations show that the CO_2 flux does vary linearly with the CO_2 partial pressure difference between air and sea when the other variables are held constant. This is demonstrated in Figure 7.9 for a range of dew-point temperatures, with air and sea temperatures fixed at 288 K, a renewal time of 0.1 s, and the same stagnant-layer thickness as in Figures 7.3 and

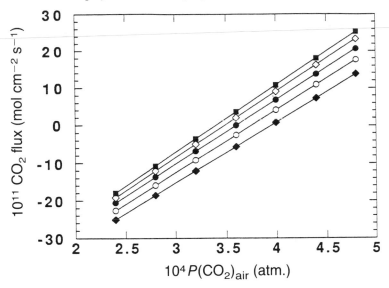

Figure 7.9. Air–sea CO_2 flux as a function of the partial pressure of CO_2 in air, for fixed sea partial pressure (3.6×10^{-4} atm.) and fixed air and sea temperatures (both 288 K). Renewal time (0.1 s) and stagnant layer thickness (10^{-3} cm) as in Figures 7.3 and 7.6. Lines correspond to dew-point temperatures of 278, 283, 288, 293 and 298 K, reading from top to bottom.

7.6. The flux is *proportional* to the air–sea partial pressure difference, as implied by Equation 7.1, only when the dew-point and air temperatures are identical with the sea temperature.

The effect of varying the thickness of the stagnant air layer is shown in Figure 7.10 for three different combinations of air, sea and dew-point temperatures. These results were calculated for a renewal time of 0.1 s and for the same air and sea partial pressures of CO_2 as in Figures 7.2–7.4 and 7.6–7.8. When the air and dew-point temperatures are the same as the sea temperature (open circles in Figure 7.10: all three temperatures 288 K) the CO_2 flux is found to be independent of the thickness of the stagnant air layer for thicknesses less than about 1 mm (a very large value), and does not begin to fall off markedly until the thickness exceeds 3 mm. This is consistent with the usual view that the main resistance to CO_2 exchange is to be found on the sea side of the interface. Thus the *direct* effect on the CO_2 flux of varying the thickness of the stagnant air layer is negligible. However, as we have seen, indirect effects can occur as a result of variations in either the latent or sensible heat fluxes to the surface.

When the dew-point is markedly less than the air temperature (filled diamonds in Figure 7.10: dew-point 278 K, other temperatures 288 K),

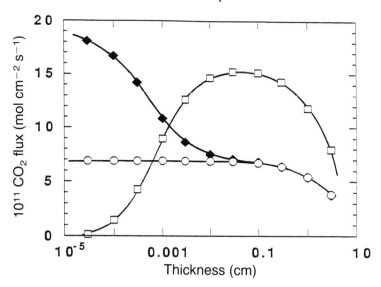

Figure 7.10. Air–sea CO_2 flux as a function of the thickness of the stagnant air layer, for a renewal time of 0.1 s, with air and sea partial pressures of CO_2 equal to 4.0 and 3.6×10^{-4} atm., respectively, and a bulk sea temperature of 288 K. Unfilled circles: air temperature and dew-point both 288 K. Filled diamonds: dew-point 278 K, air temperature 288 K. Unfilled squares: air temperature 298 K, dew-point 288 K.

varying the thickness of the stagnant air layer produces a corresponding variation in the flux of water vapour and so alters the temperature of the surface layer, provided the surface layer is not so thick that the latent heat flux is negligible. Thus, for a thin stagnant layer (less than about 0.1 mm) the CO_2 flux is enhanced by cooling of the surface, and the enhancement is greater for a thinner stagnant air layer.

When the air temperature is markedly different from the sea temperature (unfilled squares in Figure 7.10: air temperature 298 K, sea and dew-point temperatures both 288 K), varying the thickness of the stagnant air layer varies the extent to which the temperature of the surface layer of liquid is altered by conduction, and so affects the flux of CO_2. In Figure 7.10, for a thin layer (again less than about 0.1 mm), the flux into the sea is reduced because of heating of the surface. The flux calculated for a thick layer, with an air temperature of 288 K, is greater than that found for the other two sets of data because of the greater thermodynamic driving force, but the fall-off with increasing stagnant layer thickness still sets in at a thickness of about 3 mm.

The effect of radiative cooling of the surface has been investigated for the surface-renewal model by assuming a typical value of 10 μm (P.

Schluessel, pers. commun.) for the *e*-folding distance for absorption of radiation in the liquid and adding the appropriate terms to the heat flux at different depths (Phillips, 1994c). The effect of the radiation is to increase the cooling of the liquid surface and so increase the gas flux into the liquid or decrease the flux from liquid to gas. The relative magnitudes of the effects of surface cooling by varying the dew-point and varying the intensity of long-wavelength radiation are compared in Figure 7.11. The range of values of radiation intensity is considerably larger than even the short-term (5-minute) averages found experimentally (Schluessel *et al.*, 1990), but it should be borne in mind that the timescale of these calculations is much shorter than the timescale of their observations (surface-renewal time 0.1 s) so these values are probably not too unreasonable. The results in Figure 7.11 show that the effect of radiative cooling on the CO_2 flux is linear, that it is likely to be indistinguishable in practice from the effect of cooling by evaporation of water, and that the radiative effects are small enough to be negligible in most cases.

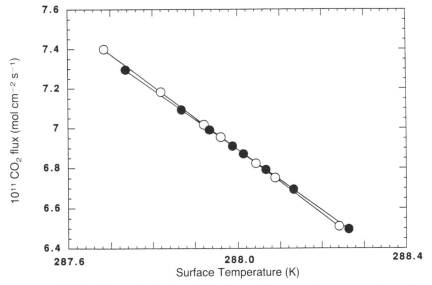

Figure 7.11. Variation of calculated CO_2 flux with sea-surface temperature, the surface temperature being varied either by changing the dew-point of the air (unfilled circles) or by varying the intensity of radiation (filled circles). Air and sea temperatures both 288 K, air and sea partial pressures of CO_2 3.6 and 3.2×10^{-4} atm., surface renewal time 0.1 s, thickness of stagnant air layer 0.1 mm. Filled circles are for dew-point 288 K and integrated radiative fluxes -2000, -1000, -500, -100, $+100$, $+500$, $+1000$, and $+2000$ W m^{-2}, reading from left to right on the plot. Unfilled circles, zero radiation flux and dew-point temperatures of 278, 283, 286, 287, 289, 290 and 293 K, reading from left to right.

Kinetics: 2. Incorporation of a turbulent air layer

The model used in this section includes an upper layer of turbulent air in which the profiles of temperature and gas concentration are given as a function of height z by the standard equation of 'K-theory' (see, for example, Panofsky and Dutton, 1984):

$$(q-q_0)/q^* = (1/k_a)\{Ln\ (z/z_0) - \Psi(z/L)\} \tag{7.19}$$

where q is the value of the particular intensive variable, q^* is equal to the flux associated with the profile of q divided by u^*, u^* being the friction velocity, with an additional divisor $\varrho\sigma$ (density times specific heat) when q is temperature, k_a is Von Karman's constant, which we set equal to 0.4, L is the Monin–Obukov length, z_0 is the roughness length, and the quantities on the right-hand side of Equation 7.19 are assumed to be the same for heat and matter fluxes. For a stable temperature profile ($T_{air} > T_{sea}$) we use, following Panofsky and Dutton,

$$\Psi(z/L) = -5z/L \tag{7.20}$$

and for an unstable profile

$$\Psi(z/L) = 2\ Ln\ [0.5 + 0.5(1 - 16z/L)^{1/2}] \tag{7.21}$$

Equation 7.19 becomes invalid when the value of $-z/L$ is greater than about 2.

Beneath the turbulent layer, within which the bulk concentrations and temperature are measured, is a thin, stagnant layer of air where transport is controlled by an ordinary diffusion coefficient or a thermal diffusivity. Equation 7.19 is used to relate the concentration or temperature measured in the turbulent region to the value at the top of the stagnant layer. The height z_{1q} at which a transition from turbulent to molecular transport occurs for the flux associated with variable q is given implicitly by

$$D_q = k_a u^*\ Z_{1q}/\phi_m(z_{1q}) \tag{7.22}$$

where D_q is either a diffusion coefficient or a thermal diffusivity and $\phi_m(z)$ is either $(1-16z/L)^{-1/2}$, in unstable air (negative L), or $(1+5z/L)$ in stable air. (Note that Equation 7.22, which is obtained from Panofsky and Dutton's equations 6.9.4 and 6.9.5 by assuming that D_q and the coefficient for eddy diffusion become equal at the transition height z_{1q}, fails for stable air when $5D_q/k_a u^* L$ exceeds unity, making z_{1q} negative. Note also that Equation 7.22 does not imply that Fick's law of diffusion applies to gas transfer across the stagnant air layer except in the special case where there is no conductive heat flux.) Hence we define the quantities Ω_q

$$\Omega_q = (q_a - q_1)/q^* = (1/k_a) \{Ln \ (z_a/z_{1q}) - \Psi(z_a/L) + \Psi(z_{1q}/L)\} \quad (7.23)$$

which relate the intensive variable q_a measured at height z_a in the turbulent region to q_1 at the transition height z_{1q}.

To complete the model, the temperature of the liquid surface is assumed to be fixed by the steady-state fluxes of sensible and latent heat at the surface, thermal and vapour-pressure equilibrium are assumed for the thin film of gas immediately adjacent to the surface and the fluxes in the liquid layer immediately beneath the surface are controlled by the coefficients for either molecular or eddy diffusion together with the values of the intensive variables q in the bulk liquid and at the surface. The discussion which follows will be given in terms of the exchange of carbon dioxide but applies as well to other gases.

The flux of sensible heat above the liquid surface is given by

$$J_{ha} = -\varkappa_a \ T_m' \quad (7.24)$$

and the flux of water vapour by

$$J_w = -D_w w_m \ (w_m'/w_m + Q_w^* \ T_m' \ / \ R \ T_m^2) \quad (7.25)$$

where the symbols are as defined above and Equation 7.25 is the water-vapour counterpart of Equation 7.15.

Combining the two versions of Equation 7.23 which apply to the profiles of temperature and water vapour concentration gives

$$w_a - w_1 = f_r \varrho\sigma \ \Omega_w(T_a - T_1)/\Omega_T \quad (7.26)$$

where f_r is the flux ratio J_w/J_h, $\varrho\sigma$ is the product of density and specific heat for air, w_a and T_a are the water vapour concentration and temperature, as measured at a height which was 10 m in the eddy-correlation studies of Smith and coworkers (Smith and Jones, 1986; Smith *et al.*, 1991) and 17 cm in the wind tunnel studies of Liss *et al.* (1981), and w_1 and T_1 are the values at the transition heights z_{1W} and z_{1T}, respectively.

The ratio of the gradients of water vapour concentration w_m and temperature T_m on the air side of the interface is given approximately by

$$w_m' \ / \ T_m' = (w_m - w_1) \ z_{1T}/(T_m - T_1) \ z_{1W} \quad (7.27)$$

The approximation is expected to be a very good one when the stagnant layer is thin or in the long-time limit when the profiles are linear. In view of Equations 7.24 and 7.25, this ratio of gradients is also given by

$$w_m{}' / T_m{}' = f_r x_a / D_w - w_m Q_w{}^*/R\,T_m{}^2 \qquad (7.28)$$

If we combine Equations 7.27 and 7.28 and then use Equation 7.26 to eliminate w_1, the following expression for T_1 is obtained:

$$T_1 - (w_a - w_m + \mu T_m - \lambda\,T_a)/(\mu - \lambda) \qquad (7.29)$$

where we have written λ in place of $\varrho\sigma f_r \Omega_w/\Omega_T$ and μ in place of $(z_{1w}/z_{1T})(f_r x_a/D_w - w_m Q_w{}^*/RT_m{}^2)$. It is a helpful feature of Equation 7.29 that the quantities Ω_q and z_{1q}, and the fluxes of heat and water vapour, enter the equation only as ratios. In the rare cases where the calculation of z_{1q} by Equation 7.22 fails, we make the approximation of setting the ratios Ω_w/Ω_T and z_{1w}/z_{1T} equal to unity in Equation 7.29.

On the water side of the interface we have for the heat flux

$$J_{hs} = J_{ha} + Q_w J_w = - x_S\,T_{sm'} \qquad (7.30)$$

and for the flux of carbon dioxide

$$J_{cs} = -D_{cs} C_{sm'} \qquad (7.31)$$

where the subscript s identifies values on the water side of the interface and again the subscript m indicates that we are considering gradients that obtain right at the surface. (Note that any heating or cooling effect of long-wavelength radiation can be ignored in Equation 7.30. We aim to scale the CO_2 flux to the conductive heat flux immediately below the surface of the liquid. The conductive flux is the total heat flux minus the radiative flux, and is equal to the sum of the sensible and latent heat fluxes above the surface, which again is the total heat flux minus the radiative flux.) If ordinary diffusion is involved, D_{cs} is the diffusion coefficient of carbon dioxide in water. In the case of eddy diffusion, D_{cs} is set equal to D_{xs}, the thermal diffusivity, given by $x_s/\varrho\sigma$, where ϱ and σ are now the density and specific heat of liquid water. (Note that in practice the actual values of eddy diffusion coefficients do not enter the calculations because the total heat flux J_{hs} provides a calibration.)

If the profiles of temperature and concentration on the water side of the interface are assumed to be linear and to refer to the same thickness of water above a mixed layer, the gradients of temperature and carbon dioxide concentration at the surface are related by

$$C_{ms}{}' / T_{ms}{}' = (C_{ms} - C_s)/(T_m - T_s) \qquad (7.32)$$

where C_s and T_s are the bulk values of CO_2 concentration and temperature for the water and C_{ms} is the surface concentration of CO_2 in the water. Alternatively, if the profiles are assumed to be of the form

$$q = q_s + (q_{ms} - q_s) \, erfc[z_s/(2D_{qs}\tau)^{1/2}] \qquad (7.33)$$

as would be appropriate for a surface-renewal model with an average renewal time τ, where z_s is measured *into* the sea, the right-hand side of Equation 7.33 has an additional factor $(D_{xs}/D_{cs})^{1/2}$ where D_{xs} is the thermal diffusivity of water. For a boundary layer model this additional factor becomes $(D_{xs}/D_{cs})^{2/3}$ (Hasse, Chapter 4, this volume).

The initial motivation for the calculations with turbulence included was a desire to avoid considering the effects of such variables as wind speed, nature of the wave field, extent of surface contamination, etc., by calculating the surface temperature T_m from the measured air and sea temperatures together with experimental values of the sensible heat and water vapour fluxes. This could be done in principle by using Equation 7.19 to calculate the temperature T_1 at the height z_{1T} obtained from Equation 7.22, and then using the value of z_{1T} together with the known temperature gradient (calculated from the conductive heat flux through the stagnant layer and the thermal conductivity of air) to calculate the surface temperature T_m. The surface temperature fixes the surface value of the solubility α, which in turn fixes the ratio of liquid and gas phase concentrations at the surface. Hence, given the surface temperature, it becomes possible to calculate the flux of the trace gas by scaling the dissolved gas flux (via the ratio of thermal and molecular diffusivities and the ratio of temperature and concentration gradients given by Equation 7.32) to the heat flux in the liquid layer immediately below the surface, and then adjusting the surface concentration iteratively until the fluxes of CO_2 given by Equation 7.31 on the water side, and by Equation 7.15 on the gas side, are equal, which gives the steady-state CO_2 flux through the interface. (In practice, because the flux of CO_2 is unknown, the profile of temperature above the surface is used initially to scale the CO_2 pressure as a function of height in order to obtain C_1, rather than using the CO_2 analog of Equation 7.25. In cases where the calculated value of T_1 is outside the range between T_m and T_a the value of C_1 is initially set equal to C_a.) Thus, in principle, it should be possible to calculate the CO_2 flux from a knowledge of the heat and water vapour fluxes and the bulk air and water temperatures and concentrations, together with u^* and the known functions Ψ and Φ. Unfortunately, in practice this procedure does not work very well with the available experimental data, because the extrapolation to z_{1T} using Equation 7.22 is too long to yield an accurate value of T_1, and so a different approach is required. The alternative approach that has been followed is to use measured values of transfer velocity to select the best value of the surface temperature T_m and then use this value of surface temperature to calculate the corresponding values of temperature and water vapour concentration at the transition heights z_{1T} and z_{1w}. These transition values are of no great interest in themselves, but they do provide criteria for deciding that a calculation has failed, as happens when the calculated value of water vapour concentration at z_{1w} is negative, or when the value of T_1 is outside the range between T_m and the air temperature T_a by more than some arbitrary amount (normally 3 K). The range of possible surface temperatures is usually defined quite closely by the directions of the heat and water vapour fluxes and the air and water temperatures. In cases where the allowable range is open-ended above or below, the range is artificially limited to values within 3 K of the bulk water temperature.

For a given set of experimental data, the range of possible surface temperatures is restricted as closely as possible by consideration of the directions of the sensible and total heat fluxes and the water vapour flux. The surface temperature is then stepped slowly through the available range, values of T_1 and w_1 being calculated for each value of T_m by means of Equations 7.28 and 7.25, respectively. A value of T_m is rejected if the corresponding value of w_1 is negative or if the value of T_1 is outside the range between T_m and T_a by more than the specified allowance.

In cases where no acceptable values of surface temperature can be found, or where K-theory does not work because the value of $-z_a/L$ exceeds 2 (this case was not encountered in the present work), the following alternative procedure is available. First, the quantities T_1, w_1 and C_1 are set equal to the bulk values T_a, w_a and C_a on the air side of the interface. Then, for each value of T_m in the allowed range, an effective thickness dz_{air} for the stagnant air layer is calculated from the known value of T_m' and the temperature difference $T_a - T_m$. This fixes the value of w_m' as $(w_m - w_a)/dz_{air}$ and so enables the water vapour flux J_w to be calculated by Equation 7.24. Values of T_m are then rejected if the calculated value of the flux ratio f_r differs from the experimental value by more than a factor of 2.

For each acceptable value of surface temperature, the flux J_c of the gas of interest is calculated by the iterative procedure outlined above and a best value of surface temperature is selected as the one which gives a value of J_c and/or K_w in best agreement with the experimental data. Any deficiency of the model should then be reflected in either an inability to calculate T_1 or an inability to reproduce the experimental value of transfer velocity.

The computer program which performs the above calculations auto- matically tests a range of options which include setting either Q_w^* or both Q_w^* and Q_c^* to zero in order to turn off the coupling of heat and matter fluxes, having either eddy or molecular diffusion on the water side of the interface, and having gradients of temperature and concentration at the water side of the interface that scale according to boundary-layer, surface-renewal (*erfc*) or linear (the long time limit of an *erfc* profile) profiles. For 'eddy diffusion', the diffusion coefficient of the dissolved gas is set equal to the thermal diffusivity. This amounts to assuming that transfer of heat and dissolved gas occurs by bulk turbulence in the liquid rather than by molecular diffusion or conduction, an assumption which will often be correct for air–sea exchange in a surf zone.

We have now reached the point where comparison of theory with experiment is possible. Of the numerous sets of measurements of gas– water exchange rates in the literature, only three have been found in which simultaneous fluxes of heat and water vapour were also measured. These are the wind tunnel measurements of oxygen exchange by Liss *et al.* (1981) and the eddy-correlation measurements of air–sea exchange of carbon dioxide by Smith *et al.* (1991) and Smith and Jones (1985). The results presented here differ in detail from those given elsewhere (Phillips, 1994a), because of an error in the solubility function for oxygen which was not discovered until after that paper had gone to press, but the general conclusions remain unaltered. The most significant difference from the results already published is that the data of Liss *et al.* no longer favour a linear model over a surface-renewal model, and that a boundary- layer model is found to fit the data equally well.

Comparison of theory and experiment

General comments

The poor agreement with equation 7.1 that is shown by the data of Figure 7.1 might well be regarded as evidence in support of the present theory. More direct evidence is provided by the work of Liss *et al.* (1981), who performed a careful series of experiments in which gas–water fluxes of heat, water vapour and oxygen were measured simultaneously in a wind tunnel. They reported that the rate of transfer of oxygen from air to water was reduced by more than 30% under condensing conditions, and referred to an observation by Hoover and Berkshire (1969) that air–water transfer of CO_2 in a wind tunnel was effectively inhibited when the dew-point of the air was higher than the water temperature. Liss *et al.* showed that the origin of this effect was not an enhancement of gas transport by increased mixing in the surface layer in the presence of evaporative cooling, as proposed by Quinn and Otto (1971), and instead attributed the effect to a reduction in transfer rate due to the inability of turbulence to propagate close to the interface under conditions of stable stratification. This is equivalent to a statement, in terms of the present theory, that the thickness of the stagnant air layer was greater under condensing conditions. However, the present surface-renewal calculations indicate that the flux does not depend directly on the thickness of the stagnant layer, and it would seem far more likely that the rate of transfer was less under condensing conditions simply because of heating of the surface layer by condensation. This conjecture has been verified by detailed calculations for the model which includes gas-phase turbulence.

Eddy correlation measurements of air–sea CO_2 fluxes by Smith and Jones (1985) gave many negative flux values, and average fluxes close to zero, despite constant supersaturation of the seawater with CO_2. These observations can be understood, at least qualitatively, in terms of the present theory, but not on the basis of Equation 7.1. Their discussion of the results in terms of bubble injection leading to supersaturation was strongly criticized, probably with justification, by Broecker *et al.* (1986), but Broecker *et al*'s. discounting of all eddy-correlation measurements over the ocean, mainly because of the high instantaneous values of CO_2 flux found by Jones and Smith (1977), Wesely *et al.* (1982), Bingham (1982), and Smith and Jones (1985), does not seem justified. As was pointed out by Smith and Jones (1986) and Wesely (1986), high instantaneous values are not necessarily incompatible with low global-average values. It is interesting that the eddy-correlation results of Smith and Jones were thought to be affected by the presence of a nearby surf zone. It is also interesting to note that, although air–sea partial-pressure differences were not measured in the earlier work, large CO_2 fluxes out of the ocean were found under conditions where the ocean surface was warming and, in the case of the measurements by Jones and Smith (1977), in the presence of high humidity or fog.

Results of improved eddy-correlation measurements by Smith *et al.* (1991) show both a poor correlation of the transfer velocity with wind speed and a clear tendency for smaller or negative CO_2 fluxes to be found under conditions of larger positive latent heat flux (the flux of 'sensible heat' was fairly constant during these measurements). The results include one data point for which the direction of the flux is clearly opposed to the negative partial pressure gradient and several for which the flux is close to zero, despite a large value of the product of wind speed and partial pressure difference. As mentioned earlier during discussion of the nature of the steady state, their observation of an apparent negative correlation between CO_2 flux and the hourly rate of change of wind speed might be related to the need for a finite time to elapse before a steady-state regime is established, and warrants further investigation.

Detailed comparison with the data of Liss *et al.* *(1981)*

Results of the present calculations for the wind tunnel data of Liss *et al.*, using the model which incorporates a turbulent air layer and assuming *erfc* profiles on the sea side of the interface, are summarized in Table 7.1. The first column of Table 7.1 contains the run numbers assigned by Liss *et al.*, and the next three columns contain the experimental values of air temperature, water temperature and dew point. Column 5 contains the measured values of transfer velocity K_w. The sixth and seventh columns give the values of T_m and T_1 for which the calculated transfer velocity for oxygen was equal to the experimental K_w. The last column refers to notes which are given in the caption to Table 7.1.

The main points that can be made about this table are as follows:

1. No calculation was possible for run 30 because the measured flux of water vapour was negative, despite the air and water temperatures being equal and below the dew-point of the air, so that water vapour should have been condensing rather than evaporating. For all of the remaining experiments it was possible to obtain acceptable values of T_m and T_1 and to reproduce the measured transfer velocity exactly on the basis of the present theory, with coupling of heat and matter fluxes turned on, with ordinary diffusion on the water side of the interface, and with *erfc* or boundary layer profiles of temperature and oxygen concentration on the water side of the interface. With linear profiles on the water side of the interface it was possible to reproduce the experimental transfer velocity for all except run 2. When eddy diffusion was assumed on the sea side of the interface the experimental value of K_w could not be reproduced for runs 1, 2 or 12.

Table 7.1 *Calculated surface temperature* (T_m) *and transition temperature* (T_1)
to fit data of Liss et al. *(1981), assuming ordinary diffusion and* erfc *profiles of
temperature and concentration on the sea side of the interface*

Run	T_a	T_s	Dew pt.	K_w	T_m	T_1	Notes
1	288.25	288.05	283.85	1.89	287.977	287.467	4
2	286.95	287.35	283.85	2.02	287.288	288.522	3, 4
3	299.55	288.35	291.85	1.78	290.302	290.785	
4	299.75	288.25	292.55	1.90	288.461	288.397	2, 4
5	279.95	288.25	278.05	2.01	287.849	286.395	2
6	280.95	287.45	278.25	2.02	287.100	285.628	2
7	287.45	289.05	287.15	2.06	288.978	288.932	
8	294.95	288.25	293.75	1.72	288.665	290.372	1
9	298.15	288.75	296.75	1.59	289.455	292.420	1
10	281.65	287.55	279.45	3.04	287.056	285.545	2
11	284.55	287.85	283.95	2.92	287.560	286.620	2
12	287.95	288.15	287.15	2.96	288.114	288.502	4
13	297.15	287.65	291.75	2.57	288.249	289.833	1
14	301.45	287.45	294.75	2.31	288.568	290.966	1
15	302.35	287.95	296.25	2.10	289.326	292.291	1
16	298.35	286.95	295.75	3.02	287.965	290.911	1
17	282.95	288.05	280.95	11.3	287.868	286.690	1
18	281.95	287.75	280.65	9.10	287.515	286.183	2
19	288.75	288.45	288.95	10.3	288.468	288.635	
20	287.95	287.35	288.85	9.18	287.397	287.605	2
21	305.06	286.95	296.15	5.45	287.947	290.955	1
22	303.15	288.05	297.85	6.45	288.843	291.330	1
23	301.05	287.65	296.95	7.29	288.294	291.728	1
24	282.55	287.15	281.85	26.6	287.062	285.728	2
25	284.25	288.25	283.75	23.5	288.160	287.124	2
26	288.25	288.15	288.15	25.6	288.151	288.156	1
27	300.15	289.45	293.25	23.2	289.612	291.137	1
28	296.85	288.25	296.05	24.8	288.450	291.054	1
29	283.55	288.45	282.95	38.2	288.365	286.898	1
30	288.35	288.35	288.75	37.9	—	—	5
31	302.35	289.25	295.95	38.4	289.428	291.201	1

Units of K_w, cm h^{-1}. *Notes:* 1. Cannot calculate T_1 without flux coupling for
water. 2. Cannot reproduce K_w without flux coupling for oxygen. 3. Cannot
reproduce K_w exactly with ordinary diffusion and *linear* profiles. 4. The
calculated value of T_1 does not lie between T_a and T_m. 5. T values inconsistent
with water vapour flux.

2. For 15 of the experiments, including most of those in which there was a
   sizeable gas–water temperature difference (runs 8, 9, 13–17, 21–23,
   26–29 and 31), it was not possible to obtain an acceptable value of T_1
   unless coupling of heat and matter fluxes was included for water
   vapour (note 1). Apart from run 30, for which the experimental data

are not self-consistent, there are no runs for which the calculation of T_1 failed when flux coupling was included.

3. For nine of the remaining 15 experiments (runs 4–6, 10, 11, 18, 20, 24 and 25), although the calculation of T_1 was deemed to be successful (i.e., T_1 was not too far outside the range from T_m to T_a) without coupling of the heat and water vapour fluxes, it still was not possible to duplicate the measured value of K_w unless flux coupling was turned on for oxygen (note 2).

4. For four of the experiments (runs 1, 2, 4 and 12) the value of T_1 given by Equation 7.29 does not lie between the values of T_m and T_a. (However, when coupling of heat and matter fluxes was omitted the calculated value of T_1 was outside the range between T_m and T_a for *every* experimental run.) The situation for water vapour is less satisfactory than for temperature: the calculated value of w_1 is outside the range from w_m to w_a for 11 of the 30 useable runs. Evidently, there is still room for improvement in the theory.

6. The best value of surface temperature T_m is often significantly different from the temperature T_s of the bulk liquid. Extreme cases are run 3, in which the surface was heated by nearly 2 K, and run 10, in which the surface was cooled by 0.5 K. Previously (Phillips, 1994a), these differences were over-estimated because of an error in the solubility function for O_2, of which more later. As noted above, cooling the surface tends to increase the gas flux into the water, while heating the surface has the opposite effect.

Detailed comparison with data of Smith and Jones (1985)

The results of the calculations of T_1 and T_m for this data set are given in Table 7.2, in similar format to Table 7.1, with the major difference that eddy diffusion was assumed on the sea side of the interface. Smith and co-workers report measured gas fluxes rather than transfer velocities. For our present purpose the fluxes have been converted to transfer velocities via Equation 7.1. The following points can be made about the results in Table 7.2:

1. The range of air and sea temperatures is rather narrow in these field measurements. For many of the runs the calculated value of T_1 lies outside the range between T_m and T_a and for six of the most interesting experimental runs, those leading to negative K_w values, the calculations initially gave K_w values with the wrong sign. The most likely source of this disagreement is thought to be the experimental values of

L. F. Phillips

Table 7.2 *Calculated surface temperature* (T_m) *and transition temperature* (T_1) *to fit data of Smith and Jones (1986), with assumptions of eddy diffusion and linear temperature and concentration profiles, on the sea side of the interface*

Run	T_a	T_s	Dew pt.	K_w	T_m	T_1	Notes
8	282.75	283.45	278.63	38.0	283.327	284.476	1, 4
9	283.35	283.45	281.19	75.8	283.429	285.121	1, 4
12	280.95	282.75	278.32	89.1	282.678	284.135	1, 4
13	280.05	282.75	280.28	27.2	282.584	282.233	1
14	280.25	282.75	279.91	38.8	282.512	282.545	4
15	277.85	280.85	274.43	−136	277.090	277.095	2, 5
16	279.35	280.85	279.94	−23.5	278.341	278.353	2, 5
17	279.85	280.95	275.48	−173	279.012	279.015	2, 5
18	280.35	281.05	276.96	−11.8	279.573	279.587	2, 5
19	280.25	281.35	278.50	−75.4	279.875	279.876	2, 5
20	281.15	281.15	278.96	−30.7	280.874	280.875	2, 4
21	281.15	280.95	278.12	39.0	280.843	277.852	1, 3, 4
22	281.15	280.85	277.90	50.6	280.775	279.746	4

Units of K_w, cm h^{-1}. *Notes:* 1. Cannot calculate T_1 without flux coupling for water. 2. Cannot reproduce K_w without flux coupling for carbon dioxide. 3. Cannot reproduce K_w exactly with eddy diffusion and *erfc* profiles. 4. The calculated value of T_1 does not lie between T_a and T_m. 5. Calculated values of K_w had the wrong sign: changed sign of sensible heat flux.

sensible heat flux. For runs 15 to 20, the air was very dry and the surface should therefore have been strongly cooled by evaporation. The negative K_w values correspond to fluxes of carbon dioxide into the sea, in the opposite direction to the air–sea concentration gradient, and this is consistent with low values of surface temperature. However, the measured heat fluxes are all in the direction from sea to air, which limits the range of possible surface temperatures to values above T_a and results in the program finding only positive K_w values. With the sign of the (small) sensible heat fluxes arbitrarily reversed, it was found possible to reproduce the experimental K_w values for these runs. The agreement was dependent on including coupling of heat and matter fluxes for water and CO_2. No agreement could be obtained (K_w very small and/or of the wrong sign) with flux coupling turned off (note 2). Also, without the sign of the heat flux reversed it was not possible to calculate values of T_1 which lay between T_m and T_a.

2. Of the other runs, there are five (8, 9, 12, 13 and 21) for which the calculation of T_1 requires coupling of heat and matter fluxes for water vapour (note 1) and none for which the calculation of T_1 or K_w is more successful without flux coupling.

Table 7.3 *Calculated surface temperature (T_m) and transition temperature (T_l) to fit data of Smith* et al. *(1991), with assumptions of eddy diffusion and linear temperature and concentration profiles on the sea side of the interface*

Run	T_a	T_s	Dew pt.	K_w	T_m	T_l	Notes
4	285.70	283.70	285.94	264	283.732	283.639	4
5.1	285.92	283.70	285.94	223	283.713	283.360	2, 4
5.2	285.99	283.70	285.37	90.0	283.765	283.514	2, 4
6	285.66	283.70	283.65	−129	284.177	284.182	2
8	285.49	283.70	280.79	21.5	283.499	280.912	1, 4
9	285.36	283.66	279.22	77.4	283.624	285.360	1, 4
10	284.57	283.65	280.66	1230	283.650	282.222	1, 3, 4
11	283.54	283.50	281.84	−3900	—	—	5
12	280.71	281.80	278.18	59.9	281.027	279.654	1, 3, 4
14	281.23	282.03	277.55	26.7	281.331	283.288	3, 4
15	281.81	282.13	277.87	2.45	281.491	279.491	2, 3, 4
17.1	282.11	282.45	277.60	188	281.627	279.752	3, 4
18	282.52	282.60	278.19	150	281.716	280.619	2, 4

Units of K_w, cm h^{-1}. *Notes:* 1. Cannot calculate T_1 without flux coupling for water. 2. Cannot reproduce K_w without flux coupling for carbon dioxide. 3. Cannot reproduce K_w exactly with eddy diffusion and *erfc* profiles. 4. The calculated value of T_1 does not lie between T_a and T_m. 5. All calculated values of K_w have the wrong sign.

3. The results in this data set are slightly in favour of eddy diffusion on the sea side of the interface. With eddy diffusion the experimental value of K_w is reproduced exactly for all except run 21; without eddy diffusion the calculations fails to reproduce the experimental value of K_w for runs 19 and 21 (boundary-layer model), for runs 15, 17, 19 and 20 (*erfc* profile), and for runs 15–17, 19 and 20 (linear profiles).

Detailed comparison with data of Smith et al. *(1991)*

The results of the calculations of T_1 and T_m for this data set are given in Table 7.3, in a similar format to Table 7.2, and again with the assumption of eddy diffusion on the sea side of the interface. The following points can be made about these results:

1. The range of values of air, sea and dew-point temperature again is rather narrow, which reduces the accuracy of the calculation of the transition temperature T_1 by Equation 7.29 and results in most of the T_1 values lying outside the range between T_m and T_a.

2. There are three runs for which it was not possible to calculate a T_1 value without coupling of heat and matter fluxes for water vapour (note 1), and none for which the calculation of T_1 worked better without flux coupling.

3. There are four runs for which the experimental value of K_w could be reproduced with coupling of heat and matter fluxes for carbon dioxide but not without such coupling, and again there are no counter-examples.

4. For eight of the 13 runs it was possible to reproduce the experimental value of K_w exactly with the present theory, and this includes one run (run 6) for which the measured carbon dioxide flux was clearly in the opposite direction to the flux predicted by Equation 7.1. The large negative value of K_w for run 11 can be disregarded because it is the result of an extremely small air–sea concentration difference and so has a very large uncertainty. For runs 10, 14 and 17.1 the best calculated value was about half the experimental value.

5. Of the runs for which the experimental value of K_w could be reproduced exactly, runs 6 and 18 required the assumption of eddy diffusion on the sea side of the interface and only for run 15, for which the transfer velocity is so low that it seems to be in a different class from the others, could the experimental K_w value not be reproduced with the assumption of eddy diffusion. Thus, overall, the data slightly favour the assumption of eddy diffusion rather than use of *erfc* or boundary-layer profiles. This, together with the similar conclusion from Table 7.2, is consistent with earlier suggestions (Broecker *et al.*, 1986; Smith and Jones, 1986; Wesely, 1986) that eddy-correlation measurements of air–sea gas exchange made close to the shore are likely to be affected by the presence of a surf zone. In such a zone the effective diffusivity of a dissolved gas is equal to the effective thermal diffusivity, instead of being much smaller. (There are, of course, several unanswered questions here, including the basic question of how close must the measurements be made to a surf zone in order for the results to be affected by it. Such questions can only be answered by experiment.)

The method of calculation used here has been criticised by Doney (1995b) on the grounds that the bulk-surface temperature differences given originally in Table 7.1 were too large and that there are serious errors in the theoretical treatment of the transition from the turbulent region to the stagnant air layer. Doney is correct about the bulk-surface temperature differences, and this error has now been corrected. His assertion that the derivation of Equation 7.29 is not self-consistent is mistaken, but the choice of transition height is indeed somewhat arbitrary, and for this reason one of the options provided by the computer program is to vary the choice of transition height. Calculations have been performed with Z_{1q} set at the height at which the eddy diffusion coefficient is equal to one quarter of the molecular coefficient and with Z_{1q} set at the height at which the eddy diffusion coefficient is four times the molecular coefficient. Such variations do have some effect on the best value of surface temperature for a particular gas flux, but have negligible effect on the relative ability of the program to find acceptable values of surface and transition temperatures with or without Q^* set to zero. This result can probably be understood by reference to Figure 7.10, which shows that the flux for a given sea-surface temperature is not sensitive to variations in the thickness of the stagnant air layer.

The conclusion that it is necessary to include coupling of heat and matter fluxes according to Equation 7.15, with Q^* as given by Equation 7.12, is seen to be remarkably robust, being unaffected by either a major change in the definition of the transition height or a ludicrous error in the solubility function for oxygen (because of a typing mistake, the program called the solubility function for CO_2 instead!). A better theoretical treatment of the transition between the turbulent and stagnant layers is certainly desirable, but the dependence of the present results on the choice of transition height is so slight as to make it extremely unlikely that such an improved treatment would invalidate the present conclusions about the coupling of heat and matter fluxes at a gas–liquid interface. The programs used in this work are now believed to be free of significant error; nevertheless, further, independent testing would be helpful.

Conclusions

Comparison of the results of the present calculations with the available experimental data provides very strong evidence for the importance of coupling of heat and matter fluxes in gas–liquid exchange, as expressed in Equation 7.15. Considering the wind-tunnel experiments of Liss *et al.*, we find that 24 of the 30 useable runs in Table 7.1 provide unequivocal support for Equation 7.15 with inclusion of flux coupling for either oxygen or water vapour. The runs for which flux coupling is not required are mostly those with rather small air–water temperature differences, so that the effects of flux coupling would in any case be expected to be small.

The field measurements by Smith and co-workers, though necessarily less well-controlled than Liss *et al.*'s wind-tunnel measurements, also provide strong support for Equation 7.15.

The calculations also give a clear demonstration of the importance of the surface temperature T_m in controlling the magnitude and sometimes even the direction of the gas flux. Without strong cooling of the sea surface it is not possible to account for the negative values of transfer velocity measured by Smith and co-workers. The sensitivity of the exchange rates to surface temperature implies that, in future experiments, temperature measurements should preferably be made with a precision of 0.01 K or better.

The present results lend some support to an explanation of the generally high CO_2 transfer velocities found by the eddy correlation measurements of Smith and co-workers in terms of turbulent mixing of the surface layer of water in a surf zone and, by implication, to the proposal of Monahan and Spillane (1984) and Kitaigorodskii and Donelan (1984) that oceanic whitecaps play an important role in air–sea gas exchange. As an extension of this idea, we may note that the margins of the ocean are likely to be relatively important for air–sea gas exchange, both because of the presence of zones of turbulent mixing and because the differences between air and sea values of temperature, partial pressures of trace gases, and water vapour pressure are likely to be greatest for air which is newly arrived over the ocean.

The practical utility of the present theory for calculating the air–sea exchange rate of a gas such as carbon dioxide is severely limited by the inability of the calculations to yield a definite value of the sea-surface temperature on the basis of measured heat and water vapour fluxes. This might be overcome in practice by making a simultaneous measurement of the flux of another gas whose flux is relatively easy to measure. Either a natural component such as oxygen or an artificial tracer such as sulphur hexafluoride might be suitable. Comparison of measured and calculated fluxes for this gas would enable the correct value of the surface temperature T_m to be selected, and so allow the flux of any other gas of interest to be calculated from its bulk air and sea concentrations. This procedure would probably be easier than making direct eddy-correlation measurements for a trace gas such as CO_2 and would represent an advance over the dual-tracer method of Watson *et al.* (1991), which model calculations (Phillips, 1995) show to be reliable under conditions where a surface-renewal model applies and only slightly less reliable (good to within about $\pm 30\%$) when a boundary-layer model applies, *provided strong surface heating or cooling effects are absent.* Another possible alternative would be to determine the true surface temperature (i.e. the temperature of the top micron or less of liquid) directly by, for example, observing the infrared spectrum of a suitable surface species with grazing-incidence radiation, but it would still be necessary to measure at least the fluxes of sensible and latent heat and the bulk air and sea temperatures and concentrations. In any event, the accuracy of the results would be enhanced by reducing the length of the extrapolation through the turbulent air layer, i.e. by making measurements of air temperature and composition as close as possible to the water surface, and also by making concentration and temperature measurements over a range of heights in order to obtain more accurate profiles through the turbulent layer.

It is apparent that many more direct, experimental measurements of air–sea fluxes of CO_2 and other gases are needed. For the purpose of improving our understanding of the phenomenon of air–sea exchange, as opposed to obtaining data of relevance to the solution of contemporary geophysical problems, the measurements should preferably not be confined to periods of 'favourable onshore winds', as in the work of Smith *et al.* (1991). Also, for the purpose of testing and refining the theory, direct measurements of gas fluxes, by eddy correlation or other means (see, for example, Anderson and Johnson, 1992) should be combined with measurements of vertical temperature profiles of similar quality to or better than those presented by Azizyan et al. (1984). Indirect measurements of CO_2 flux, such as have in the past relied on Equation 7.1 and might in future be based on the linear dependence shown in Figure 7.9, are obviously much easier to carry out on a routine basis than direct measurements. For the purpose of facilitating future indirect measurements of CO_2 flux, values of the flux would have to be obtained as a function of the air and sea temperatures and the relative humidity or dewpoint, as well as wind speed and the air and sea partial pressures of CO_2, and the possible effects of other variables, such as the nature of the wave field and extent of contamination of the sea surface, should not be neglected. In view of all these complicating factors, a method based on the simultaneous measurement of fluxes of heat, water vapour and one other gas may appear relatively attractive. Such a method would have the advantage that fluxes of all trace gases of interest could be calculated directly from their bulk air and sea concentrations, once the three calibration fluxes and the relevant bulk concentrations were known.

L. F. Phillips

Acknowledgments

I am grateful to Pieter Tans and David Erickson for helpful discussions, and to Stuart Smith for instructive correspondence on the subject of atmospheric turbulence. Much of the work described in this paper was carried out at the National Center for Atmospheric Research in Boulder, Colorado. The hospitality of NCAR and support by the Visiting Scientist Program of the Atmospheric Chemistry Division are gratefully acknowledged. NCAR is sponsored by the National Science Foundation.

References

Anderson, M. L. and Johnson, B. D. (1992). Gas transfer: a gas tension method for studying equilibration across a gas-water interface. *J. Geophys. Res.*, **97**, 17 899–904.

Atkins, P. W. (1994). *Physical Chemistry*, 5th edn. Oxford: Oxford University Press.

Azizyan, G. V., Volkov, Yu. A. and Soloviev, A. V. (1984). Experimental investigation of thermal structure of thin boundary layers of the atmosphere and over ocean. *Izvestiya, Atmos. Ocean. Phys.*, **20**, 482–8.

Bingham, G. E. (1982). *Integrated Area CO_2 Flux Studies: A Progress Report*. Livermore, Ca: Lawrence Livermore Laboratory Report UCLR-1478.

Broecker, W. S. and Peng, T.-H. (1992). Interhemispheric transport of carbon dioxide by ocean circulation. *Nature*, **356**, 587–9.

Broecker, W. S., Ledwell, J. R., Takahashi, T., Weiss, R., Merlivat, L., Memery, L., Peng, T-H., Jähne, B. and Munnich, K. O. (1986). Isotopic versus micrometeorologic ocean CO_2 fluxes: a serious conflict. *J. Geophys. Res.*, **91**, 10 517–27.

Coantic, M. (1986). A model of gas-transfer across air–water interfaces with capillary waves. *J. Geophys. Res.*, **91**, 3925–43.

Danckwerts, P. V. (1970). *Gas–Liquid Reactions*. New York: McGraw-Hill.

de Groot, S. R., and Mazur, P. (1962). *Non-equilibrium Thermodynamics*. Amsterdam: North-Holland.

Denbigh, K. G. (1951). *The Thermodynamics of the Steady State*. London: Methuen.

Doney, S. C. (1994). Irreversible thermodynamic coupling between heat and matter fluxes across a gas/liquid interface. *J. Chem. Soc. Faraday Trans.*, **90**, 1865–74.

Doney, S. C. (1995a). Irreversible thermodynamics and air–sea exchange. *J. Geophys. Res.*, **100**, 8541–53.

Doney, S. C. (1995b). A comment on interfacial processes and air–sea gas exchange. *J. Geophys. Res.*, **100**, 14 347–50.

Erickson, D. J. III (1993). A stability-dependent theory for air–sea gas exchange. *J. Geophys. Res.*, **98**, 8471–88.

Etcheto, J. and Merlivat, L. (1988). Satellite determination of the carbon dioxide exchange coefficient at the ocean-atmosphere interface: A first step. *J. Geophys. Res.*, **93**, 15 669–78.

Etcheto, J., Boutin, J. and Merlivat, L. (1991). Seasonal variation of the CO_2 exchange coefficient over the global ocean using satellite wind speed measurements. *Tellus*, **43B**, 247–55.

Ewing, G. C. and McAlister, E. D. (1960). On the thermal boundary layer of the ocean. *Science*, **131**, 1374–6.

Hasse, L. (1990). On the mechanism of gas exchange at the air–sea interface. *Tellus*, **42B**, 250–3.

Hoover, T. E., and Berkshire, D. C. (1969). Effects of hydration on carbon dioxide exchange across an air–water interface. *J. Geophys. Res.*, **74**, 456–464.

Jähne, B., Munnich, K. O., Bosinger, R., Dutzi, A., Huber, W. and Libner, P. (1987). On parameters influencing air–water gas exchange. *J. Geophys. Res.*, **92**, 1937–49.

Jones, E. P., and Smith, S. D. (1977). A first measurement of air–sea CO_2 flux by eddy correlation. *J. Geophys. Res.*, **82**, 5990–2.

Kitaigorodskii, S. A. and Donelan, M. A. (1984). Wind–wave effects on gas transfer. In *Gas Transfer at Water Surfaces*, ed. W. Brutsaert and G. H. Jirka, pp. 147–70. Higham MA: D. Reidel.

Liss, P. S. and Merlivat, L. (1986). Air–sea gas exchange rates: introduction and synthesis. In *The Role of Air–Sea Exchange in Geochemical Cycling*, ed. P. Buat-Menard, pp. 113–27. Dordrecht: D. Reidel.

Liss, P. S., Balls, P. W., Martinelli, F. N. and Coantic, M. (1981). The effect of evaporation and condensation on gas transfer across an air–water interface. *Oceanol. Acta*, **4**, 129–38.

Monahan, E. C. and Spillane, M. C. (1984). The role of oceanic whitecaps in air–sea gas exchange. In *Gas Transfer at Water Surfaces*, ed. W. Brutsaert and G. H. Jirka, pp. 495–503. New York: D. Reidel.

Panofsky, H. A. and Dutton, J. A. (1984). *Atmospheric Turbulence*. New York: John Wiley & Sons.

Paulson, C. A. and Simpson, J. J. (1981). The temperature difference across the cool skin of the ocean. *J. Geophys. Res.*, **86**, 11 044–54.

Phillips, L. F. (1991a). CO_2 transport at the air–sea interface: effect of coupling of heat and matter fluxes. *Geophys. Res. Lett.*, **18**, 1221–4.

Phillips, L. F. (1991b). Steady-state heat and matter exchange at a phase interface. *J. Chem. Soc. Faraday Trans.*, **87**, 2187–91.

Phillips, L. F. (1992). CO_2 transport at the air–sea interface: numerical calculations for a surface-renewal model with coupled fluxes. *Geophys. Res. Lett.*, **19**, 1667–70.

Phillips, L. F. (1994a). Experimental demonstration of coupling of heat and matter fluxes at a gas-liquid interface. *J. Geophys. Res.*, **99**, 18 577–84.

Phillips, L. F. (1994b). Steady-state thermodynamics of transfer through a gas-liquid interface, treated as a limiting case of thermo-osmosis. *Chem. Phys. Lett.*, **228**, 533–8.

Phillips, L. F. (1994c). Theory of air–sea gas exchange. *Accounts Chem. Res.*, **27**, 217–23.

Phillips, L. F. (1995). CO_2 transport at the air-sea interface: numerical test of the dual-tracer method. *Geophys. Res. Lett.*, **22**, 2597–600.

Quinn, J. A. and Otto, N. C. (1971). Carbon dioxide exchange at the air-sea interface: flux augmentation by chemical reaction. *J. Geophys. Res.*, **76**, 1539–49.

Robertson, J. E. and Watson, A. J. (1992). Thermal skin effect for the surface ocean and its implications for CO_2 uptake. *Nature*, **358**, 738–40.

Sarmiento, J. L. and Sundquist, E. T. (1992). Revised budget for the oceanic uptake of anthropogenic carbon dioxide. *Nature*, **356**, 589–93.

Sarmiento, J. L., Orr, J. C. and Siegenthaler, U. (1992). A perturbation simulation of CO_2 uptake in an ocean general circulation model. *J. Geophys. Res.*, **97**, 3621–45.

Schluessel, P., Emery, W. J., Grassl, H. and Mammen, T. (1990). On the bulk-skin temperature difference and its impact on satellite remote sensing of sea surface temperature. *J. Geophys. Res.*, **95**, 13 341–56.

Smith, S. D. and Jones, E. P. (1985). Evidence for wind-pumping of air–sea gas exchange based on direct measurements of CO_2 fluxes. *J. Geophys. Res.*, **90**, 869–75.

Smith, S. D. and Jones, E. P. (1986). Isotopic and micrometeorological ocean CO_2 fluxes: different time and space scales. *J. Geophys. Res.*, **91**, 10 529–32.

Smith, S. D., Anderson, R. J., Jones, E. P., Desjardins, R. L., Moore, R. M., Hertzman, O. and Johnson, B. D. (1991). A new measurement of CO_2 eddy flux in the nearshore atmospheric surface layer. *J. Geophys. Res.*, **96**, 8881–7.

Tans, P. P., Fung, I. Y. and Takahashi, T. (1990). Observational constraints on the global atmospheric CO_2 budget. *Science*, **247**, 1431–8.

Turner, D. R. (1990). Air–sea exchange. *Discussion paper for IUPAC Workshop: Assessment of Uncertainties in the Projected Concentrations of CO_2 in the Atmosphere*. Petten, Netherlands.

Upstill-Goddard, R. C., Watson, A. J., Liss, P. S. and Liddicoat, M. (1990). Gas transfer velocities in lakes measured with SF_6. *Tellus*, **42B**, 364–77.

Volkov, Yu. A. and Soloviev, A. V. (1986). Vertical structure of the temperature field in the atmospheric layer near the surface of the ocean. *Izvestiya, Atmos. Ocean. Phys.*, **22**, 697–700.

Wanninkhof, R. (1992). Relationship between wind speed and gas exchange over the ocean. *J. Geophys. Res.*, **97**, 7373–82.

Watson, A., Upstill-Goddard, R. and Liss, P. (1991). Air–sea exchange in rough and stormy seas, measured by a dual-tracer technique. *Nature*, **349**, 145–7.

Weiss, R. F. (1974). Carbon dioxide in water and seawater: the solubility of a non-ideal gas. *Mar. Chem.*, **2**, 203–15.

Wesely, M. L., Cook, D. R., Hart, R. L. and Williams, R. M. (1982). Air–sea exchange of CO_2 and evidence for enhanced upward fluxes. *J. Geophys. Res.*, **87**, 8827–32.

Wesely, M. L. (1986). Response to 'Isotopic versus micrometeorologic ocean CO_2 fluxes: a serious conflict' by W. Broecker *et al.*, *J. Geophys. Res.*, **91**, 10 533–5.

8

The sea-surface microlayer and its effect on global air–sea gas transfer

WILLIAM ASHER

Abstract

Laboratory measurements of air–water gas transfer rates for cleaned and film-covered water surfaces have shown that the presence of soluble and insoluble surfactants can inhibit air–water gas fluxes. Naturally occurring surface-active material is known to concentrate in the marine surface microlayer and form films and slicks. It is reasonable that oceanic slicks and films may lower *in situ* gas transfer rates compared with air–sea gas exchange through a clean ocean surface. Here, a simple model of gas transfer through clean and surfactant-influenced water surfaces is used to develop parameterizations of liquid-phase, and gas-phase, rate-controlled gas transfer velocities through clean and surfactant-influenced ocean surfaces. The parameterization for liquid-phase, rate-controlled processes is used to estimate the effect of naturally occurring surface films on the net global flux of carbon dioxide. The gas-phase, rate-controlled relations are used to study the impact of films on the flux of ammonia from the central Pacific Ocean.

By relating the fractional area coverage of surfactant-influenced sea surface to a global map of net synthetic primary production, the model shows that surface films can increase or decrease the net global oceanic carbon dioxide flux, depending on the regional film coverage.

Introduction

Motivation and purpose

There is little doubt that under laboratory conditions both soluble and insoluble surfactants suppress gas–liquid mass transfer rates. It is also clear that marine phytoplankton can exude surface-active compounds, and it is thought that this material concentrates in a sea-surface microlayer. Although it is reasonable to assume that naturally occurring surfactants in the microlayer could play some role in mediating air–sea

251

gas exchange, there is little firm evidence for the effect of surface films on oceanic gas fluxes. Using available data for air–water transfer through clean and surfactant-influenced water surfaces and accepted models for gas–liquid mass transfer, this review will explore the possible impact of surfactants in the microlayer on the net global air–sea flux of carbon dioxide (CO_2) and the net flux of ammonia (NH_3) from the central Pacific Ocean. It will be shown, by estimating the fraction of the ocean surface that is surfactant-influenced from primary productivity rates and using best available estimates for gas transfer rates in the presence of surfactants, that the microlayer could have a substantial effect on gas fluxes.

It is understood that the model calculations showing the effects of surfactants in the marine surface microlayer on global air–sea gas fluxes presented here contain assumptions that are speculative in nature. Despite the lack of verifying field measurement, however, there are enough relevant laboratory and wind-tunnel data for gas exchange through clean and slicked water surfaces to show the mechanisms by which surfactants in the microlayer could affect oceanic gas exchange. This chapter will discuss these processes, provide experimental results (when available) demonstrating the magnitude of the effect, and develop one possible method for parameterizing the effects of surfactants on air–water mass transfer so that they can be included in a simple model for calculating global air–sea gas fluxes.

One major source of uncertainty in the calculations is that the simple parameterization for air–sea gas transfer in the presence of surface films that is developed has not been tested under oceanic conditions. A second source of uncertainty is that the extent of the ocean surface that is influenced by naturally occurring surfactants is not known. Also unresolved at present is whether the surfactant-influenced fraction of the ocean surface can be estimated from biological productivity. The fact that it is not definitively known how to estimate oceanic gas fluxes through clean ocean surfaces also causes some uncertainty in the calculated fluxes. Because the uncertainties in the calculation presented here are likely to be quite large, they are not an answer to the question: 'What is the global air–sea gas flux of a particular gas?'. Rather, these results are an attempt, using the best available experimental data and reasonable, although unproven, assumptions based on currently accepted hypotheses of the chemical composition and behaviour of the ocean surface microlayer, to answer the question: 'From the standpoint of estimating global air–sea gas fluxes, is there any reason to study surfactants in the sea–surface microlayer?'.

In this regard, it should be stressed that in the case of CO_2 the fluxes estimated here are not an attempt in and of themselves to resolve issues concerning global carbon budgets. However, these model results do show that the effects of surfactants could be quite large and that it is possible that these effects are not uniform over the ocean surface. It is therefore possible, though not proven, that continued research on the role of the sea-surface microlayer and its role in air-sea gas exchange could improve the accuracy of estimates of global air–sea gas fluxes.

Background information on surfactants in the microlayer

It is known that visible slicks composed of biogenically produced surface-active material can form on the ocean surface (Williams, 1986). In the absence of visible slicks, it is believed that this surface-active material forms a surface microlayer (Williams, 1986). Research has shown that the microlayer and slicks are a heterogeneous mixture containing humic material, carbohydrates, lipids, proteins (Hunter and Liss, 1981; Barger and Means, 1985; Williams *et al.*, 1986; Frew and Nelson, 1992a,b; Hunter, Chapter 9, this volume) and a substantial fraction of surface-active material (Frew and Nelson, 1992a,b). However, the concentrations and identities of these surfactants are not well known.

Coherent slicks formed on the ocean surface have been shown to increase wave damping (Broecker, Peterman and Siems, 1978; Hühner-fuss, Lange and Walter, 1982, 1984; Hühnerfuss *et al.*, 1983; Jähne *et al.*, 1984; Bock and Frew, 1993), and extracted microlayer material has been shown to decrease air–water gas transfer velocities (Frew *et al.*, 1990). Brockman *et al.* (1982) showed that an artificial slick of oleic acid on the sea surface decreased CO_2 transfer rates by 30% compared with an unslicked sea surface. Laboratory measurements of gas transfer velocities in the presence of natural and synthetic, and soluble and insoluble, surfactants have shown that they decrease gas transfer rates compared with transfer rates measured at clean water surfaces (Davies, Kilner and Ratcliff, 1964; Springer and Pigford, 1970; Moo-Young and Shoda, 1973; Broecker *et al.*, 1978; Lee, Tsao and Wankat, 1980; Jähne *et al.*, 1984; Asher and Pankow, 1986; Goldman, Dennett and Frew, 1988; Frew, Bock and Goldman, 1992). Furthermore, the experiments of Frew *et al.*, (1990) suggest that visible slicks need not be present for the microlayer to suppress gas transfer. The results from the gas exchange studies mentioned above show that it is reasonable to assume that the presence of an ocean surface microlayer and/or coherent slicks formed from it will inhibit exchange rates compared with gas exchange through a clean air–sea interface (see Frew, Chapter 5, this volume for an up-to-date review). However, quantification of this reduction under oceanic conditions remains troublesome.

One of the major problems in determining the effect of the microlayer on gas exchange is a lack of field measurements of fluxes under known microlayer conditions. This has resulted in part from the absence of a method for measuring relative levels of surfactant concentrations at the air–sea interface in real-time. However, with the development of the self-contained underway microlayer sampler (SCUMS) (Carlson, Cantey and Cullen, 1988), the computer-aided surface truthing optical recorder (CASTOR) (Bock and Frew, 1993), and nonlinear laser spectroscopic techniques (Frysinger *et al.*, 1992, Korenowski, Frysinger and Asher, 1993), *in situ* measurements are now possible. These sampling techniques, combined with remote sensing, have revealed the patchiness of ocean surface slicks (Moum, Carlson and Cowles, 1990; Frysinger *et al.*, 1992; Ochadlik, Cho and Evans-Morgis, 1992; Bock and Frew, 1993), indicating that accurate determination of microlayer effects on gas transfer requires flux measurements with high spatial (on the order of 10 km²) and temporal (less than 4 h) resolution. Development of the deliberate dual-tracer method (Watson, Upstill-Goddard and Liss, 1991b; Wanninkhof *et al.*, 1993) for measuring oceanic gas transfer rates could provide a technique with this capability. The controlled flux method of Jähne *et al.* (1989) holds promise for resolving transfer processes on very short timescales and small spatial scales.

The impact of surfactants in the microlayer on net global air–sea fluxes of radiatively important trace gases will be studied using a direct air–sea flux model. Gas fluxes in surfactant-influenced and clean ocean regions will be estimated using a parameterization developed from considerations presented below. The net global air–sea flux of CO_2 and the flux of NH_3 from the central Pacific Ocean will be studied as case examples for different scenarios of fractional microlayer coverage.

Processes affecting gas transfer

Air–sea gas exchange background

The air–sea flux, F_X (mol m^{-2} s^{-1}), of a gas X depends on a 'kinetic rate' or its overall transfer velocity, K_{OA} (cm s^{-1}) and its thermodynamic driving potential, which is the product of solubility, K_h (mol m^{-3} atm.$^{-1}$), and partial pressure difference across the air–sea interface, ΔP_X (atm.). Using the two-film model (Whitman, 1923; Liss & Slater, 1974) and ignoring chemical enhancement of transfer, F_X can be expressed as

$$F_X = K_{OA} K_h \Delta P_X \tag{8.1}$$

where K_{OA}^{-1} is the total resistance to transfer (Liss & Slater, 1974), ΔP_X is

$$\Delta P_X = P_G - P_L \tag{8.2}$$

and P_G and P_L are the bulk gas- and liquid-phase partial pressures. P_L can also be written in terms of the concentration of the gas in the bulk liquid, C_L (mol m^{-3}), as $P_L = C_L/K_h$. If the total transfer resistance is the sum of a gas-phase interfacial boundary layer resistance, r_G and an liquid-phase interfacial boundary layer resistance, r_L, K_{OA} can be written as

$$\frac{1}{K_{OA}} = \frac{1}{k_L} + \frac{K_h R T_w}{k_G} \tag{8.3}$$

Here, k_L $(= 1/r_L)$ and k_G $(= K_h R T_w/r_G)$ are defined as the liquid- and gas-phase transfer velocities, respectively, R is the gas constant (m^3 atm. mol^{-1} K^{-1}), and T_w is the water temperature (K). For sparingly soluble non-reactive gases, r_L is much greater than r_G and transfer is liquid-phase, rate-controlled with K_{OA} approximately equal to k_L. For reactive gases and/or gases with high solubility, r_G is much greater than r_L and transfer is gas-phase, rate-controlled with K_{OA} approximately equal to $k_G/K_h R T_w$.

The transfer velocity depends on diffusivity, D (m^2 s^{-1}), kinematic viscosity, v (m^2 s^{-1}), and the aqueous-phase turbulence. Brumley and Jirka (1988), among others, have shown

$$k_L = A (v/D)^{-n} f_1(Q,L) \tag{8.4}$$

where A is a constant, the ratio v/D is the Schmidt number, Sc, of the gas, the Schmidt number exponent, n, is determined by the turbulence and interfacial conditions, and $f_1(Q,L)$ represents the functional dependence on the turbulence velocity, Q, and length scale, L. Models and experiments have shown n to be in the range 0.7 to 0.5 when bubbles are absent (Jähne et al., 1987). If significant, a bubble-mediated gas flux will decrease the dependence of k_L on Sc (Asher et al., 1992; Keeling, 1993) and also cause k_L to depend on K_h (Memery and Merlivat, 1985; Woolf and Thorpe, 1991; Keeling, 1993).

The dominant near-surface turbulence generation mechanism in the ocean is the wind stress acting through momentum transfer and waves (Jähne et al., 1987). Although parameters related to the wave field may be useful in estimating k_L (Jähne et al., 1984, 1987; Back and McCready, 1988; Daniil and Gulliver, 1991; Jähne, 1991), the ease in measuring wind speed, U (m s^{-1}), has led to many attempts to develop a relation between k_L and U (Liss and Merlivat, 1986; Wanninkhof, 1992). Using k_L data from wind tunnels, lakes, and the ocean, Liss and Merlivat (1986) proposed a three-regime, piece-wise linear relation given by

$$k_L = 2.778\times10^{-6}(0.17U(Sc/600)^{-2/3}) \qquad U \leq 3.6 \text{ m s}^{-1} \qquad (8.5)$$

$$k_L = 2.778\times10^{-6}(2.85U - 9.65)(Sc/600)^{-1/2} \quad 3.6 < U \leq 13 \text{ m s}^{-1} \quad (8.6)$$

$$k_L = 2.778\times10^{-6}(5.9U - 49.3)(Sc/600)^{-1/2} \qquad 13 < U \text{ m s}^{-1} \qquad (8.7)$$

for k_L in m s^{-1}. The piece-wise, linear parameterization is supported by the wind-tunnel data of Broecker et al. (1978) and Broecker and Siems (1984). Wanninkhof (1992) has proposed a quadratic relation between U and k_L given by

$$k_L = 2.778\times10^{-6}(0.31U^2(Sc/660)^{-1/2}) \qquad (8.8)$$

(k_L in m s^{-1}). The quadratic dependence of k_L on U for a clean water surface is supported by the wind-tunnel data of Wanninkhof and Bliven (1991) and Frew et al. (1992). For comparison, the Liss and Merlivat and Wanninkhof relations and data of Wanninkhof and Bliven are shown in Figure 8.1 and the data of Frew et al. (1992) for clean distilled water are shown in Figure 8.2. Although the Frew et al. clean-surface data clearly show $k_L \propto U^2$, direct comparison of them with Equations 8.5, 8.6, and 8.7 or Equation 8.8 is uninformative because of differences in wind/wave regime between the small circular wind tunnel used by Frew et al., the larger linear wind tunnel used by Wanninkhof and Bliven and the ocean (Liss and Merlivat).

Liss (1973) has shown k_G for water is linearly related to U. From this, and assuming k_G for different species is a function of their gas-phase molecular diffusivities, k_G for a gas X can be estimated from $k_G(H_2O)$ from (Liss and Slater, 1974)

$$k_G(X) = (5.2\times10^{-5} + 3.2\times10^{-3}U)\sqrt{\frac{M_{H_2O}}{M_x}} \qquad (8.9)$$

(k_G in m s^{-1}) where the constants were determined using the data of Liss (1973) and M_X is the molecular weight of species X. Although there are

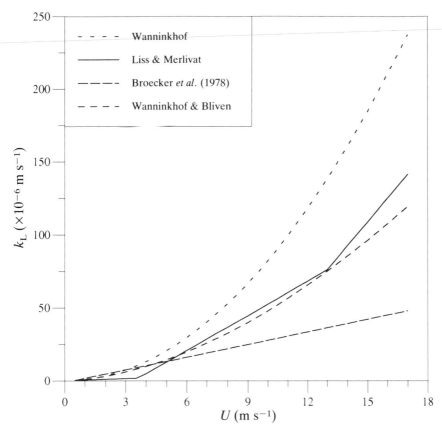

Figure 8.1. k_L *versus* U: Wanninkhof (1992) model; Liss and Merlivat (1986) model; data from Broecker *et al.* (1978) and Wanninkhof and Bliven (1991).

limited experimental data available, it can be assumed that both constants in Equation 8.9 are functions of air temperature, T_A (K). Bubbles will have little effect on a gas-phase, rate-controlled process except in the case of droplet production by bursting bubbles. These droplets could have a significant role in the case for transfer of water vapour (Ling and Kao, 1976).

   In order to understand the microlayer's effect on air–sea exchange, it is necessary to examine how the presence of a film can affect ΔP_X, k_L, and k_G. The major mechanisms are discussed in the next two sections, and the means for the inclusion of the mechanisms in a parameterization of k_L in terms of U is determined.

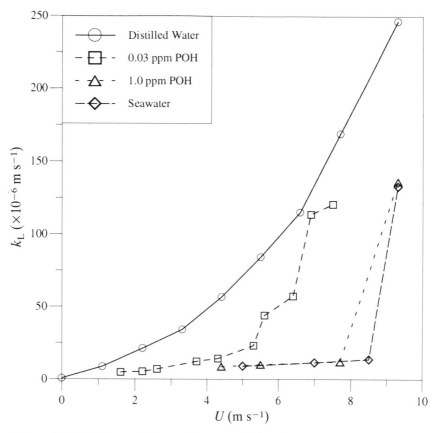

Figure 8.2. Wind tunnel data of Frew *et al.* (1992) for clean and surfactant-influenced surfaces. The surfactant used was polyethyleneoxide (POH) and the concentrations listed are bulk aqueous-phase values (ppm by volume). Seawater was taken from Vineyard Sound, Massachusetts.

Barrier effects and changes in ΔP_X, Sc, and K_h

Laboratory studies using soluble and insoluble surfactants have shown that surface films do not substantially increase the total transfer resistance of a liquid-phase, rate-controlled transfer process by adding an interfacial resistance term (Martinelli, 1979; Liss, 1983). In contrast, experiments on gas-phase, rate-controlled processes have shown that monomolecular films composed of close-packed insoluble surfactants or oil slicks can form a barrier to H_2O and sulphur dioxide (SO_2) exchange (Jarvis, 1962; Jarvis and Kagarise, 1962; Liss and Martinelli, 1978). It is believed that these results show that a coherent, close-packed surface film may have a significant effect on any nonreactive, gas-phase, rate-controlled gas by increasing the interfacial resistance to transport (Liss, 1983).

Despite the known sensitivity of soluble gases to the presence of coherent surface films, it is unlikely that the marine microlayer will affect k_G through these barrier effects. Williams *et al.* (1986) state that there is insufficient surfactant material to form an unbroken monolayer unless compacted by wind stress or current shear, and previous studies have shown that total film coverage is essential for suppression of gas transfer (Liss and Martinelli, 1978). Furthermore, Adamson (1990) reports that mixed-composition films and films of biologically produced surfactants do not form barriers to exchange. As the microlayer contains a variety of surfactants, it is unlikely that single-component films will form from it.

Oxidizing soluble gases, such as ozone and hydrogen peroxide, or radical species will react with the organic compounds in the microlayer. Because most of these species also have net flux into the ocean (Thompson and Zafiriou, 1983), films will not inhibit their transfer through interfacial mechanisms.

From Equation 8.1, the thermodynamic driving potential of the transport process is expressed by $K_h \Delta P_X$. The microlayer also can change F_X by altering the physical or chemical properties of the air–sea interface that control ΔP_X. Sea-surface temperature, T_S (K), and bulk aqueous phase concentration are the two major factors in determining P_L. For most gases, decreases in T_S will decrease P_L, and depending on P_G, ΔP_X will either increase or decrease. Due to evaporation, T_S is typically a few tenths of a degree cooler than T_W (Ewing and McAlister, 1960; Katsaros, 1980; Robinson, 1985; Schluessel *et al.*, 1990) and this is enough to have a significant effect on the net global air–sea flux of CO_2 (Robertson and Watson, 1992). If the microlayer suppressed evaporation, T_S would increase and ΔP_X would be affected. For the reasons discussed above, however, the surface microlayer will have little effect on ΔP_X by retarding evaporation and raising T_S.

Although variations in T_S on the order of 0.3 K are significant in terms of ΔP_X, Sc for CO_2 increases by only 1.4% at 293 K (variations in T_S and D will have no effect on k_G). The power-law dependence of k_L on Sc through n (Equation 8.4) with n in the range $0.5 < n < 0.7$ implies this increase in Sc will cause a 1% decrease in k_L. Goldman *et al.* (1988) report that D is relatively insensitive to surfactant concentration (Grief, Kornet and Kappesser, 1972; Ju and Ho, 1986); therefore, there is little reason to believe that it will be affected by the presence of a microlayer or coherent slick.

It has been hypothesized that for water surfaces covered by a thick film, the concentration of surfactant material could be high enough that K_h at the interface will be affected (F. Macintyre, in discussion of Liss and Martinelli, 1978). Although this could be possible in situations such as methane transfer through an oil slick, changes in K_h due to high surfactant concentrations at the interface will probably be unimportant on global scales. Furthermore, in the absence of information on the concentrations of naturally occurring surfactants in the microlayer, changes in K_h are impossible to quantify at the present time.

Hydrodynamic effects

In the case of a liquid-phase, rate-controlled transfer process, the dependence of F_X on near-surface hydrodynamics is contained in k_L. Surface films are known to decrease turbulence at the interface (Lee et al., 1980), increase the concentration boundary-layer thickness (Asher and Pankow, 1991), and damp wind-generated waves (Broecker et al., 1978; Hühnerfuss et al., 1982, 1984; Jähne et al., 1984; Bock & Frew, 1993). Because of the central role wind stress plays in generating near-surface turbulence, it is logical to assume that the effect of films will modify the functional relationship between k_L and U. In support of this, wind-tunnel experiments have shown that at a given U, surfactants decrease k_L compared with k_L at a clean surface (Broecker et al., 1978; Brockman et al., 1982; Jähne et al., 1984; Frew et al., 1992). Stirred-tank experiments have shown similar reductions in k_L in the presence of a variety of soluble and insoluble surfactants (Davies et al., 1964; Springer and Pigford, 1970; Moo-Young and Shoda, 1973; Lee et al., 1980; Jähne et al., 1984; Asher and Pankow, 1986; Goldman et al., 1988; Frew et al., 1990). More importantly, as discussed by Frew et al. (1990) and shown in the unpublished data from Asher presented below, monomolecular films or coherent slicks need not be present for surfactants to have an effect on k_L.

Broecker et al. (1978) measured k_L for CO_2 in a wind–wave tunnel for a clean water surface and for one covered by an oleyl alcohol film. They observed a four- to five-fold reduction in k_L at $U=12.5$ m s^{-1} between the film-covered and clean water surface. In contrast to the quadratic relation between k_L and U proposed by Wanninkhof (1992) or the segmented linear dependence of Liss and Merlivat (1986), Broecker et al. (1978) found k_L increased linearly with U following

$$k_L = 2.778 \times 10^{-6} (1.04U - 0.12) \left(\frac{Sc}{600} \right)^{-1/2} \quad U \leq 12.5 \text{ m s}^{-1} \quad (8.10)$$

(k_L in m s^{-1}) to the limit of their measurements at $U=12.5$ m s^{-1}. Using a small circular wind tunnel, Frew et al. (1992) measured k_L for O_2 in the presence of synthetic and naturally occurring surfactants. They found that the relationship between k_L and U was piece-wise linear, as proposed by Liss and Merlivat (1986), or strictly linear, as observed by Broecker et al. (1978). Frew et al. (1992) also report that the line segment slopes and 'critical' U values where the slope changed were a function of surfactant type and concentration. Figure 8.1 shows the Liss and Merlivat (1986) and Wanninkhof (1992) k_L–U models and the Broecker et al. (1978) k_L–U data. The results of Frew et al. (1992) for clean and surfactant-influenced surfaces are shown in Figure 8.2.

In addition to decreasing k_L as a function of U, experiments have shown that the presence of surfactants will increase the Sc exponent n (see Equation 8.4). It has been shown that in the absence of bubbles, n increases from $n=1/2$ for a clean water surface to $n=2/3$ when a surface film is present (Jähne et al., 1984, 1987). Assuming that the increase in n does not affect the parameterization with respect to U, Equation 8.10 becomes

$$k_L = 2.778 \times 10^{-6} (1.04U - 0.12) \left(\frac{Sc}{600} \right)^{-2/3} \quad U \leq 12.5 \text{ m s}^{-1} \quad (8.11)$$

and indicates that changes in Sc will cause larger changes in k_L.

There are no oceanic measurements of k_L versus U in the presence of known slick coverage. However, all available experimental results from stirred tanks and wind tunnels and the results of Brockman et al. (1982), who measured k_L under an artificial slick of oleic acid on the ocean, indicate that the presence of surfactants will cause a decrease in k_L at a given U. Therefore, it is likely that Equation 8.11 represents the maximum reduction of k_L in terms of possible parameterizations of the effect of oceanic surfactants on k_L. The results of Frew et al. (1992) demonstrate that more complicated functionality of k_L with respect to U than the strict linear dependence of Equation 8.11 can exist in the presence of surfactants. Lacking a better $k_L U$ relationship or field data, Equation 8.11 will be used with the understanding of the difficulty in extrapolating wind tunnel results to open ocean conditions.

The dynamic viscosity, η (kg m^{-1} s^{-1}), of microlayer water samples and seawater with exudates from marine phytoplankton cultures added to it has been shown to increase relative to η measured in clean seawater

(Jenkinson, 1986; Carlson, 1987). Although it is possible that changes in η will affect the governing hydrodynamics associated with air–water gas transfer, it is unclear how this would be included in a wind-speed parameterization of k_L. However, changes in η will also affect k_L by changing Sc. Carlson (1987) measured η for sampled microlayer water and saw it increased by 2.15×10^{-4} kg m^{-1} s^{-1} (at 294.2 K) compared with η measured for concurrently sampled sub-surface water. Assuming density is constant in the microlayer, an increase in Sc from 630 to 720 would be expected. If $k_L \propto$ Sc$^{-2/3}$, as shown in Equation 8.11, then it will decrease by 14% at a given U as a result of this change in η.

Slicks may also affect k_G for gas-phase, rate-controlled processes because they increase wave damping (Hühnerfuss, et al., 1982, 1984; Alpers and Hühnerfuss, 1989; Bock and Frew, 1993). The increased damping will decrease the mean-square wave slope, δ, in turn decreasing the aerodynamic roughness length, z_0 (m) (Jacobs, 1989). Decreases in z_0 will decrease the drag coefficient, C_D (Geernaert, Katsaros and Richter, 1986), which can be considered as the bulk aerodynamic transfer coefficient for momentum. More generally though, the bulk aerodynamic transfer coefficient of a gas in the atmospheric boundary layer can be written as (Smith, 1988)

$$C_{EN} = \frac{\varkappa^2}{\ln(z/z_0) \ln(z/z_e)} \tag{8.12}$$

where the transfer coefficient for water vapour, C_E, is written in Equation 8.12. In the following discussion, neutral stability will be assumed and the subscript N notation will be omitted. In Equation 8.12, z (m) is a reference height, z_e (m) is the effective roughness length for water vapour, and $\varkappa$ is the von Karman constant. Although Liu, Katsaros and Businger (1979) predict that z_e will be a function of z_0, the data of Large and Pond (1981) suggest that z_e is constant. The functionality of C_E with U displayed by the wind tunnel gas-exchange data of Ocampo-Torres et al. (1994) also is consistent with Equation 8.12 up to $U=6$ m s^{-1}. For the sake of simplicity, it is assumed that z_e is constant and that only z_0 is affected by surface films.

In principle, δ can be used to estimate z_0 using the method proposed by Jacobs (1989)

$$z_0 = z_1 \exp\left[\frac{1}{2}\left(1 - \frac{\varepsilon c}{U_*}\right)\left(\frac{\delta\varkappa}{\varepsilon}\right)^2 \right] \tag{8.13}$$

where U_* (m s^{-1}) is the friction velocity, L is the Monin–Obukhov length (m), c (m s^{-1}) is the mean-square phase speed, and ε is defined by

$$\varepsilon = \frac{\varkappa}{\ln(L/z_1)} \qquad (8.14)$$

with z_1 (m) defined by

$$z_1 = \frac{v_a}{U_*}\exp(-5.5\varkappa) \qquad (8.15)$$

where v_a is the kinematic viscosity of air. The mean-square wave slope is computed from wave slope spectra, $S(k)$ (cm), where k (rad cm^{-1}) is wavenumber. $S(k)$ from clean and slick-covered ocean surfaces, such as shown in Figure 8.3, can be used to calculate changes in δ. From this, changes in z_0 can be estimated from Equations 8.13, 8.14, and 8.15 and, in turn, the effect on C_E can be calculated using Equation 8.12.

Unfortunately, ancillary data necessary for processing the wave slope spectra shown in Figure 8.3 are unavailable at present, so the effect of a

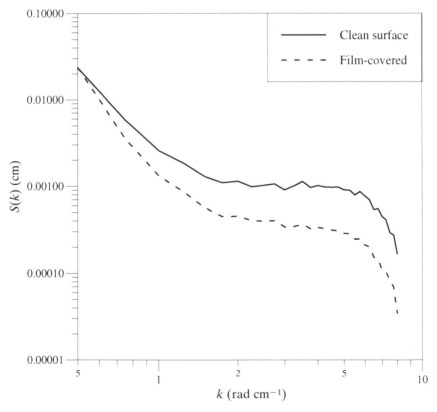

Figure 8.3. Wave slope spectra for clean and slicked ocean surfaces from the unpublished data of Bock and Hara.

surface film on z_0 will be estimated from wind-tunnel measurements of U_* made by Lange and Hühnerfuss (1978).

It must now be shown that changes in C_E can be related to changes in k_G. Following Liu et al. (1979), Smith (1988), and others, the flux of water vapour, F_{H_2O}, can be written as

$$F_{H_2O} = \frac{\varrho}{M_{H_2O}} C_E U (Q_s - Q) \qquad (8.16)$$

where ϱ is air density, and Q_s and Q are the specific humidities at the sea surface and reference height, respectively. Comparison of Equation 8.16, where ΔP_X has been substituted for $(Q_s\text{-}Q)$, with Equation 8.1, where k_G has been substituted for K_{OA}, shows that k_G is proportional to $C_E U$. This shows that changes in k_G will be proportional to changes in C_E at constant U. The decrease in k_G caused by a slick can be estimated from the corresponding decrease in C_E calculated using measured wave slope spectra and Equations 8.12 through 8.15. Therefore,

$$k_G(\text{slick}) = \frac{C_E(\text{slick})}{C_E(\text{clean})} k_G(\text{clean}) \qquad (8.17)$$

where (slick) and (clean) refer to microlayer-affected and clean ocean surfaces, respectively.

Using the data of Lange and Hühnerfuss (1978), Alpers and Hühnerfuss (1989) report $U_*=0.053U_{ref}$ at a clean water surface and $U_*=0.042U_{ref}$ at a monolayer-covered surface (U_{ref} is the wind speed at a 0.5 m reference height). U_* can be used to estimate z_0 from the Charnock and Businger relations (Smith, 1988) as

$$z_0 = 0.011 \frac{U_*^2}{g} + 0.11 \frac{v_a}{U_*} \qquad (8.18)$$

where g (m s^{-2}) is the acceleration of gravity and $v_a=1.5\times10^{-5}$ kg m^{-1} s^{-1}. At $U_{ref}=8$ m s^{-1}, z_0 (clean) and z_0 (slick) are 2.56×10^{-4} m and 1.7×10^{-4} m, respectively. Using the relations of Liu et al. (1979), z_e is estimated to be 2.67×10^{-5} m, and assuming $\varkappa=0.4$ shows that C_E decreases from 2.0×10^{-3} at a clean surface to 1.9×10^{-3} at a monolayer-covered surface.

Although the 5% decrease in C_E calculated above would be masked by current experimental uncertainty in measurement of bulk aerodynamic transfer coefficients, it is most likely an underestimate of the effect of slicks on a gas-phase, rate-controlled process. Mass transfer through the atmospheric surface layer occurs strictly through turbulent diffusion. In contrast, momentum can also be transferred by pressure forces on gravity waves (Liu *et al.*, 1979). As shown by the wave slope spectra of Bock and Hara (unpublished data) in Figure 8.3, surface films are more effective in damping capillary waves (large k) than gravity waves (small k). Therefore, a surface film may substantially reduce the transport of mass through the boundary layer by suppressing capillary waves, but momentum transport can still occur through pressure forcing by the largely undamped gravity waves.

Lange and Hühnerfuss (1978) measured the decrease in U_*, which is related to momentum transport. The observed decrease in U_* was likely due to capillary wave damping by the films, but momentum transfer was still occurring through pressure forcing by the less-damped gravity waves. Therefore, the decrease in C_E at a given U may be larger than that observed for U_* because pressure forces have little effect on mass transfer (Liu *et al.*, 1979). The result given above may underestimate the impact of the films on C_E. However, this hypothesis has not been tested using laboratory or field data.

It is acknowledged that the approach outlined in Equations 8.12 through 8.18 for estimating the effect of slicks on gas-phase, rate-controlled air–water transfer has not been tested by laboratory or field experiments. However, Deacon (1979) has reported that coral mucus reduces C_D over coral reefs. Although C_D is not equal to C_E (Liu *et al.*, 1979; Smith, 1988), the relations in Smith (1988) show that both are a function of z_0. Therefore, a similar reduction in C_E might be expected. The Deacon data suggests that the proposed relationship between slicks, C_E, and k_G captures enough of the essential physics to provide a reasonable estimate of the order of magnitude of the effect.

Changes in aqueous-phase viscosity will affect momentum coupling between the air and water, which could affect z_0, C_E, and k_G. However, these changes cannot be predicted reliably and will be ignored for the sake of simplicity.

Breaking waves and the microlayer

It is commonly accepted that visible slicks are dispersed by breaking waves. Less is known about the stability of the microlayer when whitecaps are present, although it is likely that its less-soluble components are mixed into the surface water by breaking waves. However, the microlayer could contain soluble organic material that is surface active (Frew *et al.*, 1990). Films composed of soluble surfactants, once disrupted by turbulence, reform more quickly than films composed of insoluble

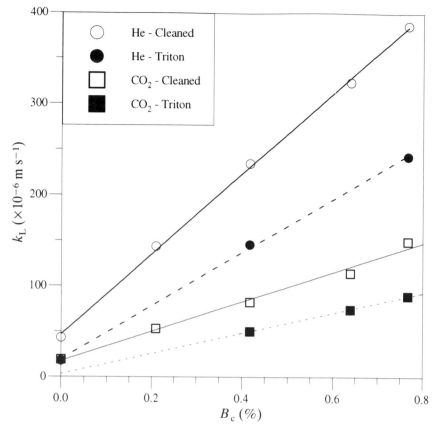

Figure 8.4. Whitecap simulation tank data for transfer of He and CO_2 through clean and surfactant-influenced surfaces showing effect of a soluble surfactant on k_L as a function of gas type and B_c. From Asher *et al.* (1996).

surfactants (Springer and Pigford, 1970). It is also known that the presence of soluble surfactants can decrease gas transfer rates (Springer and Pigford, 1970; Lee *et al.*, 1980; Goldman *et al.*, 1988; Frew *et al.*, 1990). It could be that soluble surfactants inhibit gas exchange even when waves break. This hypothesis is supported by the laboratory data presented below.

Figure 8.4 shows k_L measured by Asher *et al.*, (1996) in a whitecap simulation tank (WST) as a function of fractional area bubble coverage, B_c (per cent, area bubble plume/total tank surface area), for helium (He) and CO_2 evasion through cleaned and surfactant-influenced seawater surfaces at 293 K. The seawater used was filtered through a 5 μm spun-fibre filter and sterilized by UV irradiation. In the case of the cleaned-surface experiments, the water surface was vacuumed prior to the start of each experiment to remove contaminants. For the surfactant-influenced experiments, 2.0×10^{-6} m³ of Triton X-100 was added to the 1.8 m³ of water in the WST resulting in a surfactant concentration of 1.1 mg l⁻¹, which is similar to dissolved organic carbon (DOC) concentrations in oceanic surface water (Williams *et al.*, 1986). Further details concerning the gas exchange measurements and operation of the WST are provided in Asher *et al.* (1992, 1996) and Asher and Farley (1995).

The data in Figure 8.4 show that the effect of the soluble surfactant was to reduce k_L measured at the surfactant-influenced interface as compared with a cleaned surface at a constant B_c. This reduction in k_L was observed for relatively soluble (CO_2) and insoluble (He) gases even at the highest B_c studied. This demonstrates that soluble surfactants can affect gas transfer even when the water surface is disrupted by bubble plumes. Although these laboratory results cannot be directly related to oceanic conditions, the size and depth dependences of bubble concentrations measured in the WST are similar to oceanic bubble populations (Asher and Farley, 1995). Furthermore, a B_c of 0.77% corresponds to a wave breaking every 30 s. This frequency of whitecapping would be found at U equal to 18 m s⁻¹ (Monahan, 1993), which is not unreasonable in terms of oceanic whitecap coverage. Therefore, soluble surfactants in ocean surface waters might reduce k_L even in the presence of breaking waves at relatively high wind speeds. However, this does not imply that the microlayer and surface films or slicks are stable in the presence of breaking waves.

Models for bubble-mediated gas transfer indicate that in the presence of bubbles, k_L will not scale as the simple power-law function with respect to Sc described by Equation 8.4. As discussed in Memery and Merlivat (1985), Woolf and Thorpe (1991), Keeling (1993) and Woolf, Chapter 6 this volume, k_L will be a more complicated function of both Sc and K_h and k_L for invasion will be larger than k_L for evasion. The functionality of k_L with respect to Sc and K_h and the magnitude of the invasion/evasion asymmetry are dependent on the bubble population, sub-surface bubble dynamics including the rise velocities of individual bubbles, and the gas transfer velocity of an individual bubble (Memery and Merlivat, 1985). Naturally occurring surfactants could affect the dependence of k_L on Sc and K_h by decreasing bubble rise velocities, decreasing the transfer velocity of an individual bubble, changing the dependence of the bubble transfer velocity on Sc, and altering the bubble population (Asher and Farley, 1995). Although these effects could be predicted through detailed modelling of the bubble-driven gas flux, this type of modelling is outside the scope of this paper and is discussed elsewhere in this volume (Woolf, Chapter 6, this volume). Therefore, the effect of surfactants on bubble-mediated gas fluxes will be examined briefly using the data of Asher *et al.* (1996).

The more complicated functionality of k_L with respect to Sc and K_h is demonstrated by the gas transfer data shown in Figure 8.5 that were collected in the WST using the surfactant Triton X-100. In Figure 8.5, $\log(k_L)$ values for k_L measured for SF_6, He, and O_2 at $B_c=0.0\%$ and k_L measured for SF_6, He, O_2, and CO_2 at $B_c=0.77\%$ are plotted versus $\log(Sc)$. In the absence of bubbles ($B_c=0.0\%$), $\log(k_L)$ is linearly correlated with $\log(Sc)$, showing Equation 8.4 is applicable and k_L is proportional to Sc^{-n} where $-n$ is the slope of the solid line in Figure 8.5 (for the data shown, $n=0.63$). In contrast, the data in Figure 8.5 for $B_c=0.77\%$ show no obvious correlation between $\log(k_L)$ and $\log(Sc)$ demonstrating k_L is not proportional to Sc^{-n}.

Keeling (1993) hypothesized that the gas flux to or from the large bubbles dominates the bubble gas flux in the ocean. It should be noted that the Keeling (1993) model ignores contributions to the gas flux from small bubbles and that in seawater this flux could be significant. However, the Keeling model is useful as a rough approximation describing the dependence of k_L on K_h and Sc. According to Keeling (1993), k_L is proportional to $(RT_wK_h)^{-m}Sc^{-n}$ and as a result $\log(RT_wK_hk_L)$ will be linearly related to $\log(RT_wK_hSc^{-n})$. Figure 8.6 shows $\log(RT_wK_hk_L)$ plotted versus $\log(RT_wK_hSc^{-1/2})$ for k_L measured in the WST for SF_6, He, O_2, and CO_2 evasion at $B_c=0.77\%$ with cleaned water surfaces and $\log(RT_wK_hk_L)$ plotted versus $\log(RT_wK_hSc^{-2/3})$ for the same gases at $B_c=0.77\%$ with surfactant-influenced surfaces. For both cleaned and surfactant-influenced surfaces, the linear correlation between $\log(RT_wK_hk_L)$ and $\log(RT_wK_hSc^{-n})$ is good and the slopes of the lines for both data sets are similar and less than one. A slope that is different from one demonstrates that bubbles cause k_L to be a function of K_h and also decrease the overall dependence on Sc. Furthermore, the similarity between the slopes of the cleaned and surfactant-influenced data shows that the presence of surfactants does not significantly affect the dependence of k_L on K_h.

The WST data for transfer through cleaned and film-covered surfaces in the presence of bubbles show that the surfactant reduces k_L, but does not drastically change the functional dependence of k_L on K_h. If bubble plumes generated by breaking waves have a similar effect on gas

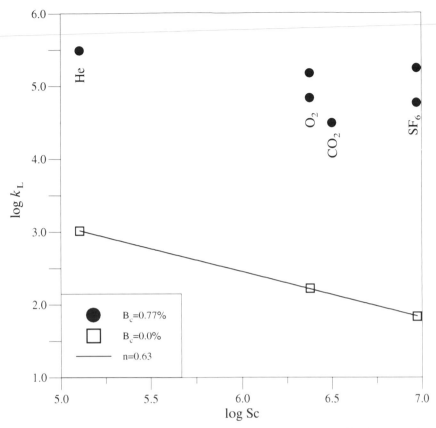

Figure 8.5. Whitecap simulation tank data showing the effect of bubbles on the functionality of k_L with respect to Schmidt number, Sc, for transfer through surfactant-influenced surfaces. In the absence of bubbles (B_c=0.0%), k_L for invasion of a gas is equal to k_L for evasion and the two values have been averaged together when available. When bubbles are present (B_c=0.77%), k_L for invasion of a gas is greater than k_L for evasion and each value has been plotted separately.

exchange in the ocean, it will further complicate parameterizing the effect of the microlayer on air–sea gas exchange. It is not clear presently how best to incorporate this effect into a simple model for estimating k_L from U, and, as a result, changes in dependence on Sc and K_h caused by bubbles have been ignored in the calculations presented below.

W. Asher

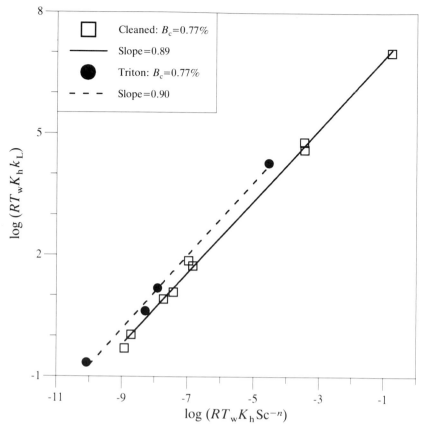

Figure 8.6. Comparison of the dependence of k_L on Sc and K_h, for bubble-mediated transfer through cleaned and surfactant-influenced surfaces. The value of n used to calculate the value of $\log(RT_wK_hSc^{-n})$ for the x-axis was ½ for the cleaned-surface data and ⅔ for the surfactant-influenced data.

Microlayer effects on global fluxes

Description of model

The effect of the surface microlayer on air–sea gas transfer was estimated for the net global flux of CO_2 and the flux of NH_3 from the Pacific Ocean. In each case, regional values for ΔP_X, K_h, and k_L or k_G were used to calculate F_x using Equation 8.1. Yearly averaged regional U and T_W were estimated from the maps of Tchernia (1980). Transfer velocities for CO_2 were calculated from U using Equation 8.8 for clean-surface regions or Equation 8.11 for surfactant-influenced regions. ΔP_{CO_2} was obtained from the map of Takahashi (1989) and K_h for CO_2 was calculated from T_W (DOE, 1994) assuming a salinity of 35 psu. k_L was normalized to Sc=660 using values calculated from T_W and the relation of Wanninkhof (1992). Transfer velocities for NH_3 were calculated using Equation 8.9 for clean regions and Equation 8.17 for microlayer-covered areas. ΔP_{NH_3} from 50° N to 11° S at 170° W was calculated from seawater and atmospheric NH_3 concentrations of Quinn *et al.* (1990). K_h for NH_3 was calculated from thermodynamic data provided by Pankow (1991) and Stumm and Morgan (1981).

The long-term oceanic uptake of atmospheric CO_2 produced by combustion of fossil fuels is rate-controlled by intermediate/deep-water ventilation, not air–sea transfer (Sarmiento, Orr and Siegenthaler, 1992). Therefore, estimating global CO_2 fluxes using this type of procedure is not strictly correct and could overestimate the importance of changes in k_L. In fact, global carbon cycle models have shown that oceanic CO_2 uptake is relatively insensitive to changes in air–sea transfer (Sarmiento *et al.*, 1992). However, these modelling studies have used constant increase or decrease in k_L for the entire ocean, and microlayer effects could reduce k_L in a specific ocean region rather than over the entire ocean surface. As will be shown in the next section, this leads to interesting effects on net global CO_2 fluxes calculated using the direct flux approach. However, it is understood that a more accurate estimate of oceanic CO_2 uptakes would be obtained using a global carbon cycle model.

Whether a region was chosen as surfactant-influenced or clean was determined by its average primary production. It is acknowledged that biological activity alone is insufficient information to predict accurately the extent, coverage, or concentration of naturally occurring surfactants and their effect on k_L (Goldman *et al.*, 1988). However, it is known that phytoplankton can produce large amounts of surface-active material (Zutic *et al.*, 1981; Williams *et al.*, 1986; Frew *et al.*, 1990) and given the coarse regional-scale model used, it is not clear whether there is another globally mapped parameter that would be a better indicator of the presence of surfactants. The global map of yearly-averaged synthetic primary production (SPP) from Berger and Herguera (1992) was used.

In the case of CO_2, it is known that ΔP_{CO_2} has a strong seasonal component in some regions of the ocean. It is also known that SPP can have a strong seasonal variation, depending on latitude. Although both of these seasonal cycles could affect the calculated fluxes, neither were included and, for the sake of simplicity, yearly averaged values were used for all variables. Given the assumptions and uncertainties already contained in the model, it was decided that while including seasonal cycles in ΔP_{CO_2} and SPP would make the model more 'realistic', it is not clear that it would be of further help in determining whether surfactants in the ocean microlayer could affect global air–sea gas fluxes. Future research will undoubtedly provide a more accurate parameterization for k_L at clean and surfactant-influenced sea surfaces. Combined with a more complete understanding of how to relate surfactant concentrations and surface film coverage to SPP, one of the next steps necessary in modelling the effect of the microlayer on gas fluxes would be to include the effects of seasonal variations in this type of model. This would be essential in estimating meaningful global gas fluxes. Finally, an additional justification for not using seasonally resolved SPP data for calculating NH_3 fluxes is that seasonally averaged NH_3 concentrations in the central Pacific are not yet available.

The data of Frew *et al.* (1990) and the results presented in Figure 8.4 suggest that visible slicks need not be present for soluble surfactants to reduce k_L. Therefore, it was assumed that regions with SPP above an arbitrary threshold were surfactant-influenced. This could be a bad assumption for gas-phase rate-control, however, because wave damping and changes in δ are seen mainly in slicked regions (Bock and Frew, 1993) and the ratio of slicked to unslicked area is not well known. Even for liquid-phase rate-control, the relation between primary production and surfactant effects is not known. Because of these unknowns, the four microlayer-coverage cases simulated were as follows: no surfactant influence (Baseline Case); Total Coverage, and three in which the coverage was related to SPP. In Case 1, regions having a maximum SPP>90 g-C m^{-2} y^{-1} were defined as microlayer-covered. In Case 2, regions having a maximum SPP>40 g-C m^{-2} y^{-1} were defined as microlayer-covered. Case 3 used the same microlayer-covered regions as Case 1, but it was assumed Sc increased in the surfactant-influenced regions due to increased microlayer viscosity (Jenkinson, 1986; Carlson, 1987). Table 8.1 lists the approximate latitudes and longitudes of the regions modelled for CO_2, and each region's yearly averaged T_W, U, and ΔP_{CO_2} and whether the region was clean (C) or microlayer-affected (S) under Case 1 or Case 2. Table 8.2 gives the equivalent information for NH_3.

Global air–sea CO_2 flux

Table 8.3 gives the calculated net global CO_2 fluxes for the Baseline, Total Coverage, and Case 1, 2 and 3 scenarios described above. When the microlayer is ignored (Baseline Case), the global flux is -1.9 Gt-C y^{-1} (giga-tons of carbon per year) where the minus sign denotes flux into the ocean. Although this results agrees with current estimates of the yearly oceanic CO_2 uptake (Sarmiento and Sundquist, 1992), it must be stressed that it should not be interpreted as a new 'best-estimate' of the net global ocean CO_2 flux. It is to be used solely for comparison purposes in this study and, as a rough check on the input data, indicates that the ΔP_{CO_2}, U, and T_W values used are not grossly inaccurate. Finally, because of the inherent uncertainties in these types of calculation, the somewhat fortuitous agreement between this current estimate and previous results does not imply that surface films are unimportant in determining the actual global air–sea CO_2 flux.

If the entire ocean surface is assumed to be affected by a microlayer (Total Coverage Case), the global flux is calculated to be -0.32 Gt-C y^{-1}, which is a significant decrease from the Baseline Case. However, because the assumption that the entire ocean surface is surfactant-influenced is not realistic, the Total Coverage case represents the maximum effect of surfactants in the microlayer on a liquid-phase, rate-controlled transfer process.

The global flux for the Case 1 scenario, where microlayer effects were limited to very high SPP regions, increased to -2.6 Gt-C y^{-1} from -1.9 Gt-C y^{-1} in the Baseline Case. This increase results from suppression of CO_2 evasion in the equatorial Pacific (Region 14, Table 8.1). For Case 2, with microlayer effects expanded to waters with intermediate SPP, the net global flux decreases to -0.78 Gt-C y^{-1}. This decrease is due mainly to suppression of CO_2 invasion in the north Atlantic (Region 15, Table 8.1). The net global flux calculated for Case 3 is -2.7 Gt-C y^{-1}, which is not significantly different from Case 1.

Figure 8.7 shows the fractional flux reduction/increase for CO_2 plotted as a function of fractional area microlayer coverage. The most interesting feature of the CO_2 data is that increases in microlayer coverage can increase the estimated net flux of CO_2 into the ocean. This occurs because gas transfer from the ocean to the atmosphere is suppressed in regions with $\Delta P_{CO_2} > 0$ in the Case 1 scenario. This causes a net increase in the estimated flux. In the Case 2 scenario, gas transfer is decreased in both regions with $\Delta P_{CO_2} > 0$ and $\Delta P_{CO_2} < 0$ with the net result a decrease of the CO_2 flux.

Table 8.1 Global ocean regions and physical parameters for CO_2 flux

Region	Latitude	Longitude	U (m s^{-1})	T_w (°C)	ΔP_{CO_2} (µatm.)	Case 1	Case 2
1	20°N–10°N	50°E–80°E	2.6	25	+25	S	S
2	20°N–10°N	60°E–100°E	9.9	25	+25	S	S
3	10°N–10°S	50°E–115°E	4.8	26	+10	C	S
4	10°S–40°S	30°E–125°E	5.7	25	–10	C	C
5	40°S–55°S	180°W–180°E	9.8	15	–21	C	S
6	55°S–60°S	180°W–180°E	11	5	–33	S	S
7	60°S–70°S	180°W–180°E	9.8	0	–5	C	S
8	50°N–40°N	150°E–180°E	8.6	8	+27	S	S
9	40°N–10°N	130°E–180°E	3.4	22	–9	C	C
10	10°N–10°S	125°E–165°E	1.7	27	+28	S	S
11	10°S–40°S	150°E–70°W	6.2	24	–16	C	C
12	60°N–40°N	180°W–130°W	12	12	+4	S	S
13	40°N–10°N	180°W–110°W	6.9	20	+9	S	S
14	10°N–10°S	165°E–80°W	6.2	25	+72	S	S
15	80°N–55°N	60°W–0°W	9.0	6	–37	C	S
16	55°N–40°N	65°W–10°W	5.6	10	–22	C	C
17	40°N–10°N	80°W–15°W	6.1	20	–15	C	C
18	10°N–10°S	40°W–25°E	2.6	26	+39	S	S
19	10°S–40°S	50°W–25°E	3.5	17	–7	C	C

Regions correspond to those defined by the ΔP_{CO_2} map of Takahashi (1989). U and T_w are yearly averages from the global wind field and temperature maps of Tchernia (1980). Microlayer-affected regions are designated by S, clean-surface regions by C.

Table 8.2 *Pacific ocean regions and physical parameters for NH_3 flux*

Region	Latitude	Longitude	U (m s^{-1})	T_w (°C)	ΔP_{NH_3} (µatm. ×10^3)	Case 1	Case 2
1	50°N–40°N	180°W–170°W	13	11	+3.5	S	S
2	40°N–30°N	180°W–170°W	10	19	+5.4	C	S
3	30°N–20°N	180°W–170°W	8.5	24	+6.1	C	C
4	20°N–10°N	180°W–170°W	4.4	26	+4.4	C	C
5	10°N–0°N	180°W–170°W	4.8	28	+4.1	C	S
6	0°S–10°S	180°W–170°W	2.7	28	+8.0	C	S
7	10°S–20°S	180°W–170°W	4.0	27	+11	C	C
8	50°N–40°N	170°W–160°W	13	11	+3.5	S	S
9	40°N–30°N	170°W–160°W	10	19	+5.4	C	S
10	30°N–20°N	170°W–160°W	8.5	24	+6.1	C	C
11	20°N–10°N	170°W–160°W	4.4	25	+4.4	C	S
12	10°N–0°N	170°W–160°W	4.8	25	+4.1	C	S
13	0°S–10°S	170°W–160°W	2.7	25	+8.0	C	S
14	10°S–20°S	170°W–160°W	4.0	25	+11	C	C

Regions are 10° E–W, 10° N–S intervals along 170°W. ΔP_{NH_3} data are from Quinn *et al.* (1990). U and T_w are yearly averages from the global wind field and temperature maps of Tchernia (1980). Microlayer-affected regions are designated by S, clean-surface regions by C.

Table 8.3 *Flux calculation results for* CO_2

Simulation scenario	SPP limit (g-C m^{-2} y^{-1})	Estimated flux (Gt-C y^{-1})
Baseline	∞	-1.9
Total coverage	0	-0.32
Case 1	90	-2.6
Case 2	40	-0.78
Case 3	90	-2.7

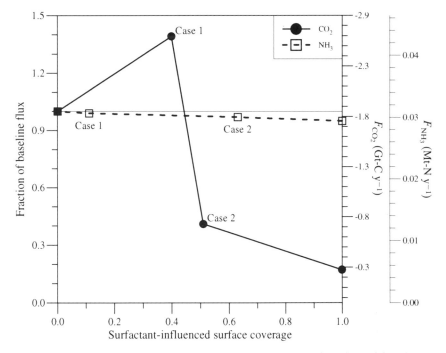

Figure 8.7. Fractional reduction/increase in net flux as a function of fractional area surfactant-influenced surface coverage for CO_2 and NH_3.

Comparison of the Case 1 and Case 2 fluxes show that microlayer effects can cause a change of nearly 2 Gt-C y^{-1} in the net CO_2 flux, which is a significant difference. Based on the results of the simple model presented here, it can be concluded that the sea-surface microlayer can have a large impact on the global air–sea flux of liquid-phase, rate-controlled gases. Furthermore, microlayer coverage of relatively small areas of the ocean can have a large effect on the global flux, and there is not a simple relation between fractional area microlayer coverage and flux reduction or increase. This result could be of great importance given the observed spatial variability in slicks coverage (Moum *et al.*, 1990; Ochadlik *et al.*, 1992; Bock and Frew, 1993) and ΔP_{CO_2} (Watson *et al.*, 1991a). This implies that the effect of the microlayer will be a function of microlayer coverage and the global map of the air–sea partial pressure difference of the gas. The results also indicate that increased v will have little effect on the global flux of a liquid-phase, rate-controlled gas.

NH_3 flux in the Pacific Ocean

Table 8.4 gives the average yearly flux of NH_3 from the central Pacific Ocean for the Baseline, Total Coverage, and Case 1 and 2 scenarios. Case 3 was not run for NH_3 because changes in Sc will not affect k_G. The net flux from the ocean to the atmosphere for the Baseline Case is 0.0309 Mt-N y^{-1} (mega-tons of nitrogen per year). On an area-averaged basis, this number agrees with fluxes estimated in Quinn *et al.* (1990). Total microlayer coverage results in a net flux of 0.0294 Mt-N y^{-1}. Comparison of these two results shows that the maximum estimated effect of the microlayer on gas-phase, rate-controlled exchange is a 5% reduction in the flux. This arises from the simple proportionality between k_G in slicked and clean regions given in Equation 8.17.

The flux for the Case 1, high-productivity scenario is 0.0306 Mt-N y^{-1} compared with 0.0302 Mt-N y^{-1} for the Case 2, low-productivity scenario. In contrast to CO_2, for which microlayer effects can either increase or decrease the global flux depending on regions covered, the net flux of NH_3 out of the central Pacific decreases regardless of the microlayer coverage. Furthermore, the decrease in the NH_3 flux is proportional to the fraction of the surface assumed to be microlayer-affected. The relation between coverage and flux is monotonically decreasing because ΔP_{NH_3} measured by Quinn *et al.* (1990) is always greater than zero over the region studied. Therefore F_{NH_3} is always from sea to air, and films can only decrease the net flux.

Also shown in Figure 8.7 is the fractional flux reduction for NH_3 plotted as a function of fractional area microlayer coverage. In contrast to the results for CO_2, the net flux of NH_3 decreases monotonically with increasing microlayer coverage. If ΔP_X for a gas-phase, rate-controlled species changed sign from region to region, its net flux could increase or decrease compared with its baseline value, as was observed for CO_2. It is

Table 8.4 *Flux calculation results for NH₃*

Simulation scenario	SPP limit (g-C m^{-2} y^{-1})	Estimated flux (Mt-N y^{-1})
Baseline	∞	+0.0309
Total coverage	0	+0.0294
Case 1	90	+0.0306
Case 2	40	+0.0301

unfortunate that this effect cannot be demonstrated for NH₃ with the data of Quinn *et al.* (1990), but the NH₃ simulation results for the Baseline and Total Coverage Cases do show that the microlayer can have up to a 5% reduction in the net flux. However, the microlayer-coverage scenarios used in Cases 1 and 2, which show a modest reduction of 1% to 3%, could be more realistic.

These results show, based on the assumptions used here, that the sea-surface microlayer has a minimal effect on suppressing gas-phase, rate-controlled air–sea fluxes. As discussed above, larger changes in k_G could be possible if decreases in capillary wave slope are more important in controlling air-phase interfacial mass transport than in governing momentum transport. This would increase the effect of surface films on estimated air–sea fluxes.

Conclusions, implications for further research

In summary, all available experimental evidence suggests that the major effect of surfactants associated with the ocean surface microlayer on air–sea gas exchange will be to decrease k_L or k_G at a particular wind speed compared with a clean ocean surface. In the absence of breaking waves, it is also probable that the microlayer and/or slicks will increase the dependence of k_L on Sc. The interaction among the microlayer, breaking waves, and gas transfer under oceanic conditions is not well known and cannot be resolved at present. It is unlikely, given current understanding of the chemical composition of the microlayer and the interaction of mixed-compositions films with gas exchange, that the microlayer will act as a barrier to gas transfer. Similarly, it is believed that changes in K_h will be unimportant on global scales. The simple, air–sea flux calculations performed have shown that changes in viscosity will have little effect on global fluxes.

There are obvious limitations to the modelling approach used here, in light of the patchiness of ocean slicks, the largely unknown effect of the microlayer on k_L and k_G, and the coarse estimates of U and T_W used. Especially in the case of CO_2, ocean general circulation models (GCMs) are more accurate at calculating net global oceanic gas uptakes because CO_2 storage in the ocean is a function of deep-water formation and biological productivity (Sarmiento *et al.*, 1992). Therefore, these results should not be used to support or refute estimates of the global ocean CO_2 sink (Tans, Fung and Takahashi, 1990; Sarmiento and Sundquist, 1992). The results presented here do illustrate that using the best available parameterizations for k_L or k_G in terms of U for clean and surfactant-influenced sea surfaces, the microlayer could have a significant effect on global or regional air–sea gas fluxes. However, it is understood that the assumptions used here concerning the extent of the area covered by films on the ocean surface and the effect of these films on k_L and k_G and the Sc dependence of k_L have not been tested. Therefore, accurate estimation of the effect of films on global gas fluxes remains problematic because of a lack of field data on gas fluxes in the presence of surface slicks and unslicked, but microlayer-covered, sea surfaces.

Although there has been a great advance in understanding the marine microlayer and its effect on near-surface ocean processes (see, for example, the sea-surface microlayer issue of *J. Geophys. Res.*, **97, C4**), there remain many unknowns in estimating the effect of the microlayer on air–sea gas fluxes. Further research is required to resolve the parameterization of k_L and k_G in terms of U, Sc, and K_h for film-covered and clean ocean surfaces. Of particular use would be methods resolving gas transfer at small spatial and short temporal scales. It is hoped that continued refinement of dual-tracer and passive-flux techniques will simplify these measurements. Additionally, it must be determined whether visible slicks need be present for the microlayer to suppress gas transfer. It is clear that synthetic aperture radar can determine ocean slick coverage (Ochadlik *et al.*, 1992), but there is currently no remote-sensing method for determining whether an ocean region is surfactant-influenced in the absence of a visible slick. If, as suggested by Frew *et al.* (1990), visible slicks need not be present for suppression of gas transfer, methods for remote determination of the presence of a microlayer must be developed.

Finally, if further field and laboratory studies provide evidence that the microlayer has a significant effect on global air–sea fluxes, microlayer-induced changes in k_L, k_G, and/or ΔP_X must be parameterized for inclusion in geochemical models including oceanic and atmospheric GCMs. This will require accurate knowledge of how the microlayer affects the parameterization of the transfer velocities with wind speed or other environmental forcing functions and a method for predicting whether an ocean region will be affected by a microlayer from model-derived variables.

Acknowledgements

Erik Bock of Woods Hole Oceanographic Institute (WHOI) and Tetsu Hara of University of Rhode Island helped formulate the relation between wave slope and C_E and provided the wave slope spectra. Nelson Frew of WHOI provided his wind tunnel data for K_L/U analysis. Their help is greatly appreciated. The WST data were collected and analyzed by Lisa Karle, Bruce Higgins, and Paul Farley of Pacific Northwest Laboratory/Marine Sciences Laboratory and Ira Leifer of Georgia Institute of Technology. This work was supported by the US Department of Energy, Office of Health and Environmental Research, Enironmental Sciences Division under Contract DE-AC06–76RLO 1830. Pacific Northwest Laboratory is operated for DOE by Battelle Memorial Institute.

References

Adamson, A. W. (1990). *Physical Chemistry of Surfaces*, 5th edn., New York: Wiley. 777 pp.

Alpers, W. and Hühnerfuss, H. (1989). The damping of ocean waves by surface films: a new look at an old problem. *J. Geophys. Res.*, **94C**, 6251–65.

Asher, W. E. and Farley, P. J. (1995). Phased-Doppler anemometer measurement of bubble concentrations in laboratory-simulated breaking waves. *J. Geophys. Res.*, **100C**, 7045–56.

Asher, W. E. and Pankow, J. F. (1986). The interaction of mechanically generated turbulence and interfacial films with a liquid phase controlled gas/liquid transport process. *Tellus* **38B**, 305–18.

Asher, W. E. and Pankow, J. F. (1991). The effect of surface films on concentration fluctuations close to a gas/liquid interface. In *Air–Water Mass Transfer*, ed. S. E. Wilhelms and J. S. Gulliver, pp. 68–80. New York: ASCE American Society of Civil Engineers.

Asher, W. E., Farley, P. J., Wanninkhof, R., Monahan, E. C. and Bates, T. S. (1992). Laboratory and field measurements concerning the correlation of fractional area foam coverage with air/sea gas transport. In *Precipitation Scavenging and Atmosphere–Surface Exchange, Vol. 2. The Semonin Volume: Atmosphere–Surface Exchange Processes*, ed. S. E. Schwartz and W. G. N. Slinn, pp. 815–28. Washington DC: Hemisphere.

Asher, W. E., Karle, L. M., Higgins, B. J., Farley, P. J., Leifer, I. S. and Monahan, E. C. (1996). The influence of bubble plumes on air–seawater gas transfer velocities. *J. Geophys. Res.*, **101C**, 12 027–41.

Back, D. D. and McCready, M. J. (1988). Effect of small wavelength waves on gas transfer across the ocean surface. *J. Geophys. Res.*, **93C**, 5143–52.

Barger, W. R. and Means, J. C. (1985). Clues to the structure of marine organic material from the study of physical properties of surface films. In *Marine and Estuarine Geochemistry*, ed. A. C. Sigleo and A. Hattori, pp 47–67. Chelsea, MI: Lewis.

Berger, W. H. and Herguera, J. C. (1992). Reading the sedimentary record of the ocean's productivity. In *Primary Productivity and Biogeochemical Cycles in the Sea*, ed. P. G. Falkowski and A. D. Woodhead, pp. 455–86. New York: Plenum Press.

Bock, E. J. and Frew, N. M. (1993). Static and dynamic response of natural multi-component oceanic surface films to compression and dilation: laboratory and field observations. *J. Geophys. Res.*, **98C**, 14 599–617.

Brockman, U. H., Hühnerfuss, H., Kattner, G., Broecker, H-Ch. and Hentzschel, G. (1982). Artificial surface films in the sea near Sylt. *Limnol. Oceanogr.*, **27**, 1050–8.

Broecker, H-Ch. and Siems, W. (1984). The role of bubbles for gas transfer from water to air at higher wind speeds. In *Gas Transfer at Water Surfaces*, ed. W. Brutsaert and G. H. Jirka, pp. 229–36. Hingham, MA: Reidel.

Broecker, H-Ch., Peterman, J. and Siems, W. (1978). The influence of wind on CO_2 exchange in a wind wave tunnel, including the effects of monolayers. *J. Mar. Res.*, **36**, 595–610.

Brumley, B. B. and Jirka, G. H. (1988). Air–water transfer of slightly soluble gases: turbulence, interfacial processes and conceptual models. *Physico-chem. Hydrodynam.*, **10**, 295–319.

Carlson, D. J. (1987). Viscosity of sea-surface slicks. *Nature*, **329**, 823–5.

Carlson, D. J., Cantey, L. L. and Cullen, J. J. (1988). Description of and results from a new surface microlayer sampling device. *Deep Sea Res.*, **35A**, 1205–13.

Daniil, E. I. and Gulliver, J. S. (1991). Influence of waves on air–water gas transfer. *J. Environ. Engn.*, **117**, 522–40.

Davies, J. T., Kilner, A. A. and Ratcliff, G. A. (1964). The effect of diffusivities and surface films on rates of gas absorption. *Chem. Engng. Sci.*, **19**, 583–90.

Deacon, E. L. (1979). Role of coral mucus in reducing wind drag over coral reefs. *Boundary Layer Meteorol.*, **17**, 517–21.

DOE (1994). *Handbook of Methods for the Analysis of the Various Parameters of the Carbon Dioxide System in Sea Water, Version*, 2, ed. A. G. Dickson and C. Goyet, 156 pp. ORNL/CDIAC-74.

Ewing, G. C. and McAlister, E. D. (1960). On the thermal boundary layer of the ocean. *Science*, **131**, 1374–6.

Frew, N. M. and Nelson, R. K. (1992a). Isolation of marine microlayer film surfactants for ex situ study of their surface physical and chemical properties. *J. Geophys. Res.*, **97C**, 5281–90

Frew, N. M. and Nelson, R. K. (1992b). Scaling of marine microlayer film surface pressure–area isotherms using chemical attributes. *J. Geophys. Res.*, **97C**, 5291–5300.

Frew, N. M., Goldman, J. C., Dennett, M. R. and Johnson, A. S. (1990). Impact of phytoplankton-generated surfactants on air–sea gas exchange. *J. Geophys. Res.*, **95C**, 3337–52.

Frew, N. M., Bock, E. J. and Goldman, J. C. (1992). Surfactant films: an unconstrained parameter in wind speed–gas exchange relationships. *Eos Trans. AGU, Suppl.*, **73(43)**, 245.

Frysinger, G. S., Asher, W. E., Korenowski, G. M., Barger, W. R., Klusty, M. A., Frew, N. M. and Nelson, R. K. (1992). Study of ocean slicks by nonlinear laser processes. 1. Second harmonic generation. *J. Geophys. Res.*, **97C**, 5253–69.

Geernaert, G. L., Katsaros, K. B. and Richter, K. (1986). Variation of the drag coefficient and its dependence on sea state. *J. Geophys. Res.*, **91C**, 7667–79.

Goldman J. C., Dennett, M. R. and Frew, N. M. (1988). Surfactant effects on air/sea gas exchange under turbulent conditions. *Deep-Sea Res., A*, **35**, 1953–71.

Grief, R., Cornet, I. and Kappesser, R. (1972). Diffusivity of oxygen in a non-newtonian saline solution. *Int. J. Heat Mass Transfer*, **15**, 593–7.

Hühnerfuss, H., Lange, P. and Walter, W. (1982). Wave damping by mono-molecular surface films and their chemical structure. Part I: Variation of the hydrophobic part of carboxylic acid esters. *J. Mar. Res.*, **40**, 209–25.

Hühnerfuss, H., Alpers, W., Garrett, W. D., Lange, P. A. and Stolte, S. (1983). Attenuation of capillary and gravity waves at sea by mono-molecular organic surface films. *J. Geophys. Res.*, **88C**, 9809–16.

Hühnerfuss, H., Lange, P. and Walter, W. (1984). Wave damping by mono-molecular surface films and their chemical structure. Part II: Variation of the hydrophobic part of the film molecules including natural substances. *J. Mar. Res.*, **42**, 737–59.

Hunter, K. A. and Liss, P. S. (1981). Organic sea surface films. In *Marine Organic Chemistry*, ed. K. Duursma and R. Dawson, pp. 259–268. Amsterdam, Netherlands: Elsevier.

Jacobs, S. J. (1989). Effective roughness length for turbulent flow over a wavy surface. *J. Phys. Oceanogr.*, **19**, 998–1010.

Jähne, B. (1991). New experimental results on the parameters influencing air–sea gas exchange. In *Air–Water Mass Transfer*, ed. S. E. Wilhelms and J. S. Gulliver, pp. 582–592. New York: ASCE.

Jähne, B., Huber, W., Dutzi, A., Wais, T. and Ilmberger, J. (1984). Wind/wave tunnel experiments on the Schmidt number and wave field dependence of air–water gas exchange. In *Gas Transfer at Water Surfaces*, ed. W. Brutsaert and G. H. Jirka, pp. 303–10. Hingham, MA: Reidel.

Jähne, B., Munnich, K. O., Bosinger, R., Dutzi, A., Huber, W. and Libner, P. (1987). On the parameterization of air–water gas exchange. *J. Geophys. Res.*, **92C**, 1937–49.

Jähne B., Libner, P., Fischer, R., Billen, T. and Plate, E. J. (1989). Investigating the transfer process across the free aqueous viscous boundary layer by the controlled flux method. *Tellus*, **41B**, 177–95.

Jarvis, N. L. (1962). The effect of monomolecular films on surface temperature and convective motion at the water/air interface. *J. Colloid Sci.*, **17**, 512–22.

Jarvis, N. L. and Kagarise, R. E. (1962). Determination of the surface temperature of water during evaporation studies: a comparison of thermistor with infrared radiometer measurements. *J. Colloid Sci.*, **17**, 501–11.

Jenkinson, I. R. (1986). Oceanographic implications of non-newtonian properties found in phytoplankton cultures. *Nature*, **323**, 435–7.

Ju, L. and Ho, C. S. (1986). The measurement of oxygen diffusion coefficients in polymeric solutions. *Chem. Engng. Sci.*, **41**, 578–89.

Katsaros, K. B. (1980). The aqueous thermal boundary layer. *Boundary Layer Meteorol.*, **18**, 107–27.

Keeling, R. F. (1993). On the role of large bubbles in air–sea gas exchange and supersaturation in the ocean. *J. Mar. Res.*, **51**, 237–71.

Korenowski, G. M., Frysinger G. S. and Asher, W. E. (1993). Noninvasive probing of the ocean surface using laser-based nonlinear optical methods. *Photogramm. Engng. Remote Sensing*, **59**, 363–369.

Lange, P. A. and Hühnerfuss, H. (1978). Drift response of monomolecular slicks to wave and wind action. *J. Phys. Oceanogr.*, **8**, 142–50.

Large, W. G. and Pond, S. (1981). Open ocean momentum flux measurements in moderate to strong winds. *J. Phys. Oceanogr.*, **11**, 324–36.

Lee, Y. H., Tsao, G. T. and Wankat, P. C. (1980). Hydrodynamic effect of surfactants on gas–liquid oxygen transfer. *Am. Inst. Chem. Engng. J.*, **26**, 1008–1012.

Ling, S. C. and Kao, T. W. (1976). Parameterization of the moisture and heat transfer process over the ocean under whitecap sea states. *J. Phys. Oceanogr.*, **6**, 306–15.

Liss, P. S. (1973). Processes of gas exchange across the air–sea interface. *Deep-Sea Res.*, **20**, 221–38.

Liss, P. S. (1983). Gas transfer: experiments and geochemical implications. In *Air–Sea Exchange of Gases and Particles*, ed. P. S. Liss and W. G. N. Slinn, pp. 241–98. Hingham, MA: Reidel.

Liss, P. S. and Martinelli, F. N. (1978). The effect of oil films on the transfer of oxygen and water vapour across an air–water interface. *Thalassa Jugosl.*, **14**, 215–20.

Liss, P. S. and Merlivat, L. (1986). Air–sea gas exchange rates: introduction and synthesis. In *The Role of Air–Sea Exchange in Geochemical Cycling*, ed. P. Buat-Menard, pp. 113–27. Hingham, MA: Reidel.

Liss, P. S. and Slater, P. G. (1974). Fluxes of gases across the air–sea interface. *Nature*, **247**, 181–4.

Liu, W. T., Katsaros, K. B. and Businger, J. B. (1979). Bulk parameterization of air–sea exchanges of heat and water vapor including molecular constraints at the interface. *J. Atmos. Sci.*, **36**, 1722–35.

Martinelli, F. N. (1979). The effect of surface films on gas exchange across the air–sea interface, PhD. Thesis, University of East Anglia, Norwich, United Kingdom.

Memery, L. and Merlivat, L. (1985). Modeling of the gas flux through bubbles at the air–water interface. *Tellus*, **37B**, 272–85.

Monahan, E. C. (1993). Occurrence and evolution of acoustically relevant subsurface bubble plumes and their associated, remotely monitorable, surface whitecaps. In *Natural Physical Sources of Underwater Sound*, ed. B. R. Kerman, pp. 503–17. Dordrecht, Netherlands: Kluwer.

Moo-Young, M. and Shoda, M. (1973). Gas absorption at the free surface of a flowing water stream: effects of a surfactant and of surface baffles. *Ind. Engng. Chem. Proc. Dev.*, **12**, 410–14.

Moum, J. N., Carlson, D. J. and Cowles, T. J. (1990). Sea slicks and surface strain. *Deep Sea Res.*, **37**, 767–75.

Ocampo-Torres, F. J., Donelan, M. A., Merzi, N. and Jia, F. (1994). Laboratory measurements of mass transfer of CO_2 and H_2O vapour for smooth and rough flow conditions. *Tellus*, **46B**, 16–32.

Ochadlik, A., Cho, R. P. and Evans-Morgis, J. (1992). Synthetic aperture radar observations of currents colocated with slicks. *J. Geophys. Res.*, **97C**, 5325–30.

Pankow, J. F. (1991). *Aquatic Chemistry Concepts*. Chelsea, MI: Lewis. 683 pp.

Quinn, P. K., Bates, T. S., Johnson, J. E., Covert, D. S. and Charlson, R. J. (1990). Interaction between the sulfur and reduced nitrogen cycles over the central Pacific Ocean. *J. Geophys. Res.*, **95D**, 16 405–16.

Robertson, J. E. and Watson, A. J. (1992). Thermal skin effect of the surface ocean and its implication for CO_2 uptake. *Nature*, **358**, 738–40.

Robinson, I. S. (1985). *Satellite Oceanography*. Chichester, England: Ellis Horwood. 455 pp.

Sarmiento, J. L. and Sundquist, E. T. (1992). Revised budget for the oceanic uptake of anthropogenic carbon dioxide. *Nature*, **356**, 589–93.

Sarmiento, J. L., Orr, J. C. and Siegenthaler, U. (1992). A perturbation simulation of CO_2 uptake in an ocean general circulation model. *J. Geophys. Res.*, **97C**, 3621–46.

Schluessel, P., Emery, W. J., Grassl, H. and Mammen, T. (1990). On the bulk-skin temperature difference and its impact on satellite remote sensing of sea surface temperature. *J. Geophys. Res.*, **95C**, 13 341–56.

Smith, S. D. (1988). Coefficients for sea surface wind stress, heat flux, and wind profiles as a function of wind speed and temperature. *J. Geophys. Res.*, **93C**, 15 467–72.

Springer, T. G. & Pigford, R. L. (1970). Influence of surface turbulence and surfactants on gas transport through liquid interfaces. *Ind. Engng. Chem. Fundament.*, **9**, 458–65.

Stumm, W. & Morgan, J. J. (1981). *Aquatic Chemistry*, 2nd edn. New York: Wiley. 780 pp.

Takahashi T. (1989). The carbon dioxide puzzle. *Oceanus*, **32**, 22–30.

Tans, P. P., Fung, I. Y. and Takahashi, T. (1990). Observational constraints on the global atmospheric carbon dioxide budget. *Science*, **247**, 1431–8.

Tchernia, P. (1980). *Descriptive Regional Oceanography*. New York: Pergamon Press. 253 pp.

Thompson, A. M. and Zafiriou, O. C. (1983). Air–sea fluxes of transient atmospheric species. *J. Geophys. Res.*, **88C**, 6696–708.

Wanninkhof, R. (1992). Relationship between wind speed and gas exchange over the ocean. *J. Geophys. Res.*, **97C**, 7373–82.

Wanninkhof, R. and Bliven, L. F. (1991). Relationship between gas exchange, wind speed, and radar backscatter in a large wind–wave tank. *J. Geophys. Res.*, **96C**, 2785–96.

Wanninkhof, R., Asher, W. E., Weppernig, R., Chen, H., Schlosser, P., Langdon, C. and Sambrotto, R. (1993). Gas transfer experiment on Georges Bank using two volatile deliberate tracers. *J. Geophys. Res.*, **98C**, 20 237–48.

Watson, A. J., Robinson, C., Robinson, J. E., Williams, P. J. le B. and Fasham, M. J. R. (1991a). Spatial variability in the sink for atmospheric carbon dioxide in the north Atlantic. *Nature*, **350**, 50–3.

Watson, A. J., Upstill-Goddard, R. C. and Liss, P. S. (1991b). Air–sea gas exchange in rough and stormy seas measured by a dual-tracer technique. *Nature*, **349**, 145–7.

Whitman, W. G. (1923). Preliminary experimental confirmation of the two-film theory of gas absorption. *Chem. Metall. Engng.*, **29**, 148–57.

Williams, P. M. (1986). The chemical composition of the sea-surface micro-layer and its relation to the occurrence and formation of natural sea-surface films. In *ONRL Workshop Proceedings: The Role of Surfactant Films on the Interfacial Properties of the Sea Surface, Report C-11–86*, ed. F. Herr and J. Williams, pp. 79–110. London: US Office of Naval Research.

Williams, P. M., Carlucci, A. F., Henrichs, S. M., Van Vleet, E. S., Horrigan, S. G., Reid, F. M. H. and Robertson, K. J. (1986). Chemical and micro-biological studies of sea surface films in the southern Gulf of California and off the west coast of Baja California. *Mar. Chem.*, **19**, 17–98.

Woolf, D. K. and Thorpe, S. (1991). Bubbles and the air–sea exchange of gases in near-saturation conditions. *J. Mar. Res.*, **49**, 435–66.

Zutic, V., Cosovic, B., Marcenko, E. and Bihari, N. (1981). Surfactant pro-duction by marine phytoplankton. *Mar. Chem.*, **10**, 505–20.

9

Chemistry of the sea-surface microlayer

KEITH A. HUNTER

Abstract

The last 15 years has seen a considerable increase in our understanding of the chemical composition of the sea-surface microlayer. However, many new developments in methods of chemical analysis have yet to be applied systematically to the microlayer. The development of continuous microlayer samplers coupled to UV absorbance or fluorescence detectors now allows much greater temporal and spatial resolution to be achieved in field measurements, and will have great application with the development of new chemical sensing technology. These techniques, and a greater range of studied environments, indicate that microlayer enrichment of the major classes of organic compounds (protein, carbohydrate) or of organic parameters such as dissolved organic carbon, is less than earlier studies might have indicated. Enrichment factors larger than 1.5 are relatively rare, and depletion is often observed. Improvements in the analysis of specific organic compounds, including better sample blanks, sensitivity and improved identification of individual compounds, have all been seen. Recently-reported concentrations of PCB and organochlorine insecticides in low latitude regions are extremely low, although microlayer samples enriched in these components are found at high latitudes. Organotin species have also been reported in microlayers. For trace elements passively associated with surface-layer organic material such as trace metals, a confusing picture exists. Some coastal, polluted waters appear to be enriched with trace metals in the microlayer, but for other regions reported microlayer measurements are in error owing to spurious sample contamination during sample collection. Particulate inorganic species in the microlayer are better understood. Bubble flotation, mixing processes and atmospheric deposition control microlayer concentrations of these components to differing extents depending on geographic location. Overall, particulate species are the most consistently enriched in the surface microlayer because of stabilization of particles at the air–sea interface through surface tension forces.

Prologue

The sea-surface microlayer has fascinated and perplexed environmental scientists for most of the last three decades. Despite the fact that there are still many gaps in our understanding of the physical, chemical and biological properties of the microlayer, its environmental significance has remained widely appreciated. Indeed, it is the effects of the microlayer on environmental processes that make it such an important focus of study. Some examples of these effects are:

- Sea-to-air transfer of chemical substances and organisms through bubble bursting and aerosol formation.
- Influence on air–sea gas exchange rates.
- Formation of slicks and foams which can be aesthetically displeasing when they come ashore on beaches.
- Possible harmful effects on neuston, especially neustonic larval stages of marine organisms.
- Sea-surface smoothing through damping of capillary waves.

Perhaps the first recorded speculation about the existence of the microlayer was that made by Alec Wilson, who in 1959 reported the presence of both albuminoid nitrogen compounds and an excess of potassium (compared with seawater) in snow samples collected at high altitude in New Zealand (Wilson, 1959). He concluded that:

It is difficult to avoid the conclusion that the source is the ocean itself . . . In order to explain these facts, it is necessary to speculate that the upper, very thin layer of the ocean has a different composition from that of the rest of the ocean.

A similar observation, whose significance in terms of microlayer effects on sea-to-air mass transport was pointed out by MacIntyre in his 1974 review, is 'Dean's Recipe' for reproducing the rainwater that falls on coastal Taita in New Zealand (Dean, 1963):

Seawater	0.5 ml
Dried algae and plankton	4 mg
Distilled water	to make 1 l

The combination of 4 mg marine organic material with only 0.5 ml seawater represents an increase over the concentration of organic matter in bulk seawater of at least three orders of magnitude, clearly suggesting a significant enrichment process during the formation of sea-spray.

The 1960s witnessed an enormous growth in scientific interest in the air–sea exchange of gases, particularly those, such as CO_2, implicated in global warming. Thus, the microlayer took on a new significance, not only through possible direct effects on gas transfer rates, but also through indirect effects as a result of controlling the surface physical state through wave damping.

The formation of slicks and foams is an obvious and direct manifestation of the microlayer. The general public usually link these phenomena to pollution effects, but in fact both slicks and foams occur quite naturally in unpolluted waters and probably represent an important, if transient, habitat and food source for many species.

However, natural sea-surface films and foams are a substrate in which pollutants can accumulate from both sub-surface water and atmospheric deposition (Liss, 1975). Thus, in recent years there have been increasing reports claiming enhanced toxic effects and negative biological impacts in the microlayer. For example, surface microlayer contamination has been suggested as a cause of reduced viability of fish eggs, increases in chromosomal abnormalities in developing embryos, larval abnormalities and mortality (e.g. Cross *et al.*, 1987; Hardy *et al.*, 1987; Kocan *et al.*, 1987), although direct evidence for these effects is difficult to obtain.

The influence of organic surfactants in the microlayer on the damping of capillary waves was recognized very early on (Garrett, 1967). This process leads to the characteristic smooth surface appearance of visible slicks. In recent years, this role of the microlayer has assumed a new importance because of the development of remote sensing technology that is affected by the sea-surface state, including nonlinear laser techniques, passive microwave detectors and other radar-based methods (e.g. Onstott and Rufenach, 1992; Frysinger *et al.*, 1992).

These introductory examples (which are by no means exhaustive in scope) illustrate the wide range of scientific aims and interests that converge on the topic of the sea-surface microlayer. But what is the microlayer itself? In this review I will adopt a broad phenomenological definition:

The sea-surface microlayer is that microscopic portion of the surface ocean which is in contact with the atmosphere and which may have physical, chemical or biological properties that are measurably different from those of adjacent sub-surface waters.

It can be seen that this definition encompasses the variety of processes that are important at the air–sea interface, and their ranges and time-scales of action. It also avoids the question of defining the physical thickness of the microlayer, one which has been particularly vexing (Liss, 1975). Methods for sampling the microlayer, which impose a practical definition of microlayer thickness, are discussed in the next section, as well as in Hardy, Chapter 11, this volume.

Methods for sampling the microlayer

The screen sampler

A variety of techniques have been used to collect samples of the sea-surface microlayer for chemical or biological analysis. The first practical device was the well-known screen sampler described by Garrett (1965). This comprises a metallic mesh, typically with mesh wires about 0.2–0.3 mm in diameter and 60–70% void space, held in a suitable frame. In use, the screen is held horizontally and dipped briefly into the sea surface, after which it is withdrawn and the seawater and surface film material entrapped by the mesh spaces allowed to drain into a sample bottle. Plastic versions of the screen device have also been used for study of inorganic constituents of the microlayer such as trace metals. The thickness of the screen microlayer sample can be calculated from the surface area of the screen and the volume of the water collected. Typical results are in the range 150–400 μm, and depend mainly on the diameter of the mesh wires, but also on how the sampler is used. The efficiency of Garrett's original device (made of Monel metal) for the sampling of oleic acid monolayers was found to be 75% down to surface film pressures of $10^{-3}\,\mathrm{N\,m^{-2}}$ (Garrett, 1965). The less than 100% efficiency arises because on the first dip of the screen, oleic acid is adsorbed to the screen material itself. This initial adsorption is considered to deactivate the Monel metal towards further such adsorption, so that on subsequent dipping the screen removes a sample equivalent to its void area.

The glass plate sampler

Another very simple device is the glass plate sampler described by Harvey and Burzell (1972). A clean glass plate (typically 20 × 20 cm) with an attached handle is immersed vertically through the sea surface and withdrawn at a steady rate of about 20 cm s^{-1}. Because clean glass is a high-energy solid surface that is easily wetted, a water film of some 40–80 μm thickness drawn from the sea surface adheres to the plate after it is withdrawn. The film is then removed into a sample bottle by means of a wiper blade. Although this technique superficially resembles the mono-layer plate sampling technique of Blodgett (1934, 1935), the Harvey–Burzell plate is much larger and its rate of withdrawal much faster.

Sample retention takes place largely through viscous retention, rather than by adsorption to the glass surface, as in the Blodgett technique.

Carlson (1982a) carried out a comparison of the plate and screen samplers. His measurements included DOC, POC, PON, UV absorbance, chlorophyll-*a* and ATP. Plate samples were always thinnest (52 ± 2 μm), and their sample thickness depended on surface wave state and water temperature. Carlson (1982a) concluded that because the plate and screen devices sample the microlayer in different ways, it was not advisable to compare too closely the data obtained by the two methods.

A related technique is the funnel sampler devised by Morris (1974). A large polythene funnel (20 cm diameter) is immersed in seawater beneath the surface with its stem plugged and then slowly withdrawn vertically through the water surface, thus isolating a section of the sea surface (about 300 cm^2 in area). The plug is then removed and the seawater allowed to slowly drain away, leaving the microlayer sample adhering to the inside surface of the funnel. The sample is then removed by rinsing the walls of the funnel with a small volume of suitable solvent. Morris (1974) used a chloroform/methanol mixture in his study of microlayer lipids. A problem with this sampler is that it is not possible to measure easily the thickness of the sample collected. However, it is possible to calculate the amount of material present per unit area of surface sampled.

Hydrophobic samplers

Several investigators have employed microlayer samplers that use a hydrophobic, non-wettable surface material in contact with the microlayer to minimize the amount of sub-surface water collected as well. Larsson *et al.* (1974) used a Teflon plate that is placed on the surface of the water, allowing microlayer materials to adsorb. It was necessary to perforate the plate with a number of conical holes in order to reduce the water–air contact area, so that this device collects a reasonable quantity of entrained water in the holes, yielding an effective sample thickness of 60–100 μm. Garrett and Barger (1974) have described a similar hydrophobic adsorption sampler made of thin Teflon sheet that is clipped to a circular metal holder. The sheet is touched to the water surface and removed, entraining very little sub-surface water. After removal from the frame, the Teflon sheet can be cut up into strips and placed in a Soxhlet extraction apparatus for removal of organic materials. Like the Morris funnel sampler, it is not possible to measure easily the thickness of the sample collected using the Teflon sheet.

Rotating drum samplers

All of the samplers described so far are operated manually, and while it is possible to obtain a small-to-moderate-sized sample for analysis without too much effort, they are all relatively tedious to use and do not easily provide for large sample volumes. Accordingly, much effort has been devoted to the development of larger-volume microlayer samplers. The first, described by Harvey (1966), consists of a rotating drum with a hydrophilic ceramic surface mounted on floats so that it is half-immersed in the water. As the device is moved through the water, a scraper blade adjacent to the top of the drum scrapes off the water film adhering to the drum into a sample bottle. The rotation axis of the drum is perpendicular to the direction of movement through the water, so that the device superficially resembles a half-submerged steam roller slowly moving forward through the water. The Harvey drum sampler operates on similar viscous retention principles as the glass plate, and collects a surface film of approximately 60–100 μm thickness. Under average conditions, the sample collection rate is about 300 ml/min., making it quite simple to collect large sample volumes.

A Harvey-type drum sampler which incorporates a SeaStar sampler for *in situ* filtration and extraction of microlayer samples has been used by Nicholls and Espey (1991) for characterization of organic matter at the air–sea interface near the Malabar sewage outfall in Sydney, Australia.

A self-contained underway microlayer sampler (with the apt acronym SCUMS) of a similar general principle was described by Carlson *et al*. (1988). This is illustrated in Figure 9.1. Like the Harvey drum sampler, it uses a rotating hydrophilic surface (in this case, glass). However, the drum is a hollow cylinder with its rotation axis *parallel* to the direction of forward movement, rather than perpendicular. This means that the forward cross-section of the drum is very small, minimizing disturbance to the microlayer. The SCUMS device is controlled and powered remotely from the mother ship via a cable. The microlayer sample scraped off the drum is pumped directly to this vessel for collection and/or analysis. By coupling the sample flow to a suitable detector (e.g. UV or fluorescence detectors), this ingenious device allows continuous, almost real-time measurements of important microlayer properties to be made concurrently with measurements of hydrographic and atmospheric conditions.

As an example, Figure 9.2 shows a trace of the ratio of UV absorbance for microlayer and sub-surface (10 cm depth) waters. This dramatically

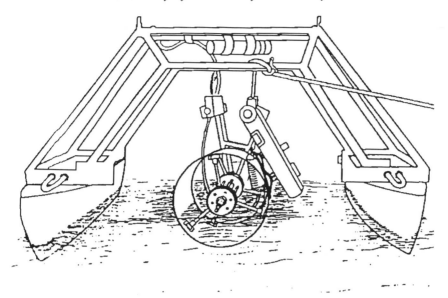

Figure 9.1. Schematic front-end view of the SCUMS microlayer sampling device (reproduced from Carlson *et al.*, 1988).

reveals the spatial and temporal variability of microlayer enrichment of UV-absorbing organic materials. With the development of new real-time chemical sensors, this sampling technique clearly offers tremendous potential for future microlayer studies.

The prism-dipping method

Most of the samplers described so far collect relatively thick microlayer samples (40–400 μm). Some attempts have been made to collect much thinner layers that may be more representative of the depth range over which microlayer enrichment of chemical and biological materials occurs. In surface chemistry research, the classical approach to the collection of surface monolayers of hydrophobic material, such as a fatty acid spread over a water surface, is that devised by Blodgett (1934). A small plate of glass or any other high-energy hydrophilic surface (e.g. a microscope slide) is immersed into the water and withdrawn vertically through the surface *very slowly* so that the plate emerges quite dry. As a result of this process, a monomolecular layer of the surface-active material adheres to the plate with its hydrophobic hydrocarbon chains extending outwards (Figure 9.3). This renders the plate surface hydrophobic, and prevents

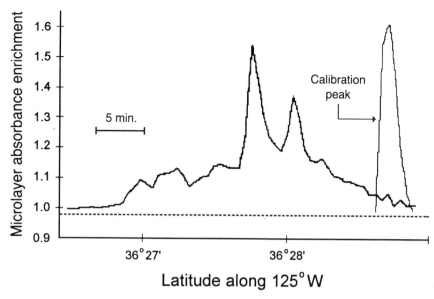

Figure 9.2. Example of nearly real-time measurement of microlayer enrichment of UV absorbance using the SCUMS sampling device. The solid line indicates the microlayer enrichment of UV absorbance (280 nm), calculated as the ratio of absorbance in microlayer samples to that of sub-surface water from 10 cm depth. The results are scaled arbitrarily so that the absorbance ratio for clean (unslicked) areas is 1.00. The dotted line shows the mean absorbance ratio for clean, unslicked areas for the whole tow. Data presented are 40 second averages of measurements made every 4 seconds (reproduced from Carlson et al., 1988).

the monolayer from coming off if the plate is immersed a subsequent time. Thus with careful work, a series of up to 100 monomolecular films may be built up by successive dipping (Blodgett, 1935). Instead of removing the plate so slowly that the film emerges dry, it is also possible to consolidate the deposited film by drying it before the next dip.

Baier (1972) and Baier et al. (1974) have employed this elegant technique, mostly with only a single dipping, to collect surface film samples on germanium prisms. After collection, the sample can then be examined by multiple internal reflection infrared spectroscopy to identify the major functional groups of the organic molecules. In addition, the surface film thickness and refractive index can be measured using ellipsometry. Finally, the contact angles of a range of standard organic liquids may be determined by placing small drops on the prism, allowing calculation of the critical surface tension intercept. The latter property is highly characteristic of the chemical nature of the surface and its surface energy (Zisman, 1964).

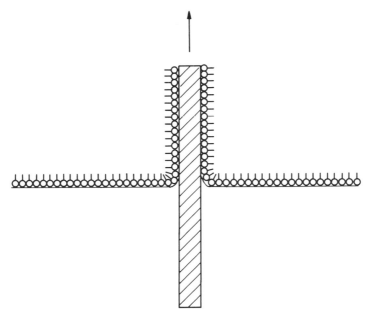

Figure 9.3. The Blodgett technique for sampling a monomolecular film of hydrophobic material on a small solid plate

The bubble microtome

A further sampling technique for thin layers is based on the bubble microtome effect described by MacIntyre (1974). The jet drop ejected into the atmosphere from an air bubble bursting at a film-covered surface consists of material that was originally located within a distance of the interface equivalent to 0.05% of the bubble diameter. For a 1-mm diameter bubble, this corresponds to a surface-layer slice of 0.5 μm thickness, considerably thinner than the samples taken by most of the microlayer samplers in common use. In principle, therefore, microlayer samples collected using the bubble microtome effect should exhibit much greater enrichment factors for material that is enriched within about 1 μm of the surface. A sea-going implementation of this effect, known as the bubble interfacial microlayer sampler (BIMS), was described by Fasching *et al.* (1974).

Although the BIMS undoubtedly produces samples that are more enriched than microlayers collected by the screen, plate and drum techniques, it has one major disadvantage. The scavenging of surface-active materials from sub-surface waters and their subsequent transport to the air–sea surface by attachment to the surfaces of rising bubbles is a well-documented process (e.g. Blanchard and Syzdek, 1974; Blanchard, 1975; Wallace and Duce, 1978a,b; Tseng *et al.*, 1992). Therefore, it is not clear whether the BIMS device samples material that is enriched in the upper 0.5 1 μm of the sea surface at the time bubbles arrive there, or whether the BIMS actively transports sub-surface material to the micro-layer and thence into the atmosphere above. It seems more likely that the BIMS is best regarded as a practical model of sea-to-air transport of surface-active material by rising bubbles in the ocean itself (Liss, 1986).

Enrichment factors and microlayer thickness

Enrichment factor EF

For chemical constituents, we are normally interested in the extent to which the microlayer is enriched relative to sub-surface water. This can be expressed in a dimensionless manner using the *enrichment factor (EF)*, which is the ratio of the microlayer concentration to that of the sub-surface water. For a component X having concentrations $[X]_\mu$ in the microlayer and $[X]_b$ in the sub-surface water

$$EF = \frac{[X]_\mu}{[X]_b} \tag{9.1}$$

Thus, constituents that are selectively accumulated in the microlayer sample will have EF values greater than unity, while those that are not accumulated will have $EF = 1$. Examples also exist where microlayer constituents are *depleted* relative to bulk water, i.e. $EF < 1$ (e.g. Hunter, 1980a).

One difficulty with the enrichment factor concept is that EF depends on the physical thickness of the microlayer sample, making it difficult to compare results obtained with different sampling methods. As we saw in the previous section, the actual dimensions of microlayers sampled from the sea can vary by at least two orders of magnitude depending on the type of sampler used.

Surface excess concentration

The results of different sampling methods can be compared in a simple way if it can be assumed that the depth region obtained with any particular sampler *completely contains* that depth region of the microlayer in which concentrations differ from those of sub-surface water. If this condition is met, we can calculate the *surface excess concentration* (Γ)

$$\Gamma = ([X]_\mu - [X]_b) \cdot d \tag{9.2}$$

where d is the thickness of the microlayer sample obtained with a particular device. The surface excess concentration is seen to correspond to the excess amount of the substance of interest per unit area of surface.

As pointed out by Hunter and Liss (1981a), the surface excess concentration is a well-defined concept in surface chemistry. It is tempting to think of a still water surface as a molecular plane dividing water and air phases, and to consequently envision enrichment of molecules in the microlayer as resulting from their attachment to such a plane through attractive forces such as hydrophobic interactions.

In reality, a single water molecule at the air–water interface experiences about 3×10^5 thermal collisions per second from other molecules in the air and water phases. At a molecular level, the surface is in a state of immense disturbance. Adamson (1976) has suggested that the effects of these collisional disturbances may be manifested in the solution as far as 10 nm below the interface. Under sufficient magnification, the interface would appear as a fuzzy blur with the properties of water, air and adsorbed surface-active species smeared out over at least several molecular diameters.

Although the surface excess allows comparison of the results of different microlayer samplers, it does make it difficult to compare the enrichment of species whose bulk concentrations differ appreciably.

Organic chemical composition of the microlayer

A wide variety of organic compounds, compound classes and parameters related to organic material such as dissolved and particulate organic carbon (DOC, POC) have been studied in the microlayer. Recent research has been characterized by two important improvements:

- An increased variety of sample sites and conditions, which has allowed a better appreciation of the 'typical' degree of microlayer enrichment.
- Improvements in analytical methods, particularly from the application of HPLC and from greatly improved techniques of single-compound identification.

Lipids and hydrocarbons

Much early work on the organic composition of the microlayer focused on the measurement of lipid materials such as fatty acids, fatty alcohols and hydrocarbons. These materials are well-known for their surface activity and/or high insolubility in seawater. Hydrocarbons have a natural relevance to slicks derived from oil seepage and spills. In almost all cases, these lipid materials are found to be enriched in the microlayer, particularly in the presence of obvious surface slicks. As discussed by several earlier reviewers (Liss, 1975; Hunter and Liss, 1977, 1981a; Hardy, 1982), lipids make up only a small percentage of the organic material found in either the microlayer or sub-surface water. This has been confirmed by more recent work in which DOC and other fractions of the organic carbon pool have been analyzed along with lipids (Williams *et al.*, 1986; Garabetian *et al.*, 1993). As is found to be the case for most microlayer organics, particulate lipids and hydrocarbons are generally much more enriched that their dissolved counterparts.

Although they are only minor components of their respective DOC/POC pools, many lipid compounds and hydrocarbons have particular use as specific biological source markers. For example, Marty *et al.* (1988) used specific lipid markers to demonstrate the importance of atmospheric input, flotation from sub-surface waters and lateral input from microlayer organisms as joint sources of chemical substances in the microlayer.

Dissolved and particulate organic carbon

DOC has been widely measured in microlayer samples. Carlson (1983) compiled available data comparing microlayer and sub-surface water concentrations. His results (Figure 9.4) show that microlayer DOC values are for the most part similar to those of the corresponding sub-surface waters, with enrichment factors EF > 1.5 being the exception rather than the rule. Much of the DOC in seawater is hydrophilic in nature and is capable of solubilizing more hydrophobic materials, which form surface films. This is more likely to occur under conditions where the sea surface is not calm and the flexing of surface films by wave and ripple motion squeezes material out of the surface.

Williams *et al.* (1986) have reported a very comprehensive set of microlayer and sub-surface water measurements in the southern Gulf of California and off the west coast of Baja California. Mean enrichment

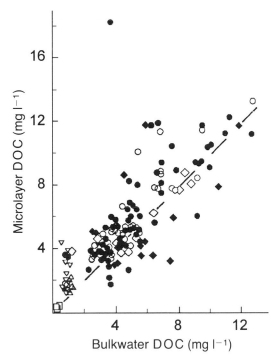

Figure 9.4. Comparison of microlayer and sub-surface DOC measurements compiled by Carlson (1983). The broken line corresponds to equal microlayer and sub-surface values.

factors were 1.1–2.4 for DOC, DON, urea, carbohydrate, lipids, and the soluble inorganic nutrients NH_4^+, NO_2^-, NO_3^-, PO_4^{3-} and SiO_3^{2-}; 1.3–2.0 for ATP, chlorophyll-a, microplankton and bacteria; and 1.1–3.7 for POC, PON, and dissolved and particulate protein. Particulate and dissolved carbon were the only measured constituents that were never depleted in microlayer samples relative to sub-surface waters. Systematic correlations between the measured parameters were few, illustrating the complexity of processes which form and maintain the chemical composition of the films.

In confirmation of the conclusions of earlier reviews, lipid was not the major component of the DOC in microlayer and sub-surface waters, averaging 18% of the POC and 2.5% of the DOC. Film samples had a greater protein : carbohydrate ratio than the sub-surface waters, as did the particulate material. This finding suggests that the microlayer may be selectively accumulating micro-particulate (i.e. colloidal) material from the DOC pool.

The generally greater microlayer enrichment of POC compared with DOC (and of PON compared with DON) has been widely observed. Results compiled by Hunter and Liss (1981a) show that EF values for POC and PON are generally in the range 6–10. Similarly, Carlson's (1983) summarized data show enrichments of POC ranging from 1.36 to 38.4. Hunter (1980a) suggested that organic particles are readily stabilized at the air–sea interface through surface-tension forces. Moreover, surface-active organics readily adsorb from seawater onto the surface of suspended particles (Hunter, 1980b), which is likely to make them adhere to the air–water interface.

Plant pigments

De la Giraudiere *et al.* (1989) have described the application of HPLC analysis to chlorophyll pigments in microlayer samples. By use of this method, it is possible to obtain an accurate picture of the pigments and their breakdown products present in the sample, and to deduce the origin of phytoneustonic cells. POC and ATP measurements were incorporated into this study as well. The pigment results obtained did not provide evidence for a specific photochemical response of each chlorophyll form in the microlayer. For example, the enrichment factor for chlorophyll-*a* was greater than that for chlorophyll-*c*, even though the latter pigment is considered to be more photochemically stable. Little difference was observed between the relative pigment compositions of microlayer and bulk samples, suggesting that the main source of microlayer pigments is sub-surface waters. Finally, the high proportion of chlorophyll-*a* and correspondingly low proportions of its breakdown products found in the microlayer is indicative that microlayer organisms are in an impeded metabolic state.

Organic pollutants

Many organic pollutants have been found in the microlayer (Hardy, 1982). A number of early papers report enrichment of chlorinated insecticides and polychlorinated biphenyls (PCBs) in microlayer samples (Seba and Corcoran, 1969; Bidelman and Olney, 1972; Duce *et al.*, 1972; Bidelman *et al.*, 1975; Ofstad *et al.*, 1979). The analysis of these compounds has undergone considerable improvement since the 1970s, both through the better control of method blanks and sample contamination, and through positive identification of individual compounds afforded by mass spectrometric detection methods. One might have expected that this improvement in analytical capability would have led to a rush of new data showing the contamination of the sea-surface microlayer with PCBs and chlorinated insecticides. There are some recent reports of enrichment of PCBs and chlorinated hydrocarbons of the pesticide group in microlayer samples collected in polluted coastal areas (Picer and Picer, 1992). However, a comprehensive study made by Sauer *et al.* (1989) in open ocean waters of the US east coast and in the Gulf of Mexico found only two microlayer samples (out of a total of 27) in which PCBs were above the detection limits (0.1–1.0 ng/l). Furthermore, none of the chlorinated pesticides was detected in any samples, either from the microlayer or sub-surface waters. These findings, while environmentally encouraging, are not easy to interpret unequivocally. Have microlayer and sub-surface concentrations of these compounds declined substantially since the 1970s, to the extent that they are no longer detectable using better methods, even in the microlayer? Or were the earlier results compromized by inadvertent analytical errors arising from high blanks, sample contamination or the identification of the wrong compounds?

Another common class of environmental contaminant, the polynuclear aromatic hydrocarbons (PAHs), have also been reported in microlayer samples. Unlike the chlorinated hydrocarbons, PAHs are widely distributed materials and arise from a variety of emissions: gasoline and diesel powered vehicles, refuse incineration and coal-fired power stations. Hardy *et al.* (1990) found substantial enrichments of particulate hydrocarbons, including PAHs, in microlayer samples collected in Chesapeake Bay. Some samples had PAH levels 200–400 times higher than sub-surface seawater, although the study included only a single sub-surface water sample. Microlayer samples consistently included high levels of fluorene, pyrene and chrysene; indicating a common source of PAHs. These compounds accounted for 77% of the total PAHs. The concentrations of PAHs found in about half of the microlayer samples taken from Chesapeake Bay were equal to, or exceeded, concentrations at which demonstrated toxic effects to developing fish embryos are seen.

Other, less common, organic contaminants that have been found enriched in the microlayer regime are silicone polymers (Batley and Hayes, 1991) and tributyl tin (Gucinski, 1986; Cleary and Stebbing, 1987; Maguire and Tkacz, 1987).

UV absorption and fluorescence

Considerable use has been made of simple physical measurements that are indicative of broad organic chemical properties in the study of microlayer organic materials. As already mentioned, these techniques allow nearly real-time microlayer measurements (Carlson *et al.*, 1988). Both UV absorbance and fluorescence are associated with aromatic structures in the DOC pool. Carlson and Mayer (1982) compared UV absorbance (280 nm) of microlayer and sub-surface water samples with DOC measurements and analysis of phenolic materials using the Folin–Ciocalteau method. A glass plate sampler (thickness 51 ± 2 μm) was used for microlayer sample collection. A high correlation was found between UV_{280} absorbance and the concentration of phenolics (expressed as the equivalent concentration of phloroglucinol, PGE), suggesting that most of the UV absorbing materials are phenolic in nature. Comparison of results with macroalgal exudates showed that the latter had a quite different A_{280} : PGE ratio, suggesting that terrestrially derived phenolic material contributes to the samples studied. DOC results did not show consistent microlayer enrichments, whereas A_{280} and PGE were consistently higher in microlayer samples, especially in marine samples.

Using the same techniques, Carlson (1982b) studied the relationship between A_{280} and DOC microlayer enrichment factors and the occurrence of surface slicks. The samples were classified into one of four types corresponding to slick prevalence: 1. *certified slicks* obviously defined by boundaries; 2. *slick-influenced areas*, without a definable boundary but with evidence of surface foam or floating debris; 3. *mixed areas* in which some dips of the microlayer sample were slick-affected and others were not and 4. *clean surface areas* with no evidence of surface films. The EF results obtained are summarized in Table 9.1.

These results show a definite relationship between A_{280} enrichment factor and the degree of slick presence. Although there are not many DOC results for comparison, there is a poor relationship between DOC enrichment factors and slick occurrence. Similar findings were reported by Carlson (1983). The mean DOC enrichment factor was only 1.33, or 1.17 if only coastal data were included. By contrast, the mean EF for A_{280} was 1.27 in non-slicked samples.

We can conclude from these exhaustive studies that, given its simplicity and application to real-time measurements at sea, the UV_{280} technique provides a relatively sensitive measure of the microlayer concentration of a major fraction of film-forming materials.

Table 9.1 *EF values of A_{280} and DOC materials in various classes of sticks (for definition of terms see text)*

	A_{280}	DOC
Certified	2.5 (30)	1.4 (8)
Influenced	1.7 (24)	1.7 (6)
Mixed	2.0 (13)	0.81 (2)
Clean	1.27 (97)	1.13 (4)

The number of samples is shown in parentheses

Electrochemical methods

Electrochemical techniques that selectively measure the concentration of surface-active substances have also been applied to microlayer studies. These techniques are based on the use of the mercury–water interface as an analogy for the air–water interface. Adsorption of hydrophobic materials on the surface of a liquid mercury electrode induces changes in interfacial properties that can be monitored in a variety of ways. The first of these techniques applied to microlayer samples was the Kalousek commutator method which was compared with the classical methylene blue absorption method by Cosovic *et al.* (1977). The so-called *polarographic maximum technique* described by Hunter and Liss (1981b) has also been applied to microlayer samples. In this method, a standard dropping mercury electrode (DME) is induced to spin at the electrode tip by operating it at the reduction wave of Hg^{2+}. The stirring of the solution caused by this electrode motion enhances the mass transport of electroactive species through the interfacial boundary layer to the electrode surface, giving rise to a very large increase in observed current (the so-called 'polarographic maximum'). Adsorption of surface-active species on the electrode surface acts to damp the spinning motion, thus decreasing the observed current. The technique is calibrated using standard solutions of a known surfactant, usually the polymer Triton X-100 (polyoxyethylenated tert-octylphenol). For water samples, the results are expressed as the Triton X-100 concentration that has the same suppression effect on the polarographic maximum as the natural organic surfactants in the sample. This measure is known as the *surfactant activity* (*SA*). Hunter and Liss (1981b) showed that the surfactant activity of a large number (160) of riverine, estuarine and coastal seawater samples was quite well correlated with DOC measurements.

Table 9.2 *Microlayer and sub-surface surfactant activity (SA) and calculated EF values for various samples from the Alde estuary, Suffolk, UK*

Description of sample	Microlayer SA (mg l^{-1} TX-100)	Sub-surface SA (mg l^{-1} TX-100)	EF
Icy surface slick	11.6	10.0	1.16
	13.0	9.5	1.37
Patchy films	5.9	4.7	1.26
	6.8	5.9	1.15
Sheltered slick	12.0	11.4	1.05
Near dense foam	11.7	11.0	1.06
	12.3	11.4	1.08
Inside dense foam	13.3	11.7	1.14
	15.7	11.4	1.37
	12.7	12.3	1.03
Slight slick patches	12.0	11.7	1.03
	13.0	8.7	1.49
	7.5	7.0	1.07
Smooth slick	8.7	7.0	1.24

A series of microlayer samples were collected for study from the Alde estuary, Suffolk using a Monel metal screen sampler as described by Garrett (1965). Samples were obtained during periods when surface slicks and wind-blown scums were very much in evidence. The results obtained are summarized in Table 9.2 (the precision of replicate measurements is ± 0.3 mg l^{-1} Triton X-100). It is seen that all microlayer samples have higher SA values than their corresponding sub-surface waters, although the enrichment factors are never very large (maximum 1.49). There does not appear to be much correlation between the observed intensity of slicks and the EF values. Given the high correlation between DOC and the SA measurement, this is in accordance with Carlson's findings mentioned earlier. Clearly the polarographic max-imum technique is not especially selective for hydrophobic materials and must measure a substantial portion of the more hydrophilic surface-active compounds in seawater having less affinity for the air–water interface than their hydrophobic counterparts.

More recently, the Yugoslavian group pioneered the use of a.c. polarographic techniques for surfactant measurement (Cosovic and Vojvodic, 1982, 1987, 1989). These methods take advantage of the fact that adsorption of surface-active molecules on electrodes cause measurable changes in the electrode double layer capacity. The decrease in either the double layer capacity, or the capacity current, as surfactant is adsorbed by the electrode is measured at a selected constant potential, and the time required for adsorption can be related to the surfactant activity of the solution. Like the polarographic maximum technique, this method is calibrated using standard solutions of a known surfactant such as Triton X-100. Cosovic and Vojvodic (1989) and Marty *et al.* (1988) showed that the a.c. polarographic technique is more selective for hydrophobic materials, and consequently shows higher EF values for microlayer samples, than the polarographic maximum method and DOC. This is consistent with the results of Hunter and Liss (1981b) for the Alde estuary mentioned above.

Trace elements in the microlayer

Dissolved trace elements

Trace elements have also been widely studied in the sea-surface microlayer, starting with early studies by Barker and Zeitlin (1972) and Piotrowicz *et al.* (1972) which demonstrated microlayer enrichment of a range of trace metals. As with chlorinated hydrocarbons, there have been vast improvements in analytical methodology for trace elements in seawater since the 1970s, with the consequence that most, if not all, of the concentration data published for dissolved trace metals in seawater prior to the establishment of the so-called 'clean' methods of sampling and analysis have been rendered meaningless by contamination artifacts (e.g. Patterson and Settle, 1976).

Having grappled with this problem personally and having finally succeeding in producing good-quality reliable data for bulk seawater samples (e.g. Frew and Hunter, 1992), I find it almost impossible to believe that *any* of the conventional microlayer sampling devices collect anything but spurious contamination, except perhaps in considerably polluted waters. In other words, I cannot erase residual doubt about the freedom from contamination artifacts of existing microlayer sampling techniques.

Hunter and Liss (1981c) presented a rather elaborate model that attempted to account for dissolved trace metal enrichments in the microlayer. Although that discussion may be based on erroneous data, the general conclusions drawn from their discussion are still valid. The basic question posed in the analysis by Hunter and Liss (1981c) is whether surface enrichment of dissolved metals can be reasonably accounted for by metal-ion complexation with organic ligands enriched in the microlayer. Their conclusion was that the literature microlayer surface excess concentrations of trace metals could be explained provided the hypothetical ligand unit was able to bind one trace metal ion for every 140 carbon atoms. The authors considered this to be a reasonable likelihood. Since sample contamination effects almost certainly mean that microlayer surface excess concentrations of trace metals are even lower, this requirement becomes more easily met. However, the required stability constant for binding of the metal ion to the hypothetical surface ligand must increase at lower trace metal concentration. As pointed out by Hunter and Liss (1981c), this poses problems with competition by abundant complexing ions like Mg^{2+} and Ca^{2+}.

Major cations of seawater

There have been some reports of very small microlayer enrichments of the seawater major cations in microlayer samples. Barker and Zeitlin (1972), using a glass-plate sampler, found EF = 1.03 for Na, 1.07 for K, 1.01 for Mg, 1.08 for Ca and 1.13 for Sr. These values are very difficult to explain by means of complexation with microlayer organics (notwithstanding the unlikely occurrence of Na^+ and K^+ complexes). For Na, K, Mg and Ca the results given convert to surface excess concentrations that are much greater than that for carbon atoms, based on DOC results presented earlier. The values are also too large to be explained by counter-ion effects using a negatively charged surface film. The surface excess concentration of Na^+ alone is equivalent to an electrical charge density at the surface of about 200 C m^{-2}, whereas a value of about 0.1 C m^{-2} would be more typical of an organic film.

MacIntyre (1974) has presented a very thorough analysis of physical mechanisms that might lead to small enrichments of major ions, but none of the mechanisms he advances is able to produce surface excess concentrations of the magnitude suggested by the Barker and Zeitlin (1972) results. It seems likely, therefore, that these enrichments of major ions are experimental artifacts. More recently, Savenko (1990) reports that small enrichments of major ions and salinity in microlayer samples can arise from water evaporation, both *in situ* and as the sample is collected.

Particulate trace metals

It is considerably simpler to collect, handle and analyze a microlayer sample for trace elements in particulate form than the corresponding dissolved fraction. Consequently, I believe that there are some reliable numbers describing microlayer enrichments of particulate trace metals.

Figure 9.5 illustrates the transport processes affecting particulate materials in the microlayer. Sources for the microlayer are deposition from the atmosphere above and both bubble flotation and surface hydrodynamic renewal from sub-surface water below. Since particles are macroscopic in nature, their transport through the microlayer must take place through Brownian motion (all directions), sedimentation (downwards) and capture by aggregation (all directions). As argued by Hunter (1980a), these processes will dominate within the hydrodynamic boundary layer adjacent to the air–sea interface. The thickness of this layer depends to a great extent on surface conditions and wind speed, but 50 μm can be taken as a typical figure.

In a seminal paper, Hoffman *et al.* (1974) compared the rates of atmospheric deposition of particulate trace metals with their microlayer concentrations in oceanic regions off West Africa within the influence of the Saharan dust plume. They showed that the deposition rates of Fe, Mn

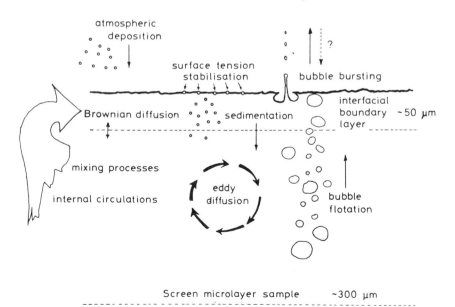

Figure 9.5. Conceptual model for processes affecting particulate material in the microlayer (reproduced from Hunter, 1980a).

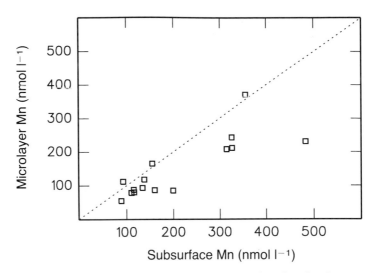

Figure 9.6. Comparison of particulate Mn concentrations in microlayer and subsurface water samples from the North Sea. The line indicates equal concentrations in both phases (drawn using data presented by Hunter, 1980a).

and V contained in relatively large, soil-sized particles were sufficiently great to account for their observed enrichments in the microlayer. On this basis, transit times across the microlayer as a consequence of sedimentation were calculated to be of the order of 2 seconds.

The West African data of Hoffman *et al.* (1974) show that processes mixing particles out of the hydrodynamic boundary layer (surface film renewal, replacement of surface eddies) must have a characteristic timescale that is significantly longer than a few seconds, otherwise the enrichments of Fe, Mn and V should not have been observed. In support of this notion, Hunter (1980a) found consistent depletion of particulate Fe (mean EF 0.84, 15 samples) and Mn (mean EF 0.74, 15 samples) in microlayer samples collected at a coastal site in the North Sea. The Mn results are summarized in Figure 9.6, from which it is seen that most of the microlayer samples plot well below the line of equal concentrations (EF = 1) and none is significantly above it.

In this case, the principal source of Fe- and Mn-containing particles is resuspension of bottom sediments and fluvial input, *beneath* the microlayer. To account for these results, Hunter (1980a) argued that if processes mixing soil-sized particles into and out of the microlayer are intrinsically slower than their sedimentation rates, then if the particles come from below the microlayer, depletion must result. He argued further that if it could be assumed that the hydrodynamic boundary layer contained almost no particles compared with the sub-surface waters, then the (negative) surface excess of Fe and Mn could be used to estimate the minimum boundary-layer thickness. A result of 72 μm was obtained, which is a reasonable result compared with estimates based on more conventional methods such as gas exchange.

In areas of the Atlantic Ocean outside of the Saharan plume, Hoffman *et al.* (1974) concluded that atmospheric deposition rates were too low to account for the observed microlayer enrichment of particulate metals. Thus, another mechanism must be invoked to explain the latter enrichments. Similarly, Hunter (1980a) found enrichments of particulate Cu, Ni, Cd, Zn and Pb in the same North Sea microlayer samples in which Fe and Mn were depleted, while Lion *et al.* (1979) reported a range of enriched particulate metals in estuarine salt marsh microlayers. As mentioned already, Hunter (1980a) argued that this enrichment arises from the stabilization of non-wettable particles at the sea surface by surface-tension forces. Non-wettable particles having a low-energy, hydrophobic surface are more stable attached to the interface than submerged beneath it.

Brügmann *et al.* (1992) reported microlayer enrichments of particulate Ba, Cd, Cu, Pb and Zn in samples collected in the Baltic and North Seas. Interestingly, no enrichment of Al, Ca, Mg, Mn, Ni and V was found, and the authors reported that there was even 'some tendency towards slightly higher metal contents in samples taken from 0.2 m depth'. Since most of the latter group of elements are closely associated with clay minerals in coastal sediments, this finding may be another example of the Fe and Mn depletion reported by Hunter (1980a).

Physico-chemical properties of sea-surface films

Sea slicks, capillary waves and film pressures

Sea slicks comprise a coherent surface film of organic matter on the sea surface and are visible because capillary waves (wavelength < 2 cm) are rapidly damped out. This leads to changes in the surface reflectance properties of the ocean. Under these circumstances, a slick can be seen as a floating patch with a silvery sheen, particularly if it is observed very near the water surface. Slicks usually have well-defined boundaries and are commonly observed in the wake of a large ship, often many kilometres behind its present position. Surface slicks are particularly evident during rain because the capillary ripples caused by falling raindrops are rapidly damped out within the slick (Blanchard, 1963). Slicks not caused by ship wakes are often aligned in parallel streaks. Some may mark regions of surface convergence driven by internal waves, and can be aligned in any particular direction, while others are wind-aligned (Ewing, 1950).

Garrett (1967) has shown that insoluble monolayer films will damp out capillary waves on water with an exponential decay of wave amplitude $a(x)$ with distance x from its source:

$$a(x) = a_o \exp(-kx)$$

where a_o is the initial amplitude. The damping coefficient k goes through a maximum at a film pressure of about 10^{-3} N m^{-1} and is much smaller at lower film pressures.

Measurements of surface film pressure can be made at sea using the spreading drop technique, originally described by Adam (1937). Drops of surfactant/paraffin oil mixtures of precisely known surface film pressure are placed on the water surface and observed closely. If the oil drop spreads, the ambient film pressure must be lower than that of the applied drop. In this case, successive oil mixtures having lower film pressures would be tried until one is found that does not spread, thus bracketing the ambient film pressure value. In the field, the technique has a reported resolution of about 10^{-3} N m^{-1}, although this depends mainly on the composition range of the spreading oils used.

Using this technique, Barger et al. (1974) found that 150 out of 168 measurements made at Mission Beach, San Diego had film pressures below 10^{-3} N m^{-1}, too low to cause capillary wave damping and obvious slick presence. Few measurements are available for open ocean situations, but it seems unlikely that film pressures exceeding 10^{-3} N m^{-1} are common in situations where wind speeds exceed a few metres per second. Studies by Goldacre (1949) on lake films suggest that the film pressure gradient developed across an ambient slick through wind will be sufficiently great to collapse the film by lateral wind pressure into a scum or foam at this sort of wind speed.

Recently, Peltzer et al. (1992) have described a high-resolution, automated version of the Adam spreading drop technique. The apparatus comprises an instrumented catamaran that is used to apply spreading oils remotely to the sea surface by command from the mother ship. The effect of oil drop addition is monitored using a video camera mounted aboard the catamaran. The device, which carries the acronym STEMS (Surface Tension Measuring System), has so far been applied to the study of film formation and persistence in the wake of US Navy vessels. Very sharp changes in film pressure are observed at the boundaries of the wake, with reductions in seawater surface tension of as much as $0.010 - 0.015$ N m^{-1} in some cases. Ambient sea slicks also appear in the results presented by the authors. A film pressure resolution of 1.5×10^{-3} N m^{-1} is claimed, this being achieved by selection of a wide range of closely similar spreading oil mixtures. An example of the results obtained is given in Figure 9.7, which shows that both ambient surface slicks and those created by the wake of a passing ship are made very evident by this technique.

Wei and Wu (1992) have reported *in situ* measurements of surface tension and wave damping properties of the sea surface. They used a bottomless tank mounted on a catamaran which incorporated a plunger-type 12 Hz wave-making device and paired laser beams to measure surface refraction. The authors were able to calculate both surface tension and capillary wave damping rate from their optical data. Significantly, they found variations in wave damping not only in visibly slicked areas, but also in 'clean' surface areas.

As mentioned in the Introduction, the influence of capillary wave damping by slicks on the optical reflectance of the sea surface clearly has considerable importance to remote imaging of surface sea state, including nonlinear laser techniques, passive microwave detectors and other radar-based methods (e.g. Onstott and Rufenbach, 1992; Frysinger *et al.*, 1992). This is a very rapidly expanding area that may lead to new insights into the occurrence and frequency of slicks in more remote areas of the ocean, for which few data are presently available. The passive microwave detector (Onstott and Rufenbach, 1992) has been recently used in

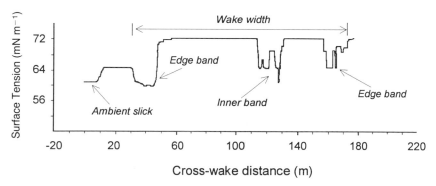

Figure 9.7. Cross-wake surface tension distribution measured at 4637 m aft of a Navy frigate using the STEMS system (re-drawn from Peltzer *et al.*, 1992).

association with the SCUMS automated microlayer sampler. Remote-sensing instruments such as these, and STEMS, may help to remove the subjectivity associated with reporting the presence of slicks and their intensity when chemical or microbiological examination of microlayer samples is being carried out.

Film pressure–area measurements

The film pressure–area properties of surface films can often be characteristic of the type of surface-active material present in natural films. The capabilities of this technique and typical results have been summarized by Hunter and Liss (1977, 1981a). In this approach, microlayer samples are placed in a film pressure apparatus and the resultant film pressure is recorded as the film is compressed using a moveable barrier. Today, the apparatus is most often automated, with results accumulated by microcomputer.

Figure 9.8 compares film pressure–area plots for a number of surface microlayer and atmospheric aerosol samples spread on clean seawater, and illustrates the type of information obtained from these measurements. These films from two different sources have considerable similarity in the degree of surface elasticity, which may indicate a similar general composition.

Van Vleet and Williams (1983) compared film pressure–area curves of surface microlayer samples with those of a wide range of standard organic substances. They concluded from their results that natural microlayer films most closely resembled the film pressure–area behaviours of proteins, polysaccharides, humic materials and waxes. Their results also suggest relatively small amounts of free fatty acids, free fatty alcohols and triglycerides in the microlayer films. All of these observations would seem to accord with the findings of chemical analysis discussed earlier.

Frew and Nelson (1992) isolated surface-active organic matter from microlayer samples for film pressure–area studies using C_{18} reversed phase adsorbents. Materials collected by this technique were re-spread onto clean seawater surfaces for measurement. The reconstituted surface films had pressure–area isotherms and static elasticities closely approximating those measured for films formed by allowing surface-active materials to diffuse to the surface from the untreated microlayer sample. The authors indicate that the hydrophobic materials extracted using C_{18} were the major class of compounds responsible for the microlayer film characteristics as measured, even though other more hydrophilic materials (that would not be adsorbed by C_{18}) may have been enriched in

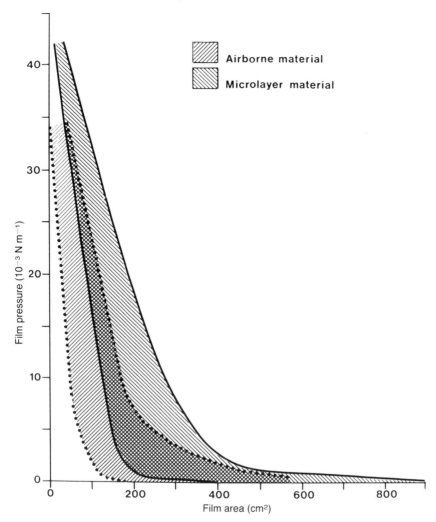

Figure 9.8. Comparison of film pressure–area plots for a number of surface microlayer and atmospheric aerosol samples spread on clean seawater (reproduced from Hunter and Liss, 1977).

the microlayer samples. Against this assertion must be set the observations made by Hunter and Lee (1986) that significant interactions occur between hydrophobic and hydrophilic surface-active species when they are present together (at least in freshwater). They observed that the surfactant activity of separated fractions was considerably greater than that of the original sample. This suggests that conclusions about the quantitative importance of hydrophobic or hydrophilic components of surface-active organic matter in producing film properties cannot be made if they are separated from each other by adsorption techniques.

Epilogue

Considerable progress has been made in many areas of our understanding of the sea-surface microlayer over the last 15 years. In particular, we have seen the application of improved measurement and sample collection techniques to a variety of chemical substances. I am sure that much remains to be done with specific organic compounds that act as markers or indicators of biochemical processes. The plant pigment studies of de la Giraudiere *et al.* (1989) have already been mentioned as an example of this approach.

Although much has been written of the effects of the microlayer as a region of chemical transformation, there are relatively few clear and direct demonstrations of this role as yet. However, interesting results are emerging from new studies. I believe that one of the most promising areas is the study of photochemical transformations in the microlayer. The microlayer is a region of concentration anomalies, especially for organic compounds. It also receives sunlight unattenuated by superjacent water, and is a region where diffusional transport by eddy processes is largely damped, leading to enhanced residence times. Finally, studies on the water column have given us a new appreciation of the importance of transient species, including those involved in photochemical processes, for example H_2O_2 (Moffett and Zafiriou, 1993), Cu(I) and Fe(II) (Moffett and Zika, 1987, 1988).

New spectroscopic techniques developed over the last decade may also be useful in examining microlayer film materials directly without the need to remove them onto a solid substrate such as the germanium prism used by Baier (1972). The vibrational spectrum of surface film compounds can now be determined by Fourier transform infrared (FTIR) reflection spectroscopy (e.g., Dluhy and Cornell, 1985). Similarly, where the surface molecules contain a chromophore, the added sensitivity of resonance Raman spectroscopy can be exploited (Takenaka and Fukuzaki, 1979). This may prove useful for studying plant pigments in the microlayer. Photon correlation spectroscopy in which a laser beam incident on a liquid surface is scattered by surface thermal waves can be used to monitor the degree of surface coverage by film-forming materials (e.g. Chen *et al.*, 1986). This topic is developed further by Korenowski. (Chapter 15, this volume).

References

Adam, N. K. (1937). A rapid method for determining the lowering of surface tension of exposed water surfaces, with some observations on the surface tension of the sea and some inland waters. *Proc. R. Soc. London*, **B122**, 134–9.

Adamson, A. W. (1976). *Physical Chemistry of Surfaces*. New York: Wiley, 698 pp.

Baier, R. E. (1972). Organic films on natural waters: their retrieval, identification and modes of elimination. *J. Geophys. Res.*, **77**, 5062–75.

Baier, R. R., Goupil, D. W., Perlmutter, S. and King, R. (1974). Dominant chemical composition of sea surface films, natural slicks and foams. *J. Rech. Atmos.*, **8**, 571–600.

Barger, W. R., Daniel, W. H. and Garrett, W. D. (1974). Surface chemical properties of banded sea slicks. *Deep-Sea Res.*, **21**, 83–9.

Barker, D. R. and Zeitlin, H. (1972). Metal ion concentrations in the sea surface microlayer and size-separated atmospheric aerosol samples in Hawaii. *J. Geophys. Res.*, **77**, 5076–86.

Batley, G. E. and Hayes, J. W. (1991). Polyorganosiloxanes (silicones) in the aquatic environment of the Sydney region. *Austr. J. Mar. Freshwater Res.*, **42**, 287–93.

Bidelman, T. F. and Olney, C. E. (1972). Chlorinated hydrocarbons in the Sargasso Sea atmosphere and surface water. *Science*, **183**, 513–18.

Bidelman, T. F., Rice, C. P. and Olney, C. E. (1975). High molecular weight chlorinated hydrocarbons in the air and sea: rates and mechanisms of air/sea transfer. In *Marine Pollutant Transfer*, ed. H. L. Windom and R. A. Duce, pp. 323–51. Lexington, MA: Lexington Books.

Blanchard, D. C. (1963). The electrification of the atmosphere by particles from bubbles in the sea. *Progr. Oceanogr.*, **1**, 73–202.

Blanchard, D. C. (1975). Bubble scavenging and the water-to-air transfer of organic material in the sea. In *Applied Chemistry at Protein Interfaces*, ed. R. E. Baier, *Advances in Chemistry Series*, **145**, 360–87.

Blanchard, D. C. and Syzdek, L. D. (1974). Importance of bubble scavenging in the water to air transfer of organic material and bacteria. *J. Rech. Atmos.*, **8**, 529–40.

Blodgett, K. B. (1934). Monomolecular films of fatty acids on glass. *J. Am. Chem. Soc.*, **56**, 495.

Blodgett, K. B. (1935). Films built by depositing successive monomolecular layers on a solid surface. *J. Am. Chem. Soc.*, **57**, 1007–22.

Brügmann, L., Bernard, P. C. and van Grieken, R. (1992). Geochemistry of suspended matter from the Baltic Sea. 2. Results of bulk trace metal analysis by AAS. *Mar. Chem.*, **38**, 303–23.

Carlson, D. J. (1982a). A field evaluation of plate and screen microlayer samplers. *Mar. Chem.*, **11**, 189–208.

Carlson, D. J. (1982b). Surface microlayer phenolic enrichments indicate sea surface slicks. *Nature*, **296**, 426–9.

Carlson, D. J. (1983). Dissolved organic materials in surface microlayers: Temporal and spatial variability and relation to sea state. *Limnol. Oceanogr.*, **28**, 415–31.

Carlson, D. J. and Mayer, L. M. (1982). Enrichment of dissolved phenolic material in the surface microlayer of coastal waters. *Nature*, **286**, 482–3.

Carlson, D. J., Cantey, J. L. and Cullen, J. J. (1988). Description and results from a new surface microlayer sampling device. *Deep-Sea Res.*, **35**, 1205–13.

Chen, Y., Sano, M., Kawaguchi, M., Yu, H. and Zographi, G. (1986). Static and dynamic properties of pentadecanoic acid monolayers at the air–water interface. *Langmuir*, **2**, 338–41.

Cleary, J. J. and Stebbing, A. R. D. (1987). Organotin in the surface microlayer and subsurface waters of southwest England. *Mar. Pollut. Bull.*, **18**, 238–46.

Cosovic, B. and Vojvodic, V. (1982). The application of a.c. polarography to the determination of surface-active substances in seawater. *Limnol. Oceanogr.*, **27**, 361–9.

Cosovic, B. and Vojvodic, V. (1987). Direct determination of surface-active substances in natural waters. *Mar. Chem.*, **22**, 363–75.

Cosovic, B. and Vojvodic, V. (1989). Adsorption behaviour of the hydrophobic fraction of organic matter in natural waters. *Mar. Chem.*, **28**, 183–98.

Cosovic, B., Zutic, V. and Kozarac, Z. (1977). Surface-active substances in the sea-surface microlayer by electrochemical methods. *Croatica Chim. Acta*, **50**, 229–41.

Cross, J. N., Hardy, J. T., Hose, J. E., Hershelman, G. P., Antrim, L. D., Gossett, R. W. and Crecelius, E. A. (1987). Contaminant concentrations and toxicity of sea-surface microlayer near Los Angeles, California. *Mar. Environ. Res.*, **23**, 307–23.

Dean, G. A. (1963). The iodine content of some New Zealand drinking waters with a note on the contribution from sea spray to the iodine in rain. *N. Z. J. Sci.*, **6**, 208–14.

de la Giraudiere, I., Laborde, P. and Romano, J. C. (1989). HPLC determination of chlorophylls and breakdown products in surface microlayers. *Mar. Chem.*, **26**, 189–204.

Dluhy, R. A. and Cornell, D. G. (1985). In situ measurements of the infrared spectrum of insoluble monolayers at the air–water interface. *J. Phys. Chem.*, **89**, 3195–7.

Duce, R. A., Quinn, J. G., Olney, C. E., Piotrowicz, S. R., Ray, B. J. and Wade, T. L. (1972). Enrichment of heavy metals and organic compounds in the surface microlayer of Narragansett Bay, Rhode Island. *Science*, **176**, 161–3.

Ewing, G. (1950). Slicks, surface films and internal waves. *J. Mar. Res.*, **9**, 161–87.

Fasching, J. L., Courant, R. A., Duce, R. A. and Piotrowicz, S. R. (1974). A new surface microlayer sampler utilising the bubble microtome. *J. Rech. Atmos.*, **8**, 649–52.

Frew, R. D. and Hunter, K. A. (1992). The cadmium-phosphate properties of Southern Ocean waters. *Nature*, **360**, 144–6.

Frew, N. M. and Nelson, R. K. (1992). Isolation of marine microlayer film surfactants for ex-situ study of their surface physical and chemical properties. *J. Geophys. Res.*, **97**, 5281–90.

Frysinger, G. S., Asher, W. E., Korenowski, G. M., Barger, W. R., Klusty, M. A., Frew, N. M. and Nelson, R. K. (1992). Study of ocean slicks by non-linear laser processes: 1. Second-harmonic generation. *J. Geophys. Res.*, **97**, 5253–69.

Garabetian, F., Romano, J. C., Paul, R. and Sigoillot, J. C. (1993). Organic matter composition and pollutant enrichment of sea surface microlayer material inside and outside of slicks. *Mar. Chem.*, **35**, 323–39.

Garrett, W. D. (1965). Collection of slick-forming materials from the sea surface. *Limnol. Oceanogr.*, **10**, 602–5.

Garrett, W. D. (1967). Damping of capillary waves at the air–sea interface by oceanic surface-active material. *Deep-Sea Res.*, **14**, 221–7.

Garrett, W. D. and Barger, W. R. (1974). *Sampling and Determining the Concentration of Film-forming Constituents of the Air–Water Interface.* Naval Research Laboratory Memorandum Report 2852, 13 pp.

Goldacre, R. J. (1949). Surface films on natural bodies of water. *J. Anim. Ecol.*, **18**, 36–9.

Gucinski, H. (1986). The effect of sea surface microlayer enrichment on TBT transport. In *Proceedings of the Oceans '86 Conference, Marine Technology Society*, **4**, 1266–74.

Hardy, J. T. (1982). The sea-surface microlayer: biology, chemistry and anthropogenic enrichment. *Progr. Oceanogr.*, **11**, 307–28.

Hardy, J. T., Kiesser, S. L., Antrim, L. D., Stubin, A. I., Kocan, R. and Strand, J. A. (1987). The sea-surface microlayer of Puget Sound: Part I. Toxic effects on fish eggs and larvae. *Mar. Environ. Res.*, **23**, 227–49.

Hardy, J. T., Crecelius, E. A., Antrim, L. D., Kiesser, S. L., Broadhurst, V. L., Boehm, P. D. and Steinheauer, W. G. (1990). Aquatic surface contamination in Chesapeake Bay. *Mar. Chem.*, **28**, 333–51.

Harvey, G. W. (1966). Microlayer collection from the sea surface: a new method and initial results. *Limnol. Oceanogr.*, **11**, 608–14.

Harvey, G. W. and Burzell, L. A. (1972). A simple microlayer method for small samples. *Limnol. Oceanogr.*, **17**, 156–7.

Hoffman, G. L., Duce, R. A., Walsh, P. R., Hoffman, E. J., Ray, B. J. and Fasching, J. L. (1974). Residence time of some particulate trace metals in the oceanic surface microlayer: significance of atmospheric deposition. *J. Rech. Atmos.*, **8**, 745–59.

Hunter, K. A. (1980a). Processes affecting particulate trace metals in the sea surface microlayer. *Mar. Chem.*, **9**, 49–70.

Hunter, K. A. (1980b). Microelectrophoretic properties of natural surface-active organic matter in coastal seawater. *Limnol. Oceanogr.*, **25**, 807–22.

Hunter, K. A. and Lee, K. C. (1986). Polarographic study of the interaction between humic acids and other surface-active agents in river waters. *Water Res.*, **20**, 1489–91.

Hunter, K. A. and Liss, P. S. (1977). The input of organic material to the oceans: air–sea interactions and the organic chemical composition of the sea surface. *Mar. Chem.*, **5**, 361–79.

Hunter, K. A. and Liss, P. S. (1981a). Organic sea surface films. In: *Marine Organic Chemistry*, ed. E. K. Duursma and R. Dawson, pp. 259–298. Amsterdam: Elsevier.

Hunter, K. A. and Liss, P. S. (1981b). Polarographic measurement of surface-active material in natural waters. *Water Res.*, **15**, 203–15.

Hunter, K. A. and Liss, P. S. (1981c). Principles and problems of modelling cation enrichment at natural air–water interfaces. In: *Atmospheric Pollutants in Natural Waters*, ed. S. J. Eisenreich, pp. 99–127. Ann Arbor: Ann Arbor Science.

Kocan, R. M., Westernhagen, H. V., Landolt, M. L. and Furstenberg, G. (1987). Toxicity of sea-surface microlayer: II. Effects of hexane extract on Baltic herring (*Clupea harengus*) and Atlantic cod (*Gadus morhua*) embryos. *Mar. Environ. Res.*, **23**, 291–305.

Larsson, K. Oldham, G. and Sodergren, A. (1974). On lipid films on the sea: I. A simple method for sampling and studies of composition. *Mar. Chem.*, **2**, 49–57.

Lion, L. W., Harvey, R. W., Young, L. Y. and Leckie, J. O. (1979). Particulate matter, its association with microorganisms and trace metals in an estuarine salt marsh microlayer. *Environ. Sci. Technol.*, **13**, 1522–7.

Liss, P. S. (1975). The chemistry of the sea surface microlayer. In: *Chemical Oceanography*, Vol. 1, ed. J. P. Riley and G. S. Skirrow, pp. 193–243. London: Academic.

Liss, P. S. (1986). The chemistry of near-surface seawater. In *Dynamic Processes in the Chemistry of the Upper Ocean*, ed. J. D. Burton, P. G. Brewer and R. Chesselet, pp. 41–51. New York: Plenum.

MacIntyre, F. (1974). Chemical fractionation and sea-surface microlayer processes. In *The Sea*, Vol. 5, ed. E. D. Goldberg, pp. 245–299. New York: Wiley.

Maguire, R. J. and Tkacz, R. J. (1987). Concentration of tributyltin in the surface microlayer of natural waters. *Water Pollut. Res. J. Can.*, **22**, 227–33.

Marty, J. C., Zutic, V., Precali, R., Saliot, A., Cosovic, B., Smodlaka, N. and Cauwet, G. (1988). Organic matter characterization in the northern Adriatic Sea with special reference to the sea surface microlayer. *Mar. Chem.*, **25**, 243–63.

Moffet, J. W. and Zafiriou, O. C. (1993). The photochemical decomposition of hydrogen peroxide in surface waters of the Eastern Caribbean and Orinoco River. *J. Geophys. Res.*, **98**, 2307–13.

Moffett, J. W. and Zika, R. G. (1987). The reaction kinetics of hydrogen peroxide with copper and iron in seawater. *Environ. Sci. Technol.*, **21**, 804–10.

Moffett, J. W. and Zika, R. G. (1988). Measurement of Cu(I) in surface waters of the subtropical Atlantic and Gulf of Mexico. *Geochim. Cosmochim. Acta*, **52**, 1849–57.

Morris, R. J. (1974). Lipid composition of surface films and zooplankton from the eastern Mediterranean. *Mar. Pollut. Bull.*, **5**, 105–09.

Nicolls, P. D. and Espey, Q. I. (1991). Characterization of organic matter at the air–sea interface, in subsurface water, and in bottom sediments near the Malabar sewage outfall in Sydney's coastal region. *Aust. J. Mar. Freshwater Res.*, **42**, 327–48.

Ofstad, E. B., Lunde, G. and Drangsholt, H. (1979). Chlorinated organic compounds in the fatty surface film on water. *Int. J. Environ. Anal. Chem.*, **6**, 119–31.

Onstott, R. and Rufenach, C. (1992). Shipboard active and passive microwave measurement of ocean surface slicks off the Southern Californian coast. *J. Geophys. Res.*, **97**, 5315–23.

Patterson, C. C. and Settle, D. M. (1976). The reduction of orders of magnitude errors in lead analysis of biological materials and natural waters by controlling external sources of industrial Pb contamination introduced during sample collection, handling and analysis. In *Reliability in Trace Analysis*, ed. D. M. La Fleur, pp 321–343. Special Publication 422. Washington, DC: National Bureau of Standards.

Peltzer, R. D., Griffin, O. M., Barger, W. R. and Kaiser, J. A. C. (1992). High-resolution measurement of surface-active film redistribution in ship wakes. *J. Geophys. Res.*, **97**, 5231–52.

Picer, N. and Picer, M. (1992). Inflow, levels and the fate of some persistent chlorinated hydrocarbons in the Rijeka Bay area of the Adriatic Sea. *Water Res.*, **26**, 899–909.

Piotrowicz, S. R., Ray, B. J., Hoffman, G. L. and Duce, R. A. (1972). Trace metal enrichment in the sea surface microlayer. *J. Geophys. Res.*, **77**, 5243–54.

Sauer, T. C., Durell, G. S., Brown, J. S., Redford, D. and Boehm, P. D. (1989). Concentrations of chlorinated pesticides and polychlorinated biphenyls in microlayer and seawater samples collected in open ocean waters of the US east coast and in the Gulf of Mexico. *Mar. Chem.*, **27**, 235–7.

Savenko, V. S. (1990). Determination of the salinity of the thin surface microlayer of seawater. *Oceanol. Acad. Sci. USSR*, **30**, 289–92.

Seba, D. B. and Corcoran, E. F. (1969). Surface slicks as concentrators of pesticides in the marine environment. *Pestic. Monit.*, **3**, 190–3.

Takenaka, T. and Fukuzaki, H. (1979). Resonance Raman spectra of insoluble monolayers spread on a water surface. *J. Raman Spectrosc.*, **8**, 151–4.

Tseng, R. S., Viechnicki, J. T., Skop, R. A. and Brown, J. W. (1992). Sea-to-air transfer of surface-active organic compounds by bursting bubbles. *J. Geophys. Res.*, **97**, 5201–6.

Van Vleet, E. S. and Williams, P. M. (1983). Surface potential and film pressure measurements in seawater systems. *Limnol. Oceanogr.*, **28**, 401–14.

Wallace, G. T. and Duce, R. A. (1978a). Open-ocean transport of particulate trace metals by bubbles. *Deep-Sea Res.*, **25**, 827–35.

Wallace, G. T. and Duce, R. A. (1978b). Transport of particulate organic matter by bubbles in marine waters. *Limnol. Oceanogr.*, **23**, 1155–67.

Wei, Y. and Wu, J. (1992). In situ measurements of surface tension, wave damping and wind properties modified by natural films. *J. Geophys. Res.*, **97**, 5307–17.

Williams, P. M., Carlucci, A. F., Henrichs, S. M., Van Fleet, E. S., Horrigan, S. G., Reid, F. M. H. and Robertson, K. J. (1986). Chemical and microbiological studies of sea-surface films in the southern Gulf of California and off the west coast of Baja California. *Mar. Chem.*, **19**, 17–98.

Wilson, A. T. (1959). Surface of the ocean as a source of air-borne nitrogenous material and other plant nutrients. *Nature*, **184**, 99–101.

Zisman, W. A. (1964). Relation of the equilibrium contact angle to liquid and solid constitution. In *Contact Angle: Wettability and Adhesion*, ed. R. F. Gould, *Advances in Chemistry Series*, **43**, 1–51.

10

Biophysics of the surface film of aquatic ecosystems

MICHAIL I. GLADYSHEV

Abstract

The surface film of water (the laminar layer about 0.5 mm thick and the intermediate layer about 5 mm thick) is regarded as a 'bottleneck' for heat and mass exchange between the atmosphere and natural water bodies. The dependence of the surface film temperature on air and water temperature and humidity under laboratory conditions is described. As demonstrated, replacing a 'warm' by a 'cold' surface film results in the oxygen transfer rate increasing by 8%.

The surface film of natural water bodies is inhabited by specific neuston organisms. The freshwater zooneuston of large Siberian reservoirs is described, and their general similarity with the marine neuston of the Sea of Japan is shown. The statement is made that there are two ways in which the biota influence the properties of surface films: 1. mechanical – by providing turbulence in the laminar layer by the swimming action of small zooneuston organisms; and 2. chemical – by the influence of biogenic surfactants on the water film. Experimental evidence of the biotic influence is demonstrated.

Introduction

From the hydrophysical or thermophysical point of view the surface film of water is envisaged as consisting of a laminar layer through which heat and mass transfer takes place by molecular diffusion, and an intermediate layer in which the rates of diffusion increase from molecular to turbulent scales (Figure 10.1). On the open surface of lakes and seas the laminar layer average thickness is about 0.2–2.0 mm, and the intermediate layer is about 10 mm (Khundzhua et al., 1977; Panin, 1985; Ilyin et al., 1986). These layers exist due to suppression of turbulence by the water surface, due to molecular interactions. The rate of molecular diffusion is known to be five to six orders of magnitude slower than that of turbulent diffusion. Thus, the surface film represents a 'bottleneck' for heat and mass exchange between the atmosphere and natural water bodies.

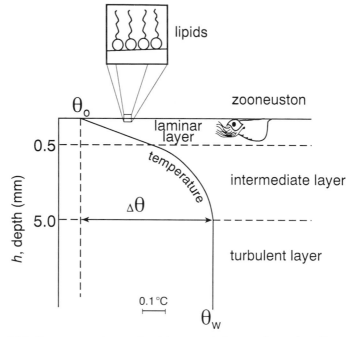

Figure 10.1. Structure and ways that biota may influence the surface film of water. The zooneuston *Scapholeberis* is drawn to scale.

The exchange rates depend on the effective thickness of the surface film and the temperature profile in the film. These surface properties in turn depend on many parameters, including wind velocity, heat balance, and the presence of surface-active materials. The wind influence is known to be the major determining factor for heat and mass exchange between the atmosphere and natural water bodies. An especially important effect is wave breaking under high winds, which produces a dramatic change in heat exchange. But under low wind conditions, with $U_{10} < 3$ m s^{-1} (U_{10} refers to the wind velocity 10 m above the water surface), the state of the surface film of water is not significantly affected. Thus, in such situations the film properties become the determining factor for the heat and gas exchange (Malevskii-Malevich, 1974; Panin, 1985; Brekhovskikh, 1988; Savenko, 1990). Damping of capillary waves by slicks decreases the wind component in the exchange and also increases the role of the surface film properties. Thus, at low winds the detailed physical properties of the surface film of water can be of great ecological importance.

On the open surface of natural water bodies the surface film of water is clearly affected by biotic processes. While many authors remark on the importance of investigating the biotic influence on the surface film properties, this subject remains almost unexplored (MacIntyre, 1974; Katsaros, 1980; Lebedev, 1986). There are two main ways that the aquatic biota affect the detailed properties of the surface film (Figure 10.1):

1. the chemical influence of biogenic surfactant films on the water film; and
2. the mechanical influence of turbulence in the laminar layer, caused by the swimming motion of small zooneuston organisms.

Before considering biotic effects, it is necessary to determine the primary detailed physical properties of the film.

Many researchers consider that the temperature difference between the surface film and the turbulent layer (bulk water), $\Delta\theta$, (Figure 10.1) is one of the important physical characteristics (Malevski-Malevich, 1974; Panin, 1985; Brekhovskih, 1988; Savenko, 1990). This difference is a result of the heat balance of the water surface:

$$\Delta\theta = \theta_0 - \theta_w = -h/k \, (R + P + LE) \tag{10.1}$$

where θ_0 is the surface temperature, θ_w is the bulk water temperature at depth h (Figure 10.1), k is the average heat conductivity in the layer $(0-h)$ and R, P and LE are heat fluxes due to radiation, convection and evaporation respectively (Malevski-Malevich, 1974). The laminar and intermediate layers are also sometimes designated as a 'cold film' because their temperature in oceans and lakes during summer is often lower than that of the turbulent layer, as is shown in Figure 10.1 (Malevski-Malevich, 1974; Khundzhua *et al.*, 1977). However, the surface film can also be warmer than the turbulent layer, depending on relationships between the radiation, convection and evaporation heat fluxes.

Experimental studies

The dependence of the difference of the temperatures of the surface film and the bulk water, $\Delta\theta$, on the convective heat flux, i.e. the difference between air and water temperature, $\theta_a - \theta_w$, was investigated in a laboratory experiment. This experiment was carried out in a small cylindrical vessel of 0.72 l volume and 36 cm^2 of open water surface, placed in a hygrostat to maintain a given air humidity. Such vessels were constructed for subsequent investigations of biotic effects. The temperature of the water surface film was measured by a microthermoresistor that had a glass-covered bead of diameter 0.7 mm. Thus, it gave the integrated average temperature of the surface film down to a depth of about 1 mm. This temperature is designated below as $\bar{\theta}_0$. A detailed description of the measurement techniques and experimental devices has previously been published (Gladyshev, 1988). It was found that under usual room humidity of 21% (Figure 10.2), a cold film ($\bar{\theta}_0 - \theta_W < 0$) existed even when the air temperature was warmer than the bulk temperature ($\theta_a - \theta_W > 0$). In this case the convective flux was positive, i.e. directed from air to water, but it was compensated for by evaporation. If the evaporation was suppressed by enhanced humidity (Figure 10.2; 88%), a cold film only existed when there was a negative difference between the air and water temperature, i.e. under negative convective flux. The radiation flux in this experiment was negligible.

In general, results of the experiment were comparable qualitatively

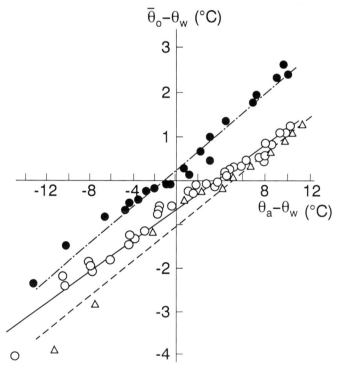

Figure 10.2. The dependence of the temperature differences between the surface film and bulk water ($\theta_a - \theta_w$) on the temperature differences between the air and water ($\theta_0 - \theta_w$) in a small experimental vessel under three relative humidity conditions: 88% (filled circles and dashed-dotted line), 21% (open circles and solid line), and 15% (triangles and dashed line).

and quantitatively with results of similar measurements made in large pools and natural reservoirs under low wind conditions (Melnikov and Mineev, 1979; Ilyin *et al.*, 1986), so it appears that the small size of the laboratory vessel did not influence the investigated principal properties of the surface film. Therefore, similar vessels were used in subsequent ecological experiments.

The three data sets obtained under different humidity conditions (Figure 10.2) were approximated by the linear equation:

$$\bar{\theta}_0 - \theta_w = \alpha\,(\theta_a - \theta_w) + \beta \qquad (10.2)$$

Empirical coefficients α and β were extracted by least squares fit (Table 10.1). There were no statistically significant differences between the values of slope coefficients α for these three graphs, with the implication that variation in intensity of evaporation caused by changing the air humidity did not affect the hydrodynamic properties of the surface film. This fact appears to be of importance for subsequent determination of biotic effects on the surface film. Values of the intercepts, i.e. β, differed significantly at a confidence level of 0.99 (Table 10.1) and were satisfactorily approximated by a dimensionless parameter of evaporation which was equal to the ratio of the humidity deficit to the atmospheric pressure (Gladyshev, 1988). This provided an opportunity to determine the relevant equation which described the dependence of $(\bar{\theta}_0 - \theta_w)$ on air and water temperature and humidity, and, by varying these parameters, to create experimentally cold and warm films of a given temperature.

Cold and warm films differ not only quantitatively but qualitatively as well. A cold film has gravitational instability at water temperatures $>$ 4 °C. It grows due to cooling of the surface, then reaches a critical size and its lower part collapses, sinking into the bulk water, which is warmer and therefore less dense than the film (Ginsburg, Zatsepin and Fedorov, 1977; Katsaros, 1980). This process is accompanied by fluctuations of the surface film temperature. In contrast, a warm film is naturally stable and its temperature does not change with time under steady-state conditions. These two different kinds of film may be of different ecological importance, because it is clear that thermal instability provides enhanced rates of transfer between the water and the atmosphere and accordingly could also influence considerably the rates of air–water exchange of gases (Savenko, 1990).

Due to the opportunity to change experimentally the film properties by varying the parameters described above, a comparison of oxygen transfer through the warm and cold surface films of equal temperature was carried

Table 10.1 *Values of the empirical coefficients of Equation 10.2 and their standard errors* (SE) *at three relative humidities*

Humidity (%)	α	SE	β	SE
88	0.204	0.007	0.259	0.047
21	0.194	0.009	−0.724	0.062
15	0.220	0.011	−1.060	0.087

out. The oxygen transfer coefficient, K_L (cm h^{-1}), was determined experimentally. As is known,

$$K_L = D/\delta \qquad (10.3)$$

where D is the molecular diffusivity for oxygen and δ is the thickness of the molecular diffusion layer. A complete description of this experiment has been given by Gladyshev (1991a). It was found that the oxygen transfer coefficients for cold films were in general higher than those for warm films (Figure 10.3). This difference was statistically significant by the Student's t-test at a confidence level of 0.99. The average oxygen transfer coefficient value for the cold film was 108% of that for the warm film. Thus, the state of the surface film of the water may be of considerable importance for aquatic ecosystems.

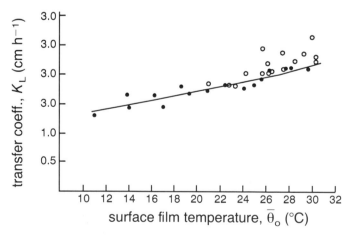

Figure 10.3. Dependence of the oxygen transfer coefficient on the temperature of the surface film of water. Open circles, cold film; filled circles, warm film; solid line, least-squares fit for the warm film.

Effects of surface-living organisms

As mentioned above, neustonic organisms may influence the surface film of water mechanically, e.g. by providing turbulence in the laminar layer by swimming into it. To estimate the large-scale effect of zooneuston mechanical influences on air–water exchange processes, one needs to know two things: 1. the mechanism of the neuston influence on the relevant properties of the film and quantitative values of the effect; and 2. the distribution of the neuston density in oceans, seas and lakes.

The neuston is known to be a specific group of organisms inhabiting the surface film of water. One of the recent classifications of neustonic living forms and ecological groups was made by Gladyshev and Malyshevski (1982b). At the beginning of this century, the neuston was regarded to exist only in small pools. Then the marine zooneuston was discovered and studied (Zaitsev, 1971). Following Zaitsev's initial work, a paradoxical situation developed: zooneuston species were considered to live only in seas and small freshwater bodies, because all known freshwater zoo-neuston species were limited to smooth, undisturbed water surfaces (Kiselev, 1969; Berezina, 1973; Rivier, 1975; Konstantinov, 1979). Then large Siberian reservoirs were investigated using the methods of marine neustonology, and freshwater zooneuston tolerant to wind and wave influences were identified (Gladyshev, 1980, 1986, 1994). For the following discussion of the biophysics of the mechanical effect of zooneuston on the surface film and its potential global importance, several aspects of neustonology should be emphasized.

The zooneuston of large freshwater reservoirs and the marine zoo-neuston have similar general ecological features. For example, both in a marine bay and in a freshwater reservoir bay the primary fraction (more than 80%) of the zooneuston biomass consists of benthoneustonic organisms which move very actively (Gladyshev and Malyshevski, 1982a; Gladyshev, 1994). The dynamics of daily vertical migrations of these marine and freshwater benthoneuston species are similar in detail (Figure 10.4). In addition to the benthoneuston migrants, most of the freshwater zooneuston organisms from the large reservoirs appeared to be universal contourobionts, i.e., they can inhabit successfully all types of interfaces: water–air and water–solid substrata (both benthic and periphytonic). The term 'contourobiont' was coined by Zaitsev (1986) for marine organisms inhabiting interfaces. However, an important difference for freshwater environments is that it is much easier for individual organisms to move from one interface to another. Thus, the same species, at the same

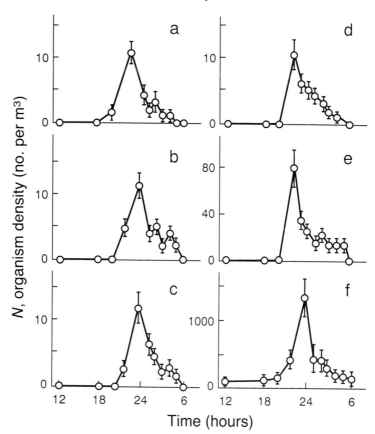

Figure 10.4. Dynamics of the density of the benthoneuston in the surface-layer 0–5 cm: (a) *Caprella excelsa*; (b) *Pontogeneia rostrata*; (c) *Diastylis alaskensis*; (d) *Campilaspis glabra*; (e) *Isopoda* (was not identified) in Vostok Bay of the Sea of Japan, July–September, 1978; (f) *Cricotopus gr. silvestris* in the Mokhovskii pool of the Krasnoyarsk reservoir, June–August, 1979.

stage of development, for example, *Chydorus sphaericus* (Crustacea, Cladocera), can occupy all the interface habitats.

Initially, attempts were made to carry out biophysical experiments to determine the effect of freshwater zooneuston on the surface film temperature in small laboratory vessels using the dominant benthoneuston species, *Cricotopus gr. silvestris* (larvae of Diptera, Chironomidae) (Figure 10.4f). Larvae collected from the Krasnoyarsk reservoir were placed in the experimental vessel. Under daylight conditions they concentrated on the bottom, and under darkness they floated up toward the surface. But after reaching the surface they soon contacted the edge of the vessel, attached to it and stopped moving. Thus, this benthoneuston organism did not swim in the surface film of small experimental vessels, and their effect on the film could not be measured. Because of this, in the main series of experiments the euneustonic species, *Scapholeberis mucronata* (Crustacea, Cladocera, Figure 10.1), was used. The primary difficulties in these experiments were that *S. mucronata* preferred the surface film, but from time to time they left it and, because of their contourobiont nature, concentrated on the other interfaces – the walls and bottom of the experimental vessel. It was believed that they might leave the film if food on the surface of the experimental water volume was insufficient. There-fore, natural water with suitable film material was collected and after observation that individual zooneuston concentrated on the surface, the relevant measurements were carried out.

The experiment was done under steady-state conditions simultan-eously in the experimental vessel with natural water containing the zooneuston (30 individuals per 33 cm^2) and in the control vessel without zooneuston; for further details, see Gladyshev (1991b). The surface film temperature in the experimental vessel was lower than that in the control vessel, and the difference was statistically significant (Table 10.2). This meant that zooneuston movements had enhanced evaporation and cooled the surface film. Further, the difference between the surface film and the water bulk temperature ($\bar{\theta}_0 - \theta_w$) was also significantly lower in the experimental vessel than in the control vessel (Table 10.2). This meant that the zooneuston generated turbulence in the laminar layer and decreased its effective thickness, thus enhancing the intensity of the exchange between the water and air. The change of one of the primary characteristics of the surface film, $\Delta\theta$, caused by the mechanical influence of the zooneuston in the experiment was more than 25%. Moreover, by decreasing the effective thickness of the laminar layer, the zooneuston also led to an increase in the rate of oxygen transfer according to Equation 10.3. Thus, the effect of small neustonic organisms on the heat and mass exchange between water and air may be comparable with that of a moderately strong wind.

Some data concerning the details of the mechanical influence of the zooneuston on the surface film have now been obtained. The experiment was carried out with freshwater zooneuston, but the general ecological similarity of the freshwater and marine zooneuston mentioned above suggests that the mechanical influence of the marine neuston would be similar. To estimate the potential global effect of this influence one needs information on the zooneuston density in lakes and seas. There are few

Table 10.2 *Temperature (°C) of the surface film* $(\bar{\theta}_o)$, *bulk water* (θ_w) *and their difference* $(\bar{\theta}_o - \theta_w)$ *in the experimental vessel (exp.) containing zooneuston* Scapholeberis mucronata *and in the control vessel during the five hours experiment*

	θ_w		$\bar{\theta}_o$		$\bar{\theta}_o - \theta_w$	
	exp.	control	exp.	control	exp.	control
	25.34	25.85	24.59	24.63	−0.74	−1.22
	25.57	26.16	24.79	25.01	−0.77	−1.15
	25.41	25.92	24.68	24.72	−0.72	−1.20
	25.67	26.19	24.68	24.95	−0.99	−1.24
	25.50	25.94	24.61	24.81	−0.89	−1.13
	25.71	26.23	24.77	25.02	−0.94	−1.20
	25.48	25.92	24.46	24.59	−1.02	−1.33
	25.69	26.14	24.55	24.90	−1.14	−1.23
Mean	25.55	26.04	24.64	24.83	−0.90	−1.21
SE	0.048	0.053	0.039	0.060	0.053	0.021
t-test	6.83		2.66		5.43	

data on the zooneuston density distribution. Only some particular preliminary suggestions may be made. Data on the large freshwater Siberian reservoirs of the Yenisei River are given in Table 10.3. In the experiment described above the zooneuston biomass density was about 80 mg m^{-2}. Thus, it was of the same order of magnitude as the maximum biomass density in the littoral of the Krasnoyarsk reservoir (Table 10.3). At sea, high densities of zooneuston may also occur; for example, Hardy *et al.* (1987) reported zooneuston abundances of up to 35 individuals m^{-2}. Although biomass data were not published, the species comprised relatively large organisms (such as Hyperiids, crab larvae, etc.), so that their biomass density could be of comparable magnitude to that of the freshwater neuston. Thus, further quantitative information on zooneuston abundance in different environments is very desirable and necessary.

The other way in which biota can influence the state of the surface film is chemical (Figure 10.1). The biogenic surface-active materials exuded in the process of metabolism by aquatic organisms inhabiting the water column accumulate at the water surface due to floatation and adsorption. They modify the physical properties of the surface film of water so that, besides damping capillary waves, they may increase the temperature of the surface film and retard evaporation. However, some authors report a dual effect of surfactants on the temperature of the surface film, i.e. such films can both increase and decrease the surface temperature (Jarvis,

Table 10.3 *Density (N, individuals m^{-2}) and biomass (B, mg m^{-2}) of the zooneuston species in freshwater reservoirs in the Yenisei River region (Siberia, Russia) in June–September of 1977–1981*

Water body	Number of samples	Species	Maximum		Average	
			N	*B*	*N*	*B*
Forest pool (Biostation of Krasnoyarsk State University)	6	*Scapholeberis mucronata*	150	1.35	68	0.61
		Cricotopus gr. silvestris	16	1.96	9	1.10
Krasnoyarsk reservoir, littoral	82	*Scapholeberis mucronata*	11	0.10	1	0.01
		Cricotopus gr. silvestris	330	40.42	17	2.08
Krasnoyarsk reservoir, deepwater	15	*Chydorus sphaericus*	435	3.13	69	0.50
		Alona costata	14	0.08	5	0.03
Sayano-Shushenskoye reservoir, deepwater	9	*Chydorus sphaericus*	165	1.19	45	0.32
		Alona costata	12	0.07	4	0.02

1962; Bogorodskii and Kropotkin, 1984; Ivanov, Kolomeev and Chekryzhov, 1989). Therefore, a special experiment was carried out in the small laboratory vessel to clarify this problem of duality. Despite the primary components of the natural organic surface film being regarded as high molecular weight compounds, e.g. 'glycoproteins' (Baier *et al.*, 1974, Hunter and Liss, 1981), it has been demonstrated experimentally by Dragcevic and Pravdic (1981) that free fatty acid (FFA) monolayers are, with respect to some surface physico-chemical features, similar to the natural organic films collected at sea. Thus, a monomolecular condensed film of stearic acid was used as a model biogenic surfactant in our experiments. For details of the experimental methods and materials used, see Gladyshev and Sushchik (1994).

We found that values of the difference between the surface film and bulk water temperature for clean water, $\Delta\theta$ (clean), were lower than that for the water covered with the monolayer, $\Delta\theta$ (acid), in the case of the warm film, i.e. ($\Delta\theta$ (clean) $- \Delta\theta$ (acid)) < 0 if ($\theta_a - \theta_w$) > 0, but in the case of the cold film the opposite took place: ($\Delta\theta$ (clean) $- \Delta\theta$ (acid)) > 0 if ($\theta_a - \theta_w$) < 0 (Figure 10.5). That is, the acid monolayer increased the

M. I. Gladyshev

temperature of the warm film and decreased the temperature of the cold film, in comparison with the clean water surface. This effect cannot be explained by retardation of evaporation, because that would only result in increasing the surface film temperature, as described above (Figure 10.2). The following explanation of this phenomenon is suggested. The monolayer of fatty acid first stabilized the surface film, i.e. it increased its effective thickness. Therefore, the monolayer influenced the convective flux between the surface film and bulk water, rather than evaporation. Thus, under positive downward convective heat flux ($\theta_a - \theta_w > 0$) it retarded the heat conduction from the warm film to the cold bulk and the film became warmer. Under negative upward convective heat flux ($\theta_a - \theta_w < 0$) it slowed the conduction of heat from the warm bulk to the cold film and the film became cooler. Thus, the dual effect of a thin surfactant layer on the temperature of the surface film of water would seem to be explained.

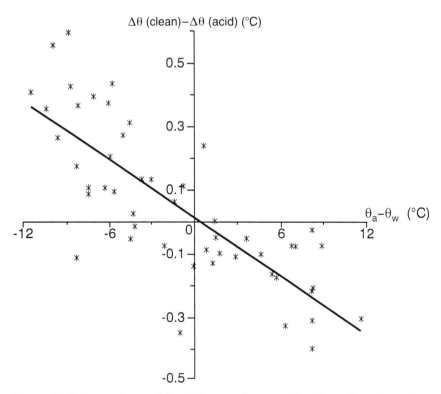

Figure 10.5. Dependence of the differences between the $\Delta\theta$ gradients in surface films of clean water and those covered with a monolayer of stearic acid on the differences between the air and water temperature. The solid line is the least squares fit.

As mentioned above, natural organic films consist for the most part of complex polymeric material ('wet' surfactants). These surfactants appear to stabilize the laminar layer even more effectively than 'dry' FFAs, although they do not retard evaporation so well. Thus, the dual effect found in the experiment described above may also apply under natural conditions.

It is clear that to estimate the large-scale effect of biogenic surfactants on water–atmosphere heat and mass exchange, information is needed about the distribution of surfactant films of different quality on the surface of seas and lakes, as well as a better understanding of the mechanisms involved.

It is worth pointing out that these two ways by which biota influence the physical properties of the surface film of water (Figure 10.1) produce opposite effects: the zooneuston decrease the surface film thickness and increase rates of transfer, while the surfactant increases the film thickness and decreases rates of transfer. On this basis, a speculative hypothesis about the potential ability of aquatic biota to affect heat exchange processes between a reservoir and the atmosphere may be suggested. For example, in spring the surface film is the warmest part of the water column. In it the zooneuston develop intensively, generate turbulence in the laminar layer and thus accelerate the heating of the water. In contrast, in autumn the abundant plankton exude large quantities of biogenic surfactants, that float to the surface, stabilize the surface film and so retard the cooling of the water.

The physical role of surface-active materials adsorbed on the water surface is noticeable, but besides this they are considered to have informative value for ecological monitoring as well. Biogenic surface-active materials, for instance lipids, including such well-known bio-markers as free fatty acids, are exuded by aquatic organisms inhabiting the entire water column. These biomarkers are accumulated in the surface film. Thus, the surface film seems to be a natural integrator of information about ecological processes taking place throughout the water column. Decoding of this information may be very promising for ecological monitoring of extended areas because it gives a chance to reduce the three-dimensional problem of analysis of aquatic ecosystems to two dimensions.

A successful attempt to investigate the surface film as a mirror of metabolism in the water column has been made in the large freshwater Krasnoyarsk reservoir. The FFA composition of the surface film was used as an ecological biomarker (Gladyshev, Kalachova and Sushchik,

1993b). During two summers, samples of the surface film and water column phytoplankton samples were taken at the same time. Three usual phases of phytoplankton seasonal dynamics took place: a spring domination by diatoms; a bloom (exponential growth) of blue-green algae; and autumn equilibrium of blue-green algae and diatoms. Using cluster analysis in conjunction with analysis of variance, it was found that the specific composition of FFAs in the surface film corresponded to each phase of the seasonal succession. The inverse problem was also solved, i.e. FFA-indicators were identified that indicated whether algal biomass was increasing or decreasing in the investigated area on the basis of a single FFA sample. This may be especially useful for monitoring cruises through large areas, because in this case one cannot investigate the dynamics of the phytoplankton biomass over several days at every sampling station.

The enhancement and decline of phytoplankton biomass (i.e. the sign of the derivative of the biomass dynamics) seem to be a very important characteristic of the integral functional properties of an entire ecosystem. For example, it was found that, in freshwater systems, the kinetics of the self-purification from phenol closely corresponded with the sign of the derivative of the phytoplankton biomass dynamics (Gladyshev, Gribovskaya and Adamovich, 1993a). Thus, using the FFA-identifier mentioned, it might be possible to estimate the potential self-purification ability of the aquatic ecosystem without long-time experiments, but on the basis of a single surface film sample.

A new approach to the investigation of integral functions of aquatic ecosystems is now being developed on the basis of a 'type of ecosystem' (Gladyshev *et al.*, 1993a; Gitelson *et al.*, 1993). It is supposed that a finite number of discrete patterns (or types) of aquatic ecosystems exists in time and space. It is likely that for every type of ecosystem particular values of kinetic characteristics, such as self-purification rate, may be attributed. Similarly, to every species of organism, values of maximum specific growth rate, half-saturation constants, etc., are attributed. To realize such an approach one needs to obtain some indications of the ecosystem types, and the FFA composition of the surface film of water seems to be one of the best indicator characteristics.

In conclusion, special monitoring of the surface film of water is likely to be very important for studying aquatic ecosystems, both for estimation of water–atmosphere exchange and for biomarker information analysis. This monitoring seems to be very promising, especially if one takes into account the potential opportunity of applying remote sensing approaches

(Robinson, Chapter 16, this volume), and in particular laser second-harmonic generation techniques (Frysinger *et al.*, 1992; Korenowski, Frysinger and Asher, 1993; Korenowski, Chapter 15, this volume) for investigating the quantity and composition of surfactants in the surface film of water.

References

Baier, R. E., Goupil, D. W., Perlmutter, S. and King, R. (1974). Dominant chemical composition of sea-surface films, natural slicks and foams. *J. Rech. Atmos.*, **8**, 571–600.

Berezina, N. A. (1973). *Hydrobiology*. Moscow: Pishchevaia Pro Myshlennost (in Russian).

Bogorodskii, V. V. and Kropotkin, M. A. (1984). Effect of oil pollutions on processes taking place in water bodies. *Vodn. Resur.*, **1**, 161–8 (in Russian, translated into English).

Brekhovskikh, V. F. (1988). *Hydrophysical Factors of Forming of Oxygen Status of Water Bodies*. Moscow: Nauka (in Russian).

Dragcevic, D. and Pravdic, V. (1981). Properties of the sea-water interface. 2. Rates of surface film formation under steady conditions. *Limnol. Oceanogr.*, **26**, 492–9.

Frysinger, G. S., Asher, W. E., Korenowski, G. M., Barger W. R., Klusty, M. A., Frew, N. M. and Nelson R. K. (1992). Study of ocean slicks by nonlinear laser processes: 1. Second-harmonic generation. *J. Geophys. Res.*, **97**(C4), 5253–69.

Ginsburg, A. I., Zatsepin, A. G. and Fedorov, K. N. (1977). Detailed structure of thermal boundary layer in water near the water–air interface. *Izvestiya AN SSSR, Seriya Fisiki Atmosfery i Okeana*, **13**, 1268–77 (in Russian, translated into English).

Gitelson, I. I., Gladyshev, M. I., Degermendzhy, A. G., Levin, L. A. and Sid'ko, F. Ya. (1993). Ecological biophysics and its role in investigation of aquatic ecosystems. *Biofizika*, **38** (6), 1069–1078 (in Russian, translated into English).

Gladyshev, M. I. (1980). Neuston biogeocenosis of the Krasnoyarsk reservoir. In *Biological Processes and Self-purification of the Krasnoyarsk Reservoir*, ed. Z. G. Gold, A. A. Vyshegorodtsev, N. S. Abro-sov and V. M. Gold, pp. 92–103. Krasnoyarsk: Krasnoyarsk State University (in Russian).

Gladyshev, M. I. (1986). Neuston of continental water bodies (review). *Gidrobiol. Z.*, **22**(5), 12–19 (in Russian, translated into English).

Gladyshev, M. I. (1988). Investigations of the water surface film temperature in the laboratory vessel of small volume. *Vodn. Resur.*, **4**, 69–73 (in Russian, translated into English).

Gladyshev, M. I. (1991a). Comparison of rates of oxygen transference through the 'warm' and the 'cold' surface film of water. *Izvestiya AN SSSR, Seriya Fisiki Atmosfery i Okeana*, **27**(11), 1251–5 (in Russian, translated into English).

Gladyshev, M. I. (1991b). Experimental study of effect of zooneuston organisms on the temperature of the surface film of water. *Doklady Akademii Nauk SSSR*, **320**(6), 1489–90 (in Russian, translated into English).

Gladyshev, M. I. (1994). Zooneuston of the Yenisei reservoirs. *Gidrobiol. Z.*, **30**(3), 3–15 (in Russian, translated into English).

Gladyshev, M. I. and Malyshevski, K. G. (1982a). Daily vertical migrations of the benthoneuston in the Vostok Bay, Sea of Japan. *Biol. Moria*, **2**, 20–3 (in Russian, translated into English).

Gladyshev, M. I. and Malyshevski, K. G. (1982b). The concept of 'Neuston'. *Gidrobiol. Z.*, **18** (3), 19–23 (in Russian, translated into English).

Gladyshev, M. I. and Sushchik, N. N. (1994). Effect of monomolecular layer of fatty acid on temperature of the surface film of water. *Izvestiya RAN, Seriya Fisiki Atmosfery i Okeana*, **1**, 74–7 (in Russian, translated into English).

Gladyshev, M. I., Gribovskaya, I. V. and Adamovich V. V. (1993a). Disappearance of phenol in water samples taken from the Yenisei river and the Krasnoyarsk reservoir. *Water Res.*, **27**(6), 1063–70.

Gladyshev, M. I., Kalachova, G. S. and Sushchik N. N. (1993b). Free fatty acids of surface film of water in the Sydinsky bay of the Krasnoyarsk reservoir. *Int. Rev. Gesamten Hydrobiol.*, **78**(4), 575–87.

Hardy, J., Kiesser, S., Antrim, L., Stubin, A., Kocan, R. and Strand, J. (1987). The sea-surface microlayer of Puget Sound: Part 1. Toxic effects on fish eggs and larvae. *Mar. Environ. Res.*, 23, 227–49.

Hunter, K. A. and Liss, P. S. (1981). Organic sea surface films. In *Marine Organic Chemistry*, ed. E. K. Duursma and R. Dawson, pp. 259–98. Amsterdam: Elsevier.

Ilyin, Yu.A., Panin, G. N., Popov, N. N., Skorohvatov, N. A., Tserevetinov, F. O., Chernyshev, O. A., Shevchenko, V. I. and Grigoriev, V. T. (1986). Experimental investigations of thermal structure of subsurface layer of water body. *Vodnyie Resursy*, **2**, 97–101 (in Russian, translated into English).

Ivanov, V. V., Kolomeev, M. P. and Chekryzhov, V. M. (1989). Effect of thin oil films on radiation temperature of water surface and rate of gas exchange. *Izv. Akad. Navk. SSSR, Ser. Fis. Atmos. Okeana*, **25** (2), 202–6 (in Russian, translated into English).

Jarvis, N. L. (1962). The effect of monomolecular films on surface temperature and convective motion at the water/air interface. *J. Colloid Sci.*, **17**(6), 512–22.

Katsaros, K. B. (1980). The aqueous thermal boundary layer. *Boundary-layer Meteorol.*, **18**, 107–27.

Khundzhua, G. G., Gusev, A. M., Andreev, E. G., Gurov, V. V. and Skorokhvatov, N. A. (1977). About the structure of the surface cold film of ocean and heat exchange between ocean and atmosphere. *Izv. Akad. Nauk. SSSR, Ser. Fis. Atmos. Okeana*, **13** (7), 753–8 (in Russian, translated into English).

Kiselev, I. A. (1969). *Plankton of Seas and Continental Reservoirs*, Vol. 1. Leningrad: Nauka (in Russian).

Konstantinov, A. S. (1979). *General Hydrobiology*. Moscow: Vysshaia Shkola (in Russian).

Korenowski, G. M., Frysinger, G. S. and Asher, W. E. (1993). Noninvasive probing of the ocean surface using laser-based nonlinear optical methods. *Photogr. Eng. Remote Sensing*, **59** (3), 363–9.

Lebedev, V. L. (1986). *Interfaces in the Ocean*. Moscow: Moscow State University Press (in Russian).

MacIntyre, F. (1974). The top millimeter of the ocean. *Sci. Am.*, **230**, 62–77.

Malevskii-Malevich, S. P. (1974). Peculiarities of the temperature distribution in the surface water layer. In *Processes of Exchange near the Ocean–Atmosphere Boundary Surface*, ed. A. S. Dubov, pp. 135–61. Leningrad: Gidrometeoisdat (in Russian).

Melnikov, G. S. and Mineev, E. N. (1979). Detection of surface temperature gradient in the liquid boundary layer of the system 'sea–atmosphere'. In *Optical Methods of Investigation of Oceans and Inland Reservoirs*, ed. G. I. Galazii, K. S. Shifrin and P. P. Sherstyankin, pp. 212–19. Novosibirsk: Nauka (in Russian).

Panin, G. N. (1985). *Heat and Mass Exchange between a Reservoir and Atmosphere at Natural Conditions*. Moscow: Nauka (in Russian).

Rivier, I. K. (1975). Zooplankton and neuston. In *Methods of Investigation of Biogeocenosis of Inland Reservoirs*, ed. G. G. Vinberg, pp. 138–58. Moscow: Nauka (in Russian).

Savenko, V. S. (1990). *Chemistry of Water Surface Microlayer*. Leningrad: Gidrometeoizdat (in Russian).

Zaitsev, Y. P. (1971). *Marine Neustonology*. Washington, DC: National Marine Fisheries Service, NOAA and NSF.

Zaitsev, Y. P. (1986). Contourobionts in ocean monitoring. *Environ. Monitor. Assess.*, **7**, 31–8.

11

Biological effects of chemicals in the sea-surface microlayer

JOHN T. HARDY

Abstract

The upper metre of the ocean is a transition zone between the atmosphere and deeper water and can be subdivided into strata (nanolayer, microlayer, millilayer and centilayer) with different chemical and biological characteristics. A variety of techniques has been used to collect samples from different depths for chemical and biological analysis. Autotrophic and heterotrophic neuston (surface dwelling biota) range in size from less than 2 µm (piconeuston) to a metre or more (macroneuston) and are represented by, perhaps, several thousand species worldwide. They occur in much greater densities than their sub-surface counterparts, the plankton and nekton.

Anthropogenic organic compounds and metals frequently occur in greater concentrations in the microlayer (upper 10^{-6} m) than in deeper layers. These surface enrichments originate from a variety of sources, but in offshore and in some coastal areas atmospheric deposition is particularly important. Enrichment factors (microlayer concentration/bulk-water concentration) for Pb, Cd, Cu, Zn, some radionuclides and aromatic hydrocarbons are typically 10 to 10^2 and for chlorinated organics may be 10^3 or more.

Increases in global sea-surface contamination and/or ultraviolet radiation could threaten important sea-surface biological communities and processes. Contaminated films deposited on intertidal beaches during receding tides could negatively impact shellfish and infauna. The majority of marine fish has floating eggs or larvae which could be adversely affected. Although more study is needed, some evidence suggests that microneuston may be important in the sea-surface biogeochemical cycles linking the atmosphere and ocean.

The aquatic surface layer

Important physical, chemical and biological processes occur at interfaces between different environments. The upper metre of the ocean (surface layer), covering 71% of the earth's surface, is a transitional gradient where physical and biological conditions change more rapidly near the

339

surface than below. The chemistry and the biology of the surface layer differs greatly from that of the water column. The sea surface is a habitat for a great diversity of organisms, an area of high productivity, a vital nursery area for eggs and larvae, a collection point for natural organic compounds produced in the water column, and an area of concentration for anthropogenic toxic chemicals from the atmosphere and water column.

A general conceptual model considers that the aquatic surface layer (SL or upper 1 metre of the water surface) actually consists of a series of sublayers or microstrata of different thicknesses, each with different physical, chemical, and biological characteristics (Hardy and Word, 1986). These microstrata include the thin surface nanolayer (SNAN, $<10^{-6}$ m), highly enriched in organic compounds; the sea-surface microlayer (SMIC, $<10^{-3}$ m), with high densities of particles and microorganisms (bacterioneuston and phytoneuston); the surface milli-layer (SMIL $<10^{-2}$ m), inhabited primarily by small crustaceans and the eggs and larvae of fish and shellfish (zooneuston); and the surface centilayer (SCEN, 10^{-1} m), with larger crustaceans and floating organisms such as certain snails, jellyfish, and seaweeds (Figure 11.1). The aquatic surface layer of freshwaters shares much of the same physical, chemical, and biological characteristics as the sea surface (Gladyshev, 1986).

The use of different terminology has led to some confusion, particularly regarding the differences between the neuston and plankton communities. Gladyshev and Malyshevski (1982a) suggest that the neuston represent a unique ecosystem (biocenosis) inhabiting the neustal zone (upper 5 cm), which includes the upper (subaerial) side of the surface film inhabited by epineuston and the lower (water) side of the surface film inhabited by hyponeuston. The neustal may also contain planktonic and benthic species that migrate up into this layer from the water column and, in some cases, are only temporary inhabitants (Gladyshev and Malyshevski, 1982b).

Research on the aquatic surface layer has been limited to within-discipline rather than interdisciplinary study. Chemists have quantified the high concentrations of nutrients, biogenic organics, and anthropogenic contaminants in the upper few microns. Microbiologists have documented the abundance of bacteria and protozoa in the same layer. Phycologists have found abundant populations of microalgae in the upper 100 μm or less; zoologists have identified different metazoans that inhabit the top few centimetres of the water, and fisheries biologists point

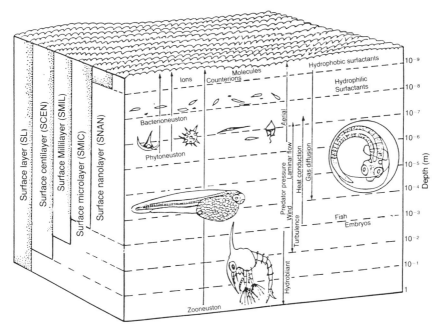

Figure 11.1. Conceptual model of the sea-surface layer (modified from Hardy and Word, 1986).

to the importance of the upper few centimetres for the early developmental stages of fish and shellfish. No investigators have sampled different biological or chemical components within the different surface-layer strata simultaneously to characterize the sea-surface ecosystem. Therefore, the basic trophic structure of the neuston community and its interactions with the plankton community remain largely unknown.

Following is a brief synopsis of the biology and chemistry of the sea surface. Additional information is provided by a number of previous and current reviews (see Zaitsev, 1971; Hempel and Weikert, 1972; Liss, 1975; Norkrans, 1980; Mileikovskii, 1981; Lion and Leckie, 1981; Hardy, 1982; Peres, 1982; Hunter, Chapter 9, this volume).

Sea-surface sample collection

Several depths within the upper metre of the water column should be sampled simultaneously using techniques appropriate for each type of biological or chemical component. The surface-layer strata can be sampled using more than 20 different techniques. Each has its own

Table 11.1 *Sea-surface sampling methods appropriate for taxa abundance, biomass, and chemistry*

Layer	Depth	Sampling method	Chemistry	Neuston			Near-surface plankton		
				Pico & nano	Micro	Meso & macro	Pico & nano	Micro	Meso & macro
Nanolayer SCAN	0 to 1 μm	Prism or remote sensing	+						
Microlayer SMIC	1 to 1000 μm	Membrane		+					
		Glass plate	+	+	+				
		Drum	+	+	+				
		Screen							
Millilayer SMIL	0.1 to 10 mm	Pump	+	+	+				
Centilayer SCEN	1 to 10 cm	Pump	+		+	+			
		Net				+			
Surface layer SL	0.1 to 1 m	Pump	+				+	+	+
		Net							+

Size ranges: Piconeuston and plankton = < 2 μm; Nanoneuston and plankton = 2 to 20 μm; Microneuston and plankton = 20 to 200 μm; Mesoneuston and plankton = 0.2 to 20 mm; Macroneuston and plankton = > 20 mm.

advantages, disadvantages and depth of sampling, from less than one micrometre to several metres (Garrett and Duce, 1980; Hühnerfuss, 1981; Hardy, 1982; Carlson, 1982c). However, six basic types of sampling gear (the membrane filter, glass plate, rotating drum, screen, net and pump) can be used effectively to collect different biological and chemical components from appropriate depth intervals of the surface layer (Table 11.1).

Membrane filter: Piconeuston in the upper 20 μm of the water surface can be sampled by the polycarbonate membrane filter technique (Sewell *et al.*, 1981). Briefly, a 47 mm diameter filter is floated on the water surface for approximately 5 seconds and then removed with sterile forceps and placed in a sterile test tube, with 3 ml of glutaraldehyde fixative.

Glass plate: A clean glass plate dipped vertically through the water surface and withdrawn collects samples of SMIC 30 to 55 μm in thickness (Harvey & Burzell, 1972; Hardy *et al.*, 1985b). It is somewhat tedious, since many repeated dips are necessary to collect a sizeable volume of sample, but it is quite simple and portable. The thickness and type of sample collected agrees closely with that of the drum (below).

Drum: Rotating drum samplers provide an efficient method for collecting large-volume samples of the SMIC, and collect a surface layer about one tenth as thick as the screen method. We developed and tested a rotating Teflon drum sampler (Hardy *et al.*, 1988). The drum collects a relatively thin layer (~34 μm), and is comparable in collection characteristics to the more time-consuming glass plate technique (Hardy *et al.*, 1985b). Nano- and microneuston can be collected using the Teflon drum sampler. Microlayer samples for chlorophyll pigments and chemical analysis can also be collected using the Teflon drum. Samples collected with the drum at Beaufort 4 sea conditions suggest that the SMIC is surprisingly stable (Hardy *et al.*, 1988). A glass drum sampler collects a similar depth sample and, at the same time, continuously collects sub-surface (10 cm depth) water and a variety of environmental data (Carlson *et al.*, 1988).

Screen: The screen sampler (Garrett, 1967) has been widely used. The screen is dipped in the water surface and then drained into a collector. Samples collected in this manner have been referred to as 'microlayer'. However, screens tend to collect substantial quantities of sub-surface water (to a depth of 300 to 400 μm) and effectively dilute the microlayer by a factor of six to eight compared with plate or drum samplers (typically 50 μm). Data on the microlayer of Baja California are typical and must be interpreted with caution, as the authors state 'enrichment of dissolved and particulate organic matter, nutrients, bacteria, etc. compared to the sub-surface waters could be up to 10^5 times greater than reported here'.

Pump: Pico-, nano- and microplankton can be collected by pumping approximately 5 to 10 litres of SCEN from depths of three or more centimetres through a 200 μm mesh screen (to remove larger organisms). Pumping has been criticized as potentially damaging to organisms, especially large watery forms. However, with the appropriate choice of gear, damage is negligible (Beers, 1978). SCEN samples for chlorophyll pigments and chemical analyses can also be collected by pumping.

Nets: Mesoneuston, macroneuston, and plankton for enumeration, and for live rate-determining experiments, can be collected by towing a two-tiered 200 μm neuston net (Schram *et al.*, 1981). The net collects mesoneuston from a depth of 0 to 5 cm in the upper compartment while simultaneously sampling mesoplankton from a depth of 40 to 60 cm in the lower compartment (Hardy *et al.*, 1987a). Samples for enumeration can be preserved in 2% buffered formalin and returned to the laboratory for determination of species and densities.

Biology of the aquatic surface layer

Recognition of a special group of organisms inhabiting the water surface (neuston) occurred early in this century (Naumann, 1917), but study of the role of these organisms in aquatic ecosystems began only in the late 1960s with research in the US (Hardy, 1971; Hardy, 1973) and in the former Soviet Union (Zaitsev, 1971 and Chapter 12, this volume). Most studies suggest that the sea surface represents a highly productive, metabolically active interface and a vital biological habitat. Organisms from most major divisions of the plant and animal kingdoms either live, reproduce, or feed in the surface layers (SNAN, SMIC, SMIL, and SCEN) (Hardy, 1982). As with the plankton, it is useful to consider neuston of different size classes.

Piconeuston

Piconeuston (<2 μm), like picoplankton (Sieburth *et al.*, 1978), are being increasingly recognized as important links in the recycling of organic matter in aquatic ecosystems. High densities of metabolically active bacterioneuston are found in the SMIC (Sieburth 1971; Bezdek and Carlucci 1972; Sieburth *et al.*, 1976; Carlucci *et al.*, 1985). For example, bacterioneuston off the coast of Baja California account for 1.4 to 5.9% of the total microbial carbon biomass and a similar percentage (1.9 to 5.1%) of the microbial carbon production in the SMIC. Also, they actively assimilate dissolved free amino acids (Carlucci *et al.*, 1986). Studies along the coast of Maine (USA) also suggest that the SMIC contains highly active heterotrophic and autotrophic populations which respond rapidly to environmental changes (Carlucci *et al.*, 1991).

Bacterial enrichments in the SMIC result, at least in part, from the greater degree of hydrophobicity of bacterioneuston compared with bacterioplankton and, thus, their adhesion to organically enriched surface microlayers (Dahlback, *et al.*, 1981). Bacteria also adhere to the air–water interface of bubbles and may be injected into the atmosphere as part of sea-salt aerosols (Blanchard, 1983). Bacterioneuston probably play an important role in degrading, not only natural organics, but also anthropogenic chemicals collected at the sea surface. In the Bay of Marseilles (France), hetertrophic activity, as represented by CO_2 production in the SMIC, is enhanced relative to sub-surface water (Garabétian, 1990).

The importance of autotrophic picoplankton is also being increasingly recognized (Johnson *et al.*, 1982). In the subarctic Pacific, 28% of the biomass (as carbon) was in the <2 μm size fraction and was almost exclusively the blue-green autotroph *Synechococcus* spp. (Booth *et al.*, 1988). Picoautotrophs in the microlayer could influence air–sea exchange of gases such as CO_2, but relevant studies are lacking.

Autotrophic nano- and microneuston

Autotrophic nano- and microneuston (2 to 200 μm in size), including dozens of species of microalgae (phytoneuston) occur in great abundance in the microlayer (Hardy and Apts, 1984; Hattori *et al.*, 1983; Hardy 1973; Hardy and Valett, 1981). Enrichment ratios (number of individuals per volume in the SMIC/numbers per volume in the SCEN) are generally 10 to 1000.

The blue-green alga (cyanobacterium) *Trichodesmium* is a common phytoneustonic organism in tropical open ocean waters. In the tropical North Atlantic Ocean it is the most important primary producer (165 mg C m^{-2} d^{-1}) and is also responsible, through nitrogen fixation, for the largest fraction of new nitrogen to the euphotic zone (30 mg m^{-2} d^{-1}) (Carpenter & Romans, 1991). Although maximum densities of *Trichodesmium* sp. occurred at near-surface (15 m) depths (Carpenter and

Romans, 1991), this may simply reflect the absence of microlayer sample collection, which might otherwise show the buoyant filaments even more concentrated in the microlayer. For example, in the Atlantic Ocean 150 km off the Florida coast, the surface microlayer enrichment ratio for total phytoneuston (dominated by *Trichodesmium* sp. and microflagellates) can reach 400 with a phytoneuston species composition very different from that of the sub-surface phytoplankton community (Hardy *et al.*, 1988).

Certain dinoflagellate species are commonly abundant in the SMIC. For example, in the North Sea, near the island of Sylt, a bloom of the dinoflagellate *Prorocentrum micans*, and several other phytoneuston species, can contribute to the formation of visible surface slicks. Micro- layer enrichment ratios for *P. micans* of more than 1000 occur (Brockmann *et al.*, 1976). The phytoneuston community off Baja California is dominated by dinoflagellates and differs greatly from the phytoplankton in species composition (Williams *et al.*, 1986). Also, in the Black Sea, continuing eutrophication has led to frequent algal blooms (including dinoflagellates) and screen-collected microlayer phytoneuston/phytoplankton enrichment ratios of 100 to 10000 (Nestrova, 1980).

Only a few measurements of photosynthetic carbon reduction in the SMIC exist. Most data focus on nearshore, productive environments: an enclosed lagoon (Hardy, 1973); a salt marsh (Gallagher, 1975); coastal estuaries (Albright 1980; Cullen *et al.*, 1989); and the Mediterranean Sea near Marseilles (De Souza Lima and Chretiennot-Dinet, 1984). One offshore study was conducted near Baja California (Williams *et al.*, 1986). All these investigators enclosed collected microlayers and their communities (microneuston) in bottles, thus disrupting the natural surface-film habitat.

Study of undisrupted SMIC, where a water column and surface is enclosed in microcosms, incubated under ambient temperature and solar energy, and the SMIC then sampled using the membrane technique, suggest significant enhancements in biomass and productivity in the SMIC compared with bulkwater (Hardy and Apts, 1989). Enrichment ratios (SMIC/SCEN concentration) in slick (S) and non-slick (NS) areas were: phytoneuston population abundance, 37 (NS) to 154 (S); total chloro- phylls, 1.3 (NS) to 18 (S); particulate carbon fixation, 2 (NS) to 52 (S); and dissolved carbon excretion by autotrophs, 17 (NS) to 63 (S). As in other studies, the species composition of the phytoneuston was distinctly different from that of the phytoplankton (Hardy and Apts, 1989).

Photoautotrophic biomass, as indicated by chlorophyll pigments, is often greater in the SMIC, especially in areas of visible slicks, although SMIC depletions sometimes occur outside slick areas (Carlson, 1982a; Hardy and Apts, 1984; Cullen *et al.*, 1989).

Most studies suggest that phytoneuston productivity, although remaining high compared with that of plankton, is inhibited by high summer light intensities. At one coastal site in the Pacific Northwest, USA, winter and spring carbon reduction rates were 20 to 150 times greater in the SMIC than in the SCEN, but in high summer light intensities, photoinhibition of 36 to 89% occurred in phytoneuston (Hardy and Apts, 1984). Photoinhibition of phytoneuston also occurred off Baja California, where photosynthesis increased 179% in neuston incubated at 1 m depth compared with incubation at the surface (Williams *et al.*, 1986). Assimilation ratios (mg C mg chl-a^{-1} h^{-1}) in the SMIL (upper 3 mm) of the Aegean Sea in November and December are only about 45% of values at 1 m depth (Ignatiades, 1990). In a coastal estuary in Maine, USA, only slight photoinhibition occurs in spring and autumn (Cullen *et al.*, 1989). However, the Maine study does not include mid-summer measurements in ultraviolet-transparent containers under natural sunlight where photoinhibition could be expected.

Heterotrophic nano- and microneuston

Microzooplankton (the heterogenous group of organisms between 2 and 200 μm) form a vital link in the planktonic food web between small bacteria, flagellates, and larger metazoans including fish. They include protozoa and the developmental stages of many pelagic and benthic animals, but few adult metazoans. Their feeding modes include particle filtration, phagotrophy, and pinocytosis (Conover, 1982). From 4 to > 70% of primary production is consumed by microzooplankton (Conover, 1982). Small (<20 μm) ciliates, at times, make up 4 to 57% of the total biomass of heterotrophic (apochlorotic) nanoplankton in diverse marine systems (Sherr *et al.*, 1986). Nano- and microheterotrophs release dissolved organics that form substrates (along with organics from phototrophs) for the growth of bacteria on which they feed. They can be thought of as 'microbial gardeners' (Davis and Sieburth, 1982). Despite their probable importance, few quantitative data are available on their abundance or trophic relationships, so that 'To be able to attach reliable rates and biomass to just one of the pathways would be a major achievement' (Conover, 1982).

Compared with the microzooplankton, even less is known about the species composition, abundance or dynamics of microzooneuston. Based on a limited number of investigations, heterotrophic microneuston inhabiting the SMIC include small ciliates (Zaitsev, 1971), protozoa (Norris, 1965) and yeasts and moulds (Kjelleberg and Hakansson, 1977). Tintinnids often form an important component of the marine neuston, probably feeding on the high densities of bacterioneuston (Hardy, 1971; Zaitsev, 1971). Also, the heterotrophic dinoflagellates *Noctiluca scintillans* and *Oxyrrhis marina*, are often abundant in neuston samples (Hardy, 1971; Zaitsev, 1971).

Mesozooneuston

Mesozooneuston (0.2 to 20 mm in size) feed on the high densities of microneuston (Zaitsev, 1971). Samples from many areas of the world's oceans indicate that copepods, often dominated by pontellid species, are abundant components of the SMIL and SCEN (Zaitsev, 1971; Hardy, 1982). In the Irish Sea (Tully and O'Ceidigh, 1986), and many other locations (Zaitsev, 1971), neustonic isopods and amphipods are abundant, often associated with floating driftweed and debris.

Numerous species of fish, including cod, sole, flounder, hake, menhaden, anchovy, mullet, flying fish, greenling, saury, rockfish, halibut, and many others, have surface-dwelling egg and/or larval stages (Zaitsev, 1971; Ahlstrom and Stevens, 1975; Brewer, 1981; Safronov, 1981; Kendall and Clark, 1982; Shenker, 1989). Crab and lobster larvae in estuarine, coastal, and shelf areas concentrate in the surface film during midday as a result of positive phototaxis (Smyth, 1980; Jacoby, 1982; Provenzano *et al.*, 1983; Tully and O'Ceidigh, 1987). Pelagic larvae of benthic invertebrates and fishes can be transported shoreward in surface slicks associated with tidally forced internal waves (Shanks, 1983).

In Puget Sound, English sole (*Parophrys vetulus*) and sand sole (*Psettichthys melanostictus*) spawn between January and April, releasing trillions of eggs that collect on the water surface. The embryos float until hatching occurs, generally 6 to 7 days after fertilization (Budd, 1940 and pers. obs.). Because of the buoyancy of their large yolk sacks, newly hatched larvae of both species often float upside-down at the surface of the water (Budd, 1940).

Although the micro-vertical distribution of zooneuston in the upper 0 to 50 cm can be disturbed by wind mixing, the effect seems to disappear quickly once the wind subsides (Champalbert, 1977). As Zaitsev (1971) noted, '. . . most of the eggs must rise to the surface film and remain suspended there. Although the specific weight of the egg increases during development, it remains low enough to hold the egg close to the surface film . . . and surface enrichments of organisms were found to be stable even at sea states of 5 to 6 Beaufort'. Fish eggs, concentrated at the sea surface, are dispersed at wave heights exceeding 1 to 2 m, while larvae and fry remain there even when waves reach heights of 3 to 4 m (Zaitsev, 1971).

Neuston net tows, from widely dispersed areas of the global ocean, indicate that floating tar and plastic debris is important as a habitat colonized by dozens of species of marine invertebrates. The distributional abundance of many neuston, including attached eggs, is positively correlated with the distribution of pelagic tar. While the tar acts as a habitat for some species, it may exert a detrimental effect on the majority of neuston (Holdway and Maddock, 1983).

Macroneuston

The larger (> 2 cm) neuston organisms have been referred to as 'pleuston', although a more consistent term might be 'macroneuston'. These, perhaps 100 species or so, inhabit the surface layer (upper metre) generally by floatation or by association with floatable seaweed (Cheng, 1975). Coelenterates include the jellyfish *Physalia* spp., which feed on small fish, and *Velella* spp., which feed on copepods and fish and invertebrate eggs. Common gastropods in the open ocean surface layer are species of prosobranchiata and nudibranchiata. Oceanic crabs and many other organisms are found associated with the floating pelagic seaweed, *Sargassum* spp. These macroneuston undoubtedly form part of the food web linking the sea surface and sub-surface layer, but the details of this food web remain unclear.

Anthropogenic chemical enrichment of the sea surface

The SMIC is a stagnant boundary layer at the air–water interface where turbulent mixing is greatly reduced compared with the layers below (Hunter, 1980). The presence of surface-active organic films (Baier *et al.*, 1974) may cause water molecules below the interface to orient into 'icelike' clathrate structures, resulting in additional stabilization of the boundary layer (Hühnerfuss and Alpers, 1983). This stable boundary (approximately the upper 50 μm) including the SNAN and SMIC serves as a concentration layer for particles, natural and anthropogenic organic compounds, nutrients, and metals. A large portion of the organics associate with suspended particles and ultimately deposit in the bottom sediments. However, compounds that have low water solubility, or that associate with floatable particles, concentrate in the SNAN and SMIC.

Hundreds of samples collected from rivers, estuaries, bays, and offshore areas, indicate that the SMIC is generally highly enriched in bacteria, microalgae and often toxic metal and organic pollutants compared with the SCEN (Hardy, 1982; Cross *et al.*, 1987; Hardy *et al.*, 1987 a,b; 1990). Concentrations of potentially toxic contaminants including PAHs, PCBs, and metals, orders of magnitude greater than US Environmental Protection Agency water quality criteria standards, occur in the SMIC of Puget Sound (Hardy *et al.*, 1985a, 1986, 1987a,b), Chesapeake Bay (Hardy, *et al.*, 1990), Southern California (Cross *et al.*, 1987) the North Sea (Hardy and Cleary, 1992), the Persian/Arabian Gulf (Hardy *at al.*, 1993), and elsewhere (review in Hardy, 1982).

The sea surface also serves as a point of photochemical and biological degradation of contaminants. For example, the industrial plasticizer di(2-ethylhexyl)phthalate (DEHP), in microcosm experiments, accumulates in the surface microlayer where bacterioneuston play an important role in biodegradation and removal of this contaminant (Davey *et al.*, 1990). Also, bacterioneuston can actively degrade aromatic and saturate bituminous hydrocarbons in laboratory microcosms (Wyndham and Costerton, 1982).

Sources of contamination

The sea-surface is contaminated from a variety of sources. Ocean dumping and spills provide large quantities of petroleum hydrocarbons and plastic, which are carried by currents and distributed globally. In coastal areas such as the Southern California Bight (Patterson and Settle, 1974), the New York Bight, the North Sea (Cambray *et al.*, 1979), the Mediterranean (WMO, 1985) and in offshore areas beyond the continental shelves (Duce *et al.*, 1991) the major source of sea-surface contamination is atmospheric deposition. For example, in the North Sea, large quantities of anthropogenic pollutants enter the SMIC from riverine input and ocean dumping (Hardy and Cleary, 1992). However, atmospheric inputs of Cu, Zn, Pb and Cd to the surface microlayer of the North Sea exceed, by one order of magnitude, the combined input of the rivers Scheldt, Rhine and Meuse (Dehairs *et al.*, 1982).

In coastal areas sewage discharge, dredge material disposal, industrial discharges, and non-point source runoff can contribute to sea-surface contamination. Secondary treated sewage contains 10 to 25% of floatable organics as well as bacteria (Word *et al.*, 1990), and sewage effluent fractions have been detected in US and Australian surface waters (Nichols and Espey, 1991; Selleck, 1975; Word and Ebbesmeyer, 1984). In Concarneau Bay in South Brittany (France), an area under the influence of wastewater discharge, SMIC/SCEN enrichments of faecal coliform bacteria range from 24 to 720 (Plusquellec *et al.*, 1991). Typical contaminated dredge material can contain a floatable fraction of up to 0.2% of the total mass. A computer model (Hardy and Cowan, 1986), along with laboratory microcosm experiments (Word *et al.*, 1987), suggest that disposal of contaminated dredge material could have toxic effects on surface biota.

Metals

Metals enter coastal areas from a large variety of sources including non-point source runoff, sewage discharge, dredge disposal, industrial point sources and atmospheric deposition. Since anthropogenic metals from combustion sources are usually associated with organically coated (floatable) particles, it is not surprising that the sea-surface microlayer is generally enriched in anthropogenic metals compared with the sub-surface water. Substantial quantities of heavy metals deposited in the microlayer from atmospheric particle deposition are soluble in seawater (Hardy and Crecelius, 1981) and potentially toxic.

High concentrations of potentially toxic metals occur in the SMIC of many coastal areas such as Puget Sound (Hardy *et al.*, 1985b; 1987b), Chesapeake Bay (Hardy *et al.*, 1990), Southern California (Cross *et al.*, 1987) and elsewhere (review in Hardy, 1982). In Puget Sound, although concentrations of microlayer and bulkwater metals (Pb, Zn, Cd and Fe) were 2 to 15 times greater in an urban than a rural bay, enrichment ratios (SMIC/SCEN concentrations) were similar in the two bays, ranging from 6 to 65 (Hardy *et al.*, 1985b). Similar enrichment ratios were found in Chesapeake Bay where, for example, a station in the Central Bay had enrichments of Cu and Pb of 9 and 43, respectively (Hardy *et al.*, 1990). In the Southern California Bight microlayer, concentrations of total metals (Ag, Cd, Cr, Cu, Fe, Mn, Ni, Pb, and Zn) generally increased from 18 μg l^{-1} offshore to 12168 μg l^{-1} in Los Angeles Harbor (Cross *et al.*, 1987).

Table 11.2 *Mean metal enrichments (microlayer concentration/sub-surface water concentration)*

Location	Sample collection depth (μm)	Enrichment			Reference
		Cu	Pb	Zn	
Narragansett Bay	150	7	6		Piotrowicz *et al.* (1972)
New York Bight	150	2	4		Piotrowicz *et al.* (1972)
North Atlantic	150	10	7		Piotrowicz *et al.* (1972)
English Channel	200	7	25	7	Dehairs *et al.* (1982)
North Sea	300		2		Hunter (1980)
Central North Pacific	150	7		20	Barker and Zeitlin (1972)
Chesapeake Bay	34	43	9	3	Hardy *et al.* (1990)
Puget Sound	48	13	45	22	Hardy *et al.* (1985b)
Florida Shelf	34	12	6	6	Hardy *et al.* (1988)
North Sea	30 to 400	5	3	2	Hardy and Cleary (1992)
Mean of all locations		12	12	10	

In the North Sea, metal concentrations in the SMIC greatly exceed those in the near-surface bulkwater at almost all stations sampled (Hardy and Cleary, 1992). Minimum and maximum enrichment ratios (SMIC/SCEN concentrations) are: Cd 0 to 3; Ni 1 to 2; Cu 2 to 9; Pb 2 to 6; Zn 1 to 5; organotins 3 to 17; and tributyl tin 2 to 11. Microlayer concentrations of TBT and Cu exceed both the UK Environmental Quality Standard value of 2 ng TBT l^{-1} and the US EPA chronic water quality limit of 2.9 μg Cu l^{-1}. These North Sea enrichments are similar to those measured in previous studies (Dehairs *et al.*, 1982) and generally agree with those predicted by an atmospheric deposition sea-surface fractionation model (Hardy *et al.*, 1985b).

Mean microlayer enrichment ratios for Cu, Pb and Zn are typically about one order of magnitude (Table 11.2), although the ratios can differ greatly both temporally and spatially. Ratios of 100 are common and, in convergence foams or slick areas, ratios can reach 1000 to 10 000 (Szekielda *et al.*, 1972). The actual depth of the metal-enriched layer is not clearly known, may exist as a depth gradient, and is probably obscured by site-to-site differences. In the North Sea, no significant differences in metal enrichments occurred between samples collected with screen, glass plate or drum samplers (Hardy and Cleary, 1992).

Little attention has been given to the residence times in the microlayer of metals which are deposited from the atmosphere to the sea surface. However, laboratory seawater microcosm studies, using natural urban air particles, suggest residence times of 1.5 to 15 hours with predicted enrichment ratios of Pb>Cu>Zn>Ag>Ni>Mn (Hardy *et al.*, 1985a). Subsequent field sampling and microlayer analysis generally validate the predictions of the laboratory-derived model (Hardy *et al.*, 1985a).

Radionuclides

Radionuclides enter the open ocean primarily through the microlayer as a result of atmospheric deposition. Therefore, neuston are likely to receive higher doses of radiation than most other marine populations (Polikarpov, 1966). Studies in the vicinity of a British coastal nuclear power plant indicate that actinides are enriched considerably in marine aerosol, probably due to enrichment of fine grained particulates carried with the aerosol. Microlayer/bulkwater (screen-collected) enrichment ratios in the particulate fractions are: $23\times$ for ^{238}Pu, $20\times$ for ^{239}Pu and $8\times$ for ^{241}Am. Aerosol/seawater particulate enrichment factors are on the order of $10^2\times$ for ^{238}Pu, ^{239}Pu, and ^{241}Am. Actinides seem to behave like other trace metals in their association with fine particles in the microlayer and in marine aerosols.

Microlayer (300 μm screen samples)/bulkwater particulate enrichments in the ratio of the radon daughters ^{210}Po/^{210}Pb are 2.4 to 8.8 in open ocean samples and closer to 1 in coastal samples (Bacon and Elzerman, 1980). Experiments in which columns of seawater are bubbled with air suggest that a significant fraction of trace metal and radionuclide enrichments in the microlayer is maintained by transport of sub-surface particulate matter to the interface (Bacon and Elzerman, 1980). Likewise, in the Western Mediterranean, there is a significant flux of ^{210}Po from bulk seawater to the microlayer and the atmosphere (Heyraud and Cherry, 1983). The degree of enrichment is positively correlated with neuston biomass per unit volume of microlayer. Neustonic copepods *Anomalocera patersoni* bioconcentrate ^{210}Po to levels of about 16 pCi/g dry biomass. The highest concentration factor for neuston with respect to bulk seawater is 1.7×10^5 for ^{210}Po.

Organics

The microlayer often contains high concentrations of anthropogenic organic chemicals, many of which cannot be detected in bulkwater samples. For example, in Puget Sound, although considerable temporal and spatial heterogeneity exists, maximum and mean microlayer concentrations are: total aromatic hydrocarbons 8030 (mean 132) µg 1^{-1}, saturate hydrocarbons 2060 µg 1^{-1}, and pesticides 44 (mean 0.5) ng 1^{-1} (Hardy *et al.*, 1987b), while such compounds in the water column are usually below detection limits. In Chesapeake Bay total polyaromatic hydrocarbon microlayer enrichments are on the order of 10 to 10^2 (Hardy *et al.*, 1990). In the Southern California Bight, microlayer concentrations of total PAHs and chlorinated hydrocarbons generally increased from about 0.01 µg 1^{-1} offshore to >38 µg 1^{-1} in harbour areas (Cross *et al.*, 1987).

Sea-surface contamination is not restricted to coastal bays and estuaries, but occurs in freshwaters and offshore marine areas. For example, high surface film concentrations of toxic organotin compounds, used in anti-fouling boat paints, occur not only in English and North American marinas (Cleary and Stebbing, 1987; Hall *et al.*, 1986), but also at sites more than 100 km from shore in the North Sea (Hardy and Cleary, 1992). Samples from several Norwegian fjords and freshwater lakes contained high concentrations of PCBs with enrichment ratios of 10^6 to 10^7 compared with the sub-surface water (Ofstad *et al.*, 1979), and the proposed source was long-distance atmospheric transport and deposition. Chlorinated hydrocarbon concentrations in the SMIC of the Niagra River (USA) are 40 times that of the sub-surface water. Following a PCB spill on the Niagara, the SMIC concentration was 6400 times greater than in the sub-surface water (Maguire and Tkacz, 1988).

Petroleum hydrocarbon concentrations are elevated throughout the world's coastal and offshore waters and have been found in measurable levels in tissues of neustonic organisms (Butler and Sibbald, 1987; Cross *et al.*, 1987; Hardy *et al.*, 1987a, 1990; Marty *et al.*, 1979; Morris, 1974). Pesticides and PCBs have not only been found in microlayers of nearshore urban bays and estuaries, but also offshore in the Sargasso Sea and the Gulf of Mexico (Bidleman and Olney, 1974; Burns *et al.*, 1985; Duce *et al.*, 1972; Eganhouse *et al.*, 1990; Ofstad and Lunde, 1978; Seba and Corcoran, 1969).

Many studies have documented high concentrations of plastic particles and debris (Wolfe, 1987) and petroleum-derived tar balls (Van Dolah *et al.*, 1980; McCoy, 1988; Knap *et al.*, 1986) in the sea surface at widely dispersed sites in the global ocean. Using data of Van Dolah *et al.*, (1980), of 1 mg of tar per 10 litres of seawater, and assuming the tar is floating within the upper centimetre of water, the concentration of tar in the surface centilayer of the southern part of the South Atlantic Bight is about 0.1 mg 1^{-1}.

Biological effects of sea-surface contamination

Despite the demonstrated importance of the sea surface as both a vital biological habitat and a concentration point for contaminants, spatial distributions or temporal trends in surface contaminants, or their impacts on the reproduction of valuable marine species, remain largely unknown. However, several recent studies link aquatic surface contamination with negative biological impacts. For example, the North Sea study (Hardy and Cleary, 1992) used a variety of microlayer toxicity tests and identified a zone of toxic surface water, with contaminant levels exceeding water quality standards, extending from the shore outward. Stations less than 100 km from Cuxhaven at the mouth of the Elbe River were generally toxic, while samples collected 284 km from Cuxhaven were not toxic.

Effects on intertidal biota

In coastal regions during tidal ebb, sea-surface films become stranded and can deposit toxic materials on beaches. This was demonstrated in Puget Sound, where a modified screen sampler was used to collect surface film deposits from sites in rural, semi-rural and urban bays (Gardiner, 1992). Collected samples were compared using three types of echinoderm toxicity test. Larvae incubated in urban bay surface film deposits had significantly higher rates of mortality, abnormality, and cytogenetic aberration than larvae in either the semi-rural or rural surface film deposits. Bulkwater responses in all tests were considerably lower than both surface film and surface deposit responses for all three embayments.

A conceptual model based on surface current, wind, demographic, and drift card data predicts that Puget Sound shorelines, which are downwind or downcurrent from contaminant sources, are especially at risk from these effects. Results suggest that in urbanized coastal regions surface film deposits on beaches could have significant impacts on the health of intertidal organisms (Gardiner, 1992).

Effects on fish eggs and larvae

Numerous laboratory and some *in situ* studies suggest that sea-surface contamination can have toxic effects on fish eggs and larvae. In the North Sea, fish embryos exposed to microlayer samples from Heligoland Harbour had reduced hatching success and an increase in larval abnormalities, while those exposed to water from 20 cm depth showed normal development (von Westernhagen *et al.*, 1987; Kocan *et al.*, 1987).

In Puget Sound, fertilized eggs of Sand Sole (*Psettichthys melanostictus*), as well as of several other species, were exposed in the laboratory to microlayer samples collected from several sites with a rotating Teflon drum. Compared with rural reference sites, samples from contaminated urban bays resulted in more chromosomal aberrations in developing sole embryos, reduced hatching success of sole larvae, and reduced growth in trout cell cultures (Hardy *et al.*, 1987a). Fertilized sole eggs, exposed *in situ* in floating enclosures, also showed increased toxic effects compared with *in situ* exposure in cleaner waters (Hardy *et al.*, 1987a). Multiple regression analyses indicated that toxic effects increased with increasing concentrations of a complex mixture of microlayer contaminant chemicals. Principal components analysis suggested that several major classes of contaminants (metals, aromatic hydrocarbons, saturate hydrocarbons, pesticides and PCBs) all contributed to some degree to the toxicity (Hardy *et al.*, 1987b).

Likewise, studies of the Southern California Bight area indicate that a variety of toxic effects (e.g. increased mortality, decreased size, increased abnormalities, and increased cytogenetic effects) are induced in kelp bass embryos and larvae exposed to collected microlayer samples. Toxic effects generally are positively correlated with concentrations of toxic chemical contaminants present in the microlayer which, in turn, increase with proximity to shore (Cross *et al.*, 1987).

Does microlayer contamination pose a significant threat to resident populations of pelagic fish? Floating embryos of plaice (*Pleuronectes platessa*), cod (*Gadus morhua*) and whiting (*Merlangius merlangus*), collected from the surface microlayer of the North Sea, have a very high incidence (5 to 26%) of abnormalities (Dethlefsen *et al.*, 1985), possibly as a result of exposure to surface microlayer contamination. Similarly, the incidence of cytogenetic and developmental abnormalities in populations of Atlantic mackerel eggs, collected from the sea surface in the vicinity of pollutant sources in the New York Bight, is greater than that in populations elsewhere (Longwell and Hughes, 1980). Densities of neuston, including sole eggs, are lower in urban bays of Puget Sound containing contaminated microlayers than in similar bays without microlayer contamination (Hardy *et al.*, 1987b). Likewise, polluted areas of the Gulf of Marseille have lower densities of zooneuston than cleaner areas (Champalbert, 1977). Based on available laboratory and *in situ* studies, it seems likely that present levels of microlayer contamination do pose a threat to the sustainability of fish populations in some areas.

Because radionuclides tend to concentrate greatly in the SMIC compared with sub-surface seawater (see above), releases (accidental or intentional) of radionuclides to the marine environment could pose a particular threat to neuston (Polikarpov, 1977). The large number of fish species whose embryos develop in contact with the SMIC could be subject to radiation-induced toxic effects (Polikarpov, 1966; LaTier, 1992). Also, radionuclides could enter organisms, including fish and birds, that feed on the sea surface.

Effects on microneuston and gas exchange

Properties of the SMIC influence the flux of gases across the air–sea interface (Liss and Slater, 1974). Microlayer contamination could negatively impact microneuston involved in biogeochemically important reactions at the sea surface. For example, deposition of natural urban air particulate matter (APM) to the surface of seawater microcosms caused a significant decrease in photosynthetic carbon assimilation of phytoneuston (Hardy and Apts, 1984). Deposition rates of APM in many coastal areas, and possibly some offshore areas, appear high enough (i.e. >1 mg m^{-2} h^{-1}) to inhibit carbon fixation in the sea-surface microlayer (Figure 11.2), (Hardy and Apts 1984). Also, polyaromatic hydrocarbons, introduced into the microlayer of microcosms, can inhibit the growth of phytoneuston (Riznyk *et al.*, 1987).

Photochemical and biological effects of ultraviolet-B radiation (UV-B) in the sea surface are discussed in detail elsewhere (Blough, Chapter 13, this volume; Hardy & Gucinski, 1989). Increased UV-B, due to stratospheric ozone depletion, is likely to result in decreased productivity and/or a change in the species composition of near-surface phytoplankton and phytoneuston (Figure 11.3) (Behrenfeld *et al.*, 1993 a,b). These effects are most likely to occur at temperate southern latitudes where both present productivity and future predicted ultraviolet increases are relatively high (Hardy *et al.*, in press).

Bioaccumulation of sea-surface contaminants by neuston could lead to their introduction into planktonic food webs. As Liss (1975) suggested, solubilization of airborne particles and contaminants at the sea surface will lead to their rapid introduction into marine food webs via the high concentrations of microorganisms found in the microlayer.

Relation of effects to visible slicks

Although microlayer toxicity is often strongly correlated with (or restricted to) areas of visible slicks, this does not diminish the overall spatial

J. T. Hardy

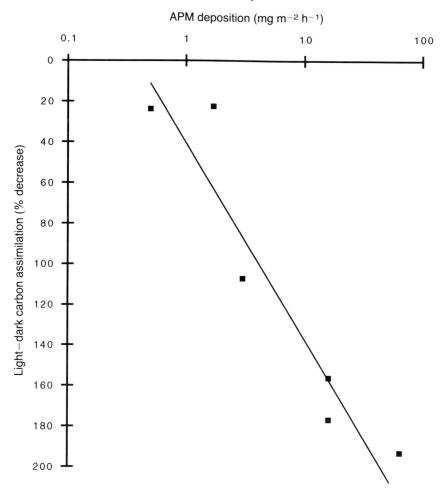

Figure 11.2. Inhibition of radiocarbon-measured primary productivity in the sea-surface microlayer (upper 20 μm) with increasing additions of metal-rich atmospheric particulate matter (APM) (drawn from data of Hardy and Apts, 1984).

importance of sea-surface toxicity. Surface films form from a complex mixture of biogenic organic exudates which collect at the sea surface after being released by plankton in the water column or nearshore (Baier *et al.*, 1974; Carlson, 1982b; Garabétian *et al.*, 1993). Wind and current patterns collapse the films into thicker visible slicks. Along the Mediterranean coast of France, visible slicks occur about 30% of the time, and, when present, cover greater than 50% of the sea surface (Garabétian *et al.*, 1993). Visible slicks in lakes also contain elevated nutrient

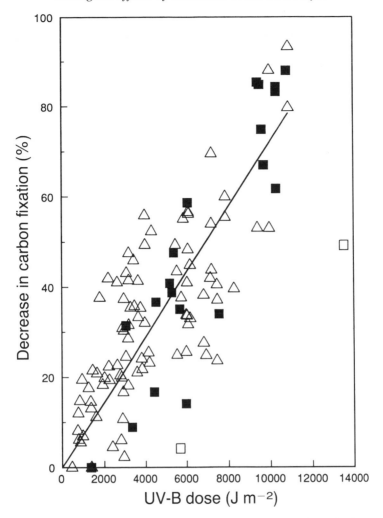

Figure 11.3. Photoinhibition of carbon fixation in marine phytoplankton as a function of total dose of UV-B radiation. $\triangle$ = 1991 data from Washington coast and data from Behrenfeld *et al.* (1993a); $\blacksquare$ = ambient treatment data from Smith *et al.* (1980) converted from photoinhibition doses to EXP_{300} doses; $\square$ = ambient data of Smith *et al.* (1980) not used in calculation of combined regression coefficient. (From Behrenfeld *et al.*, 1993a.)

and organic chemical concentrations compared with sub-surface water (Södergren, 1993).

In areas remote from contaminant sources these slicks may contain little or no contamination. However, where toxic hydrophobic contaminants are present they concentrate in natural films and toxicity increases with the relative degree of organic surface film (surface pressure, dynes cm^{-2}) (Hardy *et al.*, 1987a). A film may be moderately toxic, but not visible. *In situ*, embryos may be exposed intermittently to such drifting films or they may concentrate in, and be transported by, such films (pers. obs. and Shanks, 1983). Contaminated films, driven by winds and tides, can drift shoreward and deposit on intertidal beaches (the so-called bathtub ring effect). Such beach contamination could pose a threat to intertidal communities (Gardiner, 1992). Thus, though dense visible contaminated films may be restricted in area, their effects may be greater than their spatial coverage would suggest. Their extent needs to be routinely determined during pollution and monitoring surveys.

Global effects of sea-surface stress

The ocean is a significant sink for increasing anthropogenic carbon dioxide emissions. The sea-surface microlayer (upper 100 μm layer) serves as the primary point for exchange of materials, including carbon dioxide, between the atmosphere and hydrosphere. Although speculative at this time, it is not inconceivable that high densities of phytoneuston at the sea surface could affect the transport of carbon dioxide between the atmospheric boundary layer and the aquatic surface layer. Areas of high surface primary productivity often have low dissolved CO_2 (Watson *et al.*, 1991; Hardy *et al.*, in press) (Figure 11.4). On the other hand, microheterotrophs in the SMIC can give rise to a significant enhancement

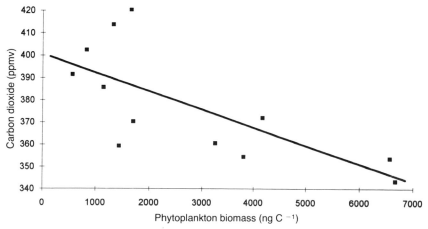

Figure 11.4. Relation between dissolved surface CO_2 and surface phytoplankton biomass. Data collected along a north–south transect from equatorial to Antarctic waters in the southeast Pacific ocean (Hardy *et al.*, in press).

of CO_2 production and diffusion across the air–water interface (Garabétian, 1990).

Neuston are part of a complex food web beginning at the sea surface. Rather than direct effects such as acute mortality, global increases in stressors such as chemical toxicants and radiation are most likely to have subtle effects on the reproduction and growth of sensitive species, while tolerant species may thrive and become dominant. Such shifts in species population abundances and food web relationships could involve species that are important in biogeochemical cycling and atmosphere–ocean exchange processes (for example, $CaCO_3$ containing coccolithophotes or other organisms important as a sink for CO_2 or in the production of organosulphur compounds). Alteration of such processes could affect global climate.

Summary

Our current concept of the sea-surface microlayer is that of a biologically and chemically active boundary layer. Compared with sub-surface water, the surface layer is generally more physically stable, environmentally stressed, and enriched in particulate matter, organic compounds, metals and biota. Increasing sea-surface pollution and ultraviolet radiation represent threats to sea-surface biota. Some surface biota may be important for fisheries recruitment and for biogeochemical processes at the air–water interface. However, major important questions, some of which were posed more than 14 years ago (Hardy, 1982), remain unresolved. For example:

1. What are the species and densities of organisms inhabiting the sea surface, and how do they vary spatially and temporally?
2. What is the functional depth distribution of physical/chemical/ biological components of the sea surface?
3. Does the abundance or type (species assemblage) of microneuston affect the air–sea exchange of nitrogen, carbon dioxide, or radiatively important trace gases?
4. How important is the microlayer as a habitat, particularly for recruitment of the early life stages of fish and invertebrates? If it is important, can it be protected by water quality standards?
5. What are the effects of sea-surface contaminants and predicted increases in ultraviolet radiation on neuston?

To understand the structure and dynamics of neuston (surface dwelling organisms), and their role in the exchange of gases and particles between the atmosphere above and the water column below, all the major biological and chemical components at a site need to be sampled using a variety of techniques. A great deal of research effort will need to be expended before we begin to understand the sea-surface ecosystem and how processes occurring there link the world's atmosphere and hydrosphere.

References

Ahlstrom, E. H. and Stevens, E. (1975). Report on neuston (surface) collections made on an extended CalCOFI cruise during May 1972. *Calif. Coop. Fish. Investig. 28*, 1 July 1973 to 30 June 1975.

Albright, L. J. (1980). Photosynthetic activities of phytoneuston and phytoplankton. *Can. J. Microbiol.*, **26**, 389–92.

Bacon, M. P. and Elzerman, A. W. (1980). Enrichment of ^{210}Pb and ^{210}Po in the sea-surface microlayer. *Nature*, **284**, 332–4.

Baier, R. E., Goupil, D. W., Perlmutter, S. and King, R. (1974). Dominant chemical composition of sea-surface films, natural slicks, and foams. *J. Rech. Atmos.*, **8**, 571–600.

Barker, D. R. and Zeitlin, H. (1972). Metal-ion concentrations in the sea-surface microlayer and size-separated atmospheric aerosol samples in Hawaii. *J. Geophys. Res.*, **77**, 5076–86.

Beers, J. R. (1978). About microzooplankton. In *Phytoplankton Manual*, ed. A. Sournia, pp. 288–96. Paris: UNESCO.

Behrenfeld, N. J., Chapman, J. W., Hardy, J. T. and Lee, H. (1993a). Is there a common response to ultraviolet-B radiation by marine phytoplankton? *Mar. Ecol. Prog. Ser.*, **102**, 59–68.

Behrenfeld, M., Hardy, J., Gucinski, H., Hanneman, A., Lee, H. and Wones, A. (1993b). Effects of ultraviolet-B radiation on primary production along latitudinal transects in the South Pacific Ocean. *Mar. Environ. Res.*, **35**, 349–63.

Bezdek, H. F. and Carlucci, A. F. (1972). Surface concentrations of marine bacteria. *Limnol. Oceanogr.*, **17**, 566–9

Bidleman, T. F. and Olney, C. E. (1974). Chlorinated hydrocarbons in the Sargasso Sea atmosphere and surface water. *Science*, **183**, 516–18.

Blanchard, D. C. (1983). The production, distribution, and bacterial enrichment of the sea-salt aerosol. In *Air–Sea Exchange of Gases and Particles*, ed. P. S. Liss and W. G. N. Slinn, pp. 407–54. NATO-ASI Series No. 108. Hingham, MA: Reidel.

Booth, B. C., Lewin, J. and Lorensen, C. J. (1988). Spring and summer growth rates of subarctic Pacific phytoplankton assemblages determined from carbon uptake and cell volumes estimated using epifluorescence microscopy. *Mar. Biol.*, **98**, 287–98.

Brewer, G. D. (1981). Abundance and vertical distribution of fish eggs and larvae in the Southern California Bight: June and October 1978. *Rapp. P. V. Reun. Cons. Int. Explor. Mer.*, **178**, 165–7.

Brockmann, U. H., Kattner, G., Hentzschel, G., Wandschneider, K., Junge, D. H. and Hühnerfuss, H. (1976). Natürliche oberflachenfilme im See-gebeit vor Sylt. *Mar. Biol.*, **36**, 135–46.

Budd, P. L. (1940). Development of the eggs and early larvae of six California fishes. State of California, Department of Natural Resources, Division of Fish and Game, Bureau of Marine Fisheries. *Fish Bull.* No. 56. 20 pp.

Burns, K. A., Villeneuve, J-P. and Fowler, S. W. (1985). Fluxes and residence times of hydrocarbons in the coastal Mediterranean: how important are the biota? *Estuarine Coastal Shelf Sci.*, **20**, 313–30.

Butler, A. C. and Sibbald, R. R. (1987). Sampling and GC-FID, GC/MS analysis of petroleum hydrocarbons in the ocean surface microlayer off Richards Bay, South Africa. *Estuarine Coastal Shelf Sci.*, **25**, 27–42.

Cambray, R. S., Jeffries, D. F. and Topping, G. (1979). An estimate of the input of atmospheric trace elements into the North Sea and the Clyde Sea. *Mar. Sci. Commun.*, **5**(2), 175–94.

Carlson, D. J. (1982a). Phytoplankton in marine surface microlayers. *Can. J. Microbiol.*, **28**, 1226–34.

Carlson, D. J. (1982b). Surface microlayer phenolic enrichments indicate sea surface slicks. *Nature*, **296**, 426–9.

Carlson, D. J. (1982c). A field evaluation of plate and screen microlayer sampling techniques. *Mar. Chem.*, **11**, 189–208.

Carlson, D. J., Cantey, J. L. and Cullen, J. J. (1988). Description and results from a new surface microlayer sampling device. *Deep Sea Res.*, **35**(7), 1205–13.

Carlucci, A. F., Craven, D. B. and Henrichs, S. M. (1985). Surface-film microheterotrophs: amino acid metabolism and solar radiation effects on their activities. *Mar. Biol.*, **85**, 13–22.

Carlucci, A. F., Craven, D. B., Robertson K. J. and Williams, P. M. (1986). Surface-film microbial populations: diel amino acid metabolism, carbon utilization, and growth rates. *Mar. Biol.*, **92**, 289–97.

Carlucci, A. F., Craven, D. B. and Wolgast, D. M. (1991). Microbial populations in surface films and subsurface waters: amino acid metabolism and growth. *Mar. Biol.*, **108**, 329–39.

Carpenter, E. J. and Romans, K. (1991). Major role of the cyanobacterium *Trichodesmium* in nutrient cycling in the North Atlantic ocean. *Science*, **254**, 1356–8.

Champalbert, G. (1977). Variations locales de la repartition verticale et de l'abondance de l'hyponeuston en fonction des situations meteorologiques. *Cah. Biol. Mar.*, **18**, 243–55.

Cheng, L. (1975). Marine pleuston-animals at the sea–air interface. *Oceanogr. Mar. Biol. Annu. Rev.*, **13**, 181–212.

Cleary, J. J. and Stebbing, A. R. D. (1987). Organotin in the surface microlayer and subsurface waters of southwest England. *Mar. Pollut. Bull.*, **18**(5), 238–46.

Conover, R. J. (1982). Interelations between microzooplankton and other plankton organisms. *Ann. Inst. Océanogr. Paris*, **58**(S), 31–46.

Cross, J. N., Hardy, J. T., Hose, J. E., Hershelman, G. P., Antrim, L. D., Gossett, R. W. and Crecelius, E. A. (1987). Contaminant concentrations and toxicity of sea-surface microlayer near Los Angeles, California. *Mar. Environ. Res.*, **23**, 307–23.

Cullen, J. J., MacIntyre, H. L. and Carlson, D. J. (1989). Distributions and photosynthesis of phototrophs in sea-surface films. *Mar. Ecol. Prog. Ser.*, **55**, 271–8.

Dahlback, B., Hermansson, M., Kjelleberg, S. and Norkrans B. (1981). The hydrophobicity of bacteria: an important factor in their initial adhesion at the air–water interface. *Arch. Microbiol.*, **128**, 267–70.

Davey, E. W., Perez, K. T., Soper, A. E., Lackie, N. F., Morrison, G. E., Johnson, R. L. and Heltshe, J. F. (1990). Significance of the surface microlayer to the environmental fate of di(2-ethylhexyl)phthalate predicted from marine microcosms. *Mar. Chem.*, **31**, 231–69.

Davis, P. G. and Sieburth, J. McN. (1982). Differentiation of phototrophic and heterotrophic nanoplankton populations in marine waters by epifluorescence microscopy. *Ann. Inst. Ocèanogr. Paris*, **58**(S), 249–60.

Dehairs, F., Dedeurwaerder, H., Dejonghe, M., Decadt, G., Gillain, G., Baeyens, W. and Elskens I. (1982). Boundary conditions for metals at the air–sea interface. *International Council for the Exploration of the Sea*, **1982/E:33**, 225–242.

De Souza Lima, Y. and Chretiennot-Dinet, M. J. (1984). Measurements of biomass and activity of neustonic microorganisms. *Estuarine, Coastal Shelf Sci.*, **19**, 167–80.

Dethlefsen, V., Cameron, P. and von Westernhagen, H. (1985). Üntersuchungen über die Häufigkeit von Missbildungen in Fischembryonen der südlichen Nordsee. *Inf. Fischwirtsch*, **32**, 22–7.

Duce, R. A., Quinn, J. G., Olney, C. E., Protowicz, S. R., Ray, B. J. and Wade, T. L. (1972). Enrichment of heavy metals and organic compounds in the surface microlayer of Narragansett Bay, Rhode Island. *Science*, **176**, 161–3.

Duce, R. A., Liss, P. S., Merrill, J. T., Buat-Menard, P., Hicks, B. B., Miller, J. M., Prospero, J. M., Arimoto, R., Church, T. M., Ellis, W. G., Galloway, J. N., Hanson, L., Jickells, T. D., Knap, A. H., Reinhardt, K. H., Schneider, B., Soudice, A., Tokos, J. J., Tsunogai, S., Wollast, R. and Zhou, M. (1991). The atmospheric input of trace species to the world ocean. *Global Biogeochem. Cycles*, **5**, 193–259.

Eganhouse, R., Gossett, R. and Hershelman, P. (1990). PCBs in Los Angeles harbor. *SCCWRP 1989–1990 Annual Report*, pp. 47–57, Westminster, CA: Southern California Coastal Water Research Project.

Gallagher, J. L. (1975). The significance of the surface film in salt marsh plankton metabolism. *Limnol. Oceanogr.*, **20**, 120–3

Garabétian, F. (1990). CO_2 production at the sea–air interface: an approach by the study of respiratory processes in the surface microlayer. *Int. Rev. Ges. Hydrobiol.*, **75**(2), 219–29.

Garabétian, F., Romano, J-C., Paul, R. and Sigoillot, J-C. (1993). Organic matter composition and pollutant enrichment of sea surface microlayer inside and outside slicks. *Mar. Environ. Res.*, **35**, 323–39.

Gardiner, W. (1992). *Shoreline Deposition of Contaminated Surface Film and its Effect on Intertidal Organisms*. Bellingham, WA: Huxley College of Environmental Studies, Western Washington University.

Garrett, W. D. (1967). The organic chemical composition of the ocean surface. *Deep-Sea Res.*, **14**, 221–7.

Garrett, W. D., and Duce, R. A. (1980). Surface microlayer samplers. In *Air–Sea Interaction*, ed. F. Dobson *et al.*, pp. 471–90. New York: Plenum Press.

Gladyshev, M. I. (1986). Neuston of inland waters (a review). *Hydrobiol. J.*, **22**(5), 1–7.

Gladyshev, M. I. and Malyshevski, K. G. (1982a). The concept of 'neuston'. *Hydrobiol. J.*, **18**(3), 7–10.

Gladyshev, M. I. and Malyshevski, K. G. (1982b). Daily vertical migration of benthoneuston in Vostok Bay, Sea of Japan. *Sov. J. Mar. Biol.*, **8**, 65–73.

Hall, L. W. Jr., Lenkevich, M. J., Hall, W. S., Pinkney A. E., and Bushong, S. J. (1986). Monitoring organotin concentrations in Maryland waters of Chesapeake Bay. In *Organotin Symposium. Proceedings, Oceans 86*, Vol. 4, pp. 1275–1279. Washington, DC: Marine Technology Society.

Hardy, J. (1971). Ecology of phytoneuston in a temperate marine lagoon. PhD Dissertation. Department of Botany, University of Washington, Seattle. 160 pp.

Hardy, J. T. (1973). Phytoneuston ecology of a temperate marine lagoon. *Limnol. Oceanogr.*, **18**, 525–33.

Hardy, J. T. (1982). The sea surface microlayer: biology, chemistry and anthropogenic enrichment. *Prog. Oceanogr.*, **11**, 307–28.

Hardy, J. T. and Apts, C. W. (1984). The sea-surface microlayer: phyto-neuston productivity and effects of atmospheric particulate matter. *Mar. Biol.*, **82**, 293–300.

Hardy, J. T. and Apts, C. W. (1989). Photosynthetic carbon reduction: high rates in the sea-surface microlayer. *Mar. Biol.*, **101**, 411–17.

Hardy, J. T. and Cleary, J. (1992). Surface microlayer contamination and toxicity in the German Bight. *Mar. Ecol. Prog. Ser.*, **91**, 203–10.

Hardy, J. T. and Cowan, C. E. (1986). Model and assessment of the contri-bution of dredged material disposal to sea-surface contamination in Puget Sound. *Final Report to US Army Corps of Engineers, Seattle, Washington*. Richland, WA: Pacific Northwest Laboratory Report No. PNL-5804.

Hardy, J. T. and Crecelius, E. A. (1981). Is atmospheric particulate matter inhibiting marine primary productivity? *Environ. Sci. Technol.*, **15**(9), 1103–5.

Hardy, J. T., and Gucinski, H. (1989). Stratospheric ozone depletion: impli-cations for the marine environment. *Oceanography*, 2(2), 18–21.

Hardy, J. T. and Valett, M. K. (1981). Natural and microcosm phytoneuston communities of Sequim Bay, Washington. *Estuarine Coastal Shelf Sci.*, **12** 3–12

Hardy, J. T. and Word, J. Q. (1986). Sea surface toxicity in Puget Sound. *Puget Sound Notes*. November, pp. 3–6. Seattle, WA: US EPA Region 10.

Hardy, J. T., Apts, C. W., Crecelius, E. A. and Bloom, N. S. (1985b). Sea-surface microlayer metals enrichments in an urban and rural bay. *Estuarine, Coastal Shelf Sci.*, **20**, 299–312.

Hardy, J. T., Apts, C. W., Crecelius, E. A. and Fellingham, G. W. (1985a). The sea-surface microlayer: fate and residence time of atmospheric metals. *Limnol. Oceanogr.*, **30**(1), 93–101.

Hardy, J. T., Crecelius, E. A. and Kocan, R. (1986). *Concentration and Toxicity of Sea-surface Contaminants in Puget Sound*. PNL-5834. Richland, WA: Pacific Northwest Laboratory.

Hardy, J. T., Kiesser, S. L., Antrim, L. D., Stubin, A. I., Kocan, R. and Strand, J. A. (1987a). The sea-surface microlayer of Puget Sound: Part 1. Toxic effects on fish eggs and larvae. *Mar. Environ. Res.*, **23**, 227–49.

Hardy, J. T., Crecelius, E. A., Antrim, L. D., Broadhurst, V. L., Apts, C. W., Gurtisen, J. M. and Fortman, T. J. (1987b). The sea-surface microlayer of Puget Sound: Part 2. Concentrations of contaminants and relation to toxicity. *Mar. Environ. Res.*, **23**, 251–71.

Hardy, J. T., Coley, J. A., Antrim, L. D., Kiesser, S. L. (1988). A hydro-phobic large-volume sampler for collecting aquatic surface microlayers: characterization and comparison to the glass plate method. *Can. J. Fish. Aquat. Sci.*, **45**, 822–6.

Hardy, J. T., Crecelius, E. A., Antrim, L. D., Kiesser, S. L., Broadhurst, J. L., Boehm, P. D. and Steinhauer, W. G. (1990). Aquatic surface con-tamination in Chesapeake Bay. *Mar. Chem.*, **28**, 333–51.

Hardy, J. T., Fowler, S., Price, A., Readman, J., Oregioni, B., Crecelius, E. and Gardiner, W. (1993). Environmental assessment of sea surface contamination in the Gulf. *Final Report on the Joint IOC/IUCN Gulf Mission August, 1992.* Marine Pollution Unit, Paris and IUCN, Gland, Switzerland: UNESCO, IOC. 29 pp.

Hardy, J. T., Hanneman, A., Behrenfeld, M. and Horner, R. (1996). Environmental biogeography of near-surface phytoplankton in the southeast Pacific Ocean. *Deep-Sea Res.* (in press).

Harvey, G. W. and Burzell, L. A. (1972). A simple microlayer method for small samples. *Limnol. Oceanogr.*, **17** (1), 156–7.

Hattori, H., Yuki, K., Zaitsev, Y. P. and Motoda, S. (1983). A preliminary observation on the neuston in Surgua Bay. *Umi-Mer.*, **21**, 11–20.

Hempel, G. and Weikert, H. (1972). The neuston of the subtropical and boreal North-eastern Atlantic Ocean: a review. *Mar. Biol.*, **13**, 70–88.

Heyraud, M. and Cherry, R. D. (1983). Correlation of ^{210}Po and ^{210}Pb enrich-ments in the sea-surface microlayer with neuston biomass. *Cont. Shelf Res.*, **1** (3), 283–93.

Holdway, P. and Maddock, L. (1983). Neustonic distributions. *Mar. Biol.*, **77**, 207–14.

Hühnerfuss, H. (1981). On the problem of sea surface film sampling: a comparison of 21 microlayer-, 2 multilayer-, and 4 selected subsurface-samplers. Part 1. *Sonderdruck aus Meerestechnik*, **12** (5) S, 136–142.

Hühnerfuss, H. and Alpers, W. (1983). Molecular aspects of the system water/ monomolecular surface film and the occurrence of a new anomolous dispersion regime at 1.43 GHz. *J. Phys. Chem.*, **87**, 5251–58.

Hunter, K. A. (1980). Processes affecting particulate trace metals in the sea-surface microlayer. *Mar. Chem.*, **9**, 49–70.

Ignatiades, L. (1990). Photosynthetic capacity at the surface microlayer during the mixing period. *J. Plankton Res.*, **12** (4), 851–860.

Jacoby, C. A. (1982). Behavioral responses of the larvae of *Cancer magister Dana* to light, pressure, and gravity. *Mar. Behav. Physiol.*, **8**, 267–83.

Johnson, P. W., Xu, H-S. and Sieburth, J. McN. (1982). The utilization of chroococcoid cyanobacteria by marine protozooplankters but not by calanoid copepods. *Ann. Inst. Oceanogr. Paris*, **58** (s), 297–308.

Kendall, A. W., Jr and Clark, J. (1982). Ichthyoplankton off Washington, Oregon, and Northern California, April–May 1980. *NWAFC Processed Report 82–11.* Seattle, WA: Northwest and Alaska Fisheries Center, NMFS, NOAA.

Kjelleberg, S. and Hakansson, N. (1977). Distribution of lipolytic, proteolytic, and amylolytic marine bacteria between the lipid film and the subsurface water. Mar. Biol., **39**, 103–9

Knap, A. H., Burns, K. A., Dawson, R., Ehrhardt, M. and Palmork, K. H. (1986) Dissolved/dispersed hydrocarbons, tarballs and the surface micro-layer: experiences from an IOC/UNEP workshop in Bermuda, December, 1984. *Mar. Pollut. Bull.*, **7**, 313–19.

Kocan, R. M., Westernhagen, H. V., Landolt, M. L. and Furstenberg, G. (1987). Toxicity of sea-surface microlayer: II. Effects of hexane extract on Baltic herring (*Clupea harengus*, and Atlantic cod (*Gadus morhua*) embryos. (In manuscript.)

LaTier, A. (1992). Survival of killifish (*Oryzias latipes*) embryos exposed to ultraviolet-b radiation. MS Thesis, Huxley College of Environmental Studies, Western Washington University, Bellingham, WA.

Lion, L. W and Leckie, J. O. (1981). The biogeochemistry of the air–sea interface. *Ann. Rev. Earth Planet. Sci.*, **9**, 449–86.

Liss, P. S. (1975). Chemistry of the sea-surface microlayer. In *Chemical Oceanography*, ed. J. P. Riley, and G. S. Skirrow, pp. 193–243. New York: Academic Press.

Liss, P. S. and Slater, P. G. (1974). Flux of gases across the air–sea interface. *Nature*, **247**, 181–4.

Longwell, A. C. and Hughes, J. B. (1980). Cytologic, cytogenetic and developmental state of Atlantic mackerel eggs from sea surface waters of the New York Bight, and prospects for biological effects monitoring with ichthyoplankton. *Rapp. P.V. Reun. Cons. Int. Explor. Mer.*, **179**, 275–91.

Maguire, R. J. and Tkacz, R. J. (1988). Chlorinated hydrocarbons in the surface microlayer and subsurface water of the Niagara River, 1985–86. *Water Pollut. Res. J. Can.*, **23** (2), 292–300.

Marty, J. C., Saliot, A., Baut-Menard, P., Chesselet, R. and Hunter, K. A. (1979). Relationship between the lipid composition of marine aerosols, the sea surface microlayer, and subsurface water. *J. Geophys. Res.*, **84**(C9), 5707–16.

McCoy, F. W. (1988). Floating megalitter in the eastern Mediterranean. *Mar. Pollut. Bull.*, **19**(1), 25–8.

Mileikovskii, S. A. (1981). A survey of Soviet investigations of the effect of the anthropogenic factor on natural communities of marine and estuary zooplankton and neuston. (English transl. from *Biologiya Morya*, No. 4, pp. 3–11, July–Aug 1981.) *Sov. J. Mar. Biol.* **7**(4), 215–22.

Morris, R. J. (1974). Lipid composition of surface films and zooplankton from the eastern Mediterranean. *Mar. Pollut. Bull.*, **5**, 105–109.

Naumann, E. (1917). Beiträge zur Kenntnis des Teichnanno-planktons. II. Über das Neuston des Süsswassers. *Biol. Cbl.*, **37**, 98–106.

Nestrova, D. A. (1980). Phytoneuston in the Western Black Sea. *Gidrobiol. Zh.*, **16**, 26–31.

Nichols, P. D. and Espey, Q. I. (1991). Characterization of organic matter at the air–sea interface, in subsurface water, and in bottom sediments near the Malabar sewage outfall in Sydney's coastal region. *Aust. J. Mar. Freshwater Res.*, **42**, 327–48.

Norkrans, B. (1980). Surface microlayers in aquatic environments. In *Advances in Microbial Ecology*, ed. M. Alexander, pp. 51–85, New York: Plenum Press.

Norris, R. E. (1965). Neustonic marine Craspedomonodales (choano-flagellates) from Washington and California. *J. Protozool.*, **12**, 589–602.

Ofstad, E. B. and Lunde, G. (1978). A comparison of the chlorinated organic compounds present in the fatty surface film of water and the water phase beneath. In *Aquatic Pollutants: Transformations and Biological Effects* ed. O. Hutzinger, I. H. Lelyveld and B. C. J. Zoeteman, pp. 461–462, New York: Pergamon Press.

Ofstad, E. B., Lunde, G. and Drangsholt, H. (1979). Chlorinated organic compounds in the fatty surface film on water. *Int. J. Environ. Anal. Chem.*, **6**, 119–31.

Patterson, C. C. and Settle, D. (1974). Contribution of lead via aersol deposition to the Southern California Bight. *J. Rech. Atmos.*, **8**, 957–60.

Peres, J. M. (1982). Specific pelagic assemblages. *Mar. Ecol.*, **5**(1), 313–72.

Piotrowicz, S. R., Ray, B. J., Hoffman, G. L. and Duce, R. A. (1972). Trace metal enrichment in the sea surface microlayer. *J. Geophys. Res.*, **77**, 5243–54.

Plusquellec, A., Beucher, M., Le Lay, C., Le Gal, Y. and Cleret, J. J. (1991). Quantitative and qualitative bacteriology of the marine water surface microlayer in a sewage-polluted area. *Mar. Environ. Res.*, **31**: 227–39.

Polikarpov, G. G. (1966). *Radioecology of Aquatic Organisms*. New York: Reinhold Book Division. 314 pp.

Polikarpov, G. G. (1977). Effect of ocean pollution on marine organisms and communities. 2. Accumulation of radionuclides by hydrobionts and its consequence. In *Biology of the Ocean* (in Russian), Vol. 2, pp. 331–2. Moscow: Nauka.

Provenzano, A. J. Jr, McConaugha, J. R., Philips, K. B., Johnson, D. F. and Clark, J. (1983). Vertical distribution of first stage larvae of the blue crab, *Callinectes sapidus*, at the mouth of the Chesapeake Bay. *Estuarine Coastal Shelf Sci.*, **16**, 489–99.

Riznyk, R. Z., Hardy, J. T., Pearson, W. and Jabs, L. (1987). Short-term effects of polynuclear aromatic hydrocarbons on sea-surface microlayer phytoneuston. *Bull. Environ. Contam. Toxicol.*, **38**, 1037–43.

Safronov, S. G. (1981). On neuston in Kamchatka Waters of the Sea of Okhotsk. *Biol. Morya.*, **4**, 73–4.

Schram, T. A., Svelle, M. and Opsahl, M. (1981). A new divided neuston sampler in two modifications: description, tests, and biological results. *Sarsia*, **66**, 273–82.

Seba, D. B. and Corcoran, E. F. (1969). Surface slicks as concentrators of pesticides in the marine environment. *Pestic. Monit. J.*, **3**(3), 190–3.

Selleck, R. E. (1975). The significance of surface pollution in coastal waters. In *Discharge of Sewage from Sea Outfalls*, ed. A. L. H. Gameson, pp. 143–52. New York: Pergamon Press.

Sewell, L. M., Bitton, G. and Bays, J. S. (1981). Evaluation of membrane adsorption-epifluorescence microscopy for the enumeration of bacteria in coastal surface films. *Microb. Ecol.*, **7**, 365–9.

Shanks, A. L. (1983). Surface slicks associated with tidally forced internal waves may transport pelagic larvae of benthic invertebrates and fishes shoreward. *Mar. Ecol. Prog. Ser.*, **13**, 311–15.

Shenker, J. M. (1989). Oceanographic associations of neustonic larval and juvenile fishes and Dungeness crab megalopae off Oregon. *Fish. Bull.*, **86**(2), 299–317.

Sherr, E. B., Sherr, B. F., Fallon, R. D. and Newell, S. Y. (1986). Small, aloricate ciliates as a major component of the marine heterotrophic nanoplankton. *Limnol. Oceanogr.*, **31**(1), 177–83.

Sieburth, J. (1971). Distribution and activity of oceanic bacteria. *Deep-Sea Res.*, **18**, 1111–21.

Sieburth, J. McN., Willis, P. J., Johnson, K. M., Burney, C. M., Lavoie, D. M., Hinga, K. R., Caron, D. A., French, F. W., Johnson, P. W. and

Davis, P. G. (1976). Dissolved organic matter and heterotrophic micro-neuston in the surface microlayers of the North Atlantic. *Science*, **194**, 1415–18.

Sieburth, J. McN., Smetacek, V. and Lenz, J. (1978). Pelagic ecosystem structure: heterotrophic components of the plankton and their relationship to plankton size-fractions. *Limnol. Oceanogr.*, **23**, 1256–63.

Smith, R. C., Baker, K. S., Holm-Hansen, O. and Olson, R. (1980). Photo-inhibition of photosynthesis in natural waters. *Photochem. Photobiol.*, **31**, 585–92.

Smyth, P. O. (1980). Callinectes (Decapoda: Portunidae) larvae in the Middle Atlantic Bight, 1975–77. *Fish. Bull. US*, **78**, 251–65.

Södergren, A. (1993). Role of aquatic surface microlayer in the dynamics of nutrients and organic compounds in lakes, with implications for their ecotones. *Hydrobiologia*, **251**, 217–25.

Szekielda, K. H., Kupferman, S. L., Klemas, V. and Polis, D. F. (1972). Element enrichment in organic films and foam associated with aquatic frontal systems. *J. Geophys. Res.*, **77**(27), 5278–82.

Tully, O. and O'Ceidigh, P. (1986). The ecology of *Idotea* spp. (Isopoda) and *Gammarus locusta* on surface driftwood in Galway Bay (west of Ireland). *J. Mar. Biol. Assoc. UK*, **66**, 931–42.

Tully, O. and O'Ceidigh, P. (1987). The seasonal and diel distribution of lobster larvae (*Homarus gammarus* L.) in the neuston of Galway Bay. *J. Cons. Int. Explor. Mer.*, **44**, 5–9.

Van Dolah, R. F., Burrell, V. G. Jr and West, S. B. (1980). The distribution of pelagic tars and plastics in the South Atlantic Bight. *Mar. Pollut. Bull.*, **11**, 352–356.

von Westernhagen, H., Landolt, M., Kocan, R., Furstenburg, G., Janssen, D. and Kremiling, K. (1987). Toxicity of sea-surface microlayer: effects on herring and turbot embryos. *Mar. Environ. Res.*, **23**, 273–90.

Watson, A. J., Robinson, C., Robertson, J. E., Williams, P. J. le B. and Fasham, J. (1991). Spatial variability in surface carbon dioxide in the North Atlantic, spring, 1989. *Nature*, **350**, 50–3.

Williams, P. M., Carlucci, A. F., Henrichs, S. M., Van Vleet, E. S., Horrigan, S. G., Reid, F. M. H. and Robertson, K. J. (1986). Chemical and micro-biological studies of sea-surface films in the southern Gulf of California and off the west coast of Baja California. *Mar. Chem.*, **19**, 17–98.

WMO (1985). *Atmospheric Transport of Contaminants into the Mediterranean Region.* Geneva, Switzerland: World Meteorological Organization. GESAMP. Reports & Studies No. 26. 47 pp.

Wolfe, D. A. (1987). *Plastics in the Sea.* Special Issue. *Mar. Pollut. Bull.*, **18** (6B), 1–365.

Word, J. Q. and Ebbesmeyer, C. C. (1984). The influence of floatable materials from treated sewage effluents on shorelines. In *Renton Sewage Treatment Plant Project, Seahurst Baseline Study, section 3.* ed. K. K. Chewand, Q. J. Stober, pp. 40–84. Seattle, WA: City of Seattle.

Word, J. Q., Hardy, J. T., Crecelius, E. A. and Kiesser, S. L. (1987). A laboratory study of the accumulation and toxicity of contaminants in the sea surface from sediments proposed for dredging. *Mar. Environ. Res.*, **23**, 325–38.

Word, J. Q., Boatman, C. D., Ebbesmeyer, C. C., Finger, R. E., Fischnaller, S. and Stober, Q. (1990). Vertical transport of effluent material to the surface of marine waters. In *Oceanic Processes in Marine Pollution*, Vol.

6. ed. D. J. Baumgartner and I. W. Duedall, pp. 133–49. Melbourne, FL: Robert E. Krieger.

Wyndham, R. C. and Costerton, J. W. (1982). Bacterioneuston involved in the oxidation of hydrocarbons at the air–water interface. *J. Great Lakes Res.*, **8**(2), 316–22.

Zaitsev, Y. P. (1971). *Marine Neustonology* (translated from Russian). Springfield, VA: National Marine Fisheries Service, NOAA and National Science Foundation, National Technical Information Service. 207 pp.

12

Neuston of seas and oceans

YUVENALY ZAITSEV

Abstract

The first investigations of marine neuston (surface-dwelling organisms) were conducted in the 1950s and focused on the taxonomic diversity and abundance of organisms. Later investigators examined the physics, chemistry and exchange processes between the atmosphere and ocean. Today, we know the ocean–atmosphere interface is important for many biogeochemical processes essential for life.

Physical, chemical and biological conditions differ greatly between the uppermost 5 cm of the ocean and the water below. The marine pleuston includes the larger siphonophores, *Physalia* and *Velella*, which float on the surface. Neuston can be divided into epineuston and hyponeuston. The epineuston includes more than 40 species of water striders, *Halobates*, inhabiting the open ocean and coastal areas. The hyponeuston are organisms in the surface centilayer including hydrozoa, molluscs, copepods, isopods, decapod crustaceans, fishes, and the seaweed *Sargassum*.

The neuston connect the sea surface and water column as larvae develop and migrate downward, and adult animals visit the surface to feed and reproduce. The sea surface has become a site of significant enrichment of pollutants from terrestrial and atmospheric sources. The spatial coincidence of the maximum pollutant concentrations and the biological sensitivity of its inhabitants creates a critical situation in the marine environment.

High densities of neustonic organisms in the sea surface can influence air–sea exchange processes (as discussed in Chapter 10, this volume). Anthropogenic enrichment of the sea surface not only threatens marine plants and animals directly, but may impact natural processes affecting global climate. The continually increasing pollution of the ocean represents one of the most significant factors accelerating global ecological changes.

Introduction

Specific organisms inhabiting the uppermost layer of fresh water, bordering with the atmosphere ('pleuston' and 'neuston'), were known many years ago (Naumann, 1917). However, then biologists believed that the near-surface living forms could exist only in small, wind-protected ponds, lakes, pools, and that the rough surface of seas and oceans was not suitable for organisms adapted to the special life conditions in the pelagic zone and atmosphere interface (Zernov, 1934). Significant discoveries were made only when marine biologists focused on the upper surface of the sea as a special ecosystem.

The first investigations were carried out in the 1950s by biologists who were struck by taxonomic diversity, abundance and specificity of inhabitants of this unique layer of the sea. Subsequently, investigations were made of the physics and chemistry of the surface microlayer, and of exchange processes between the ocean and atmosphere. These investigations did much for understanding the life conditions in the surface biotope. As far back as 1941, the Japanese scientist Miyazaki (Marumo *et al.*, 1971) suggested the necessity of further oceanographic studies in the 'zero metre layer' as a specific physicochemical habitat. But special investigations of abiotic conditions of this biotope were only begun in the 1960s.

Today, we are still not fully familiar with the biology, physics and chemistry of the upper layer of seas and oceans. Nevertheless, we now know that many important biogeochemical processes are focused at the ocean–atmosphere interface (as discussed in many chapters in this volume). It is also becoming clear that life on the planet in general depends on the interaction of the hydrosphere and atmosphere.

In the last two to three decades humans have begun to have an active influence on nature and on the properties of the upper ocean layer. Many processes involved began to depart from normal. This has had wide-reaching consequences for the biological, ecological, economic and social environment.

The extreme surface of the sea as a specific habitat

According to most physical and chemical parameters, the uppermost 5 cm of the sea differs not only from the water mass, but also from the sub-surface water (10–20 cm below the surface). The uppermost 10 cm of water 'captures' about half of the total amount of sunlight entering the sea (Rutkovskaya, 1965). The uppermost 1 cm layer of the Black Sea water absorbs 20% of the total sunlight (Boguslavsky, 1956).

Another important aspect of solar radiation concerns the spectral composition of the penetrating light, since the effect of light on the physiological and biochemical processes of marine organisms depends on the wavelength. In general, the shorter wavelengths are absorbed by the water as rapidly as the longer infrared. So, the upper 10 cm layer receives the largest quantity of the biologically active visible and UV light, but inhibition effects continue to at least 30 m (Sieracki and Sieburth, 1985, 1986).

Water temperature is intimately related to illumination, since solar radiation is the main source of heat in marine surface waters. But the thermal regime of the surface microlayer is strongly influenced by suspended particles, both living organisms and detritus. For example, high concentrations of bacteria and phytoplankton can raise surface water temperatures by 2–3 °C (Brekhovskikh, 1988). However, cooling effects are also possible if mobile neuston organisms directly contact the surface film causing microturbulence, and thereby increasing evaporation (Gladyshev, 1991).

Little is known at present about the vertical microdistribution of surface water temperature in the presence of ice. However, in view of the fact that various initial forms of loose ice occur in the uppermost 4–5 cm of the water, it can be concluded that the surface layer of marine waters differs from the underlying masses by having lower temperatures during certain seasons and in certain areas. This naturally has biological consequences.

The surface of seas and oceans is a collector of many particulate and dissolved organic substances. Organic matter originating from marine organisms is one source. Other supply routes are river runoff, important for coastal areas, and atmospheric inputs, from rain, snow and winds. Global atmospheric input to the ocean exceeds riverine input for several trace metals and some other substances and has increased in importance, due to human activities in the twentieth century. Among the largest particles of dead organic matter are land insects carried by the wind to the sea. According to rough estimates, about one billion land insects could be found on the whole water surface of the open Black Sea during the summer (Zaitsev, 1971). Other organic particles are of marine origin.

A very important ecological feature of the near-surface habitat is the relationships between inhabitants of the zone, especially those between predator and prey, consumer and food (Sieburth *et al.*, 1976).

Neustonic animals of moderate size (0.2–30 mm) may be consumed by predators from two environments: aquatic and aerial. The latter biological fate is not common for bulkwater plankton, nekton or benthos. Relevant aquatic predators include young and adult pelagic fishes, squid, and sometimes marine snakes and turtles. Aerial enemies of neustonic animals are littoral, neritic and oceanic birds, which feed by skimming, pattering, hydroplaning and surface filtering (Boaden and Seed, 1985), and even such mammals as specialized bats, *Noctilionidae*.

Although data characterizing conditions in the surface layer of the sea are still incomplete, there can be no doubt of its uniqueness as a habitat. This biotope differs from the underlying waters in a number of ways, notably, the inflow and accumulation of organic matter, the intensive solar radiation and the double pressure exerted by predators.

During recent decades, all these natural environmental factors have been complicated by a large number of man-made factors exerting pressure upon the abundance and biological diversity of the surface marine biotope.

Neuston and pleuston communities of marine organisms adapted to specific life conditions in the uppermost sea layer

Until recently, there was much confusion and the same species were ascribed sometimes to neuston and at other times to pleuston. Hardy (Chapter 11, this volume) has suggested the term 'macroneuston' for the large neuston formerly referred to as 'pleuston'.

Marine pleuston (macroneuston) includes siphonophores, *Physalia* and *Velella*, whose bodies project considerably above the water and can withstand prolonged desiccation and exposure to direct solar radiation. These features reflect the special histological structure of the pneumatophores in contrast to submerged tissues of these organisms. Generally, the marine pleustonic organisms are limited to tropical waters with surface temperatures greater than 15–17 °C (Savilov, 1961). Thus, it appears that the main limiting factor in this case is air temperature, whereas other marine organisms are not directly influenced by it.

The concept of 'neuston' has evolved considerably. At first it applied to unicellular algae and animals attached to the surface film. Now it applies to the whole complex of organisms found in the aquatic and aerial parts of the surface film of water bodies, including bacteria, protozoa and other microorganisms, as well as metazoa (Zaitsev, 1971; Tsyban, 1971; Sieburth, 1983).

Structure and taxonomic composition of the neuston

Topographically, the organisms of the neuston are divided into two groups. One group, known as hyponeuston, lives on the lower (aquatic) side of the surface film. The second group, the epineuston, consists of the inhabitants of the upper (aerial) side of the surface film. This division in the hyponeuston and epineuston is quite clear-cut except in the case of microorganisms, which are usually referred to simply as neuston. Some hyponeustonts can move briefly to the epineuston, and vice versa. For example, a diving epineustonic water strider finds itself briefly inside the water, while hyponeustonic pontellids remain in the air for short intervals while jumping.

The epineuston includes more than 40 species of water striders, *Halobates*, inhabiting the open ocean and coastal areas. They are tropical and subtropical organisms. In temperate coastal waters the epineuston in summer are represented by some Diptera like *Clunio marinus* and *C. ponticus*, and by different species of springtails, *Collembola*, with some, like *Anurida maritima*, preferring brackish waters.

The hyponeuston is composed of inhabitants of the 0–5 cm layer, conditionally considered as an aquatic neuston biotope (Zaitsev, 1968, 1971). It consists of permanently hyponeustonic organisms (euhyponeuston), and temporary inhabitants – with the latter comprising species associated with the surface biotope during the early stages of their development (merohyponeuston), and organisms which pass the day either at a considerable depth in the water or on the bottom, but migrating in the night to the near surface (plankto- and benthohyponeuston, together forming the nyctineuston).

Among euhyponeuston organisms are the hydrozoan *Porpita*, the actinia *Minyas*, the molluscs *Janthina* and *Glaucus*, most species of the copepod families *Pontellidae* and *Sapphirinidae*, some isopods (*Idothea metallica*), decapod crustaceans (*Planes* and *Portunus*), and fishes such as *Syngnathus schmidti, S. pelagicus, Antennarius*, and *Histrio*. The floating algae *Sargassum natans* and *S. fluitans* are also part of the euhyponeuston.

The merohyponeuston consists of organisms associated with the surface biotope during early stages of their development. After completing the hyponeustonic phase of their life cycle, they move to the water column or the bottom and become part of the plankton, nekton or benthos. In terms of numbers of metazoans, the merohyponeuston form the bulk of the hyponeuston. Indeed, the abundance of eggs, larvae and young forms of aquatic organisms in the 0–5 cm layer makes the hyponeuston an important 'incubator' of the sea (Zaitsev, 1960).

Most numerous among the merohyponeuston are larvae of bivalve and gastropod molluscs, polychaetes, cirripeds, copepods, decapods, echinoderms and fishes.

Benthohyponeuston is composed of many species of migratory poly-chaetes (*Eunice viridis, Nereis longissima, Platynereis dumerilii, Nephthis longicornis*), amphipods (*Nototropus guttatus, Dexamine spinosa, Corophium nobile*), cumaceans, mysids, shrimps (*Palaemon adspersus, P. elegans*) and many others.

Planktohyponeuston includes different planktonic species that migrate into surface waters at night. These include copepods, *Calanus fin-marchicus, C. ponticus, C. tonsus, C. cristatus*, hyperiids (for example, *Parathemisto japonica*), some squids, mesopelagic fishes, and *Mycto-phidae*.

Numbers of neustonic organisms

The investigations of A. V. Tsyban (1970, 1971), first in the Black Sea then in other seas and oceans (Sieburth, 1965, 1971), revealed that the density of bacteria at the sea surface (the bacterioneuston) is on average a hundred times greater than that of the bacterioplankton. For instance, in the surface film of the northwestern Black Sea, studies in the 1960s indicated that the numbers of bacteria in the neuston layer were $5-27 \times 10^6$ cells ml^{-1}. In the water below the numbers were only $25-50 \times 10^3$ cells ml^{-1}.

The marine phytoneuston and their high productivity have been inves-tigated by J. T. Hardy and his colleagues. Phytoneustonic microalgae inhabiting the surface microlayer form an important and often abundant community. Their densities can be ten to several hundred times greater than that of the phytoplankton in the bulkwater (Hardy, 1982). The most common representatives of the phytoneuston are diatoms, *Nitzschia closterium* and *Thalassiosira* sp., dinoflagellates, e.g. *Exuviaella* sp., and blue-green algae, *Trichodesmium* sp. (Hardy *et al.*, 1988).

The very first works in marine neustonology, using multistage nets, showed that the 0–5 cm layer is rich in microplankton species, par-ticularly *Noctiluca miliaris*, larvae of the polychaete family Spionidae, nauplii of cirriped *Balanus improvisus*, and the copepod *Centropages kröyeri pontica* (Zaitsev, 1961).

Large neustonic invertebrates measuring 3–20 mm and sometimes even tens of centimetres constitute the predatory link, feeding mainly on the organisms of the preceding stage and also on fish eggs and larvae. This group is relatively less numerous and consists of large copepods (*Pon-tella, Anomalocera, Labidocera* and others), isopods, zoea and megalopa stages of crabs, and some night migrants such as polychaetes and shrimps.

Their density in the 0–5 cm layer is 10–100 times greater than in the bulk water.

The highest buoyancy values for fish eggs have been determined in various species of *Engraulidae, Mugilidae, Pomatomidae, Carangidae, Mullidae, Callionymidae, Bothidae, Soleidae*, and others. Dense samples of such eggs obtained with neuston nets have provided accurate information on the time, place and conditions of spawning of different species in the Black Sea (Zaitsev, 1961, 1971; Savchuk, 1967).

In addition to fish eggs, the 0–5 cm layer contains large numbers of larvae and fry of pelagic and bottom fish species, mentioned above, and also of *Bellonidae, Exocoetidea, Atherinidae, Labridae, Blenniidae, Gobiidae, Balistidae, Syngnathidae*, and others.

Neuston as a connecting link in the biosphere

Close relationships exist between the neuston and other marine and terrestrial organisms. This is due to the position of the neuston at the boundary between sea and atmosphere, to its age composition (early stages of development predominate), quantity (high density and biomass), and wide distribution in all seas and oceans. After passing some time in the neuston, many larvae migrate to other marine habitats, where they develop into adults. On the other hand, a large number of adult animals visit the surface biotope for feeding or reproduction.

Three pathways connect the neuston to the benthic biocoenoses: settling of larvae to the bottom, ascent of adult organisms for spawning (such as the palolo worm, *Eunice viridis*, and other polychaetes), and night feeding of benthic animals in the neuston biotope.

These routes also connect the neuston to the plankton inhabiting the water column. Hyponeustonic larvae of many invertebrate species move to deeper layers, where they become part of the plankton. On the other hand, the surface biotope is visited by females arriving from deeper waters in order to lay eggs or leave their larvae (e.g. *Lucifer, Sagitta, Euphausiacea*). Finally, hyperiids and other deep water plankton also visit the surface layer for feeding. Similar relationships exist between the neuston and the nekton.

One of the most important features of the neuston is its contact with aerobionts. In addition to epineustonic insects, this contact is twofold: neustonts consume aerobionts falling on the sea surface, while aerobionts hunt for prey in the neuston. There are many examples of birds and even mammals feeding on neuston, with beaks, claws, digestive tracts and other parts of the body adapted to facilitate the capture and storage of neustonts, as well as their delivery to the progeny. The neuston is continually removed from the sea by air to the land, and accumulates as bird and bat guano after being digested by neustophagous animals. Birds not only consume neustonts, but also assist in their spread. Thus, eggs of *Artemia, Halobates*, and other organisms have been found on birds' feet and feathers.

Owing to a number of circumstances, the marine neuston has become a crossroads of different parts of the biosphere (Zaitsev, 1967; Polikarpov and Zaitsev, 1969). Through the marine neuston pass streams of matter and energy to a large variety of habitats situated below, in the sea, or beyond it, on land. This important role of the neuston is attracting increasing scientific attention in this period of increasing human impact on the environment.

Anthropogenic influences on neuston

The sea surface has become a major site of enrichment by different pollutants from land sources and atmospheric inputs. For instance, the chemical composition of the sea-surface microlayer samples, collected from sites in Puget Sound, indicated that contamination sources included atmospheric deposition, terrestrial runoff, and sewage disposal (Hardy *et al.*, 1987b). This leads to concentrations of organic substances, metals, pesticides, oil and other pollutants in the microlayer of coastal and open ocean waters that are 10 to 100 times greater than in the bulk water (Patin, 1977; Duce *et al.*, 1991; Piotrowicz *et al.*, 1972; Hardy and Cleary, 1992).

The spatial coincidence of maximum pollutant concentration at the sea surface and maximum biological sensitivity of its inhabitants (concentration of embryos and larvae) creates an acute critical situation in the marine environment. This important environmental problem needs more investigation.

A few data show high concentrations of some metals and other toxicants in surface-living organisms, as well as chromosome abnormalities in fish embryos developing in polluted areas. Laboratory exposure to surface microlayer samples, collected from polluted areas, resulted in more chromosomal aberrations in developing sole embryos (Hardy *et al.*, 1987a).

High concentrations of cadmium were found in the kidneys of neuston-eating shearwaters and in epineustonic sea skaters (*Halobates*) from the open ocean (Risebrough, 1971; Bull *et al.*, 1977). Levels of up to 13.5 mg kg⁻¹ of Hg and 375 mg kg⁻¹ of Zn were determined in the liver of the open ocean marlin, *Makaira indica* (MacKay *et al.*, 1975). Up to 20 mg kg⁻¹ of Hg were found in the neuston-eating guillemot (*Uria aalge*), which breed along rocky shores, especially at high latitudes, and the kittiwake (*Rissa tridactyla*) inhabiting arctic and subarctic areas (Gerlach, 1985).

An important impact on the neuston, especially for inland seas, is man-made eutrophication. The Black Sea is a good example. Nutrient enrichment of coastal waters leads to phytoplankton blooms and increasing numbers of dinoflagellates, protozoans and other microorganisms in surface waters, especially in the neuston biotope. In exceptional cases, a biomass of *Noctiluca miliaris* of 100–500 kg m⁻³ in the 0–5 cm layer can be found (Zaitsev *et al.*, 1988).

Very high densities of the tunicate *Oikopleura dioica* were observed in Saanich Inlet, as a cherry-pink neuston (Seki, 1973). This 'red tide' seems to have been due to the rapid multiplication of *O. dioica*, grazing on a heavy bloom of phytoplankton in the surface layer of the inlet.

The sedimentation of large amounts of dead phytoplankton and detritus can lead to the formation of large areas of hypoxia, with mass mortality of benthic organisms in shelf seas. This can cause a sharp decline in populations of the merohyponeuston. Thus, larvae of shrimps, ghost shrimps, crabs, flatfishes and other bottom-living organisms, formerly abundant, are extremely rare in the Black Sea neuston today. The megalopa of crabs *Rhitropanopeus harrisi tridentats* and *Brachinotus sexdentatus*, which were extremely abundant in the 1960s (Zaitsev, 1971), are practically absent now in the neuston of the Sea of Azov.

Some data show the negative influence of acid rains on marine bacterioneuston (Izrael and Tsyban, 1989).

Recent measurements confirm increasing levels of UV-B radiation reaching the earth's surface. Could the general decline in abundance of neustonic organisms be correlated with this increase? This question needs special investigation.

Neuston as a factor influencing the air–sea exchange

High densities of neustonic organisms in the top millimetre of the ocean can influence water–air exchange processes (MacIntyre, 1974). Experimentally it has been demonstrated that, in the presence of large amounts of suspended unicellular algae, the water temperature is 2–3 °C higher than in clean water (Brekhovskikh, 1988). On the other hand, the freshwater hyponeustonic cladoceran, *Scapholeberis mucronata*, causes microturbulence in the surface layer, enhancing the evaporation of water, and significantly reducing its surface temperature (Gladyshev, 1991).

Human impacts on the surface of seas and oceans, which lead to qualitative and quantitative changes in neuston, not only influence the reproduction of marine animals and plants, including commercially important species, but may also affect the natural mechanisms involved in the dynamics of weather and climate (Zaitsev, 1990). The ever-increasing area of eutrophicated waters on the planet may be a significant factor accelerating the advent of global ecological changes. A global 'neuston watch' could therefore be very timely.

Acknowledgements

I wish to express my gratitude to Professor Peter Liss and Dr Phillip Williamson for editing the manuscript, Professor Jack Hardy for many interesting suggestions and ideas, and Professor John Sieburth for helpful criticism. I am grateful to Dr Vera Lisovskaya for the first English translation of my text.

References

Boaden, J. P. S. and Seed, R. (1985). *An Introduction to Coastal Ecology.* Glasgow and London: Blackie, 300 pp.

Boguslavsky, S. G. (1956). Absorption of solar radiation in the sea and its direct influence on the temperature of the sea (in Russian). *Trudy Morskogo Gidrofizicheskogo Instituta*, **8**, 80–90.

Brekhovskikh, V. F. (1988). *Hydrophysical Factors Forming the Oxygen Regimen of Water Bodies* (in Russian). Moscow: Nauka, 168 pp.

Bull, K. R., Murton, R. K., Osborn, D., Ward, P. and Cheng, L. (1977). High levels of cadmium in Atlantic sea birds and sea skaters. *Nature*, **269**, 507–9.

Duce, R. A., Liss, P. S., Merrill, J. T., Atlas, E. L., Buat-Menard, P., Hicks, B. B., Miller, J. M., Prospero, J. M., Arimoto, R., Church, T. M., Ellis, W., Galloway, J. N., Hansen, L., Jickells, T. D., Knap, A. H., Reinhardt, K. N., Schneider, B., Soudine, A., Tokos, J. J., Tsunogai, S., Wollast, R. and Zhou, M. (1991). The atmospheric input of trace species to the world ocean. *Global Biogeochem. Cycles*, **5**, 193–259.

Gerlach, S. A. (1985). *Marine Pollution. Diagnosis and Therapy* (in Russian, translated from English). Leningrad: Gidrometeoizdat.

Gladyshev, M. I. (1991). Experimental study of the effect of zooneuston organisms on the temperature of the surface film of water (in Russian). *Doklady Academii Nauk SSSR*, **320**(6), 1489–90.

Hardy, J. T. (1982). The sea surface microlayer: biology, chemistry and anthropogenic enrichment. *Progr. Oceanogr.*, **11**, 307–28.

Hardy, J. T. and Cleary, J. (1992). Surface microlayer contamination and toxicity in the German Bight. *Mar. Ecol. Progr. Ser.*, **91**, 203–10.

Hardy, J. T., Kiesser, S. L., Antrim, L. D., Stubin, A. I., Kocan, R. and Strand, J. A. (1987a). The sea-surface microlayer of Puget Sound: Part I. Toxic effects on fish eggs and larvae. *Mar. Environ. Res.*, **23**, 227–49.

Hardy, J. T., Crecelius, E. A., Apts, C. W. and Gurtisen, J. M. (1987b). Sea-surface contamination in Puget Sound: Part II. Concentration and distribution of contaminants. *Mar. Environ. Res.*, **23**, 251–71.

Hardy, J. T., Coley, J. A., Antrim, L. D. and Kiesser, S. L. (1988). A hydrophobic large-volume sampler for collecting aquatic surface microlayers: characterization and comparison to the glass plate method. *Can. J. Fish. Aqua. Sci.*, **45**, 822–6.

Izrael, Yu. A. and Tsyban, A. V. (1989). *Anthropogenic Ecology of the Ocean*. Leningrad: Gidrometeoizdat, 528 pp.

MacIntyre, F. (1974). The top millimetre of the ocean. *Sci. Am.*, **230** (6), 62–77.

MacKay, N. J., Kazacos, M. N., Williams, R. J. and Leedow, M. I. (1975). Selenium and heavy metals in black marlin. *Mar. Pollut. Bull.*, **6**, 57–61.

Marumo, R. N., Taga, N. and Nakai, T. (1971). Neustonic bacteria and phytoplankton in surface microlayers of the equatorial waters. *Bull. Plankton Soc. Japan*, **18** (2), 36–41.

Naumann, E. (1917). Über das Neuston des Süsswassers. *Biol. Centralblatt*, **37** (2), 98–106.

Patin, S. A. (1977). Influence of ocean pollution on marine organisms and communities (in Russian). In *Biology of the Ocean. 2. Biological Productivity of the Ocean*. pp. 322–31 Moscow: Nauka.

Piotrowicz, S. R., Ray, B. J., Hoffman, G. L. and Duce, R. A. (1972). Trace metal enrichment in the sea-surface microlayer. *J. Geophys. Res.* **77**, 5243–54.

Polikarpov, G. G. and Zaitsev, Yu. P. (1969). *Horizons and Research Strategy in Marine Biology*. Kiev: Naukova Dumka. (In Russian.)

Risebrough, R. W. (1971). *Chlorinated Hydrocarbons. Impingement of Man on the Oceans*. New York: Wiley Interscience.

Rutkovskaya, V. A. (1965). Penetration of solar radiation into inland and marine waters (in Russian). *Trudy Instituta Okeanologii AN SSSR*, **78**, 245–75.

Savchuk, M. Ya. (1967). Migration and distribution of young grey mullet in the northwestern part of the Black Sea (in Russian). *Zoologicheskii Zhurnal*, **46** (5), 737–40.

Savilov, A. I. (1961). Distribution of ecological forms of *Velella lata* Ch. et Eys. and *Physalia utriculus* (La Martiniére) Esch. in the North Pacific Ocean (in Russian). *Trudy Instituta Okeanologii AN SSSR*, **45**, 223–38.

Seki, H. (1973). Red tide of Oikopleura in Saanich Inlet. *La Mer*, **11** (3), 153–58.

Sieburth, J.McN. (1965). Bacteriological samplers for air–water and water–sediment interfaces. In *Transactions of the Joint Conference, Marine Technology Society and American Society of Limnology and Oceanography, Washington, DC*, pp. 1064–8. Preprinted from *Ocean Science and Ocean Engineering*.

Sieburth, J. McN. (1971). An instance of bacterial inhibition in oceanic surface water. *Mar. Biol.*, **11** (1), 98–100.

Sieburth, J. McN. (1983). Microbiological and organic-chemical processes in the surface and mixed layers. In *Air-Sea Exchange of Gases and Particles* ed. P. S. Liss and W. G. N. Slinn, pp. 121–72. Dordrecht: Reidel.

Sieburth, J. McN., Willis, P-J., Johnson, K. M., Burney, C. M., Lavoie, D. M., Hinga, K. R., Caron, D. A., French III, F. W., Johnson, P. W. and Davis, P. G. (1976). Dissolved organic matter and heterotrophic microneuston in the surface microlayers of the North Atlantic. *Science*, **194**, 1415–18.

Sieracki, M. E. and Sieburth, J. McN. (1985). Factors controlling the periodic fluctuation in total planktonic bacterial populations in the upper ocean: comparison of nutrient, sunlight and predator effects. *Mar. Microbiol. Food Webs*, **1**, 35–50.

Sieracki, M. E. and Sieburth, J. McN. (1986). Sunlight-induced growth delay of planktonic marine bacteria in filtered seawater. *Mar. Ecol. Prog. Ser.*, **33**, 19–27.

Tsyban, A. V. (1970). *Bacterioneuston and Bacterioplankton of the Shelf Zone in the Black Sea* (in Russian). Kiev: Naukova Dumka. 275 pp.

Tsyban, A. V. (1971). Marine bacterioneuston. *J. Oceanogr. Soc. Japan*, **27**, 56–71.

Zaitsev, Yu. P. (1960). The existence of nueston biocoenosis in marine pelagic zone (in Ukrainian). *Naukovi Zapysky Odeskoyi Biologichnoyi Stantsiyi AN URSR*, **2**, 37–42.

Zaitsev, Yu. P. (1961). The surface pelagic biocoenosis of the Black Sea (in Russian). *Zool. Z.*, **40**, 818–25.

Zaitsev, Yu. P. (1967). Problems of marine neustonology (in Russian). *Gidrobiol. Z.*, **3** (5), 58–69.

Zaitsev, Yu. P. (1968). La neustonologie marine: objet, méthodes, réalisations principales et problèmes. *Pelagos. Bull. Inst. Océnogr., Alger.*, **8**, 1–48.

Zaitsev, Yu. P. (1971). *Marine Neustonology* (translated from Russian). Springfield, VA: National Marine Fisheries Service, NOAA and National Science Foundation, National Technical Information Service. 207 pp.

Zaitsev, Yu. P. (1990). Some possible consequences of forecasting climate changes and increases in ocean levels for the Black Sea area. Odessa, Dep. VINITI, No. 777–B90, pp. 1–17.

Zaitsev, Yu. P., Polischuk, L. N., Nastenko, E. V. and Trophanchuk, G. M. (1988). Superconcentrations of *Noctiluca miliaris* Suriray in the neuston layer of the Black Sea (in Russian). *Doklady Academyi Nauk Ukrainskoy SSR*, Seria B., **10**, 67–9.

Zernov, S. A. (1934) *General Hydrobiology* (in Russian). Moskow, Leningrad: Biomedgiz, 504 pp.

13

Photochemistry in the sea-surface microlayer

NEIL V. BLOUGH

Abstract

Due to the photochemical production and atmospheric deposition of highly reactive species at the sea surface, the microlayer could well act as a highly efficient microreactor, effectively sequestering and transforming select materials brought to the interface from the atmosphere and oceans by physical processes. However, very little is known about the optical and photochemical properties of this regime. Based on the measured enrichments of light absorbing material in the microlayer and employing photochemical quantum yields obtained for bulk waters, photochemical production rates and fluxes are estimated for the microlayer. The microlayer fluxes are generally small with respect to atmospheric deposition and the water column fluxes. This result argues that the microlayer is unlikely to act as a 'reaction barrier' to the exchange of trace gases across the interface. However, the higher photochemical production rates at the surface should lead to the more rapid oxidative turnover of materials at the interface and potentially to reactions and processes not observed in bulk waters.

Introduction

The sea-surface microlayer acts as a dynamic interface mediating the exchange of matter, heat, momentum, and electromagnetic radiation between the earth's oceans and atmosphere. The accumulation of surface-active material within this thin oceanic layer – defined operationally as the top 0.03 to 500 μm depending on the sampling method employed (Liss, 1986; Hunter and Liss, 1981; Carlson, 1982b) – has been shown by workers over the last several decades to alter the physical and chemical properties of the interface. These changes in surface properties can produce, in turn, a variety of effects that materially influence the exchange of matter and energy between the oceans and the atmosphere. These effects include: 1. a reduction in the rates of gas exchange, potentially even under conditions and in regions where visible slicks

383

(condensed monolayer films) are absent (Frew *et al.*, 1990); 2. the
suppression of wind-generated capillary and small gravity waves
(Hühnerfuss *et al.*, 1983b; Wei and Wu, 1992); 3. changes in the absorption and backscattering of microwave radiation and in the specular
reflection of sunlight from the sea surface owing to capillary wave
damping as observed remotely by active radar and sun glint measurements (Hühnerfuss *et al.*, 1983a; Soules, 1970); 4. the injection of both
surface-enriched organic and inorganic matter and microbiota into the
atmosphere through the collapse of bubbles at the air–sea interface (Liss,
1975; Duce and Hoffman, 1976; Lion and Leckie, 1981; Tseng *et al.*,
1992); and 5. the potentially accelerated transformation of materials
residing within the microlayer as a result of reactions with radicals and
other highly reactive species deposited from the atmosphere (Thompson
and Zafiriou, 1983), through photochemical processes arising from the
higher incident levels of UV-B (280–315 nm) and UV-A (315–400 nm)
radiation at the sea surface, or from elevated levels of microneuston
(Hardy, 1982).

Due to the enrichment of chemicals and biota within this harsher
(photo)chemical environment, the surface microlayer may well act as a
highly efficient microreactor, effectively sequestering and transforming
select materials brought to the interface from the atmosphere and oceans
by physical processes. Rapid photochemical, chemical, and biological
reactions within the microlayer could produce a variety of interesting
feedbacks. For example, photochemical reactions could destroy (or
produce) surface-active species, thereby altering surface wave damping
and gas exchange rates. Elevated levels of highly reactive intermediates
produced within this zone could present a 'reaction barrier' to the
transport of some chemicals and trace gases across the air–sea interface,
thus affecting their flux to the atmosphere or ocean. Further, reactions
occurring within the microlayer potentially could enhance (or deplete)
the surface concentrations of certain gases relative to those of bulk
seawater, chemically modify compounds during their traversal of the
interface, alter the redox state, speciation and biological availability of
trace metals deposited by the atmosphere to the interface, and influence
the types and distributions of materials introduced to the atmosphere by
bubble injection and to the oceans by particle settling.

Although very intriguing and potentially of great importance, in many
instances evidence supporting the existence of these processes is lacking.
While this paucity of experimental data can be assigned in part to the
difficulties involved in studying the microlayer, historically this field has

attracted fewer workers (and even less financial support). Thus, many simple experiments that could shed light on the importance of these processes have yet to be performed. Our understanding of the impact of photochemical reactions on the sea-surface microlayer represents a particularly cogent example. While a number of obvious experiments were recommended in the report of a 1984 ONR workshop on the sea-surface microlayer (Hartwig and Herr, 1984), to my knowledge these experiments have not been performed nor has a significant article appeared over the last nine years that deals specifically with photochemical reactions within the microlayer. Possible increases in ground-level UV-B radiation resulting from the destruction of stratospheric ozone may have important consequences for the microlayer and the processes it mediates, but the current lack of information does not allow a straightforward assessment of impacts.

Recent advances in bulk-water photochemistry combined with newer work on the composition and physical properties of the microlayer can provide fresh insights to the photochemical processes that may be occurring within this regime. To this end, this chapter will first review the optical and photochemical properties of the major bulk-water chromophores, and, where possible, will include pertinent information from studies of the microlayer. Data presented within this chapter will then be employed to acquire rough estimates of the rates of photochemical reactions within the microlayer. Finally, these estimates will be used to explore the possible impacts of microlayer photochemical processes on the exchange of materials across the air–sea boundary.

Optical properties of microlayer and bulk-water chromophores

A fundamental tenet of photochemistry is that light must first be absorbed for a photochemical reaction to be initiated. Thus, knowledge of the absorption spectra of chromophores residing within the microlayer is essential. Unfortunately, almost nothing is known about the optical absorption (and emission) properties of these species. Carlson (1982a, 1983; Carlson and Mayer, 1980) has shown that material absorbing at 280 nm is consistently enriched in the microlayer relative to bulk waters (Figure 13.1). Based on a correlation between light absorption at 280 nm and the response of the microlayer material to the Folin–Ciocalteu test, he concluded that dissolved phenolic material was the predominant light absorbing species. While this material may arise in part from the exudation of polyphenolic compounds from marine macroalgae, it seems more

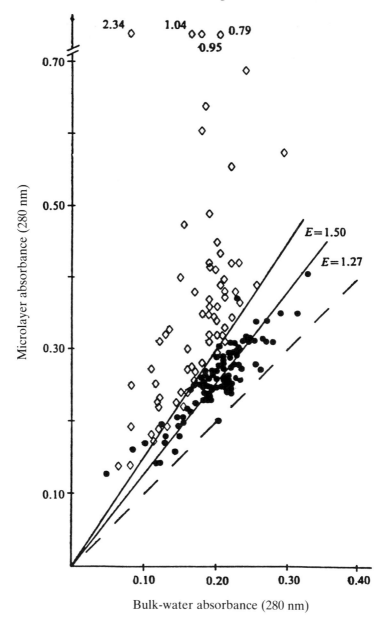

Figure 13.1. Comparison between light absorption at 280 nm by material in bulk seawater and in the microlayer. E is the microlayer enrichment factor. Absorbance values were obtained with a 10 cm optical cell. Microlayer sampling depth was ~50 μm. Points falling above the E = 1.5 line represent slicked surfaces (◇), whereas those below represent clean surfaces (●). Adapted from Carlson (1982a).

likely that it is composed principally of the plant and algae decay products commonly known as gelbstoffe, yellow substance or humic substances (Carlson and Mayer, 1980). This material, referred to here as coloured or chromophoric dissolved organic matter (CDOM), is a chemically complex and ill-defined mixture of anionic organic oligoelectrolytes known to contain phenolic moieties (Aiken *et al.*, 1985) and exhibit surface-active properties (Hayase and Tsubota, 1986).

The current lack of absorption and luminescence spectra of CDOM from the microlayer precludes, however, a more rigorous assessment of its nature and source. Furthermore, until this information is acquired, possible differences between the optical properties of the microlayer and the underlying waters cannot be defined. Differences of this sort could arise from the selective partitioning of the more hydrophobic components of bulk-water CDOM into the microlayer, potentially leading to photochemical reactions within the microlayer that differ both qualitatively and quantitatively from those of bulk waters.

CDOM is known to be the predominant light absorbing component of the dissolved organic matter (DOM) pool within bulk seawaters, far exceeding the contributions from other natural organic chromophores such as flavins (Blough and Green, 1995). Light absorption by CDOM typically decreases in an approximately exponential fashion with increasing wavelength (Zepp and Schlotzhauer, 1981; Bricaud *et al.*, 1981; Carder *et al.*, 1989; Blough *et al.*, 1993; Figure 13.2). Thus, the spectra have usually been fitted to the form

$$a(\lambda) = a(\lambda o)e^{-S(\lambda - \lambda o)} \tag{13.1}$$

where $a(\lambda)$ and $a(\lambda o)$ are the absorption coefficients at wavelength λ and reference wavelength λo, and S is a parameter that characterizes how rapidly the absorption decreases with increasing wavelength. $a(\lambda)$ is defined by the relation

$$a(\lambda) = 2.303 \, A(\lambda)/r \tag{13.2}$$

where A is the absorbance and r is the path length in metres.

$a(\lambda)$ and S vary both spatially and temporally (Blough *et. al.*, 1993; Green and Blough, 1994; Blough and Green, 1995). Values of $a(300)$ have been observed to range from <0.1 m^{-1} for very clear oligotrophic seawaters to >50 m^{-1} for coastal seawaters strongly influenced by terrestrial input. S appears to increase with decreasing $a(\lambda)$, ranging from as low as 0.012–0.013 nm^{-1} for some highly absorbing coastal waters to >0.02 nm^{-1} for weakly absorbing oligotrophic waters. These absorption

N. V. Blough

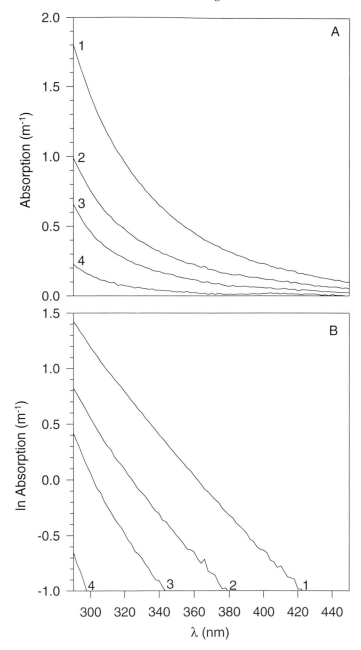

Figure 13.2. (A) Absorption spectra of 0.2 μm-filtered seawaters from the coastal/ shelf region of the Delaware Bight. Positions are 1, 38.59° N 74.77° W; 2, 38.27° N 74.36° W; 3, 38.01° N 74.05° W; 4, 37.11° N 72.94° W. (B) Log plot of the absorption coefficients versus wavelength for the samples from A.

measurements, when combined with multi-dimensional luminescence techniques, can provide information on the source and nature of the CDOM (Blough and Green, 1995). Presently, we do not know whether the magnitude and spectral dependence of the absorption and emission of light by microlayer CDOM will vary in the same fashion as that in the underlying waters or will differ due to the selective enrichment of particular components of the bulk-water CDOM.

A wide variety of other trace inorganic and organic chromophores may be present within bulk seawaters and the microlayer; these chromophores include NO_3^-, NO_2^-, transition metal complexes, metal colloids and such discrete organic compounds as flavins, phenols, and keto acids. Although these species do not usually contribute appreciably to the overall absorption of light within the oceans, many may undergo significant rates of photolysis (*vide infra*). Other than NO_3^-, and NO_2^-, which do exhibit occasional enrichments (Williams *et al.*, 1986), no inventory of trace chromophores has been reported for the microlayer. Moreover, the spectral dependence and magnitude of light absorption by particulate organic matter (POC), which is enriched significantly in the microlayer (Liss, 1986), is not known but could be determined by the filter pad method (Mitchell and Kiefer, 1988; Nelson and Robertson, 1993). Similarly, systematic tests for the presence of UV-absorbing, photoprotective chromophores (Mitchell *et al.*, 1990; Dunlap *et al.*, 1986) within the neuston have not, to my knowledge, been performed.

Photochemical reactions

Blough and Zepp (1995) have recently reviewed the methods of detection, sources and sinks, and environmental importance of photogenerated reactive oxygen species and related transient intermediates in natural waters, and so only an abbreviated account of these processes will be provided here. Instead, emphasis will be placed on reactions not covered in that review, and on available quantitative information that can be used to model photochemical processes both in the microlayer and bulk waters. To aid this discussion, major reactions are provided in Figures 13.3–13.5. Brief descriptions of the sources and sinks of major photochemical intermediates and known products are provided below. Other recent reviews and reports include those of Helz *et al.* (1993), Zepp (1991), Blough and Zepp (1990), Zafiriou *et al.* (1990) and Zika and Cooper (1987). A series of articles reporting the results of a major field experiment has also appeared recently (Zika *et al.*, 1993).

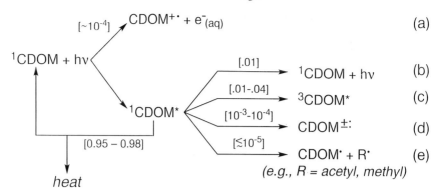

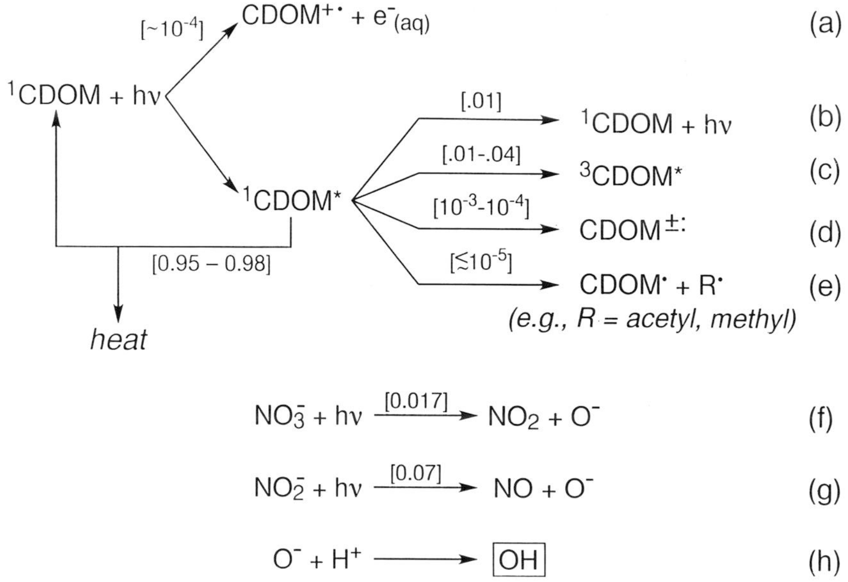

$$^1CDOM + h\nu \xrightarrow{[\sim 10^{-4}]} CDOM^{+\cdot} + e^-_{(aq)} \tag{a}$$

$$^1CDOM^* \xrightarrow{[.01]} {}^1CDOM + h\nu \tag{b}$$

$$^1CDOM^* \xrightarrow{[.01-.04]} {}^3CDOM^* \tag{c}$$

$$^1CDOM^* \xrightarrow{[10^{-3}-10^{-4}]} CDOM^{\pm :} \tag{d}$$

$$^1CDOM^* \xrightarrow{[\leq 10^{-5}]} CDOM^\cdot + R^\cdot \tag{e}$$

$$(e.g., R = acetyl, methyl)$$

$$^1CDOM^* \xrightarrow{[0.95-0.98]} heat$$

$$NO_3^- + h\nu \xrightarrow{[0.017]} NO_2 + O^- \tag{f}$$

$$NO_2^- + h\nu \xrightarrow{[0.07]} NO + O^- \tag{g}$$

$$O^- + H^+ \longrightarrow \boxed{OH} \tag{h}$$

Figure 13.3. Primary photophysical and photochemical processes thought to occur in natural waters. The values in brackets are approximate quantum yields for these reactions. Adapted from Blough and Zepp (1995).

$$^3CDOM^* + {}^3O_2 \xrightarrow{[0.25-1]} {}^1CDOM + \boxed{^1O_2} \tag{a}$$

$$^3CDOM^* + {}^3O_2 \xrightarrow{?} {}^1CDOM + {}^3O_2 + heat \tag{b}$$

$$^3CDOM^* + {}^3O_2 \xrightarrow{?} CDOM^{+\cdot} + \boxed{O_2^-} \tag{c}$$

$$+ H_2O \xrightarrow{?} CDOMH^\cdot + \boxed{OH} \tag{d}$$

$$CDOM^{\pm :} + O_2 \longrightarrow CDOM^{+\cdot} + \boxed{O_2^-} \tag{e}$$

$$CDOM^{+\cdot} + O_2 \longrightarrow ? \tag{f}$$

$$CDOM^\cdot + O_2 \longrightarrow \boxed{CDOM(O_2)} \nearrow CDOM_{OX} + \boxed{HO_2} \tag{g}$$

$$R^\cdot + O_2 \longrightarrow \boxed{RO_2^\cdot} \longrightarrow Further\ Reaction \tag{h}$$

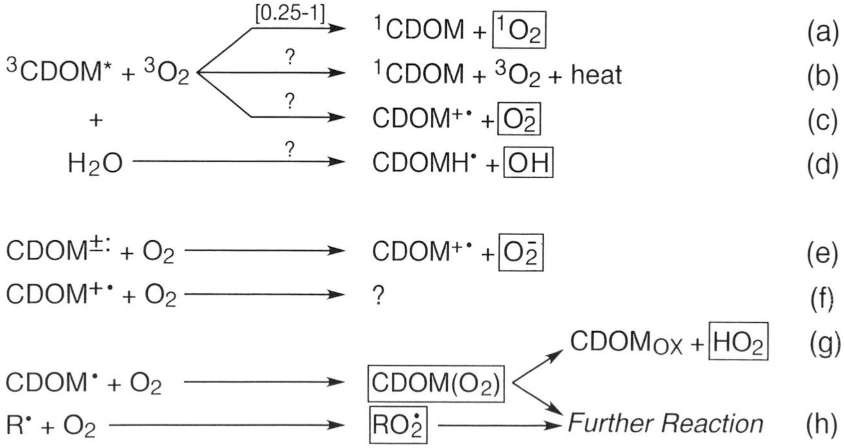

Figure 13.4. Secondary reactions producing reactive oxygen species (outlined) in natural waters. See text for additional reactions. Adapted from Blough and Zepp (1995).

$$^1O_2 \longrightarrow {}^3O_2 + \textit{heat} \tag{a}$$

$$2\,O_2^{\cdot -} + 2H^+ \longrightarrow O_2 + \boxed{H_2O_2} \tag{b}$$

$$O_2^{\cdot -} + Y \begin{cases} \longrightarrow O_2 + Y^- & \text{(c)} \\ \longrightarrow YO_2^- & \text{(d)} \end{cases}$$

$$H_2O_2 + M^{n+} \longrightarrow M^{(n+1)+} + \boxed{OH} + OH^- \tag{e}$$

$$H_2O_2 + h\nu \longrightarrow 2\,\boxed{OH} \tag{f}$$

Biological
$$\begin{cases} 2\,H_2O_2 \xrightarrow{\textit{Catalase}} O_2 + 2\,H_2O & \text{(g)} \\ H_2O_2 + XH \xrightarrow{\textit{Peroxidase}} \text{Oxidation Products} + H_2O & \text{(h)} \end{cases}$$

$$RO_2\cdot + R'H \xrightarrow[\textit{Abstraction}]{\textit{H-atom}} \boxed{RO_2H} + R'\cdot \tag{i}$$

$$2\,RO_2\cdot \xrightarrow{\textit{Termination}} \text{Non-Radical and} \tag{j}$$
$$\text{Non-Peroxidic Products}$$

$$RO_2^{\cdot} + HO_2^{\cdot} \longrightarrow \boxed{RO_2H} + O_2 \tag{k}$$

Freshwater
$$\begin{cases} OH + DOM \begin{cases} \longrightarrow DOM^{\cdot} + H_2O & \text{(l)} \\ \longrightarrow HO - DOM^{\cdot} & \text{(m)} \end{cases} \\ OH + CO_3^= \longrightarrow CO_3^{\cdot -} + OH^- & \text{(n)} \end{cases}$$

Seawater
$$\begin{cases} OH + Br^- \longrightarrow Br^{\cdot} + OH^- & \text{(o)} \\ Br^{\cdot} + Br^- \longrightarrow Br_2^{\cdot -} & \text{(p)} \end{cases}$$

Figure 13.5. Decay routes and secondary formation pathways for reactive oxygen species in natural waters. See text for additional reactions. Adapted from Blough and Zepp (1995).

Singlet dioxygen ($^1O_2$; $^1\Delta_g$)

$^1O_2$ is formed primarily through energy transfer from the excited triplet states of CDOM to ground state dioxygen, $^3O_2$ (Figure 13.4a), and wavelengths in the the UV-A and UV-B are most effective in its formation (Zepp *et al.*, 1985). Quantum yields (the fraction or percentage of absorbed photons which give rise to products) range from ~1 to 3% and generally decrease with increasing wavelength (Haag *et al.*, 1984).

Because the decay of $^1O_2$ by solvent relaxation is very rapid ($k=2.5\times10^5$ s^{-1}; Figure 13.5a), its loss is dominated by this pathway; the levels of $^1O_2$-reactive constituents in natural waters are generally too low to affect its steady-state concentration significantly. Whether this is also true in the microlayer depends on the concentration and reactivity of the materials found therein. However, even for very reactive materials (eg., $k\sim10^9$ M^{-1} s^{-1}), local concentrations of ~25 to 100 μM would be required to produce an ~10 to 70% reduction, respectively, in the steady-state concentration of $^1O_2$. Assuming a steady-state $^1O_2$ concentration of 10^{-14} M in the microlayer (*vide infra*), lifetimes for highly reactive materials ($k\sim10^9$ M^{-1} s^{-1}) would be of the order of 10^5 s, approximately five to six orders of magnitude longer than the likely residence time of (low molecular weight) materials in the microlayer. Thus, it seems unlikely that $^1O_2$ reactions could be very important in the microlayer, except in microenvironments such as those found within cells or particulate matter where locally high concentrations of $^1O_2$-reactive materials may be present (Nelson, 1993; Nelson and Roberston, 1993; Lee and Baker, 1992). Midday steady-state concentrations of $^1O_2$ range from ~10^{-13} to 10^{-15} M in surface waters, depending primarily on the chromophore content (Zepp *et al.*, 1985).

Superoxide/hydroperoxyl radical (O_2^-/HO_2)

CDOM is known to be the primary source of O_2^-, although the precise reactions producing this species remain unclear. Work by Zepp *et al.* (1987a) indicates that the hydrated electron ($e^-(aq)$) is not produced at rates sufficient to account for the production of H_2O_2 via Equations 13.3 and 13.4 (see also Figure 13.3a):

$$e^-(aq) + O_2 \rightarrow O_2^- \qquad k=2.2\times10^{10} \text{ M}^{-1} \text{ s}^{-1} \qquad (13.3)$$

$$2\,O_2^- + 2\,H^+ \rightarrow H_2O_2 \qquad k=6\times10^4 \text{ M}^{-1} \text{ s}^{-1} \text{ (pH=8)} \quad (13.4)$$

Some workers have suggested that O_2^- is formed by direct electron transfer from the excited triplet states of CDOM to O_2 (Figure 13.4c), although reduction of O_2 by radicals or radical ions produced by intramolecular electron transfer reaction reactions (Figures 13.3d, 13.4e), H-atom abstractions and/or homolytic bond cleavages (Figures 13.3e, 13.4g) is equally, if not more, plausible (Blough and Zepp, 1995).

Quantification of the photochemical production of O_2^- has been complicated in the past by the lack of simple methods to measure this species (Blough and Zepp, 1995). Recent work by our group has shown that stable nitroxide radicals act very similarly to O_2 in their reactions with one-electron reductants produced photochemically in natural waters. Rates for the anaerobic, one-electron reduction of the nitroxide, 3-aminomethyl-2,2,5,5-tetramethyl-1-pyrrolidinyloxy free radical (3-amp), to its corresponding hydroxylamine are approximately twice those recorded for the formation of H_2O_2 (Figure 13.6; see also Kieber and Blough, 1990a,b). These results are consistent with the stoichiometry of Equation 13.4 and with the 3-amp acting as a one-electron acceptor in the place of O_2. The observation of rate ratios greater than two in some cases may be due to the loss of O_2^- by pathways competitive with its dismutation to H_2O_2 (Equation 13.4), as previously reported by Petasne and Zika (1987) and Zafiriou (1990) (*vide infra*). The very similar spectral dependence of the quantum yields for formation of the hydroxylamine and H_2O_2 provides further evidence that the rates of anaerobic nitroxide reduction provide good estimates for the *primary* rates of O_2^- production in natural waters (Figures 13.7–13.9). Mid-day rates vary from lows of $\sim 10^{-11}$ M s^{-1} for oligotrophic waters to highs of $\sim 10^{-8}$ M s^{-1} for the more highly coloured coastal waters (Blough and Kieber, 1992).

In the absence of O_2^--reactive compounds, the loss of O_2^- is dominated by dismutation (Equation 13.4) with its steady-state concentration ($[O_2^-]_{ss}$) given by the relation

$$[O_2^-]_{ss} = (F/k_d)^{1/2} \tag{13.5}$$

where F is the formation rate (M s^{-1}) and k_d is the dismutation rate constant (M^{-1} s^{-1}). This equation indicates that $[O_2^-]_{ss}$ could range as high as 400 nM in some surface waters. However, transition metal complexes having one-electron reduction potentials falling between the O_2/O_2^- and O_2^-/H_2O_2 couples can catalyze dismutation through the reactions

$$M^n + O_2^- - k_r \rightarrow M^{n-1} + O_2 \tag{13.6}$$

$$M^{n-1} + O_2^- + 2H^+ - k_o \rightarrow M^n + H_2O_2 \tag{13.7}$$

where k_r and k_o are the second-order rate constants for the reaction of O_2^- with the oxidized (M^n) and reduced (M^{n-1}) metal, respectively. At steady-state, the fraction of metal ion in its oxidized form is given by

$$f_{ss} = [M^n]/[M_T] = 1/(1+r) \tag{13.8}$$

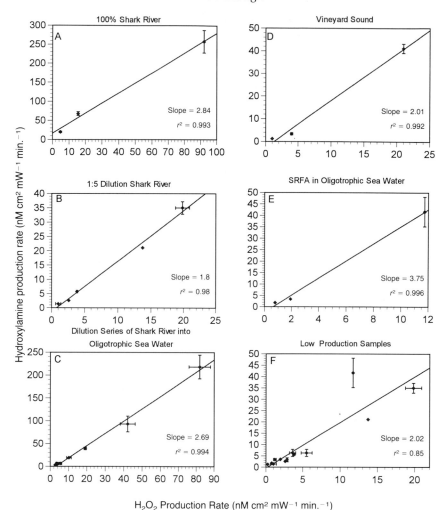

H_2O_2 Production Rate (nM cm² mW⁻¹ min.⁻¹)

Figure 13.6. Relationship between the photochemical rates for the anaerobic, one-electron reduction of 3-amp to form the hydroxylamine and the aerobic production of H_2O_2, as normalized to the incident light intensity. (A) Shark River water collected from a saline estuary draining the Everglades and irradiated at $\lambda=300\pm5$, 310 ± 5, 350 ± 10, 375 ± 10, 400 ± 10 nm. (B) 1:5 dilution of Shark River water into Sargasso seawater, irradiated at the same wavelengths as A. (C) Serial dilutions of Shark River water into Sargasso seawater irradiated at 310 ± 5 nm. (D) Vineyard Sound seawater irradiated at $\lambda=300\pm5$, 350 ± 10 and 400 ± 10 nm. (E) A solution of Suwannee River fulvic acid in Sargasso seawater irradiated at $\lambda=300\pm10$, 350 ± 10 and 400 ± 10 nm. (F) Low production samples include the less coloured coastal waters such as Vineyard Sound and all samples irradiated at $\lambda\geq350$ nm.

where $r=k_r/k_o$ and M_T is the total (electroactive) metal concentration. $[O_2^-]_{ss}$ is then given by

$$[O_2^-]_{ss} = \frac{k_o[M_T]r}{k_d(1+r)} \left\{ \sqrt{\left[1 + \frac{(1+r)^2 k_d F}{(k_o[M_T]r)^2} \right]} - 1 \right\}$$ (13.9)

This equation shows that even very low metal concentrations are capable of reducing $[O_2^-]_{ss}$ substantially, depending on the values of r and k_o (Figure 13.10).

Because of its high rate constant (Huie and Padmaja, 1993), the reaction between O_2^- and NO to form peroxynitrite (Blough and Zafiriou, 1985) is likely to be important both in surface waters and the microlayer:

$$O_2^- + NO \rightarrow {}^-OONO \quad k=6.7(\pm0.9)\times10^9 \text{ M}^{-1}\text{ s}^{-1}$$ (13.10)

During the day, this reaction may control surface [NO] as well as act as an irreversible sink of NO entering the microlayer. For example, assuming surface $[O_2^-]_{ss}= 1$ nM, the lifetime of NO would be 0.15 s. If we further assume that NO is formed only by NO_2^- photolysis in the upper water column at a midday rate of $\sim10^{-11}$ M s^{-1} (Zafiriou and True, 1979a; *vide infra*), then the surface [NO] would be limited to 1.5×10^{-12} M. Conversely, the observation by Zafiriou and McFarland (1981) that the average NO production rate and $[NO]_{ss}$ are 1.2×10^{-12} M s^{-1} and 4.8×10^{-11} M, respectively, in the central eastern Pacific argues that $[O_2^-]_{ss}$ cannot exceed 3.7×10^{-12} M in these waters. This result is surprising, since it implies that the O_2^- production rate must be $<1.2\times10^{-12}$ M s^{-1}, or that another efficient sink of O_2^- exists in these waters. Because transition metal concentrations are very low in this environment, it seems unlikely that this sink is due to metal-catalyzed dismutation (Equation 13.9; Figure 13.10).

The product of Equation 13.10, peroxynitrite, may act as a oxidant or produce strong oxidants during its decay thereby initiating secondary reactions. These reactions will be discussed in more detail below.

Another potentially important sink for O_2^- is through electron transfer and coupling reactions with phenoxy radicals (Jonnson et al., 1993; Jin et al., 1993) produced by the photoionization of phenols (Zepp et al., 1987a), or the oxidation of phenols by peroxy radicals (RO_2; Faust and Hoigné, 1987) or other photogenerated species.

The observed rate constants for O_2^-/phenoxy radical reactions are given by the sum of the rate constants for Equations 13.14 and 13.15, and range from $\sim5\times10^8$ M^{-1} s^{-1} to 3×10^9 M^{-1} s^{-1} depending on the

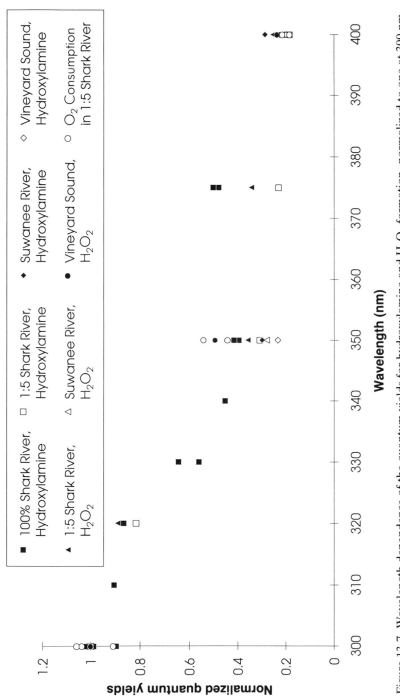

Figure 13.7. Wavelength dependence of the quantum yields for hydroxylamine and H_2O_2 formation, normalized to one at 300 nm.

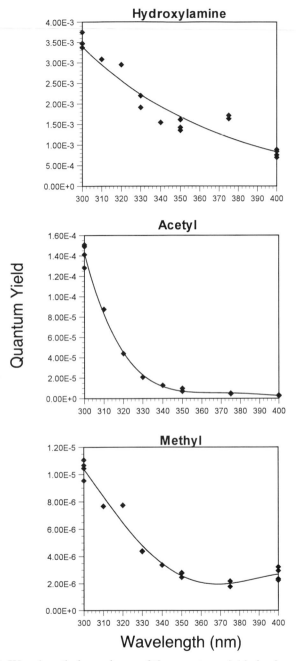

Figure 13.8. Wavelength dependence of the quantum yields for formation of the hydroxylamine and the acetyl and methyl radicals in Shark River water, as determined by the nitroxide trapping method.

$$+ \quad \text{hv} \quad \longrightarrow \quad + \quad e^-(aq) + H^+ \quad (13.11)$$

$$+ \quad RO_2 \quad \longrightarrow \quad + \quad RO_2H \quad (13.12)$$

$$(13.13)$$

$$+ \quad O_2^- \quad \xrightarrow{+H^+} \quad + \quad O_2 \quad (13.14)$$

$$\longrightarrow \quad \text{oxygenated products} \quad (13.15)$$

substituents, X (Jonnson et al., 1993). The partitioning between the electron transfer and addition pathways depends on the reduction potential of the phenoxy radical, with electron transfer becoming more favourable at higher potentials.

Warneck and Wurzinger (1988) have suggested recently that O_2^- also reacts with NO_2:

$$O_2^- + NO_2 \rightarrow NO_2^- + O_2 \quad k \approx 10^8 \ M^{-1} \ s^{-1} \quad (13.16)$$

If so, NO_2 formed by NO_3^- photolysis and through secondary reactions of NO_2^- photolysis (vide infra) may be converted rapidly to NO_2^- by this reaction.

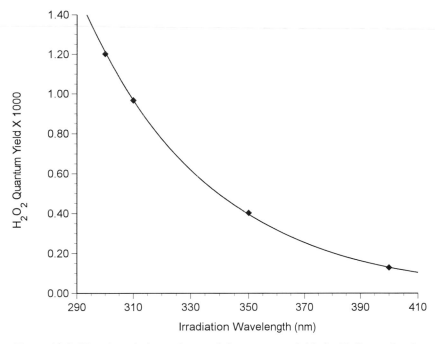

Figure 13.9. Wavelength dependence of the quantum yields for H_2O_2 production in a 1 : 5 dilution of Shark River water in Sargasso seawater. This dependence is very similar to that recently reported by Sikorski and Zika (1993).

Because the rates of peroxy radical loss by H-atom abstraction (Figure 13.5i) and disproportionation (Figure 13.5j) tend to be relatively slow in dilute solution, the reaction of HO_2/O_2^- with peroxy radicals is also likely to be important (Figure 13.5k; von Sonntag and Schuchmann, 1991). Due to the dependence of the rate constant on the redox potential of the peroxy radical, more highly oxidizing radicals such as the acetylperoxy ($k\sim10^9$ M^{-1} s^{-1}) will be affected more by this reaction than weakly oxidizing radicals such as the α-hydroxyethylperoxy ($k\sim10^7$ M^{-1} s^{-1}). Note that this reaction, as well as Equations 13.10, 13.14, 13.15 and 13.16, divert O_2^- from its path to H_2O_2, and thus may account in part for the competing sink(s) of O_2^- reported to occur in coastal seawaters by Petasne and Zika (1987). Of these reactions, only Equation 13.15 leads directly to the further oxidation of an organic species.

O_2^- is a very poor nucleophile in aqueous solutions and reacts principally as a one-electron reductant or oxidant (Sawyer and Valentine,

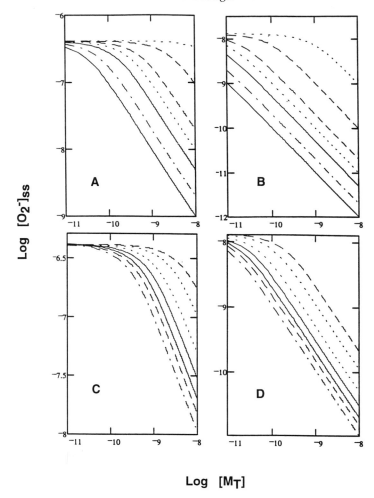

Log [M$_T$]

Figure 13.10. Dependence of the steady-state O_2^- concentration, $[O_2^-]$ (M), on transition metal concentrations, $[M_T]$ (M), for different values of r, k_o and F (Equation 13.9). (A) Parameter values are $F = 1 \times 10^{-8}\,M\,s^{-1}$, $r = 1$, $k_d = 6 \times 10^4\,M^{-1}\,s^{-1}$ and, from bottom to top, $k_o = 1 \times 10^9$, 5×10^8, 2×10^8, 1×10^8, 5×10^7, 1×10^7 and $1 \times 10^6\,M^{-1}\,s^{-1}$. (B) Same as A, except $F = 1 \times 10^{-11}\,M\,s^{-1}$. (C) Parameter values are $F = 1 \times 10^{-8}\,M\,s^{-1}$, $k_o = 5 \times 10^7\,M^{-1}\,s^{-1}$, $k_d = 6 \times 10^4\,M^{-1}\,s^{-1}$ and, bottom to top, $r = 20, 2, 1, 0.5, 0.25, 0.1$ and 0.05. (D) Same as C, except $F = 1 \times 10^{-11}\,M\,s^{-1}$.

1981). However, this species is much more reactive in organic solvents and can displace chloride ion from chlorinated hydrocarbons:

$$O_2^- + Cl\text{-}CCl_3 \rightarrow .O_2\text{-}CCl_3 + Cl^- \qquad (13.17)$$

If portions of the microlayer mimic the aprotic environment of an organic phase, reactions of this sort may be possible.

Hydrogen peroxide (H_2O_2)

Although some microorganisms do produce minor amounts of H_2O_2 in the ocean (Palenik *et al.*, 1987), this species is formed primarily through the dismutation of O_2^- (Figures 13.7–13.9). Thus, at steady-state and in the absence of competitive sinks for O_2^- (e.g., Equations 13.14–13.16), the rate of H_2O_2 formation will be half that of O_2^- (Equation 13.4).

Bacteria or small phytoplankton provide the principal sink for H_2O_2 (Lean *et al.*, 1992). Work by Moffet and Zafiriou (1990) and Cooper and Zepp (1990) indicates that two pathways are involved: a catalase-type pathway in which H_2O_2 is disproportionated to O_2 and H_2O (Figure 13.5g) and a peroxidase-type pathway in which H_2O_2 is reduced to water while an organic substrate is oxidized (Figure 13.5h). The peroxidase pathway thus leads to an additional two-electron oxidation, while the catalase pathway does not. How the relative proportions of these two pathways vary in the oceans is not well known due to the lack of measurements. H_2O_2 concentrations in surface seawaters (~50–200 nM) do not exhibit extreme variations between coastal and oligotrophic waters despite large differences in formation rates (Zika *et al.*, 1985a,b; Moore *et al.*, 1993), indicating that changes in the H_2O_2 decomposition rates largely compensate. Whether similar compensatory mechanisms operate via the biota in the microlayer is not known.

Hydroxyl radical (OH)

Although a substantial number of photochemical equivalents flow through O_2^- to H_2O_2, little of this H_2O_2 is reduced to form the strong oxidant, OH, except in waters containing higher concentrations of transition metals (*vide infra*). Instead, this species is produced primarily through other mechanisms such as the photolysis of nitrite (Zafiriou, 1974; Figures 13.3g, 13.11):

$$NO_2^- + h\nu \rightarrow NO + O^- \tag{13.18}$$

$$O^- + H^+ \rightarrow OH \tag{13.19}$$

In seawaters, OH reacts rapidly with Br^- ultimately to form Br_2^- (Zafiriou *et al.*, 1987)

$$OH + Br^- \rightarrow Br + OH^- \quad k=1.1 \times 10^{10} \ M^{-1} \ s^{-1} \tag{13.20}$$

$$Br + Br^- \rightarrow Br_2^- \quad k=1.1 \times 10^{10} \ M^{-1} \ s^{-1} \tag{13.21}$$

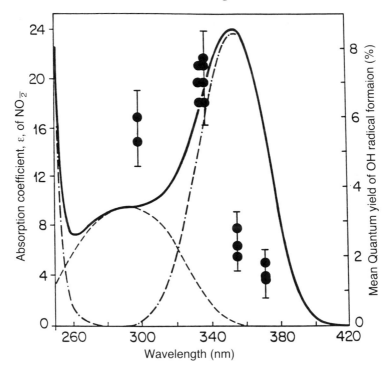

Figure 13.11. Wavelength dependence of the quantum yield for OH formation (●) from the photolysis of NO_2^-, as compared with the absorption spectrum of NO_2^- (solid line). Reprinted from Zafiriou and Bonneau (1987), copyright 1987, with kind permission from Elsevier Science Ltd.

while Br_2^- is thought to react, in turn, with carbonate (True and Zafiriou, 1987):

$$Br_2^- + CO_3^{-2} \rightarrow 2\,Br^- + CO_3^- \qquad (13.22)$$

The carbonate radical may self-terminate in competition with its oxidation of organic substrates (S):

$$2\,CO_3^- + 2\,H^+ \rightarrow 2\,CO_2 + H_2O_2 \quad k \approx 10^6 - 10^7\,M^{-1}\,s^{-1} \quad (13.23)$$

$$CO_3^- + S \rightarrow S_{ox} + CO_3^{-2} \qquad (13.24)$$

Equations 13.20–13.23 sum formally to the process

$$2\,OH \rightarrow H_2O_2 \qquad (13.25)$$

The reaction with O_2^- represents another potentially important sink for both Br_2^- and CO_3^- (Ross et al., 1992):

$$O_2^- + Br_2^- \rightarrow O_2 + 2\,Br^- \quad k=1.7\times10^8\,M^{-1}\,s^{-1} \quad (13.26)$$

$$O_2^- + CO_3^- \rightarrow O_2 + CO_3^{-2} \quad k=6.5\times10^8\,M^{-1}\,s^{-1} \quad (13.27)$$

Note that these reactions also represent an additional sink of O_2^- which does not lead to H_2O_2.

The fate of the other product of nitrite photolysis, NO, is likely to be dominated by reaction with O_2^- (Equation 13.10). The product of this reaction, peroxynitrite, is also a strong oxidant (1.4 V; Koppenol *et al.*, 1992), which undergoes the following reactions (Logager and Sehested, 1993; Yang *et al.*, 1992; Mahoney, 1970):

$$^-OONO + H^+ \rightarrow HOONO \quad pK=6.5 \quad (13.28)$$

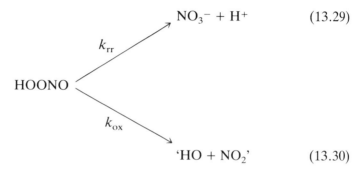

$$NO_3^- + H^+ \quad (13.29)$$

$$k_{rr}$$

$$HOONO$$

$$k_{ox}$$

$$\text{`}HO + NO_2\text{'} \quad (13.30)$$

where $k_{iso}= k_{rr} + k_{ox}= 1.0 \pm 0.2\,s^{-1}$ and the observed rate constant, k_{obs}, is given by,

$$k_{obs}= k_{iso}/\,(1 + K_a/[H^+]) \quad K_a= 1 \pm 0.3 \times 10^{-7} \quad (13.31)$$

Protonation of peroxynitrite yields peroxynitrous acid (Equation 13.28), which decomposes to form nitrate (~60% yield) in competition with the formation of HO and NO_2 or a species exhibiting very similar reactivity (~40% yield; Equations 13.29, 13.30; Yang *et al.*, 1992; Koppenol *et al.*, 1992). OH reacts further through Equation 13.20, while NO_2 is likely to be consumed through the reactions

$$2\,NO_2 \rightarrow N_2O_4 \quad k_f= 4.5 \times 10^8\,M^{-1}\,s^{-1}$$
$$\leftarrow \quad k_r= 7 \times 10^3\,s^{-1} \quad (13.32)$$

$$N_2O_4 + H_2O \rightarrow NO_2^- + NO_3^- + 2\,H^+ \quad k= 1 \times 10^3\,M^{-1}\,s^{-1} \quad (13.33)$$

$$NO_2 + O_2^- \rightarrow NO_2^- + O_2 \quad k\approx 10^8\,M^{-1}\,s^{-1} \quad (13.34)$$

$$NO_2 + S \rightarrow NO_2^- + S_{ox} + H^+ \quad k=? \tag{13.35}$$

If NO_2 acts principally as a one-electron oxidant (Equations 13.34, 13.35), up to 40% of the NO formed originally could be regenerated as NO_2^- to re-enter the photolytic cycle.

The peroxynitrite anion itself may be an important seawater oxidant in areas of high nitrite concentration. Although reactivity and rate data are as yet limited, this species is known to react with a variety of organic compounds (Yang *et al.*, 1992), including thiol containing compounds ($k = 5.9 \times 10^4\,M^{-1}\,s^{-1}$ for cysteine; Radi *et al.*, 1991). Because of its long lifetime at alkaline pH ($\sim$11 s at pH=8; Equation 13.31), steady-state concentrations of this oxidant may reach 0.1 nM, based on the mid-day rate of $10^{-11}\,M\,s^{-1}$ reported by Zafiriou and True (1979a) for nitrite photolysis. At this level, compounds exhibiting rate constants as low as that reported for cysteine could be consumed significantly over a week ($\sim$47 h lifetime under midday sun).

The photolysis of NO_3^- also generates OH as well as NO_2 (Figure 13.3f). Subsequent reactions of NO_2 can produce NO_2^- and NO_3^- (Equations 13.32–13.35), thereby coupling the dynamic cycles of NO_2^- and NO_3^- photolysis (Zafiriou and True, 1979b). The slow, net photochemical conversion of NO_2^- to NO_3^- reported by Zafiriou and True (1979b) is explainable by Equations 13.28–13.30 and 13.32–13.35, the lower quantum yield for NO_3^- (Figure 13.3f; Zafiriou and Bonneau, 1987; Zellner *et al.*, 1990), and the weaker absorption band of NO_3^- in the UV-B and far UV-A where fewer solar photons are available for reaction.

Although the oxidation of reduced metals such as Fe(II) and Cu(I) by H_2O_2 can also produce OH (Figure 13.5e; Moffett and Zika, 1987; Millero and Sotolongo, 1989; Zepp *et al.*, 1992), this process will be limited to those waters where significant levels of reduced metals are formed. The photolysis of H_2O_2 also is not a significant source of OH (Figure 13.5f) despite the high quantum yield for this reaction (0.98) due to the poor overlap between its absorption spectrum and the ground-level solar spectrum.

A major, previously unrecognized source of OH has recently been reported by Zhou and Mopper (1990) and Mopper and Zhou (1990). The OH appears to arise from the CDOM, but the mechanism of its formation remains unknown. Wavelengths in the UV-B have been reported to be the most effective in its formation (Figure 13.12).

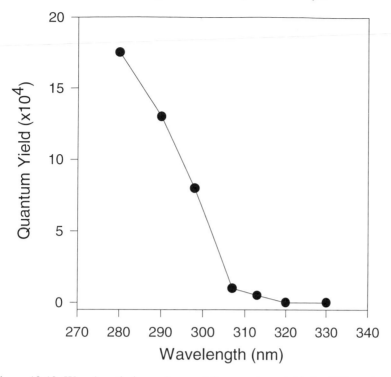

Figure 13.12. Wavelength dependence of the quantum yields for OH production in a coastal seawater. Adapted from Mopper and Zhou (1990).

Peroxy radicals (RO$_2$)

Peroxy radicals are thought to be formed by O$_2$ addition to primary (carbon-centered) radicals (R•; Figure 13.4g,h) produced photochemically from CDOM via intramolecular H-atom abstractions, electron transfer reactions (Figure 13.3d), and homolytic bond cleavages (Figure 13.3e):

$$\text{R}\cdot + \text{O}_2 \rightarrow \text{RO}_2 \qquad k \sim 10^9 \text{ M}^{-1} \text{ s}^{-1} \qquad (13.36)$$

These species can also be generated from secondary radicals produced by H-atom abstraction or addition reactions of RO$_2$ with DOM (Figures 13.4g,h, 13.5j), or one-electron oxidations of DOM by CO$_3^-$ or Br$_2^-$. Earlier work by Mill *et al.* (1980) and Faust and Hoigné (1987) provided evidence for the photochemical production of this class of radicals, but could not supply information on the specific RO$_2$ that were formed.

Indirect evidence for the formation of specific RO$_2$ has been acquired

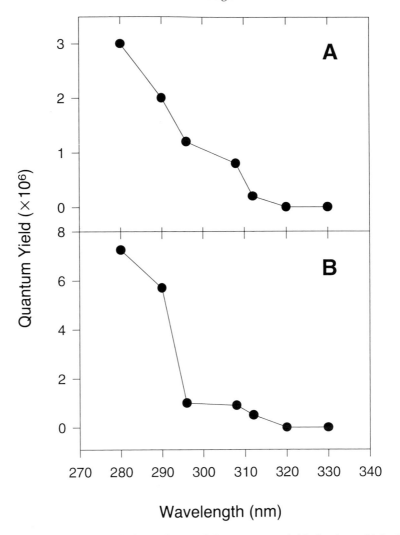

Figure 13.13. Wavelength dependence of the quantum yields for formaldehyde production in Biscayne Bay seawater (A) and Everglades freshwater (B). Adapted from Kieber, R. J. *et al.* (1990).

by employing stable nitroxide radicals to trap carbon-centered radicals, the immediate precursors to the RO_2 (Blough, 1988). Using a highly sensitive fluorescence detection scheme (Blough and Simpson, 1988) combined with high performance liquid chromatography (Kieber and Blough, 1990a,b), a number of low-molecular-weight radicals have been detected in seawaters, principally the acetyl and methyl radicals (Blough and Kieber, 1992). Evidence for the formation of radical centres on high-molecular-weight CDOM has also been obtained, although their yield appears to be significantly less than that of the low-molecular-weight species (Kieber and Blough, 1992).

Quantum yields for the formation of the major species, the acetyl and methyl radicals, exhibit a different wavelength dependence and are ~ 10- and 100-fold lower, respectively, than those for nitroxide reduction (Figure 13.8). These results imply that O_2^- production dominates RO_2 formation in surface seawaters (*vide supra*; Equation 13.36) and that different chromophores within the CDOM are responsible for the formation of these radicals. Estimated midday rates for formation of the acetylperoxy radical in natural waters range from 10^{-13} to 10^{-11} M s^{-1} as compared with 10^{-11} to 10^{-8} M s^{-1} for O_2^- (Blough and Zepp, 1995).

Because of the high rate constant, a dominant sink for the acetylperoxy radical may be the reaction with O_2^- to form peroxyacetic acid:

$$CH_3C(O)OO\bullet + O_2^- + H^+ \rightarrow$$
$$CH_3C(O)OOH + O_2 \quad k \sim 10^9 \text{ M}^{-1} \text{ s}^{-1} \quad (13.37)$$

Because the formation rate of the acetylperoxy radical is much lower than that of O_2^- it would appear that no more than $\sim 10\%$ of the O_2^- flux could be diverted to O_2 by this reaction. Given the lower reactivity of the methylperoxy radical with O_2^-, the fate of this species as well as the high-molecular-weight radical centres is less clear. Possible pathways include termination reactions to form non-radical and non-peroxidic products (Figure 13.5j) and H-atom abstractions or addition reactions to form organic peroxides and secondary radicals (Figure 13.5i).

Low molecular weight organic compounds and trace gases

The photolysis of CDOM produces a suite of low-molecular-weight (LMW) organic compounds (Kieber, D. J. *et al.*, 1989; Kieber, R. J. *et al.*, 1990; Mopper *et al.*, 1991), presumably due to radical and fragmentation reactions arising from the net oxidative flow of electrons from CDOM to O_2 (Figures 13.3–13.5). The principal products include formaldehyde, acetaldehyde, glyoxal, glyoxylate, pyruvate, acetone and methylglyoxal. Wavelengths in the UV-B are the most effective in forming these species; production rates fall off very rapidly with increasing wavelength in a fashion similar to that observed for the acetyl and methyl radicals (Figures 13.8, 13.13). Quantum yields for the formation of the likely secondary products of these radicals, the methylperoxy radical and peracetic acid, are approximately one and two orders of

magnitude larger, respectively, than the approximate yields for the
LMW organic compounds ($\leq 10^{-6}$, see Figure 13.13; Kieber *et al.*, 1990).
Thus, the levels of the methylperoxy radical and peracetic acid are more
than adequate to support the formation of the one- and two-carbon
products (eg., formaldehyde and acetaldehyde), which may arise from
further reactions of these intermediates. For example, termination of two
methylperoxy radicals can produce formaldehyde and methanol:

$$2 \, CH_3OO \rightarrow CH_3OH + CH_2O + O_2 \tag{13.38}$$

The possible source of the three-carbon products (eg., pyruvate and
acetone) is less obvious, but they could arise from other trace radicals that
are often observed by the nitroxide trapping method. Quantum yields for
O_2^- and H_2O_2 production are about three orders of magnitude larger
than those for the LMW organic compounds (Figures 13.8, 13.9, 13.13),
so that even the sum of the production rates for the known LMW
compounds is very small with respect to the flux of photochemical
equivalents from CDOM to O_2.

The LMW organic compounds are taken up by bacteria and respired to
CO_2 (Kieber *et al.*, 1989). This coupling between the photochemical
degradation of the biologically refractive CDOM and the bacterial
uptake of the LMW products represents a potentially important route for
the loss of CDOM within the oceans and may have a significant influence
on the structure of the microheterotrophic community in the upper ocean
(Kieber *et al.*, 1989).

Carbon monoxide (CO) is a major product of CDOM photolysis,
existing at supersaturated concentrations in the surface waters of most of
the earth's oceans (Conrad *et al.*, 1982; Gammon and Kelly, 1990; Jones,
1991; Jones and Amador, 1993). Recent estimates by Valentine and
Zepp (1993) indicate that up to 10–15% of the global emission of CO may
arise from CDOM photolysis in near-coastal/shelf areas. Quantum yields
are relatively high, ranging up to $\sim 2 \times 10^{-4}$ at 300 nm (Figure 13.14;
Valentine and Zepp, 1993). Interestingly and perhaps not fortuitously,
the magnitude and spectral dependence of the CO quantum yields are
similar to those for the formation of the acetyl radical over the 300–400
nm wavelength interval (Figures 13.8, 13.14). This result is suggestive of a
common photochemical origin for these species in this wavelength
regime. However, the presence of significant yields and a weaker spectral
dependence at visible wavelengths suggests that other pathways involving
different chromophores within the CDOM are also operative in CO
formation (Figure 13.14). The principal sinks for CO are biological

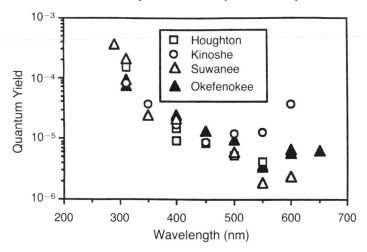

Figure 13.14. Wavelength dependence of the quantum yields for CO formation in various freshwaters. Reprinted with permission from Valentine and Zepp (1993). Copyright 1993, American Chemical Society.

oxidation to CO_2 (Jones, 1991; Jones and Amador, 1993) and release to the atmosphere, where it plays a significant role in controlling OH levels in the troposphere (Najjar *et al.*, 1994).

The photochemical formation of carbonyl sulphide (COS) in the oceans, primarily in the coastal/shelf regions, is probably the single, largest source of COS to the atmosphere, accounting for about one third of the total source strength (Khalil and Rasmussen, 1984; Andreae and Ferek, 1992). Through its oxidation in the stratosphere to form sulphate aerosol, this compound may be important in determining the earth's radiation budget and perhaps in regulating stratospheric ozone concentrations (Turco *et al.*, 1980; Rodriguez *et al.*, 1991). Work by Zepp and Andreae (1990) indicates that this species is formed by the (CDOM) photosensitized oxidation of organosulphur compounds such as mercaptans and cysteine; action spectra indicate that UV wavelengths are the most important in its formation. Not surprisingly, formation rates for COS are significantly lower than those of CO (Zepp and Andreae, 1990). The principal sinks of seawater COS are release to the atmosphere and hydrolysis (Najjar *et al.*, 1994).

Significant amounts of CO_2 may also be formed by the photolysis of CDOM (Miller and Zepp, 1992), although determining this yield in unaltered seawaters would be extremely difficult, if not impossible, due to the high pCO_2 background. Preliminary work on freshwater systems indicates that CO_2 production rates from Suwannee River water are ~20-

to 50-fold higher than that of CO (Miller and Zepp, 1992). These rates are surprisingly high considering that they must approach, if not exceed, the rate of electron flow to O_2 (Figures 13.8, 13.14).

Trace metals (Fe, Mn)

The recent suggestion that a lack of available Fe limits primary production in certain open ocean waters (Martin and Fitzwater, 1988) has intensified interest in the transport and photochemical reactions of Fe species in both seawaters (Wells *et al.*, 1991, Wells and Mayer, 1991) and atmospheric aerosols (Duce, 1986; Behra and Sigg, 1990; Zhuang *et al.*, 1990; Zhuang, 1992; Zuo and Hoigné, 1992; Zhu *et al.*, 1993). At the oxygen concentrations and pH of surface seawaters, very little biologically available Fe(III) is expected to exist due to the high thermodynamic stability of the colloidal iron (hydr)oxides. At lower pH where Fe(II) is more slowly oxidized by O_2 and H_2O_2 (Moffet and Zika, 1987; Millero and Sotolongo, 1989), evidence for the photoreductive dissolution of colloidal iron oxides by CDOM has been obtained by several groups (Waite and Morel, 1984; Sulzberger *et al.*, 1989). While photoreduction still occurs at the higher pH of seawater, the Fe(II) that is produced appears to be oxidized more rapidly than its detachment from the oxide surface, so that little if any escapes to solution (see, however, O'Sullivan *et al.*, 1991, and King *et al.*, 1991). Wells *et al.* (1991) have found that the chemical availability of iron as determined by chelation with 8-hydroxyquinoline (oxine) is strongly correlated with the growth rates of marine phytoplankton and that sunlight increases this availability. These workers have suggested that CDOM-driven cycles of light-induced reduction and rapid re-oxidation lead to the formation of amorphous Fe(III) precipitates that do serve as a source of available Fe (Waite, 1990; Wells and Mayer, 1991; Wells *et al.*, 1991).

Clear evidence for the photochemical generation of significant levels of Fe(II) in atmospheric aqueous phases (at lower pH) has been acquired by a number of groups (*vide supra*). The fate of this Fe after deposition to the sea surface is not clear. However, higher levels of CDOM within the microlayer could act to maintain or increase the biological availability of this Fe through complexation and photochemical reduction at this interface.

Manganese oxides undergo light-induced reductive dissolution in seawaters (Sunda *et al.*, 1983; Sunda and Huntsman, 1988). The product, Mn(II), is kinetically stable to oxidation in the absence of certain bacteria that are known to be photoinhibited in the upper ocean. These two effects of sunlight combine to produce a surface maximum in soluble Mn, unlike most transition metals which are depleted in surface waters due to biological removal processes. A more detailed discussion of transition metal photochemistry and its role in environmental processes is provided in Blough and Zepp (1995).

Estimated production rates and fluxes of photochemical species in the microlayer

The data presented in the previous section can be employed within a simple model to acquire estimates of photochemical formation rates and fluxes of species in the microlayer, if we assume that microlayer CDOM exhibits absorption spectra and quantum yields similar to those of the underlying bulk-waters. In this model, the downwelling solar irradiance at depth z, $E_d(\lambda,z)$, is provided by the relation

$$E_d(\lambda,z) = E_0(\lambda) \cdot e^{-K_d(\lambda) \cdot z} \tag{13.39}$$

where $E_0(\lambda)$ is the downwelling irradiance at the *surface* and $K_d(\lambda)$ is the diffuse attenuation coefficient given by

$$K_d(\lambda) = D_d \left(\sum_i a(\lambda)_i + \sum_i b_b(\lambda)_i \right) \tag{13.40}$$

where $a(\lambda)_i$ and $b_b(\lambda)_i$ are the absorption and backscattering coefficients, respectively, of the ith constituent. D_d is the distribution coefficient which accounts for the average path length of light in the water column (Preisendorfer, 1976), Neglecting skylight, $D_d \cong 1/\cos\theta$, where θ is the solar zenith angle. Although not necessary, for simplicity D_d is assumed to be one (sun directly overhead). Note that this formulation ignores decreases in $E_0(\lambda)$ due to surface reflection, which is small except for very large solar zenith angles (Zepp and Cline, 1977). This first-order model also assumes that the ocean surface is flat and that the contribution of upwelling irradiance is negligible relative to $E_d(\lambda,z)$.

The loss of photons through absorption and scattering by depth z, $L(\lambda,z)$, is given by

$$L(\lambda,z) = E_0(\lambda) - E_d(\lambda,z) = E_0(\lambda) \cdot (1 - e^{-K_d(\lambda) \cdot z}) \tag{13.41}$$

while the rate of photon loss with depth is provided by the relation

$$A(\lambda,z) = (dL/dz) = E_0(\lambda,z) \cdot K_d(\lambda) \cdot e^{-K_d(\lambda) \cdot z} \tag{13.42}$$

The wavelength dependence of the photochemical formation rate at depth z is given by the product of $A(\lambda,z)$, the fraction of photons absorbed by the photoreactive constituent, $a(\lambda)_i/K_d(\lambda)$, and the fraction of absorbed photons giving rise to products, ie. the photochemical quantum yield, $\phi(\lambda)$:

$$F(\lambda,z) = A(\lambda,z) \cdot \phi(\lambda) \cdot [a(\lambda)_i/K_d(\lambda)] = \\ E_0(\lambda) \cdot \phi(\lambda) \cdot a(\lambda)_i \cdot e^{-K_d(\lambda) \cdot z}. \quad (13.43)$$

Integration from the surface to depth z provides the photochemical flux over this interval:

$$Y(\lambda,z) = E_0(\lambda) \cdot \phi(\lambda) \cdot [a(\lambda)_i/K_d(\lambda)] \cdot (1 - e^{-K_d(\lambda) \cdot z}) \quad (13.44)$$

For the optically-thin microlayer (Figure 13.1), $K_d(\lambda) \cdot z \ll 1$, $e^{-K_d(\lambda) \cdot z} \approx 1 - Kd(\lambda) \cdot z$, and Equations 13.43 and 13.44 reduce to the form:

$$F_M(\lambda) = E_0(\lambda) \cdot \phi(\lambda) \cdot a(\lambda)_{i,M}, \quad (13.45)$$

$$Y_M(\lambda) = E_0(\lambda) \cdot \phi(\lambda) \cdot a(\lambda)_{i,M} \cdot z_M \quad (13.46)$$

where $a(\lambda)_{i,M}$ and z_M correspond to the values for the microlayer.

Integration over wavelength provides the total photochemical formation rate and flux in the microlayer:

$$F_M = \int_\lambda E_0(\lambda) \cdot \phi(\lambda) \cdot a(\lambda)_{i,M} d\lambda \quad (13.47)$$

$$Y_M = F_M \cdot z_M \quad (13.48)$$

For CDOM photoreactions, Y_M can be compared to the total water column flux ($z \to \infty$, Equation 13.44),

$$Y = \int_\lambda E_0(\lambda) \cdot \phi(\lambda) \, d\lambda \quad (13.49)$$

assuming that the absorption coefficient for CDOM, $a(\lambda)_{cm} \cong K_d(\lambda)$, which is not an unreasonable approximation for many coastal/shelf waters over the 300–400 nm wavelength range (Vodacek et al., 1994; Blough et al., unpublished observations).

Equations 13.45–13.48 indicate that F_M and Y_M are directly proportional to the absorption coefficients of the material in the microlayer. Thus, for a two-fold enrichment in absorption (Figure 13.1), F_M and Y_M will be twice as large as those obtained for an equivalent depth interval just beneath the microlayer. However, the dilution of material by sampling devices that collect over depths greater than that of the microlayer can produce significant underestimates of F_M and slight overestimates of Y_M. For example, if the true microlayer depth was 1 μm, the use of a 50 µm sampler would produce a 15-fold underestimate of F_M and a 3.3-fold overestimate of Y_M for an observed absorption enrichment of 1.4 relative to bulk water. These dilution effects can be calculated from the relation

$$a(\lambda)_M = a(\lambda)_B[1 + z_s/z_M(E-1)] \qquad (13.50)$$

where $a(\lambda)_M$ and $a(\lambda)_B$ are the microlayer and bulk absorption co-efficients, respectively, E is the apparent enrichment factor for samples collected from depth z_s, and z_M is the true microlayer depth.

$F_M(\lambda)$, F_M, and Y_M were calculated employing Equations 13.45–13.48, the fitted spectral dependencies of the quantum yield data in Figures 13.8, 13.9 and 13.14, and the downwelling solar irradiance and CDOM absorption spectra presented in Figure 13.15. $F_M(\lambda)$ for O_2^- (hydroxyl-amine formation), H_2O_2, CH_3 (CH_3OO), $CH_3C(O)$ ($CH_3C(O)OO$), and CO are presented in Figure 13.16, while the values of F_M and Y_M are presented in Table 13.1, along with Y (Equation 13.49) and the calculated fluxes for the atmospheric deposition of these species (Thompson and Zafiriou, 1983). Y_M was calculated assuming $z_M = 50$ µm. To mimic the higher levels of enrichment observed by Carlson (Figure 13.1), the CDOM absorption spectrum of a coastal water sample from the Delaware Bight (top spectrum, Figure 13.2) was multiplied by two (Figure 13.15). Note, however, that the highest levels of microlayer absorption occasionally observed by Carlson are estimated to be five- to ten-fold larger than that employed here.

This analysis indicates that atmospheric deposition of H_2O_2 and CH_3OO to the sea-surface microlayer will far outweigh their *in situ* photochemical production (Table 13.1), unless the microlayer contains material substantially more photoreactive than the bulk water CDOM on which these calculations are based. The fluxes of these species are also small with respect to the atmospheric deposition of ozone, a compound known to play a significant role in the release of volatile iodine from the sea surface (Garland and Curtis, 1981; Thompson and Zafiriou, 1983). However, the photochemical flux of O_2^- is estimated to be about an order of magnitude larger than the atmospheric flux of HO_2, and thus could be important in the microlayer due to its efficient reaction with NO (Equation 13.10), phenoxy radicals (Equations 13.14, 13.15), peroxy radicals (Equation 13.37), transition metal ions (Equations 13.6, 13.7), and possibly NO_2 (Equations 13.34) and chlorinated hydrocarbons (Equations 13.17). As previously suggested by Thompson and Zafirou (1983), the water column fluxes of these species are far larger than the sum of their atmospheric and microlayer fluxes (Table 13.1). However,

Table 13.1 *Calculated mid-day formation rates and fluxes for species generated photochemically within the microlayer and the water column, and through dry deposition from the atmosphere*

Species	F_M $(cm^{-3} s^{-1})$	Y_M $(cm^{-2} s^{-1})$[a]	Y $(cm^{-2} s^{-1})$	Air–sea flux[b] $(cm^{-2} s^{-1})$
O_2^-/HO_2	2.4×10^{11}	1.2×10^9	1.2×10^{13}	1.0×10^8
H_2O_2	5.6×10^{10}	2.8×10^8	2.4×10^{12}	1.4×10^{10}
CH_3 (CH_3OO)	4.2×10^8	2.1×10^6	1.6×10^{10}	1.1×10^8
$CH_3C(O)$ $(CH_3C(O)OO)$	2.0×10^9	1.0×10^7	5.9×10^{10}	—
CO	6.3×10^9	3.1×10^7	2.9×10^{11}	—

[a] Assumes a 50 μm microlayer.
[b] From Thompson and Zafiriou (1983); 24 hour average.

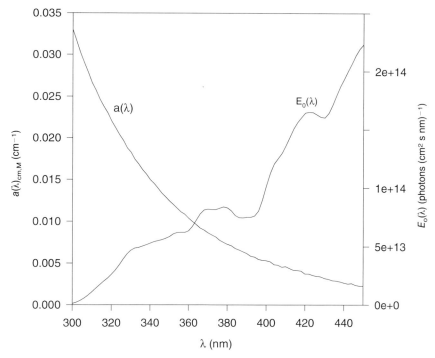

Figure 13.15. Downwelling solar irradiance and CDOM absorption spectra employed in the calculations of photochemical production rates and fluxes within the microlayer. The irradiance spectra were collected with a LICOR 1800 spectroradiometer at midday in August 1993 onboard the R/V *Cape Henlopen* offshore of Delaware. The absorption spectrum is the same as that in Figure 13.2 (spectrum 1), except multiplied by two.

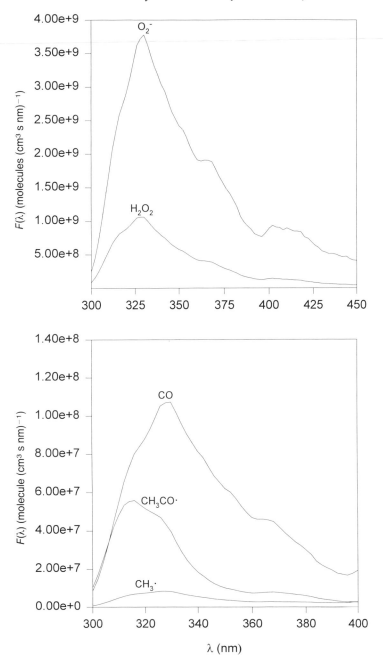

Figure 13.16. Calculated spectral dependence of the photochemical production rates for O_2^- (hydroxylamine formation), H_2O_2, CO, $CH_3(O)$, and CH_3 within the microlayer (see text).

the effectively higher 'production' rates of very reactive species within this boundary due to atmospheric deposition and *in situ* photochemistry should lead to a more rapid oxidative turnover of materials within this 'microreactor' and potentially to reactions not observed in bulk waters. While not carried out here, this analysis can be extended readily to other photochemical processes for which quantum yield data or action spectra are available (eg., production of OH, LMW compounds, NO and NO_2).

In principle, the elevated levels of highly reactive species produced photochemically within the microlayer could act as a barrier to the exchange of certain compounds across the interface. However, ignoring atmospheric deposition, the 1.5- to two-fold enrichments of light-absorbing material observed by Carlson (1982a) would produce, at most, a 1.5- to two-fold increase in the steady-state levels of reactive species in the microlayer (Equation 13.45). Thus, the levels of trace constituents passing through the microlayer would be reduced by only two-fold over that for an equivalent depth interval just under the microlayer. If z_m is significantly smaller than z_s (Equation 13.50), this loss would be reduced even further; while a 50-fold reduction in z_m for an observed enrichment of 1.4 yields a 15-fold increase in production rate (*vide supra*), the residence time (τ) for a species passing through this smaller layer is decreased 2500-fold due to the squared length dependence in the diffusion equation

$$\tau = z_M^2/2D \qquad (13.51)$$

where D is the molecular diffusion coefficient. The significantly larger fluxes of reactive species generated in the near-surface bulk waters will act as the principal determinant of the surface concentrations of these compounds. This 'barrier' will be important only if the material in the microlayer is significantly more photoreactive than that of the bulk, or if the microlayer material produces a different ensemble of reactive intermediates that exhibit selectivity for certain compounds.

Concluding remarks

The enrichment of light-absorbing materials within the microlayer is likely to increase the photochemical formation rates and fluxes of highly reactive species at the sea surface. These increased fluxes may have important impacts on the transformation and fate of materials passing through this interface as well as on the properties of the interface itself. However, the current lack of data on the spectral and photochemical properties of the microlayer presents a major barrier to a more complete understanding of these processes. Whether the major microlayer chromophore, CDOM, exhibits significantly different properties from that found in the underlying waters is a key uncertainty. If the properties are similar, currently available photochemical and spectral data should provide a reasonably good picture of the photochemical reactions occurring within this regime.

Acknowledgements

The preparation of this article was supported by grants from the ONR (NOOO14–89-J-1260), NASA through a NASA/EOS interdisciplinary investigation (NAGW-2431), and NSF (OCE-9115608). I wish to thank Sigalit Caron for her assistance in performing the photochemical experiments and calculations, and the preparation of figures.

References

Aiken, G. R., McKnight, D. M., Wershaw, R. L. and MacCarthy, P. (eds.) (1985). *Humic Substances in Soil, Sediment and Water: Geochemistry, Isolation and Characterization*. New York: Wiley Interscience, John Wiley & Sons, 691 pp.

Andreae, M. O. and Ferek, R. J. (1992). Photochemical production of carbonyl sulfide in seawater and its emission to the atmosphere. *Global Biogeochem. Cycles*, **6**, 175–83.

Behra, S. and Sigg, L. (1990). Evidence for the redox cycling of iron in atmospheric water droplets. *Nature*, **344**, 419–21.

Blough, N. V. (1988). Electron paramagnetic resonance measurements of photochemical radical production in humic substances. 1. Effects of O_2 and charge on radical scavenging by nitroxides. *Environ. Sci. Technol.*, **22**, 77–82.

Blough, N. V. and Green, S. A. (1995). Spectroscopic characterization and remote sensing of non-living organic matter. In *The Role of Non-living Organic Matter in the Earth's Carbon Cycle*, ed. R. G. Zepp and C. Sonntag, pp. 23–45. Chichester: John Wiley & Sons.

Blough, N. V. and Kieber, D. J. (1992). Photogenerated radical production in marine waters and its relationship to DOC photooxidation. Preprint extended abstracts, Division of Environmental Chemistry, ACS National Meeting, San Francisco, **32**, 73–5.

Blough, N. V. and Simpson, D. J. (1988). Chemically-mediated fluorescence yield switching in nitroxide–fluorophore adducts: optical sensors of radical/redox reactions. *J. Am. Chem. Soc.*, **110**, 1915–17.

Blough, N. V. and Zafiriou, O. C. (1985). Reaction of superoxide with nitric oxide to form peroxonitrite in alkaline aqueous solution. *Inorg. Chem.*, **24**, 3502–4.

Blough, N. V. and Zepp, R. G. (eds.) (1990). *Effects of Solar Ultraviolet Radiation on Biogeochemical Dynamics in Aquatic Environments.* Woods Hole Oceanographic Institution Technical Report, WHOI-90–09, 194 pp.

Blough, N. V. and Zepp, R. G. (1995). Reactive oxygen species in natural waters. In *Active Oxygen in Chemistry*, ed., C. S. Foote, J. S. Valentine, A. Greenberg and J. F. Liebman, pp. 280–333. New York: Chapman and Hall.

Blough, N. V., Zafiriou, O. C. and Bonilla, J. (1993). Optical absorption spectra of waters from the Orinoco River outflow: terrestrial input of colored organic matter to the Caribbean. *J. Geophys. Res.*, **98**(C2), 2271–8.

Bricaud, A., Morel, A. and Prieur, L. (1981). Absorption by dissolved organic matter of the sea (yellow substance) in the UV and visible domains. *Limnol. Oceanogr.*, **26**, 43–53.

Carder, K. L., Steward, R. G., Harvey, G. R. and Ortner, P. (1989). Marine humic and fulvic acids: their effect on remote sensing of ocean chlorophyll. *Limnol. Oceanogr.*, **34**, 68–81.

Carlson, D. J. (1982a). Surface microlayer phenolic enrichments indicate sea surface slicks. *Nature*, **296**, 426–9.

Carlson, D. J. (1982b). A field evaluation of plate and screen microlayer sampling techniques. *Mar. Chem.*, **11**, 189–208.

Carlson, D. J. (1983). Dissolved organic materials in surface microlayers: temporal and spatial variability and relation to sea state. *Limnol. Oceanogr.*, **28**, 415–31.

Carlson, D. J. and Mayer, L. M. (1980). Enrichment of dissolved phenolic organic material in the surface microlayer of coastal waters. *Nature*, **286**, 482–3.

Conrad, R., Seiler, W., Bunse, G. and Giehl, H. (1982). Carbon monoxide in seawater (Atlantic Ocean). *J. Geophys. Res.*, **87**, 8852–93.

Cooper, W. J. and Zepp, R. G. (1990). Hydrogen peroxide decay in waters with suspended soils: evidence for biologically mediated processes. *Can. J. Fish. Aquatic Sci.*, **47**, 888–93.

Duce, R. A. (1986). The impact of atmospheric nitrogen, phosphorous and iron species on marine biological productivity. In *The Role of Air–Sea Exchange in Geochemical Cycling*, ed. P. Buat-Menard, pp. 497–529. Norwell, MA: D. Reidel.

Duce, R. A. and Hoffman, E. J. (1976). Chemical fractionation at the air/sea interface. *Ann. Rev. Earth Planet. Sci.*, **4**, 187–228.

Dunlap, W. C., Chalker, B. E. and Oliver, J. K. (1986). Bathymetric adaptations of reef-building corals at Davies Reef, Great Barrier Reef, Australia. III. UV-B absorbing compounds. *J. Exp. Mar. Biol. Ecol.*, **104**, 239–48.

Faust, B. C. and Hoigné, J. (1987). Sensitized photooxidation of phenols by fulvic acid and in natural waters. *Environ. Sci. Technol.*, **21**, 957–64.

Frew, N. M., Goldman, J. C., Dennett, M. R. and Johnson, A. S. (1990). Impact of phytoplankton-generated surfactants on air–sea gas exchange. *J. Geophys. Res.*, **95**, 3337–52.

Gammon, R. H. and Kelly, K. C. (1990). Photochemical production of carbon monoxide in surface waters of the Pacific and Indian Oceans. In *Effects of Solar Ultraviolet Radiation on Biogeochemical Dynamics in Aquatic Environments*, ed. N. V. Blough and R. G. Zepp, pp. 58–60. Woods Hole Oceanographic Institution Technical Report, WHOI-90–09.

Garland, J. A. and Curtis, H. (1981). Emission of iodine from the sea surface in the presence of ozone. *J. Geophys. Res.*, **86**, 3183–6.

Green, S. A. and Blough, N. V. (1994). Optical absorption and fluorescence properties of chromophoric dissolved organic matter in natural waters. *Limnol. Oceanogr.*, **39**, 1903–15.

Haag, W. R., Hoigne, J., Gassman, E. and Braun, A. M. (1984). Singlet oxygen in surface waters. Part II. Quantum yields of its production by some natural humic materials as a function of wavelength. *Chemosphere*, **13**, 641–50.

Hardy, J. T. (1982). The sea surface microlayer: biology, chemistry and anthropogenic enrichment. *Prog. Oceanogr.*, **11**, 307–28.

Hartwig, E. O. and Herr, F. L. (1984). Chemistry and biology of the sea-surface interface: relationships to remote sensing. ONR workshop report, Office of Naval Research, Arlington, Va.

Hayase, K. and Tsubota, H. (1986) Monolayer properties of sedimentary humic acid at the air–water interface. *J. Colloid. Int. Sci.*, **114**, 220–6.

Helz, G. R., Zepp, R. G. and Crosby, D. G. (eds.) (1993). *Environmental Aspects of Surface and Aquatic Photochemistry*. Ann Arbor: Lewis Publishers. 552 pp.

Hühnerfuss, H., Alpers, W., Cross, A., Garrett, W. D., Keller, W. C., Lang, P. A., Plant, W. J., Schlude, F. and Schuler, D. L. (1983a). The modification of X and L band radar signals by monomolecular sea slicks. *J. Geophys. Res.*, **88**, 9817–22.

Hühnerfuss, H., Alpers, W., Garrett, W. D., Lange, P. A. and Stolte, S. (1983b). Attenuation of capillary and gravity waves at sea by monomolecular organic surface films. *J. Geophys. Res.*, **88**, 9809–16.

Huie, R. F. and Padmaja, S. (1993). The reaction of NO with superoxide. *Free Rad. Res. Comm.*, **18**, 195–99.

Hunter, K. A. and Liss, P. S. (1981). Organic sea surface films. *In Marine Organic Chemistry*, ed. E. K. Duursma and R. Dawson, pp. 259–98. Amsterdam: Elsevier.

Jin, F., Leitich, J. and von Sonntag, C. (1993). The superoxide radical reacts with tyrosine-derived phenoxy radicals by addition rather than by electron transfer. *J. Chem. Soc. Perkins Trans.*, **2**, 1583–8.

Jones, R. D. (1991). Carbon monoxide and methane distribution and consumption in the photic zone of the Sargasso Sea. *Deep-Sea Res.*, **38**, 625–35.

Jones, R. D. and Amador, J. A. (1993). Methane and carbon monoxide production, oxidation and turnover times in the Caribbean sea as influenced by the Orinoco river. *J. Geophys. Res.*, **98**, 2353–9.

Jonnson, M., Lind, J., Reitberger, T., Erikson, T. E. and Merenyi, G. (1993). Free radical combination reactions involving phenoxyl radicals. *J. Phys. Chem.*, **97**, 8229–33.

Khalil, M. A. K. and Rasmussen, R. A. (1984). Global sources, lifetimes and mass balances of carbonyl sulfide (OCS) and carbon disulfide (CS_2) in the Earth's atmosphere. *Atmos. Environ.*, **18**, 1805–13.

Kieber, D. J. and Blough, N. V. (1990a). Fluorescence detection of carbon-centered radicals in aqueous solution. *Free Rad. Res. Comm.*, **10**, 109–17.

Kieber, D. J. and Blough, N. V. (1990b). Determination of carbon-centered radicals in aqueous solution by liquid chromatography with fluorescence detection. *Anal. Chem.*, **62**, 2275–83.

Kieber, D. J. and Blough, N. V. (1992). Photoinitiated radical production in aquatic humic substances. Preprint extended abstracts, Division of Environmental Chemistry, ACS National Meeting, San Francisco, **32**, 84–7.

Kieber, D. J., McDaniel, J. and Mopper, K. (1989). Photochemical source of biological substrates in sea water: implications for carbon cycling. *Nature*, **341**, 637–9.

Kieber, R. J., Zhou, X and Mopper, K. (1990). Formation of carbonyl compounds from UV-induced photodegradation of humic substances in natural waters: fate of riverine carbon in the sea. *Limnol. Oceanogr.*, **35**, 1503–15.

King, D. W., Lin, J. and Kester, D. R. (1991). Determination of Fe(II) in seawater at nanomolar concentrations. *Anal. Chim. Acta*, **247**, 125–32.

Koppenol, W. H., Morens, J. J., Pryor, W. A., Ischiropoulos, H. and Beckman, J. S. (1992). Peroxynitrite, a cloaked oxidant formed by nitric oxide and superoxide. *Chem. Res. Toxicol.*, **5**, 834–42.

Lean, D. R. S., Cooper, W. J. and Pick, F. R. (1992). Hydrogen peroxide (H_2O_2) formation and decay in lakewaters. Preprint extended abstracts, Divison of Environmental Chemistry, ACS National Meeting, San Francisco, **32**, 80–83.

Lee, R. F. and Baker, J. (1992). Ethylene and ethane production in an estuarine river: formation from the decomposition of polyunsaturated fatty acids. *Mar. Chem.*, **38**, 25–36.

Lion, L. W. and Leckie, J. O. (1981). The biogeochemistry of the air–sea interface. *Ann. Rev. Earth Planet. Sci.*. **9**, 449–86.

Liss, P. S. (1975). Chemistry of the sea-surface microlayer. In *Chemical Oceanography*, ed. J. P. Riley and G. Skirrow, Vol. 2, p. 193. New York, London, San Francisco: Academic Press.

Liss, P. (1986). The chemistry of near-surface seawater. In *Dynamic Processes in the Chemistry of the Upper Ocean*, ed. J. D. Burton, P. G. Brewer and R. Chesselet, pp. 41–51. NATO Conference Series IV: Marine Sciences. New York: Plenum Press.

Logager, T. and Sehested, K. (1993). Formation and decay of peroxynitrous acid: a pulse radiolysis study. *J. Phys. Chem.*, **97**, 6664–9.

Mahoney, L. R. (1970). Evidence for the formation of hydroxyl radicals in the isomerization of pernitrous acid to nitric acid in aqueous solution. *J. Am. Chem. Soc.*, **92**, 5262–3.

Martin, J. H. and Fitzwater, S. E. (1988) Iron deficiency limits phytoplankton growth in the north east Pacific subarctic. *Nature*, **331**, 341–3.

Mill, T., Hendry, D. G. and Richardson, H. (1980). Free-radical oxidants in natural waters. *Science*, **207**, 886–7.

Miller, W. L. and Zepp, R. G. (1992). Photochemical carbon cycling in aquatic environments: Formation of atmospheric carbon dioxide and carbon monoxide. Preprint extended abstracts, Division of Environmental Chemistry, ACS National Meeting, San Francisco, **32**, 158–60.

Millero, F. J. and Sotolongo, S. (1989). The oxidation of Fe(II) with H_2O_2 in seawater. *Geochim. Cosmochim. Acta*, **53**, 1867–73.

Mitchell, B. G. and Kiefer, D. A. (1988). Chlorophyll *a* specific absorption and fluorescence excitation spectra for light-limited phytoplankton. *Deep-Sea Res.*, **35**, 665–89.

Mitchell, B. G., Vernet, M. and Holm-Hanson, O. (1990). Ultraviolet radiation in antarctic waters: particulate absorption and effects on photosynthesis. In *Effects of Solar Ultraviolet Radiation on Biogeo-chemical Dynamics in Aquatic Environments*, ed. N. V. Blough and R. G. Zepp, pp 135–6. Woods Hole Oceanographic Institution Technical Report: WHOI-90–09.

Moffett, J. W. and Zafiriou, O. C. (1990) An investigation of hydrogen peroxide chemistry in surface waters of Vineyard Sound with $H_2^{18}O_2$ and $^{18}O_2$. *Limnol. Oceanog.*, **35**, 1221–9.

Moffett, J. W. and Zika, R. G. (1987) Reaction kinetics of hydrogen peroxide with copper and iron in seawater, *Environ. Sci. Technol.*, **21**, 804–10.

Moore, C. A., Farmer, C. T. and Zika, R. G. (1993) Influence of the Orinoco river on hydrogen peroxide distribution and production in the eastern Caribbean. *J. Geophys. Res.*, **98**, 2289–98.

Mopper, K. and Zhou, X. (1990). Hydroxyl radical photoproduction in the sea and its potential impact on marine processes. *Science*, **250**, 661–4.

Mopper, K., Zhou, X., Kieber, R. J., Kieber, D. J., Sikorski, R. J. and Jones, R. D. (1991). Photochemical degradation of dissolved organic carbon and its impact on the oceanic carbon cycle. *Nature*, **353**, 60–2.

Najjar, R. G., Erickson III, D. J. and Madronich, S. (1994). Modeling the air–sea fluxes of carbonyl sulfide and carbon monoxide. In *The Role of Non-living Organic Matter in the Earth's Carbon Cycle*, ed. R. G. Zepp and C. Sonntag, pp. 107–32. Chichester: John Wiley & Sons.

Nelson, J. R. (1993) Rates and possible mechanism of light-dependent degradation of pigments in detritus derived from phytoplankton. *J. Mar. Res.*, **51**, 155–79.

Nelson, J. R. and Robertson, C. Y. (1993) Detrital spectral absorption: laboratory studies of visible light effects on phytodetritus absorption, bacterial spectral signal, and comparison to field measurements. *J. Mar. Res.*, **51**, 181–207.

O'Sullivan, D., Hanson, A. K., Miller, W. L. and Kester, D. R. (1991) Measurement of Fe(II) in surface water of the equatorial Pacific. *Limnol. Oceanogr.*, **36**, 1727–41.

Palenik, B., Zafiriou, O. C. and Morel, F. M. M. (1987) Hydrogen peroxide production by a marine phytoplankter. *Limnol. Oceanogr.*, **32**, 1365–9.

Petasne, R. G. and Zika, R. G. (1987) Fate of superoxide in coastal sea water. *Nature*, **325**, 516–18.

Preisendorfer, R. W. (1976) *Hydrologic Optics*. Washington, DC: US Dept. of Commerce, NOAA-ERL.

Radi, R., Beckman, J. S., Bush, K. M. and Freeman, B. A. (1991) Peroxy-nitrite oxidation of sulfhydryls: the cytotoxic potential of superoxide and nitric oxide. *J. Biol. Chem.*, **266**, 4244–50.

Rodriguez, J. M., Ko, M. K. W. and Sze, N. D. (1991) Role of heterogenous conversion of N_2O_5 on sulphate aerosols in global ozone losses. *Nature*, **352**, 134–7.

Ross, A. B., Mallard, W. G., Helman, W. P., Bielski, B. H. J., Buxton, G. V., Cabelli, D. E., Greenstock, C. L., Huie, R. F. and Neta, P. (1992) *NDRL-NIST Solution Kinetics Database – Version 1*. Gaithersburg, MD: NIST Standard Reference Data.

Sawyer, D. T. and Valentine, J. S. (1981) How super is superoxide? *Acc. Chem. Res.*, **14**, 393–400.

Sikorski, R. J. and Zika, R. G. (1993). Modeling mixed-layer photochemistry of H_2O_2: optical and chemical modeling of production. *J. Geophys. Res.*, **98**, 2315–38.

Soules, S. D. (1970). Sun glitter viewed from space. *Deep-Sea Res.*, **17**, 191–5.

Sulzberger, B., Suter, D., Siffert, C., Banwart, B. and Stumm, W. (1989) Dissolution of Fe(III) (hydr)oxides in natural waters: laboratory assessment on the kinetics controlled by surface chemistry. *Mar. Chem.*, **28**, 127–44.

Sunda, W. G. and Huntsman, S. A. (1988) Effect of sunlight on redox cycles of manganese in the southwestern Sargasso sea. *Deep-Sea Res.*, **35**, 1297–1317.

Sunda, W. G., Huntsman, S. A. and Harvey, G. R. (1983) Photoreduction of manganese oxides in seawater and its geochemical and biological implications. *Nature*, **301**, 234–6.

Thompson, A. M. and Zafiriou, O. C. (1983) Air–sea fluxes of transient atmospheric species. *J. Geophys. Res.*, **90**, 869–75.

True, M. B. and Zafiriou, O. C. (1987) Reaction of Br_2^- produced by flash photolysis of sea water with components of the dissolved carbonate system. In *Photochemistry of Environmental Aquatic Systems, Adv. Chem. Ser.* **327**, ed. R. G. Zika and W. J. Cooper, pp. 106–15. Washington, DC: American Chemical Society.

Tseng, R-S., Viechnicki, J. T., Skop, R. A. and Brown, J. W. (1992) Sea-to-air transfer of surface-active organic compounds by bursting bubbles. *J. Geophys. Res.*, **97**, 5201–6.

Turco, R. P., Whitten, R. C., Toon, O. B., Pollack, J. B. and Hamill, P. (1980). OCS, stratospheric aerosols and climate. *Nature*, **283**, 283–6.

Valentine, R. L. and Zepp, R. G. (1993) Formation of carbon monoxide from the photodegradation of terrestrially dissolved organic carbon in natural waters. *Environ. Sci. Technol.*, **27**, 409–12.

Vodacek, A., Green, S. A. and Blough, N. V. (1994) An experimental model of the solar-stimulated fluorescence of chromophoric dissolved organic matter. *Limnol. Oceanogr.*, **39**, 1–11.

von Sonntag, C. and Schuchmann, H-P. (1991) The elucidation of peroxyl radical reactions in aqueous solution with the help of radiation-chemical methods. *Angew. Chem. Int. Edn. Engl.*, **30**, 1229–53.

Waite, T. D. (1990) Photoprocesses involving colloidal iron and manganese oxides in aquatic environments. In *Effects of Solar Ultraviolet Radiation on Biogeochemical Dynamics in Aquatic Environments*, ed. N. V. Blough and R. G. Zepp, pp. 97–101. Woods Hole Oceanographic Institution Technical Report: WHOI-90-9.

Waite, T. D. and Morel, F. M. M. (1984) Photoreductive dissolution of colloidal iron oxides in natural waters. *Environ. Sci. Technol.*, **18**, 860.

Warneck, P. and Wurzinger, C. (1988) Product quantum yields for the 305-nm photodecomposition of NO_3^- in aqueous solution. *J. Phys. Chem.*, **92**, 6278–83.

Wei, Y. and Wu, J. (1992) In situ measurements of surface tension, wave damping and wind properties modified by natural films. *J. Geophys. Res.*, **97**, 5307–13.

Wells, M. L. and Mayer, L. M. (1991) The photoconversion of colloidal iron oxyhydroxides in seawater. *Deep-Sea Res.*, **38**, 1379–95.

Wells, M. L., Mayer, L. M., Donard, O. F. X., de Souza Sierra, M. M. and Ackelson, S. G. (1991). The photolysis of colloidal iron in the oceans. *Nature*, **353**, 248–50.

Williams, P. M., Carlucci, A. F., Henrichs, S. M., Van Vleet, E. S., Horrigan, S. G., Reia, F. M. H. and Robertson, K. J. (1986). Chemical and microbiological studies of sea-surface films in the southern Gulf of California and off the west coast of Baja California. *Mar. Chem.*, **19**, 17–98.

Yang, G., Candy, T. E. G., Boara, M., Wilken, H. E., Jones, P., Nazhat, N. B., Saadalla-Nazhat, R. A and Blake, D. R. (1992). Free radical yields from homolysis of peroxynitrous acid. *Free Rad. Biol. Med.*, **12**, 327–30.

Zafiriou, O. C. (1974). Sources and reactions of OH and daughter radicals in seawater. *J Geophys. Res.*, **79**, 4491–7.

Zafiriou, O. C. (1990). Chemistry of superoxide ion-radical (O_2^-) in seawater. I. pK^*_{asw} (HOO) and uncatalyzed dismutation kinetics studied by pulse radiolysis. *Mar. Chem.*, **30**, 31–43.

Zafiriou, O. C. and Bonneau, R. (1987). Wavelength-dependent quantum yield of OH radical formation from photolysis of nitrite ions in water. *Photochem. Photobiol.*, **45**, 723–7.

Zafiriou, O. C. and McFarland, M. (1981). Nitric oxide from nitrite photolysis in the central equatorial Pacific. *J. Geophys. Res.*, **86**, 3173–82.

Zafiriou, O. C. and True, M. B. (1979a). Nitrite photolysis in seawater by sunlight. *Mar. Chem.*, **8**, 9–32.

Zafiriou, O. C., and True, M. B. (1979b). Nitrate photolysis in seawater by sunlight. *Mar. Chem.*, **8**, 33–42.

Zafiriou, O. C., True, M. B. and Hayon, E. (1987). Consequences of OH radical reaction in sea water: formation and decay of Br_2^- ion radical. In *Photochemistry of Environmental Aquatic Systems, Adv. Chem. Ser. 327*, ed. R. G. Zika and W. J. Cooper, pp. 89–105. Washington: American Chemical Society.

Zafiriou, O. C., Blough, N. V., Micinski, E., Dister, B., Kieber, D. J. and Moffett, J. W. (1990). Molecular probe systems for reactive transients in natural waters. *Mar. Chem.*, **30**, 45–70.

Zellner, R., Exner, M. and Herrman, H. (1990). Absolute OH quantum yields in the laser photolysis of nitrate, nitrite, and dissolved H_2O_2 at 308 and 351 nm in the temperature range 278–353 K. *J. Atmos. Chem.*, **10**, 411–25.

Zepp, R. G. (1991). Photochemical conversion of solar energy in the environment. In *Photochemical Conversion and Storage of Solar Energy*, ed. E. Pelizzetti and M. Schiavello, pp. 497–515. Netherlands: Kluwer Academic Publishers.

Zepp, R. G. and Andreae, M. O. (1990). Photosensitized formation of carbonyl sulfide in seawater. In *Effects of Solar Ultraviolet Radiation on Biogeochemical Dynamics in Aquatic Environments*, ed. N. V. Blough and R. G. Zepp, p. 180. Woods Hole Oceanographic Technical Report: WHOI-90–09.

Zepp, R. G. and Cline, D. M. (1977). Rates of direct photolysis in the aquatic environment. *Environ. Sci. Technol.*, **11**, 359–66.

Zepp, R. G. and Schlotzhauer, P. F. (1981). Comparison of photochemical behavior of various humic substances in water. III. Spectroscopic properties of humic substances. *Chemosphere*, **10**, 479–86.

Zepp, R. G., Schlotzhauer, P. and Sink, M. R. (1985). Photosensitized transformations involving electronic energy transfer in natural waters: role of humic substances. *Environ. Sci. Technol.*, **19**, 74–81.

Zepp, R. G., Braun, A., Hoigné, J. and Leenheer, J. A. (1987a). Photoproduction of hydrated electrons from natural organic solutes in aquatic environments. *Environ. Sci. Technol.*, **21**, 485–90.

Zepp, R. G., Hoigné, J. and Bader, H. (1987b). Nitrate-induced photo-oxidation of trace organic chemicals in water. *Environ. Sci. Technol.*, **21**, 443–50.

Zepp, R. G., Faust, B. C. and Hoigné, J. (1992). Hydroxyl radical formation in aqueous reactions (pH 3–8) of iron(II) with hydrogen peroxide: the photo-Fenton reaction. *Environ. Sci. Technol.*, **26**, 313–19.

Zhou, X. and Mopper, K. (1990). Determination of photochemically produced hydroxyl radicals in seawater and freshwater. *Mar. Chem.*, **30**, 71–88.

Zhu, X. R., Prospero, J. M., Savoie, D. L., Millero, F. J., Zika, R. G. and Saltzman, E. S. (1993). Photoreduction of iron(III) in marine mineral aerosol solutions. *J. Geophys. Res.*, **98**, 9039–46.

Zhuang, G. R. (1992). The chemistry of iron in marine aerosols. *Global Biogeochem. Cycles*, **6**, 161–73.

Zhuang, G. R., Duce, R. A. and Kester, D. R. (1990). The dissolution of atmospheric iron in surface seawater of the open ocean. *J. Geophys. Res.*, **95**, 16 207–16.

Zika, R. G., and Cooper, W. J. (eds.) (1987). *Photochemistry of Environmental Aquatic Systems*, ACS Symposium Series Vol. 327, 288 pp. Washington, DC: American Chemical Society.

Zika, R. G., Moffett, J., Cooper, W. J., Petasne, R. and Saltzman, E. (1985a). Spatial and temporal variations of hydrogen peroxide in Gulf of Mexico waters. *Geochim. Cosmochim. Acta*, **49**, 1173–84.

Zika, R. G., Saltzman, E. S. and Cooper, W. J. (1985b). Hydrogen peroxide concentrations in the Peru upwelling area. *Mar. Chem.*, **17**, 265–75.

Zika, R. G., Milne, P. J. and Zafiriou, O. C. (1993). Photochemical studies of the eastern Caribbean: an introductory overview. *J. Geophys. Res.*, **98**, 2223–32.

Zuo, Y. and Hoigné, J. (1992). Formation of hydrogen peroxide and depletion of oxalic acid on atmospheric water by photolysis of iron(III)-oxalato compounds. *Environ. Sci. Technol.*, **26**, 1014–22.

14

Hydrocarbon breakdown in the sea-surface microlayer

MANFRED G. EHRHARDT

Abstract

Hydrocarbons, being minor constituents of dissolved organic matter in seawater, under normal conditions make up a small fraction of the organic surface film. In spill situations, however, they can become principal constituents. Depending on chemical structure, availability of nutrients, enrichment of microorganisms, and light regime, hydrocarbons in the surface microlayer are decomposed either microbially or photochemically. As most biogenic and the majority of fossil hydrocarbons are transparent to solar UV radiation at sea level, sensitizers are required for their photochemical oxidation and decomposition. Sensitizers include natural products, such as humic material, and anthropogenic compounds, such as polycyclic aromatic ketones. Photochemical decomposition products of hydrocarbons include alcohols, aldehydes, ketones, and terminal alkenes. Generation of low-molecular-weight carbonyl compounds by photochemical carbon chain fragmentation has been observed. Microbial decomposition of photo-oxidation products is often faster than that of the parent hydrocarbons.

Microlayer samplers

The term 'sea-surface microlayer' has been defined operationally as that thin layer of water adjacent to and including the air–sea interface which adheres to sampling devices such as wiremesh screens, glass plates, Teflon discs, rotating drums, or collectors for the spray generated by bursting bubbles (Liss, 1975, and references cited therein). Van Vleet and Williams (1980) found preferential uptake of different compound classes to depend upon the type of sampler and the material it is made of (see also Daumas *et al.*, 1976). Hühnerfuss (1981a,b) compared the sampling efficiencies of a large variety of different surface film samplers and recommended the use of a screen or a rotating drum sampler. More modern developments of this latter technique were described by Hardy *et al.* (1988) and Carlson *et al.* (1988). From the top down, the

425

microlayer usually includes an organic surface film, followed by what Broecker and Peng (1982) call the 'stagnant film' and, depending upon the type of sampler, a layer of variable thickness of bulk water. The stagnant film is approximately 40 μm thick and consists of water molecules, the near-order of which is influenced by the adjacent air–water interface and/or the vicinity of a film mainly composed of surface-active organic constituents of seawater. An organic surface film forms because accumulation of surface-active substances at the sea surface lowers the energy of the system. Molecules of surface-active substances are composed of hydrophobic sections such as aliphatic carbon chains and hydrophilic moieties such as alcoholic and phenolic hydroxyl, sulphonyl, and carboxyl groups which, because of their relatively high polarity, strongly interact with the polar molecules of water.

Compound groups found in the microlayer

Depending upon temperature, wind stress, composition and concentrations of its constituents, the thickness of the organic surface film (not the microlayer, which was taken to be approximately 1000 μm in thickness and is described in Hunter, Chapter 9, this volume, and Hardy, Chapter 11, this volume) may vary between a few and hundreds of nanometres (a monomolecular layer of fatty acids is ~2.5 nm thick, Decher, 1993). In a careful study, Williams *et al.* (1986) have shown that proteins (the sum of individual and combined amino acids), carbohydrates (the glucose equivalent of free plus combined mono- and polysaccharides), and lipids (the stearic acid equivalent of all hexane-soluble lipid components) on average accounted for 31% of the dissolved organic carbon (DOC) in surface films collected with screen samplers in a relatively clean, oceanic environment. Thus, hydrocarbons, which are part of the lipid fraction, were minor contributors to the surface films. Of the more than 60% which could not be ascribed to proteins, carbohydrates, and lipids, a significant fraction was assumed to be humic and fulvic substances, i.e. a complex assemblage of unidentified organic compounds many of which are surface active. Williams *et al.* (1986) quote a number of similar results obtained by other workers. This shows that the results obtained by Williams *et al.* (1986) were not unique, but represent a fairly common situation.

Hydrocarbons are not surface active, because their molecules lack polar hydrophilic groups. However, they form micellular aggregates with surface-active substances associating, as they do, with hydrophobic regions in the molecules of humic substances (Whitehouse, 1985;

Table 14.1 *Literature values for the Enrichment Factor (EF) of dissolved and particulate hydrocarbons (HC)*

Author(s)	Sea area	EF$_{dissolved\ HC}$	EF$_{particulate\ HC}$	Remarks
Wade and Quinn (1975)	Sargasso Sea	0.3–26		Unfiltered
Marty and Saliot (1976)	Western Mediterranean	27.4	17.6	n-Alkanes
		8.7	23.6	
	Roscoff	161	350	
	West Africa	20.2	10.4	
	Coastal western Mediterranean	6.3	169.1	
Marty et al. (1979)	West Africa	3–4		
Boehm (1980)	George's Bank	0.6–25.6		Total HC
Ho et al. (1982)	Western Mediterranean	0.44–3.6	1.3–3.1	Aromatics
		0.4–2.3		Total HC
Knap et al. (1986)	Bermuda inshore	5–30		Unfiltered
Hardy et al. (1988)	Puget Sound	2.6–221		Total HC
Anikiev and Urbanovich (1989)	N. Atlantic, Red Sea, S. China Sea	4–10		Aromatics

Whitehouse *et al.*, 1986) and, by implication, with fatty acids and other organic molecules which contain hydrophilic and hydrophobic moieties and thus also satisfy the requirements for surface activity. Although hydrocarbons are usually minor components of surface films, in spill situations hydrocarbons may easily exceed their solubilities in water (in the low 10^{-6} g 1^{-1} range for low-molecular-weight saturated, 10^{-4} to 10^{-3} g 1^{-1} for low-molecular-weight aromatic hydrocarbons, decreasing with increasing molecular weight; McAuliffe, 1966; Mackay and McAuliffe, 1988), so that they form a separate phase on the sea surface and mix with components of the natural surface microlayer. Not only in an actual oil spill, but also, to a lesser extent, in sea areas exposed to chronic low-level petroleum contamination, fossil hydrocarbons may become principal constituents of surface films (Morris, 1974). Surface-layer enrichment factors (EF = concentration of microlayer/ concentration of sub-surface water) reported by several investigators have been compiled in Table 14.1.

Table 14.1 shows that hydrocarbons are often enriched by factors up to the lower tens and sometimes are depleted in the microlayer relative to sub-surface water. Ancillary data to be found in the papers cited suggest that these conditions were associated with natural films. Enrichment factors in the hundreds point to multiple layers of hydrocarbons, as may be encountered in oil-contaminated waters. Here, it should be remembered that concentrations in the microlayer and, thus, derived enrichment factors depend upon the amount of water collected in addition to the organic surface film, and that this amount differs with different sampling techniques; also, that the material of which a sampling device is constructed affects the sample composition.

Sources and sinks of hydrocarbons in the microlayer

The sea-surface microlayer is a transient phenomenon in dynamical equilibrium with supply and removal of its constituent substances, among them hydrocarbons. Mechanisms must exist, therefore, not only for input of hydrocarbons, but also for their removal. Removal mechanisms are evaporation, injection as aerosols into the atmosphere by bursting bubbles (Tseng *et al.*, 1992), dissolution into sub-surface water, microbial degradation, and photochemical oxidation. The physical processes will not be discussed further in this chapter. Microbial degradation of hydrocarbons may be supported by elevated concentrations of microorganisms and enrichment of inorganic nutrients in the sea-surface microlayer

(Williams *et al.*, 1986, and references cited therein). Ho *et al.* (1982) found 9% n-alkanes in the saturated hydrocarbon fraction of a surface film in the western Mediterranean, versus 27% in sub-surface water. Likewise, Marty *et al.* (1979) measured a smaller fraction (6%) of n-alkanes in the microlayer than in sub-surface water (9%). Because the n-alkane component of hydrocarbons is most easily degraded by microorganisms, these results suggest increased microbial activity in the surface layer. However, conditions in the surface microlayer do not generally seem to be conducive to enhanced microbial degradation of hydrocarbons. In an earlier study, Marty and Saliot (1976) found, on average, 14% n-alkanes in the hydrocarbon fraction of the microlayer and 8% in the underlying water. Pengerud *et al.* (1984) noted a retardation of bacterial activity beneath illuminated relative to non-illuminated oil slicks. Larson *et al.* (1979) ascribe this effect mainly to formation of hydroperoxides. Herndl *et al.* (1993) even observed a direct reduction of approximately 40% in the activity of bacterioplankton in oceanic surface waters due to solar UV-B irradiation. Thus, although hydrocarbons do seem to be degraded in the microlayer by microorganisms, this mode of removal, despite more numerous microbial populations, is apparently not vastly more, and may even be less, effective than in bulk seawater. On the other hand, photooxidation of hydrocarbons should be promoted in the microlayer as no appreciable thickness of water protects them from solar radiation.

Most recently biosynthesized hydrocarbons that occur in seawater and marine organisms (Blumer and Thomas, 1965; Clark and Blumer, 1967; Blumer *et al.*, 1969, 1970, 1971; Lee and Loeblich III, 1971; Youngblood *et al.*, 1971; Youngblood and Blumer, 1973; Derenbach and Pesando, 1986; Carballeira *et al.*, 1988; Nichols *et al.*, 1988; Rowland and Robson, 1990) and many fossil hydrocarbons, with the exception of condensed aromatics and their alkyl derivatives, have no absorption bands in the spectral range of sunlight at the sea surface. Not being able to absorb light energy, aliphatic, alicyclic, and monocyclic aromatic hydrocarbons should not undergo photochemical oxidation and/or decomposition reactions under environmental conditions. However, dissolved and colloidal humic material (Zafiriou, 1983, and references cited therein; Hoigné *et al.*, 1989; Cooper *et al.*, 1989), riboflavin (Momzikoff and Santus, 1981), pteridins (Dunlap and Susic, 1985), and additional uncharacterized material (Momzikoff *et al.*, 1983), algal pigments, cyanocobalamine, thiamin, and biotin (Laane *et al.*, 1985, and references cited therein) are natural photosensitizers in marine and freshwaters.

Momzikoff *et al.* (1983) actually quantified what they called photo-sensitizing efficiency (PSE) by measuring the photochemical destruction rate of tryptophan added to seawater samples; they found that PSE increased with decreasing depth near the French coast of the Mediterranean. Lin and Carlson (1991) found elevated photo-sensitization in surface films sampled from coastal waters near San Diego, California, USA. Sensitizers most probably of anthropogenic origin, such as 9,10-anthraquinone, benzanthrone, benzophenone, and 9-fluorenone, have also been characterized in surface waters of the Baltic Sea, the Sargasso Sea, and the Mediterranean (Ehrhardt *et al.*, 1982; Ehrhardt and Burns, 1990; Ehrhardt and Petrick, 1993).

Sensitized photooxidation of hydrocarbons

Humic materials, flavins and pteridins, quinones and many aromatic ketones absorb in the solar spectral range at sea level, which extends to a short wavelength limit of approximately 295 nm (1.5×10^{-7} µEinstein $(cm^2 \, s \, 2.5 \, nm)^{-1}$ at 297.5 nm, 3.8×10^{-4} µEinstein $(cm^2 \, s \, 2.5 \, nm)^{-1}$ at 350 nm; Roof, 1982). Upon absorption of radiation energy, these substances are converted to their first excited singlet states, which are usually too short-lived ($\sim 10^{-9}$ s) to induce chemical reactions. However, transition is often possible with high quantum yields to the first excited triplet states which, because return to the ground state is spin-forbidden, are sufficiently long-lived for energy transfer to or chemical reaction with a non-absorbing target molecule. In bulk water, the most important partner in these reactions is dissolved oxygen, which is converted to either the $^1\Delta_g$ or the $^1\Sigma_g^+$ first excited singlet state. The $^1\Sigma_g^+$ state does not react efficiently, because it has a rather short lifetime ($\sim 10^{-9}$ s in condensed phases) before it relaxes to the $^1\Delta_g$ state, and because it correlates with an excited state of the product. Even if the relatively long-lived ($\sim 10^{-3}$ s) $^1\Delta_g$ state encounters a hydrocarbon molecule in sub-surface water, the energy gap of 22.5 kcal mol^{-1} (94.2 kJ mol^{-1}) to the ground state $^3\Sigma_g^-$ triplet oxygen is too small for breaking C–H or C–C bonds, because bond energies are from 83 to 104 kcal mol^{-1} (347–435 kJ mol^{-1}) and 66 to 88 kcal mol^{-1} (276 to 368 kJ mol^{-1}), respectively (Kerr and Trotman-Dickenson, 1971). However, the singlet $^1\Delta_g$ state of oxygen does react with molecules which easily undergo Diels–Alder reaction (enes and conjugated dienes) by addition to the double bond system (Kearns, 1971). Depending upon the starting materials, the primary products are dioxirans, allylic hydroperoxides, and endoperoxides, which are capable of various further reactions such as formation of ketones, quinones, and aldehydes. The aliphatic aldehydes observed in coastal surface water by Gschwend *et al.* (1982) may owe their existence, in part, to reactions of biogenic olefins with singlet oxygen.

In surface films, because of elevated concentrations relative to bulk water, hydrocarbons should be able to compete successfully with molecular oxygen as reaction partners of excited states of sensitizers. The great variety of natural and anthropogenic sensitizers makes it difficult to elucidate mechanisms of sensitization. However, analyses of carbonyl compounds in surface waters (Ehrhardt, 1987) and the product distribution in experiments modelling photooxidation under simulated environmental conditions with anthraquinone as sensitizer (Ehrhardt and Petrick, 1984, 1985) strongly suggest that hydrogen abstraction initiates photooxidation of phenylalkanes and even alkanes in surface films. Hydrogen abstraction may be effected by high-energy triplet sensitizers such as anthraquinone, but also by reactive intermediates generated by sunlight irradiation of dissolved organic matter (DOM) in seawater and in surface films (Hoigné *et al.*, 1989). Intermediates are the highly reactive OH• radical and DOM-derived peroxy radicals. The latter, although probably less reactive than OH• radicals, may well carry enough energy to at least abstract activated hydrogen atoms, e.g. in benzyl position. The intersystem crossing quantum yield (ICS) of anthraquinone to its first excited triplet state T_1 is 0.88 in benzene (Calvert and Pitts, 1967) and should also be high in a seawater medium. The combination of high ICS with a triplet energy approaching C–H bond strength (see Calvert and Pitts, 1967; Kerr and Trotman-Dickenson, 1971) renders anthraquinone an effective sensitizer. By analogy to the mechanism proposed by Bolland and Cooper (1954), reaction of anthraquinone's excited triplet state is used here to demonstrate hydrogen abstraction from an alkylbenzene under conditions of spin conservation (Figure 14.1). The immediate fate of the alkylaromatic and other peroxy radicals thus generated is difficult to determine, because seawater and, at elevated concentrations, the surface microlayer contain a large variety of compounds with which they could react. Organic peroxy radicals are oxidants. The redox potential of, for example, $CH_3OO•$, has been estimated to be 600–700 mV (von Sonntag and Schuchmann, 1991). Organic electron donors, or their inorganic counterparts, such as transition metal cations in low oxidation states, themselves produced photochemically (see, e.g. Sunda *et al.*, 1983; Moffet and Zika, 1987; Wells and Mayer, 1991), would reduce organic peroxy radicals to the corresponding

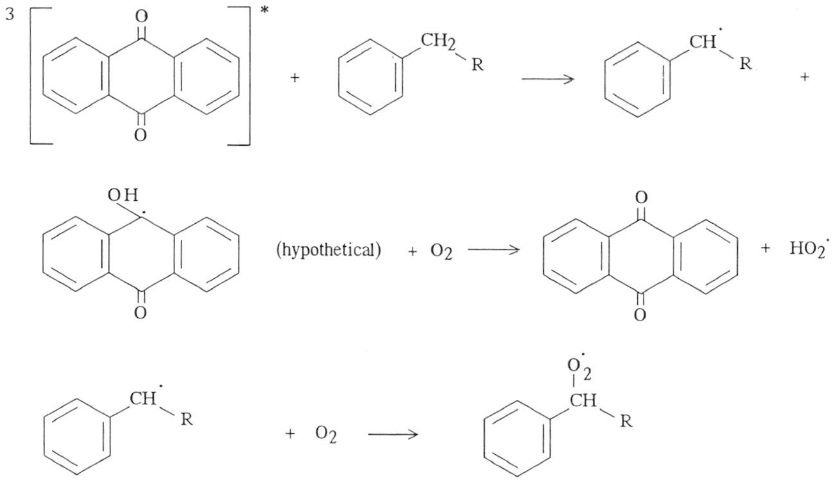

Figure 14.1. Hydrogen abstraction from an alkylbenzene by excited triplet state anthraquinone.

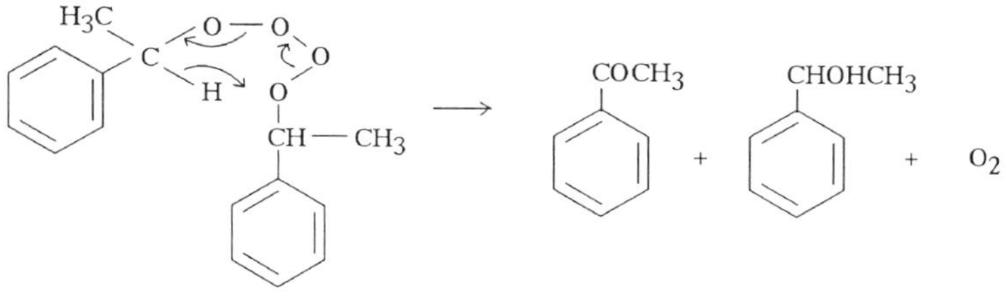

Figure 14.2. Example of the Russell mechanism.

peroxide anions. Hydrogen transfer from the hypothetical hydroxy-quinolyl radical is another possibility for hydroperoxide formation.

Photolysis of hydroperoxides yields OH• and organic oxyl radicals. However, this appears not to be an important reaction channel, because in model experiments with alkylbenzenes (Ehrhardt and Petrick, 1984) no products were found of the relatively easy hydroxylation of aromatic rings by OH• radicals. Reactions of peroxy radicals among themselves through short-lived tetroxides is more likely. This is especially so at the relatively high lipid concentrations in surface films, either via the Russell mechanism (Russell, 1957; see Figure 14.2), yielding one molecule each of the structurally related secondary alcohol and ketone and molecular oxygen, or via another concerted reaction (Zegota et al., 1984), yielding two molecules of the ketone and hydrogen peroxide.

A third possibility is fragmentation of the tetroxide into two molecules of the corresponding oxyl radical and molecular oxygen. Product analysis of photooxidation experiments with alkylbenzenes strongly suggests participation of oxyl radicals, because straight alkyl side-chains (CH_2–R in Figure 14.1) with four or more carbon atoms were fragmented between C_2 and C_3 (Ehrhardt and Petrick, 1984; Rontani *et al.*, 1987). Oxyl radicals with the oxygen functionality in the benzyl position could initiate such fragmentation reactions, but not the less reactive peroxy radicals. The mechanism of fragmentation appears to be analogous to a McLafferty rearrangement, as it leads to acetophenone and a terminal alkene with two carbon atoms less than the original side-chain. Figure 14.3 schematically depicts the alkoxy radical triggered fragmentation reaction.

A Norrish II photo-decomposition reaction of an intermediate 1-phenylalkanone could be another possibility for side-chain fragmentation yielding these products. However, exposure of a seawater solution of 1-phenylbutanone to natural sunlight did not lead to chain fragmentation (Ehrhardt and Petrick, unpublished results), most probably because dissolved oxygen or other materials present in the water effectively quenched the excited triplet state. Another argument for degradation through oxyl radicals is the photochemical generation of 1,4-diketones (Ehrhardt and Petrick, 1986; Rontani *et al.*, 1987), which must form via

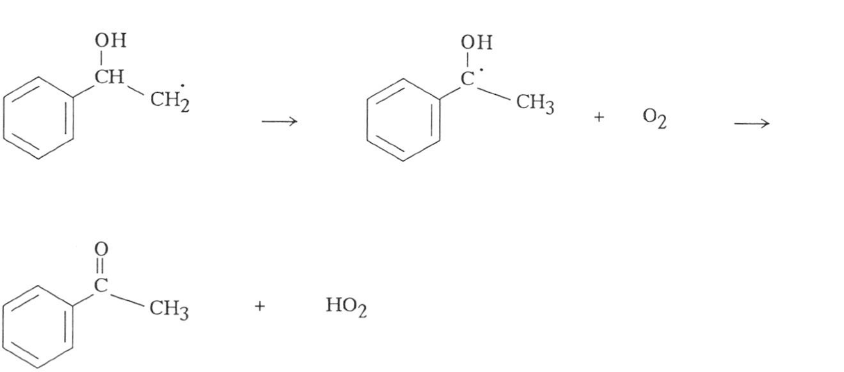

Figure 14.3. Side-chain fragmentation of a 1-phenylalkoxyl radical and formation of acetophenone after rearrangement of the primary radical.

intramolecular hydrogen abstraction by an oxyl radical involving a cyclic electron rearrangement in a six-membered ring. By intermolecular hydrogen abstraction oxyl radicals could propagate radical chains and thus initiate secondary reactions.

Benzaldehyde was an additional product of side-chain fragmentation. Again, several pathways are possible for its formation:

1. α-fragmentation of a 1-phenylalkoxyl radical;
2. a Norrish I photodecomposition of 1-phenylalkanones;
3. reaction of the primary 1-phenyl-1-hydroxyalkyl radical (see Figure 14.3) with oxygen, reduction of the resulting 1-hydroxyhydroperoxide, fragmentation between C_1 and C_2 with formation of formaldehyde and a 1-phenyl-1-hydroxymethyl radical, and its reaction with molecular oxygen.

1. would lead to benzaldehyde directly, but it involves formation of a primary alkyl radical, which is unfavourable energetically. 2. and 3. are possible pathways. However, the experimental data did not allow determination of their proportional contributions. Figure 14.4 depicts the reaction outlined under 3.

In addition to acetophenone, benzaldehyde and the 1-phenylalkanones, 1-phenylalkanols were generated by anthraquinone-sensitized photooxidation of 1-phenylalkanes. It is not clear whether they formed by fragmentation of the corresponding tetroxide via a Russell mechanism, or whether the stucturally related oxyl radical abstracted a hydrogen atom from a suitable donor. The second explanation is supported by the qualitative observation that 1-phenylethanol and ring-substituted alkyl derivatives were found in inshore waters with high organic loads, i.e. high concentration of abstractable hydrogen atoms (Ehrhardt, 1987), but not in open ocean areas (Ehrhardt and Petrick, 1993).

Not only alkylbenzenes, which, because of benzyl-activated hydrogen atoms, react under mild conditions, but also the chemically more stable saturated aliphatic hydrocarbons are subject to sensitized photooxidation by natural sunlight. After exposure of a surface film of n-pentadecane (n-$C_{15}H_{32}$) on purified seawater with anthraquinone as triplet sensitizer, Ehrhardt and Petrick (1985) found, among other products to be discussed shortly, a homologous series of aliphatic methylketones. The series extended from n-pentadecanone-2 to *n*-heptanone-2. More volatile methylketones escaped detection because of analytical limitations. n-Tetradecanone-2 was conspicuously missing from the series.

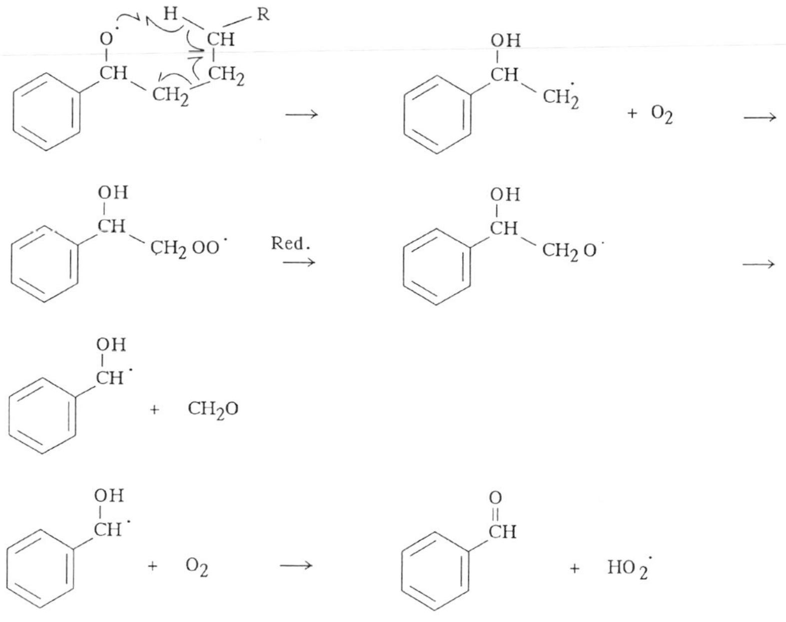

Figure 14.4. Tentative scheme for generation of benzaldehyde in photooxidation of alkylbenzenes.

Additional products were: all the other possible C_{15}-dialkylketones (but no secondary alcohols), a homologous series of terminal alkenes from n-dodecene-1 to n-decene-1 (more volatile alkenes were not seen because of experimental conditions), and all possible C_{15}-γ-diketones (Ehrhardt and Petrick, 1986).

These experimental results may be explained on the basis of abstraction, by excited sensitizer molecules (or OH radicals), of any one secondary hydrogen atom, reaction of the secondary alkyl radical with molecular oxygen, and formation (via reduction or concerted mechanisms) of secondary oxyl radicals (the missing secondary alcohols argue against a Russell mechanism). The secondary oxyl radicals can apparently be converted to dialkylketones by reaction with molecular oxygen. They can also abstract a γ-hydrogen atom with eventual formation of a γ-diketone, or undergo a McLafferty-type fragmentation reaction forming a terminal alkene with a maximum number of carbon atoms three less than contained in the original alkane, i.e. n-dodecene-1, if n-pentadecane was the starting material. Absence of n-tetradecanone-2 from the homologous series of methylketones indicates that the smallest alkene must have been ethylene. Figure 14.5 illustrates reaction of a dialkyloxyl radical ($R = CH_3 \rightarrow C_{n-2}H_{2(n-2)+1}$, $R' = C_{n-5}H_{2(n-5)+1} \rightarrow H$, $n =$ number of carbon atoms in the original n-alkane) with oxygen, its fragmentation and stabilization of a primary 2-hydroxyalkyl radical. Under the conditions of the experiment, the primary 2-hydroxyalkyl radicals rearrange (probably to some degree protected inside a cage of water molecules against reaction with oxygen) by 1,2 hydrogen shift to form the corresponding secondary 2-hydroxyalkyl radicals. Reaction with molecular oxygen then yields the homologous series of ketones and $HO_2 \cdot$. The alternative route leading to the observed reaction products, i.e. fragmentation of corresponding dialkylketones via a Norrish II mechanism, is highly unlikely. For this to occur the ketones would have to be excited by absorption of UV light. However, they only absorb at wavelengths considerably shorter than found in solar radiation at sea level.

n-Dodecene-1 (n-$C_{12}H_{24}$) was the highest-molecular-weight member of the homologous series of terminal alkenes among the photooxidation products of a surface film of n-pentadecane on seawater, i.e. it had three carbon atoms less than the original alkane substrate. The C_3-fragment which must carry the oxygen functionality was assumed to be the primary 2-hydroxypropyl radical. Rearrangement to the more stable secondary 2-hydroxypropyl radical and reaction with molecular oxygen should lead to acetone as the lowest-molecular-weight member of the homologous series of methylketones. Testing this hypothesis experimentally,

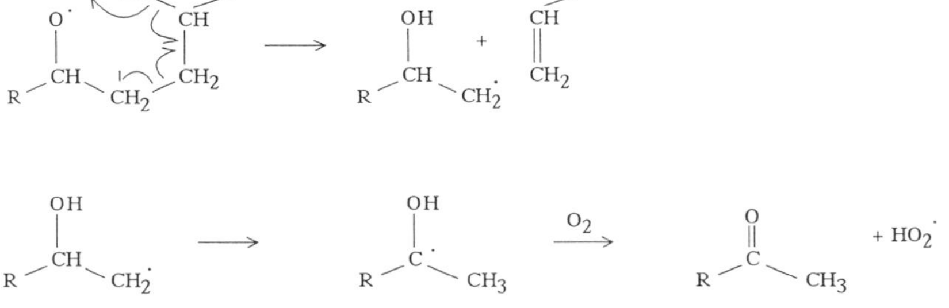

Figure 14.5. Reaction of a dialkyloxyl radical with molecular oxygen (top line), its fragmentation (middle line), and stabilization of a primary 2-hydroxyalkoxyl radical (bottom line).

Ehrhardt and Weber (1991) indeed found acetone, but also acetaldehyde and formaldehyde, as principal components among the low-molecular-weight carbonyl compounds. These results strongly suggest that the primary 2-hydroxypropyl radical also reacts with molecular oxygen before rearrangement (see Figure 14.6). The resulting 2-hydroxypropylperoxy radical will then be converted to the corresponding oxyl radical either by reduction or via cleavage of the tetroxide. Fragmentation of the 2-hydroxypropoxyl radical produces formaldehyde and the secondary hydroxyethyl radical. Reaction with oxygen yields acetaldehyde and HO_2 •. It is quite possible that under conditions other than those of the experiment, higher-molecular-weight 1-hydroxyalkyl radicals will be generated by an analogous mechanism. Reaction with molecular oxygen would then yield homologous series of aliphatic aldehydes, as have been observed by Gschwend *et al.* (1982) in coastal seawater.

The terminal alkenes generated in sensitized photo-decomposition reactions of alkylbenzenes and n-alkanes belong to a class of compounds which is biosynthesized by many algae (Blumer and Thomas, 1965; Blumer *et al.*, 1971; Youngblood and Blumer, 1973). Subterminal oxidation of n-alkanes by some bacteria and filamentous fungi yields methyl-ketones (Trudgill, 1978). Both these groups of compounds should thus

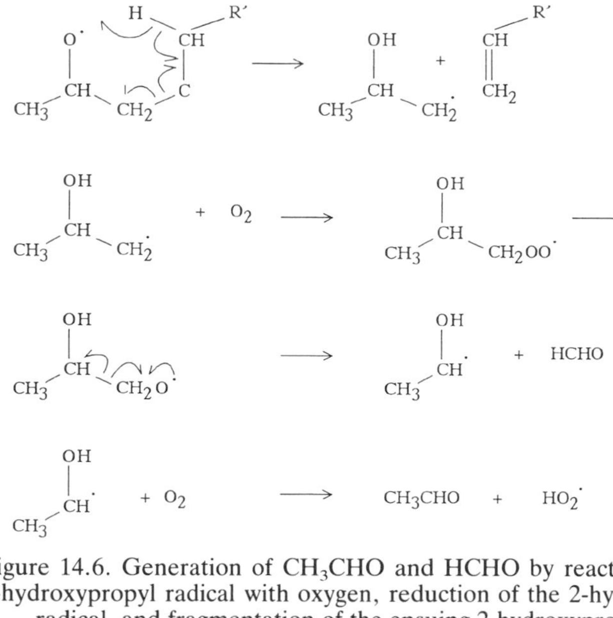

Figure 14.6. Generation of CH_3CHO and $HCHO$ by reaction of the primary 2-hydroxypropyl radical with oxygen, reduction of the 2-hydroxypropylperoxy radical, and fragmentation of the ensuing 2-hydroxypropoxyl radical.

be degraded easily by marine microorganisms, which would prevent accumulation to detectable levels of terminal alkenes and methylketones produced photochemically at low rates via the mechanism discussed above. However, under special conditions (high concentrations of hydrocarbons, high sunlight intensity, oligotrophic surface water) this may not be the case: Harvey (1988) found homologous series of terminal alkenes in surface waters of the Gulf of Mexico. Lack of an odd/even predominance within the homologous series (n-C_7H_{14} to n-$C_{15}H_{30}$) argues against a biological source and is consistent with the photo-degradation mechanisms described above.

A synergism between sensitized photooxidation and microbial decomposition was observed experimentally by Rontani *et al.* (1987). Mass spectrometric analyses proved that *Alcaligenes* sp. degraded the photooxidation products of nonylbenzene (1-phenyl-1-nonanone, 1-phenyl-1-nonanol, 1-phenyl-1,4-nonanedione) by oxidation of the functionalized side chain. Without the photochemically introduced oxgen functionality in the benzyl position, microbial decomposition was slower. Thus, ten days of irradiation with a UVGL58 lamp (UV Prod., Inc., San Gabriel, USA) equipped with a 6 W Blak-Ray® tube and a long wavelength filter (365 nm) placed 15 cm above the flask degraded 23% of the nonylbenzene substrate under sterile conditions. In the presence of *Alcaligenes* sp. PHY, but without illumination, 65% was degraded. Illumination and inoculation resulted in 84% degradation, also after ten days. More rapid elimination (by a factor of ~3) of bacterial decomposition products of nonylbenzene under sunlight illumination shows that the synergism worked both ways.

Among the various structural types of hydrocarbons, cycloalkanes are usually regarded as the most microbially recalcitrant (Trudgill, 1978), although not under all circumstances (see Leahy and Colwell, 1990). Oudot (1984) determined the following order of decreasing biodegradation rates: n-alkanes, iso-alkanes, 6,1,5, and 2-ring cycloalkanes, monocyclic and sulphur aromatics, 3- and 4-ring alkanes, 2- and 3-ring aromatics, tetra-aromatics, steranes, triterpanes, naphtheno-aromatics, penta-aromatics, asphaltenes, and resins.

In a recent study, Ehrhardt and Weber (1995) found that sensitized photooxidation converts methylcyclohexane into all possible isomeric methylcyclohexanols and methylcyclohexanones. Isomeric methylcyclohexyl hydroperoxides were also characterized by GC/MS, as well as heptan-2-one and 6-hepten-2-one generated by cleavage of the cyclohexane ring. These latter two products are additional proof that oxyl

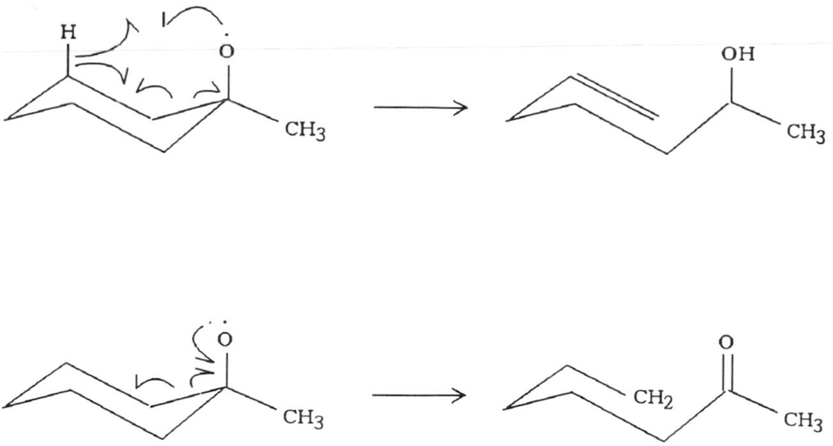

Figure 14.7. Oxyl radical induced opening of the cyclohexane ring (tentative).

radicals are intermediates in photochemical decomposition reactions of saturated hydrocarbons. A tentative mechanism rationalizing their formation is shown in Figure 14.7.

Cycloalkanols, cycloalkanones, and open-chain ketones are easier for microorganisms to degrade (Trudgill, 1978) than the parent alicyclic hydrocarbons, which again strongly suggests a synergism between sensitized photooxidation and microbial degradation in the breakdown of alicyclic hydrocarbons in the sea-surface microlayer. However, the observed formation of hydroperoxides is likely to have an antagonistic effect.

Model experiments and natural phenomena

Many of the abiotic chemical reactions of hydrocarbons in surface films discussed above have been observed in model experiments. Although natural sunlight was used as the energy source, the experimental conditions may have deviated from those in natural waters to a degree sufficient to favour formation of products of which mere traces are generated under conditions prevailing in the environment, or the lability of which prevents accumulation to detectable levels. Also, many of the reaction mechanisms are tentative, awaiting confirmation by suitable experiments. However, bearing in mind these limitations, I believe that results of model experiments carried out under suitable conditions are quite helpful to trace abiotic chemical reaction in the environment. Here,

440   *M. G. Ehrhardt*

because of much greater complexity, they are harder to observe and to interpret. I also believe that many of the reactions discussed above are not restricted to hydrocarbons, but have a wider scope and may further the understanding of abiotic chemical reactions in natural waters of more complex materials, generated with or without human participation. An example of the first is photochemical degradation of pesticides (e.g. Jensen-Korte *et al.*, 1987), of the latter, light-induced transformations of marine humic substances (e.g. Mopper *et al.*, 1991).

References

Anikiev, V. V. and Urbanovich, M. Yu. (1989). Distribution of organic pollutants in the sea surface microlayer system in some areas of the world ocean. *Geokhimiya*, **1989(5)**, 738–44.

Blumer, M. and Thomas, D. W. (1965). 'Zamene', isomeric C19 monoolefins from marine zooplankton, fishes, and mammals. *Science*, **148**, 370–1.

Blumer, M., Robertson, J. C., Gordon, J. E. and Sass, J. (1969). Phytol-derived C19 di- and triolefinic hydrocarbons in marine zooplankton and fishes. *Biochemistry*, **8**, 4067–74.

Blumer, M., Mullin, M. M. and Guillard, R. R. L. (1970). A polyunsaturated hydrocarbon (3,6,9,12,15,18-heneicosahexaene) in the marine food web. *Mar. Biol.*, **6**, 226–35.

Blumer, M., Guillard, R. R. L. and Chase, T. (1971). Hydrocarbons of marine phytoplankton. *Mar. Biol.*, **8**, 183–9.

Boehm, P. D. (1980). Evidence for the decoupling of dissolved, particulate, and surface microlayer hydrocarbons in northwestern Atlantic continental shelf waters. *Mar. Chem.*, **9**, 255–81.

Bolland J. L. and Cooper, H. R. (1954). The photosensitized oxidation of ethanol. *Proc. R. Soc.*, **225A**, 405–26.

Broecker, W. S. and Peng, T-H. (1982). The atmospheric imprint. In *Tracers in the Sea.* ed. W. S. Broecker and T-H. Peng, pp. 113–65. New York: Lamont-Doherty Geological Observatory, Columbia University.

Calvert, J. G. and Pitts, J. N., Jr (1967). *Photochemistry*. New York: John Wiley & Sons.

Carballeira, N., Negrón, V., Porras, B. and Diaz, B. A. (1988). On the hydrocarbon composition of Caribbean sponges. *Mar. Chem.*, **24**, 193–8.

Carlson, D. J., Cantey, J. L. and Cullen, J. J. (1988). Description and results from a new surface microlayer sampling device. *Deep-Sea Res.*, **35**, 1205–23.

Clark, R. C. jr. and Blumer, M. (1967). Distribution of n-paraffins in marine organisms and sediments. *Limnol. Oceanogr.*, **12**, 79–87.

Cooper, W. J., Zika, R. G., Petasne, R. G. and Fisher, A. M. (1989). Sunlight-induced photochemistry of humic substances in natural waters. In *Aquatic Humic Substances: Influence on Fate and Treatment of Pollutants*, ed. I. M. Suffet and P. MacCarthy, pp. 333–62. Washington, DC: American Chemical Society.

Daumas, R. A., Laborde, P. L., Marty, J. C. and Saliot, A. (1976). Influence of sampling method on the chemical composition of water surface films. *Limnol. Oceanogr.*, **21**, 319–26.

Decher, G. (1993). Supramolekulare Chemie: Ultradünne Schichten aus Polyelektrolyten. *Nachr. Chem. Tech. Lab.*, **41**, 793–800.

Derenbach, J. B. and Pesando, D. (1986). Investigations into a small fraction of volatile hydrocarbons. III. Two diatom cultures produce ectocarpene, a pheromone of brown algae. *Mar. Chem.*, **19**, 337–41.

Dunlap, W. C. and Susic, M. (1985). Determination of pteridines and flavins in seawater by reverse-phase, high-performance liquid chromatography with fluorimetric detection. *Mar. Chem.*, **17**, 185–98.

Ehrhardt, M. (1987). Photo-oxidation products of fossil fuel components in the water of Hamilton Harbour, Bermuda. *Mar. Chem.*, **22**, 85–94.

Ehrhardt, M. G. and Burns, K. A. (1990). Petroleum-derived dissolved organic compounds concentrated from inshore waters in Bermuda. *J. Exp. Mar. Biol. Ecol.*, **138**, 35–47.

Ehrhardt, M. and Petrick, G. (1984). On the sensitized photo-oxidation of alkylbenzenes in seawater. *Mar. Chem.*, **15**, 47–85.

Ehrhardt, M. and Petrick, G. (1985). The sensitized photo-oxidation of *n*-pentadecane as a model for abiotic decomposition of aliphatic hydrocarbons in seawater. *Mar. Chem.*, **16**, 227–38.

Ehrhardt, M. and Petrick, G. (1986). The generation of γ-diketones by sensitized photo-oxidation of *n*-alkanes under simulated environmental conditions. In *Conference Proceedings, American Chemical Society National Meeting, Miami, FL*, pp. 338–41.

Ehrhardt, M. and Petrick, G. (1993). On the composition of dissolved and particle-associated fossil fuel residues in Mediterranean surface water. *Mar. Chem.*, **42**, 57–70.

Ehrhardt, M. and Weber, R. R. (1991). Formation of low molecular weight carbonyl compounds by sensitized photochemical decomposition of aliphatic hydrocarbons in seawater. *Fresenius' J. Anal. Chem.*, **339**, 772–6.

Ehrhardt, M. G. and Weber, R. R. (1995). Sensitized photo-oxidation of methylcyclohexane as a thin film on seawater by irradiation with natural sunlight. *Fresenius' J. Anal. Chem.*, **352**, 357–63.

Ehrhardt, M., Bouchertall, F. and Hopf, H-P. (1982). Aromatic ketones concentrated from Baltic Sea water. *Mar. Chem.*, **11**, 449–61.

Gschwend, Ph. M., Zafiriou, O. C., Mantoura, R. F. C., Schwarzenbach, R. P. and Gagosian, R. B. (1982). Volatile organic compounds at a coastal site. 1. Seasonal variations. *Environ. Sci. Technol.*, **16**, 31–8.

Hardy, J. T., Coley, J. A., Antrim, L. D. and Kiesser, S. L. (1988). A hydrophobic large volume sampler for collecting aquatic surface microlayers, characterization and comparison with the glass plate method. *Can. J. Fish. Aquatic Sci.*, **45**, 822–6.

Harvey, G. R. (1988). Comments on azide and mercuric salts as seawater preservatives, and alkenes in the Gulf of Mexico. *Mar. Chem.*, **24**, 199–202.

Herndl, G. J., Müller-Niklas, G. and Frick, J. (1993). Major role of ultraviolet-B in controlling bacterioplankton growth in the surface layer of the ocean. *Nature*, **361**, 717–19.

Ho, R., Marty, J. C. and Saliot, A. (1982). Hydrocarbons in the western Mediterranean Sea, 1981. *Int. J. Environ. Anal. Chem.*, **12**, 81–98.

Hoigné, J., Faust, B. C., Haag, W. R., Scully, F. E., Jr and Zepp, R. G. (1989). Aquatic humic substances as sources and sinks of photochemically produced transient reactants. In *Aquatic Humic Substances: Influence on Fate and Treatment of Pollutants*, ed. I. M. Suffet and P. MacCarthy, pp. 363–81. Washington, DC: American Chemical Society.

Hühnerfuss, H. (1981a). On the problem of sea surface film sampling, a comparison of 21 microlayer-, 2 multilayer-, and 4 selected subsurface-samplers. Part 1. *Meerestech. Mar. Technol.*, **12**, 137–42.

Hühnerfuss, H. (1981b). On the problem of sea-surface film sampling, a comparison of 21 microlayer-, 2 multilayer-, and 4 selected sub-surface-samplers. Part 2. *Meerestech. Mar. Technol.*, **12**, 170–3.

Jensen-Korte, U., Anderson, C. and Spiteller, M. (1987) Photodegradation of pesticides in the presence of humic substances. *Sci. Total Environ.*, **62**, 335–40.

Kearns, D. R. (1971). Physical and chemical properties of singlet molecular oxygen. *Chem. Rev.*, **71**, 395–427.

Kerr, J. A. and Trotman-Dickenson, A. F. (1971) Bond strength in polyatomic molecules. In *Handbook of Chemistry and Physics*, ed. R. C. Weast, pp. F183-F186. Cleveland: The Chemical Rubber Company.

Knap, A. H., Burns, K. A., Dawson, R., Ehrhardt, M. and Palmork, K. H. (1986). Dissolved/dispersed hydrocarbons, tar balls and the surface microlayer, experiences from an IOC/UNEP workshop in Bermuda, December, 1984. *Mar. Pollut. Bull.*, **17**, 313–19.

Laane, R. W. P. M., Gieskes, W. W. C., Kraay, G. W. and Eversdijk, A. (1985). Oxygen consumption from natural waters by photo-oxidizing processes. *Neth. J. Sea Res.*, **19**, 125–8.

Larson, R. A., Bott, T. L., Hunt, L. L. and Rogenmuser, K. (1979). Photo-oxidation products of a fuel oil and their antimicrobial activity. *Environ. Sci. Technol.*, **13**, 965–9.

Leahy, J. G. and Colwell, R. R. (1990). Microbial degradation of hydrocarbons in the environment. *Microbiol. Rev.*, **54**, 305–15.

Lee, R. F. and Loeblich III, A. R. (1971). Distribution of 21:6 hydrocarbon and its relationship to 22:6 fatty acid in algae. *Phytochemistry*, **10**, 593–602.

Lin, K. and Carlson, D. J. (1991). Photo-induced degradation of tracer phenols added to marine surface microlayers. *Mar. Chem.*, **33**, 9–22.

Liss, P. S. (1975). Chemistry of the sea surface microlayer. In *Chemical Oceanography*, Vol 2, ed. J. P. Riley and G. Skirrow, pp. 193–243. London: Academic Press.

Mackay, D. and McAuliffe, C. D. (1988). Fate of hydrocarbons discharged at sea. *Oil Chem. Pollut.*, **5**, 1–20.

Marty, J. C. and Saliot, A. (1976). Hydrocarbons (normal alkanes) in the surface microlayer of seawater. *Deep-Sea Res.*, **23**, 863–73.

Marty, J. C., Saliot, A., Buat-Menard, P., Chesselet, R. and Hunter, K. A. (1979). Relationship between the lipid composition of marine aerosols, the sea surface microlayer, and subsurface water. *J. Geophys. Res.*, **84**, 5707–16.

McAuliffe, C. (1966). Solubility in water of paraffin, olefin, acetylene, cyclo-olefin, and aromatic hydrocarbons. *J. Phys. Chem.*, **70**, 1267–75.

Moffet, J. W. and Zika, R. G. (1987). Reaction kinetics of hydrogen peroxide with copper & iron in seawater. *Environ. Sci. Technol.*, **21**, 804–10.

Momzikoff, A. and Santus, R. (1981). Sur les propriétes photosensibilisatrices des pterines. Example de la biopterine. Comparison avec la riboflavine. *C. r. (hebd. Séanc.) Acad. Sci., Ser. C*, **293**, 15–18.

Momzikoff, A., Santus, R. and Giraud, M. (1983). A study on the photo-sensitizing properties of seawater. *Mar. Chem.*, **12**, 1–14.

Mopper, K., Zhou, X., Kieber, R. J., Kieber, D. J., Sikorski, R. J. and Jones, R. D. (1991). Photochemical degradation of dissolved organic carbon and its impact on the oceanic carbon cycle. *Nature*, **353**, 60–2.

Morris, R. J. (1974). Lipid composition of surface films and zooplankton from the eastern Mediterranean. *Mar. Pollut. Bull.*, **5**, 105–9.

Nichols, P. D., Volkman, J. K., Palmisano, A. C., Smith, G. A. and White, D. C. (1988). Occurrence of an isoprenoid C25 diunsaturated alkene and high neutral lipid content in antarctic sea-ice diatom communities. *J. Phycol.*, **24**, 90–6.

Oudot, J. (1984) Rates of microbial degradation of petroleum components as determined by computerized capillary gas chromatography and com-puterized mass spectrometry. *Mar. Environ. Res.*, **13**, 277–302.

Pengerud, H., Thingstad, F., Tjessem, K. and Aaberg, A. (1984). Photo-induced toxicity of North Sea crude oils toward bacterial activity. *Environ. Sc. Technol.*, **15**, 142–6.

Rontani, J. F., Bonin, P. and Giusti, G. (1987). Mechanistic study of inter-actions between photo-oxidation and biodegradation of *n*-nonylbenzene in seawater. *Mar. Chem.*, **22**, 1–12.

Roof, A. A. M. (1982). Basic principles of environmental photochemistry. In *The Handbook of Environmental Chemistry, Vol. 2, Part B, Reactions and Processes*, ed. O. Hutzinger, pp. 1–17. Heidelberg: Springer Verlag.

Rowland, S. J. and Robson, J. N. (1990). The widespread occurrence of highly branched acyclic C20, C25 and C30 hydrocarbons in recent sediments and biota: a review. *Mar. Environ. Res.*, **30**, 191–216.

Russell, G. A. (1957). Deuterium-isotope effects in the autoxidation of aralkyl hydrocarbons. Mechanism of the interaction of peroxy radicals. *J. Am. Chem. Soc.*, **79**, 3871–77.

Sunda, W. G., Huntsman, S. A. and Harvey, G. R. (1983). Photoreduction of manganese oxides in seawater and its geochemical and biological impli-cations. *Nature*, **301**, 234–6.

Trudgill, P. W. (1978). Microbial degradation of alicyclic hydrocarbons. In *Developments in Biodegradation of Hydrocarbons–1*, ed. R. J. Watkinson, pp. 47–84. London: Applied Science Publishers.

Tseng, R-S., Viechnicki, J. T., Skop, R. A. and Brown, J. W. (1992). Sea-to-air transfer of surface-active organic compounds by bursting bubbles. *J. Geophys. Res.*, **97**, 5201–6.

Van Vleet, E. S. and Williams, P. M. (1980). Sampling sea surface films, a laboratory evaluation of techniques and collecting materials. *Limnol. Oceanogr.*, **25**, 764–70.

von Sonntag, C. and Schuchmann, H-P. (1991). Aufklärung von Peroxyl-Radikalreaktionen in wässriger Lösung mit strahlenchemischen Techniken. *Angew. Chem.*, **103**, 1255–79.

Wade, T. L. and Quinn, J. G. (1975). Hydrocarbons in the Sargasso Sea surface microlayer. *Mar. Pollut. Bull.*, **6**, 54–7.

Wells, M. L. and Mayer, L. M. (1991). The photoconversion of colloidal iron oxyhydrates in seawater. *Deep-Sea Res.*, **38**, 1379–95.

Whitehouse, B. (1985). The effects of dissolved organic matter on the aqueous partitioning of polynuclear aromatic hydrocarbons. *Estuarine Coastal Shelf Sci.*, **20**, 393–402.

Whitehouse, B., Petrick, G., and Ehrhardt, M. (1986). Crossflow filtration of colloids from Baltic Sea water. *Water Res.*, **20**, 1599–601.

Williams, P. M., Carlucci, A. F., Henrichs, S. M., Van Vleet, E. S., Horrigan, S. G., Reid, F. M. H. and Robertson, K. J. (1986). Chemical and microbiological studies of sea-surface films in the southern Gulf of California and off the west coast of Baja California. *Mar. Chem.*, **19**, 17–98.

Youngblood, W. W. and Blumer, M. (1973). Alkanes and alkenes in marine benthic algae. *Mar. Biol.*, **21**, 163–72.

Youngblood, W. W., Blumer, M., Guillard, R. L. and Fiore, F. (1971). Saturated and unsaturated hydrocarbons in marine benthic algae. *Mar. Biol.*, **8**, 190–201.

Zafiriou, O. C. (1983). Natural water photochemistry. In *Chemical Oceanography*, Vol. 8, ed. J. P. Riley and R. Chester, pp. 339–79. London: Academic Press.

Zegota, H., Schuchmann, M. N., and von Sonntag., C. (1984). Cyclopentylperoxyl and cyclohexylperoxyl radicals in aqueous solution: a study by product analysis and pulse radiolysis. *J. Phys. Chem.*, **88**, 5589–93.

15

Applications of laser technology and laser spectroscopy in studies of the ocean microlayer

GERALD M. KORENOWSKI

Abstract

Laser spectroscopy and laser probes now provide previously unobtainable information on the physico-chemical structure and processes of the ocean microlayer. Methods range from laboratory techniques to *in situ* ocean microlayer probes.

In the laboratory, laser-induced fluorescence from water-soluble dye molecules is used to track aqueous layer interfacial movement. With these techniques, researchers infer turbulence in the upper several millimetres of a water surface. It is also possible to measure timescales of interfacial layer concentration fluctuations and interfacial layer penetration depths thus providing estimates of gas transfer velocities. Laser induced fluorescence methods are presently limited to laboratory studies.

Two-dimensional scanning laser slope gauges provide *in situ* measurement of ocean slope. Ocean slope is measured from the refraction of a vertical laser beam upon passing from the ocean into the air. By rapid scanning of the laser beam through a geometric pattern on the ocean surface, the researcher determines ocean wave slope at a variety of surface positions. This measurement is performed on a timescale during which the ocean surface is essentially frozen in time. From the set of slopes, an estimate is obtained of the two-dimensional capillary–gravity wave spectrum for a given instant of time and a given region of ocean surface.

Nonlinear spectroscopic processes such as reflected second harmonic generation and reflected sum frequency generation provide non-intrusive *in situ* spectroscopic probes of the ocean surface. The probes provide information on the chemical and physical structure of the top several molecular layers of the ocean surface. Chemical species identification and concentration measurements are possible even for molecules which are highly expanded monolayer films or dilute surface adsorbed species. These laser-based nonlinear spectroscopic techniques can differentiate between surface films whether of biogenic or anthropogenic origins. In addition, the techniques are employable as a remote sensing spectroscopy.

Introduction

Almost immediately after laboratory inception in 1960, the laser became a scientific probe. With technology advances, laser optics and laser spectroscopy quickly became integral tools in the physical sciences. Chemistry, physics, biology, atmospheric sciences, and many other disciplines were quick to make extensive use of the laser. In sharp contrast, ocean science has been slow to exploit this technology. This is especially true for studies of the ocean microlayer. Only recently have lasers and laser spectroscopy emerged as ocean microlayer probes. Specifically, there are three applications of lasers for direct microlayer studies and only two of these have been demonstrated for *in situ* microlayer studies.

The name 'ocean microlayer' has come to mean a number of different things to individual scientists studying the ocean surface. Universal to all these definitions is the requirement that the ocean microlayer be a thin surface layer which forms the interfacial boundary between the bulk ocean water and the atmosphere above. Researchers also define their own specific thickness for the microlayer. Here, the convention is adopted that the ocean microlayer incorporates all that is in the top one millimetre of the ocean surface. This one millimetre or less depth of the ocean surface microlayer makes it difficult to study the microlayer non-intrusively and *in situ*. Consequently, much of what is known about the physico-chemical nature of the microlayer is inferred from a combination of both laboratory and *in situ* studies. In many instances, *ex situ* analysis of microlayer samples is used to characterize this interfacial layer. Laser-based probes to study the microlayer have also been developed as laboratory experiments and in two instances the laser techniques have evolved sufficiently to permit field use. Not all of these techniques, however, may be adapted for field studies.

Turbulence in the microlayer, concentration fluctuations, interfacial layer penetration depths, and gas transfer velocities have been inferred for the microlayer using laser-induced fluorescence (LIF). Several variations of this fluorescence method have been developed. Nevertheless, all have in common the use of a water-soluble dye molecule that is made to fluoresce via laser excitation. Although a highly useful laboratory probe, it is unlikely that any of these LIF methods will function as *in situ* microlayer probes in present form.

Scanning laser slope gauges have been developed to study the effect of the surface microlayer on wind–wave formation and wave dampening. The highly directional, coherent, and monochromatic characteristics of a laser source are exploited to measure surface wave slope. The refraction of a laser beam upon traversing the water–air interface is used to determine wave slope and infer surface wave spectra. Initially developed as laboratory probes for use in wave tank studies, these scanning laser slope gauges have been successfully adapted for *in situ* use on the ocean surface.

Another successful extension of a laboratory probe for field application has taken place in the study of the chemical and structural properties of the top molecular layer of the ocean surface. In this application, the laser-based phenomena of nonlinear optics and nonlinear spectroscopic processes are exploited. The techniques of reflected second harmonic generation and reflected sum frequency generation have gained acceptance as versatile non-intrusive interfacial spectroscopic laboratory probes. It is possible with these techniques to characterize the surface-layer physical and chemical structure of a water–air interface. These techniques have recently been successfully adapted for *in situ* probing of the ocean surface. Consequently, these methods have become the first true ocean surface molecular spectroscopic probes.

In the following, the major focus will be on laser technology and laser spectroscopy, which have been developed for *in situ* studies of the microlayer. Consequently, the major emphasis of this chapter will be with the scanning laser slope gauges and the nonlinear spectroscopic probes. Nevertheless, a small overview of the laboratory-based LIF techniques is presented because of the importance of information gained using these methods.

Laser-induced fluorescence probes

Preceded and/or motivated by the work of Pankow, Asher and List (1984), several different laser-induced fluorescence probes have been developed to study processes at the water–air interface. These include extensions of the original technique by Asher and Pankow (1986, 1989, 1990, 1991), the work of Wolff, Liu and Hanratty (1990), that of Jähne (1990, 1993) and that of Münsterer and Jähne (1993 and 1994). Common to these techniques is the introduction of a fluorescent dye molecule to the aqueous phase. In each case the fluorescence of the molecule is affected by a second dissolved species. The fluorescein dyes, in the case of Asher and Pankow as well as those in the case of Jähne, and Jähne and Münsterer, are various forms of the dye molecule in which the fluorescence is pH dependent. Pyrenebutyric acid is used as the dye in the experiments of Wolff, Liu and Hanratty. For pyrenebutyric acid, the fluorescence is quenched by molecular oxygen.

In the experiments of Asher and Pankow, CO_2 is introduced into the air over the water. As CO_2 enters the air–water interface, hydration of CO_2 results and the weak acid H_2CO_3 is formed. This species can subsequently form equilibrium species (HCO_3^-, CO_3^{2-} and H_3O^+) with concentrations dependent upon the baseline pH of the aqueous medium. Any dissociation of H_2CO_3 alters the pH of the aqueous medium in that region. This in turn deactivates the fluorescent probe allowing the fluorescence quenching of the aqueous phase dye molecules to be used as a measure of mass transfer. The only restriction with the Asher and Pankow scheme is that care must be exercised so that the formation of H_2CO_3 and hence attainment of equilibrium pH not be the rate limiting step relative to mass movement in the water matrix. Considering the importance of CO_2 as a greenhouse gas, an advantage of the Asher and Pankow system is that it allows the direct study of CO_2 transfer across the air–water interface.

Although the method of Jähne also uses fluorescein dyes, it differs in that the gases introduced into the air over the water surface are either a strong acid such as HCl or a strong base such as NH_3. For HCl, the acid immediately dissociates upon entering the interface causing an immediate transfer or deactivation of the fluorescent pH indicator. Similarly, the strong base NH_3 immediately hydrates and dissociates to NH_4^+ and OH^- upon entering the water surface again deactivating the pH indicator. This method has the advantage that pH change and, consequently, fluorescence quenching occur immediately upon mass transfer of the HCl or NH_3 providing potentially better temporal resolution.

In the approach of Wolff, Liu and Hanratty, molecular oxygen is removed from the dye-containing aqueous medium. Fluorescence from the pyrenebutyric acid dye is subsequently of equal intensity throughout the oxygen-depleted solution. Molecular oxygen is then introduced into the air above the interface. As the oxygen is transported across the interface, it reduces the fluorescence lifetime of the pyrenebutyric acid from 100 nanoseconds to 65 nanoseconds. This results in a fluorescence intensity decrease for the oxygen-enriched regions of the aqueous medium.

Schematic diagrams of the three experiments are found in Figure 15.1. All of the methods use the monochromatic and spatially well-defined output of a laser to excite the fluorescence emission. In the methods employing fluorescein, the 488 nm emission line of an argon ion laser is used to induce the dye emission. The 337 nm near-UV output of a nitrogen laser is used to induce the pyrenebutyric acid emission. Where the Asher and Pankow experiment used a single detector looking at a

(a)

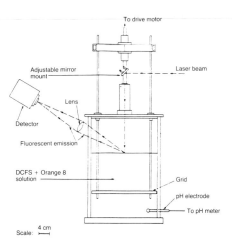

(b) (c)

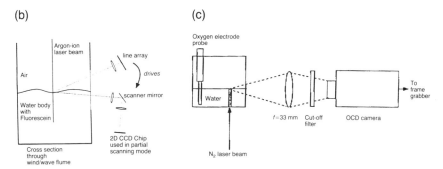

Figure 15.1. Three laser-induced fluorescence techniques for monitoring inter-facial process at the air–water interface: the apparatus of (A) Asher and Pankow (1989), (B) Münsterer and Jähne (1993) and (C) Wolff, Liu and Hanratty (1990).

single point in space, the later experiments of Jähne, Jähne and Münsterer, and Wolff, Liu and Hanratty employ a charge coupled device (CCD) to image the fluorescence emission. Use of the CCD detector allows spatial imaging of the mass transport as a function of time.

These LIF methods have demonstrated that it is possible to follow mass transport across the air–water interface, measure the timescales of inter-facial layer concentration fluctuations, and determine interfacial-layer penetration depths. It is also possible to infer turbulence in the upper several millimetres of a water surface with the LIF probes. However, it is unlikely that these techniques can be extended for *in situ* studies of the ocean surface without major design changes in the experimental schemes.

G. M. Korenowski

The scanning laser slope gauge

Knowledge of surface wave type and distribution on the ocean surface is important to our understanding of microlayer phenomena. Scanning laser slope gauges provide such information about ocean surface waves. For further information and background on the subject of ocean surface waves and the specialized terminology of this subject, the reader is directed to the text by J. R. Apel (1987).

In the following, we will focus only on those laser slope gauges which have been used *in situ* to study the microlayer effects on wind–wave formation and dampening. This reduces the review to the work of E. Bock and co-workers (Bock and Hara, 1992), (Martinsen and Bock, 1992), (Bock and Frew, 1993), (Bock and Hara, 1995), and (Hara, Bock and Lyzenga, 1994) and that of J. Wu and co-workers (Wei and Wu, 1992) and (Li *et al.*, 1993).

The basic premise of the laser slope gauge is quite simple. A laser beam passing through an interface between two media of different refractive indices is refracted according to Snell's law. Assume in a first approximation that the laser source is fixed in space so that the laser beam is normally incident on a calm water surface from below. According to Snell's law, the laser beam passing through the water–air interface would exit the interface normal to the surface. Now let the interface tilt relative to the laser beam incident from below. The new direction of the laser beam exiting the interface into the air is refracted from the surface normal according to Snell's law. Since the only variable is the tilt angle of the water–air interface, a measurement of this refraction angle directly yields the tilt angle or slope of the interface.

Positioning of the laser source below the air–water interface has two advantages. The detection system, larger than the laser source, is located above the surface reducing potential apparatus interference in the natural processes being studied. A second advantage is that the optical detection system is looking down into the ocean rather than at the sky. The latter advantage is important for any optical detection system operating in the field.

In the apparatus of Bock, the aspheric lens is designed to redirect the refracted beam exiting the interface into a beam approximately parallel to the surface normal. This light beam which is now parallel to the surface normal is incident on a diffuser. An array of silicon photodiodes is then used to determine the beam location on the detection screen providing an estimate of ocean surface slope.

In the apparatus of Wu, the Fresnel lens focuses the refracted laser beam on a position-sensitive detector (lateral-effect diode). Intermediate between the lens and the detector is a diffuser. The end result is again an estimate of the surface slope of the air–water interface.

In both systems, it was realized early on that a single point measurement of slope was insufficient to characterize the surface capillary–gravity wave spectrum. In the development of both systems, the approach was expanded to measure the surface slope over a number of spatial positions over a time interval in which the air–water interface remains essentially frozen in time. These measurements in turn may then be time averaged over any given temporal period. The two-dimensional scanning of the laser beam is accomplished via moving mirrors. Two mirrors, each mounted on a limited rotation servo motor (galvanometer), are used to move the incident laser beam in a two-dimensional pattern on the surface in the Bock apparatus. In the apparatus of Wu, the same task is accomplished by two rotating polygon mirrors. The two systems provide essentially equivalent results.

Figure 15.2 depicts a scanning laser slope gauge during ocean deployment. The typical deployment vehicle is a catamaran because it provides a somewhat stable platform. Nevertheless, the three-dimensional motion of the catamaran is deconvoluted from the data to refine further the results in both systems discussed above.

Figure 15.3 shows the wavenumber–frequency spectrum of the surface slope averaged over one hour as obtained with the apparatus of Bock and co-workers during a field deployment at the Woods Hole Oceanographic Institution (Hara, Bock and Lyzenga, 1994). During this experiment the mean wind friction velocity was 0.22 m s^{-1}. Also displayed in the figure are the capillary–gravity wave dispersion relations at surface tensions of 73 mN m^{-1} and 50 mN m^{-1} (the latter surface tension representing a slicked surface). The degree of saturation of capillary–gravity waves for the wind stress of 0.22 m s^{-1} is given in Figure 15.4.

In the above experiment, it is interesting to note that radar backscattering measurements were performed simultaneously during measurements of the capillary–gravity wavenumber–frequency spectra. Using the surface slope results, Bragg wave number spectra are plotted against the corresponding radar cross-section in Figure 15.5. This comparison shows the first strong field evidence that Bragg scattering theory provides an estimate for radar backscatter at intermediate incidence angles (vertical polarization).

With these scanning laser slope gauges now passing beyond the

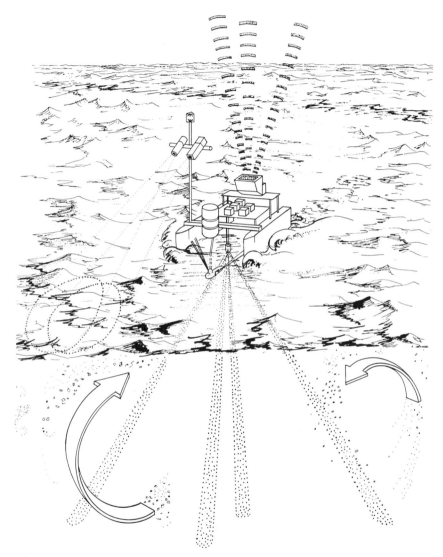

Figure 15.2. The scanning laser slope gauge of Bock and co-workers. The slope gauge is mounted on a research catamaran along with an array of support instrumentation including an acoustic anemometry package, X-band Doppler radar, attitude measuring unit, acoustic Doppler current meter, a video camera and an atmospheric acoustic profiler, the mini-SODAR (Hara, Bock and Lyzenga, 1994).

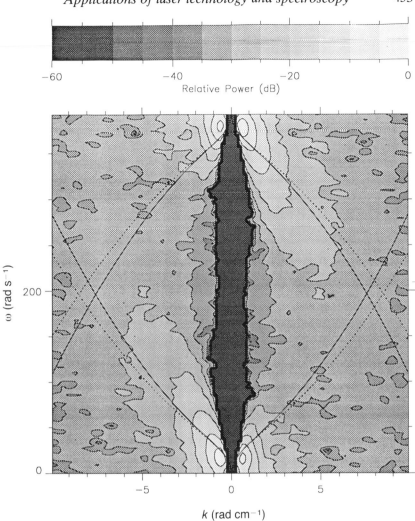

Relative Power (dB)

Figure 15.3. The wavenumber–frequency slope spectrum of capillary–gravity waves averaged over one hour (10:55 to 11:55 on 21 October 1992) as obtained with the laser slope gauge during a field experiment at the Woods Hole Oceanographic Institution (Hara, Bock and Lyzenga, 1994). The mean wind friction velocity was 0.22 m s^{-1} and the relative power 0 dB corresponds to 10^{-4} (cm^2 s). The solid dispersion line corresponds to a surface tension of 73 mN m^{-1} while the dotted dispersion line is for a surface tension of 50 mN m^{-1}. Measurements are along the wind direction. Here k is the wave number and ω is the frequency.

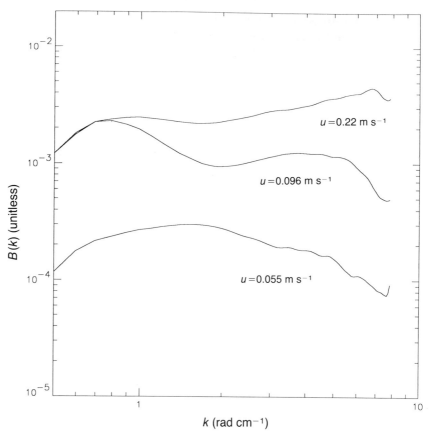

Figure 15.4. Capillary–gravity wave degree of saturation versus wave number for three wind stresses along the wind direction. Here $B(k)$ is the degree of saturation of capillary–gravity waves, k is the wave number, and u is the wind velocity (Hara, Bock and Lyzenga, 1994).

developmental stage, one can expect to perform many different *in situ* studies which were previously limited to the laboratory and large wind–wave tanks.

Nonlinear optical probes of the ocean surface

The top molecular layer of the ocean surface is a complex and as yet unexplored region of ocean science. It is this uppermost molecular layer which controls and/or dictates many oceanic physico-chemical processes. Wind–wave coupling (wave formation and dampening), gas exchange, and the reflection, transmission, and emission of electromagnetic radiation at the ocean surface are but a few of these processes. This surface layer with its large Gibbs free energy of formation also acts as a magnet for adsorption and concentration of trace chemical species from both the atmosphere and bulk ocean waters. Organic films, even as expanded monolayers, dramatically alter many of the physico-chemical properties of this interface. Consequently, methods of studying this top molecular layer *in situ* and information obtained on the interfacial chemical and physical structure would have far-ranging applications.

Over approximately the last ten years, the laser-based nonlinear optical techniques of reflected second harmonic generation (SHG) and

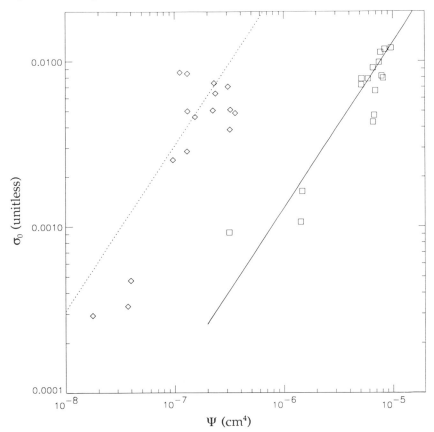

Figure 15.5. A comparison between the normalized radar scattering cross-section (σ_0) and the Bragg wave number spectrum ($\Psi = B(k)k^4$). X-band measurements are given as squares and compared with the theoretical ratio – the solid line. K-band measurements are given by diamonds, and the theoretical ratio is given by the dotted line (Hara, Bock and Lyzenga, 1994).

reflected sum frequency generation (SFG), have emerged as powerful new laboratory surface probes (Shen, 1984, 1989). As spectroscopic techniques, these surface probes are capable of obtaining electronic and vibrational spectra of the molecules at the air–water interface. In these laboratory experiments, sum frequency generation using visible and infrared radiation has been used to probe molecular vibrational resonances of molecules at interfaces. The probes are also used to determine the average spatial orientation of molecular species with respect to the interface normal. Such orientation studies can be performed even for large biological molecules or with the ensuing complexity of a copolymer surface film (Frysinger *et al.*, 1994; Barnoski, *et al.*, 1994; Barnoski, Gaines and Korenowski, 1995). It was realized early on that the technique could be used to study natural ocean surfactants on a substrate of seawater, albeit in the laboratory (Asher, Frysinger and Korenowski, 1988). It was not long before these techniques were demonstrated to be effective as *in situ* field probes, thus becoming the first ocean surface spectroscopic techniques (Korenowski *et al.*, 1989; Frysinger *et al.*, 1992; Korenowski, Frysinger and Asher, 1993). (It should be clarified that the meaning of surface here refers to the top several molecular layers if not the top molecular layer of the ocean surface.)

The basis behind these optical processes are the interactions between matter and the laser light field that are possible with intense laser sources. The electric field of the laser light induces a polarization in any material. This polarization is a macroscopic polarization extending over the dimension of the laser beam due to the coherence of the laser source. In a first approximation, the magnitude of this induced polarization is directly proportional to the electric field of the incident laser beam. However, this is just a first approximation. The macroscopic polarization is often treated as a power series expansion in the incident light field. This expansion is represented in the following equation.

$$\mathbf{P} = \chi^{(1)} \cdot \mathbf{E} + \chi^{(2)}{:}\mathbf{E}\mathbf{E} + \chi^{(3)}{:}\mathbf{E}\mathbf{E}\mathbf{E} + \cdots$$

where $\mathbf{P}$ is the macroscopic polarization vector in the material and $\mathbf{E}$ is the incident electric field vector of the laser light. The proportionality constants are the optical susceptibilities of the material. $\chi^{(n)}$ represents the nth order optical susceptibility tensor of the material. For (n) two or greater, the susceptibility tensors are referred to as the nonlinear optical susceptibilities of the material.

This macroscopic polarization, $\mathbf{P}$, oscillates at the optical frequency of the inducing light source for the linear term or at combination or multiple frequencies of this source for the higher order or nonlinear contributions. In the case that more than one laser beam is present, the nonlinear contributions to the polarization will oscillate at the sum and difference frequencies of the incident optical fields as well as at harmonics of the individual light fields. This is the basis of nonlinear optical and nonlinear spectroscopic processes.

The macroscopic polarization leads to either absorption or emission of optical radiation at the optical frequencies by the host medium. Hence, the linear polarization is responsible for well-known spectroscopic processes such as absorption and emission spectroscopy and optical effects such as refraction. The nonlinear polarizations are responsible for the nonlinear counterparts such as multiphoton absorption, stimulated Raman scattering, second harmonic generation, and so forth. These higher-order processes are collectively known as the nonlinear spectroscopic and nonlinear optical processes. We will focus here on just the second-order nonlinear events, which are the first of the nonlinear optical effects.

Electric dipole forbidden in the bulk of an isotropic medium (the bulk structure of water is isotropic over laser beam dimensions), second-order nonlinear optical processes are allowed at the sea–air interface. These processes are allowed due to the breaking of the bulk isotropy by molecular ordering at the interface. Reflected second-order nonlinear optical signals originate in the top one or two molecular layers of the sea–air interface. As surface processes, the resulting nonlinear optical signals are sensitive to the physical structure, chemical composition, and internal structure of the molecules at the surface. Optical resonances with electronic or vibrational transitions of interfacial molecules enhance the nonlinear signals and allow these processes to be used as a surface spectroscopy. Even though a surface will always yield a reflected second-order signal, tuning the wavelength of the incident laser(s) scans through the resonances of the interfacial molecules causing dramatic increases in these signals. In this manner the spectroscopic transitions of the interface molecules are recorded and species identified.

In reflected second-order processes, a single laser beam or two laser beams are made incident on the interface. As stated above, these intense coherent light sources induce a macroscopic polarization (equal in area to the incident laser beam footprint) in the interface. For two laser beams of optical frequencies ω_1 and ω_2, this surface polarization may be of frequencies ω_1 or ω_2 (i.e., a linear process and not surface selective) or at the nonlinear polarizations of $2\omega_1$, $2\omega_2$, $\omega_1+\omega_2$, $|\omega_1-\omega_2|$, $\omega_1-\omega_1=0$, and

$\omega_2-\omega_2=0$. Polarizations at $2\omega_1$ and $2\omega_2$ are examples of second-harmonic polarizations and require only one laser beam to be incident. The polarization at $\omega_1+\omega_2$ is a sum frequency polarization and that at $|\omega_1-\omega_2|$ is a difference frequency polarization. The DC polarizations at $\omega_1-\omega_1=0$ and $\omega_2-\omega_2=0$ are examples of the optical rectification process.

These surface nonlinear polarizations in turn reradiate electromagnetic energy at the induced frequencies. (This is true except for the optical rectification process which, however, will radiate electromagnetic energy if induced with a pulsed laser.) In each case the surface nonlinear polarization must conserve energy. For example, in SFG the emitted photon at $\omega_1+\omega_2=\omega_3$ equals the energy of the two photons annihilated from the incident laser beams. Momentum must also be conserved in the light field for these processes. Again using the SFG example, the sum frequency photon must possess a wave vector $\mathbf{k}_3$ which is equal to the sum of the annihilated photon wave vectors $\mathbf{k}_1+\mathbf{k}_2$ with consideration for the process of reflection $|\mathbf{k}|=n2\pi/\lambda$ where n is the refractive index at λ and on a quantum scale the momentum vector is $\mathbf{p}=\mathbf{k}h/2\pi$).

Each nonlinear polarization (second-harmonic, sum frequency, and so on) is viewed as a three-wave-mixing process where energy is exchanged between the incident and induced light fields. As such, these signals possess the same temporal profile as the incident laser pulses, a wavelength bandwidth directly related to that of the incident laser pulses, a well-defined wavelength because of energy conservation and because of momentum conservation a highly directional signal (except in the case of optical rectification). All of these properties make the nonlinear signals easy to identify and scale for quantitative measurements of species concentrations. This is best seen using the example of SHG at $\omega=\omega_1+\omega_1$.

The intensity of the SHG signal is given by

$$\mathbf{S}(\omega) = 32\pi^3\omega_1\,s^2\Theta\,(h/2\pi)^{-1}c^{-3}\,\varepsilon(\omega_1)^{-1}\varepsilon(\omega)^{-1/2}\,|\chi^{(2)}|^2\,\mathbf{I}^2\,\mathbf{A}\,\Delta t.$$

Θ is the laser incidence angle with respect to the interface normal, ε are the optical dielectric constants of the seawater surface, $\mathbf{A}$ is the laser beam footprint area, Δt is the laser pulse duration, $\mathbf{I}$ is the incident laser intensity, and $\chi^{(2)}$ is the second order nonlinear optical susceptibility of the ocean surface.

The intensity dependence of the nonlinear signal on the incident laser intensity is an additional mechanism of confirming the signal along with the bandwidth, temporal width, and direction restrictions on the signal. For the essentially non-dispersive medium air, the reflected laser beam and SHG signal are collinear. This enables one to scale the SHG return for quantitative measurements. Since the surface reflectivity of the ocean surface at the incident laser is known and the detection equipment is quantitatively calibrated, one can determine what portion of the reflected light pulse is collected in the optical detection system. This is then used to scale the SHG signal for quantitative measurements of the nonlinear signal. (The same technique is used for the SFG experiment where the two incident laser beams are collinear and normally incident on the ocean surface.)

In the absence of very strong dipole–dipole interactions between interfacial molecules, the $\chi^{(2)}$ is expressed as the number of molecules at the surface, N, multiplied by the individual molecular second-order polarizability of the molecules, $\alpha^{(2)}$.

$$\chi^{(2)} = N \, \alpha^{(2)}$$

$\alpha^{(2)}$ is molecule specific, possessing resonances characteristic of the molecular electronic and vibrational structure, the same as the linear polarizability is associated with linear spectroscopic events. Unlike a linear spectroscopy, the second-order processes exhibit a concentration squared dependence in the signal intensity making them especially sensitive to concentration variations. For a multicomponent surface (more than one molecular species), $\chi^{(2)}$ is expressed as a sum of individual susceptibilities.

$$\chi^{(2)} = \chi_a^{(2)} + \chi_b^{(2)} + \chi_c^{(2)} + \chi_d^{(2)} + \chi_e^{(2)} + \ldots$$

By tuning the incident laser wavelength (or wavelengths) to a strong characteristic resonance of a given molecular species, the signal from one particular molecule is isolated. Knowing the absolute magnitude of $\alpha^{(2)}$ for a given molecule at this wavelength, one can determine the number of these molecules per unit area of surface from a quantitative measurement of the nonlinear signal. Nevertheless, one must be careful in extracting concentrations from the nonlinear signals. For example, when the nonlinear signal is in resonance with a molecular transition of a particular molecule, molecular reorientations of this molecule at the interface can dramatically alter the nonlinear signal intensity. Such molecular reorientations can occur with variations in surface concentration of the species (Barnoski *et al.*, 1994). Also, pH dependent shifts in molecular resonances can lead to complications in extracting concentration from resonant second-harmonic and sum frequency generation measurements.

For a more extensive review of the surface-selective second-order nonlinear optical processes and how they relate to molecular structure,

(a)   (b)

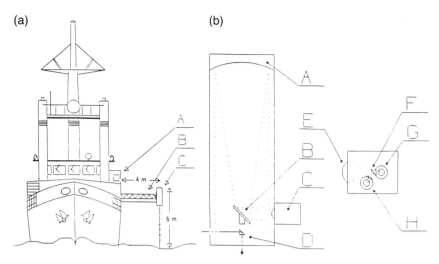

Figure 15.6. The experimental arrangement for use of the SHG and SFG probes from a research vessel. A, the van in which the laser and detection electronics were housed; B, the detector boom; and C, the delivery and collection telescope. (b) The collection and delivery telescope used showing the 0.4 m diameter primary mirror (A), secondary mirror (B), optical detector box (C), laser beam steering prism (D), the enlargement of the detector box (E), showing filters (F) and photomultiplier tubes, (G) and (H), respectively.

the reader is directed to more in-depth treatments of nonlinear optical phenomena (Shen, 1984, 1989; Frysinger *et al.*, 1992).

Both the second-harmonic and sum frequency techniques have been successfully used for *in situ* ocean studies. These probes have been successfully deployed from a Woods Hole Oceanographic Institution dock, an ocean tower, and a ship (Korenowski *et al.*, 1989; Frysinger *et al.*, 1992; Korenowski, Frysinger and Asher, 1993; Frysinger 1992).

Figure 15.6a shows a deployment of the second-harmonic apparatus from a ship during the SLIX 89 experiment. Figure 15.6b provides a schematic of the delivery optics and the collection optics, which are based on a Newtonian telescope. A near-normal incidence geometry is employed. The source laser used in all the experiments was a Q-switched Nd:YAG laser and in some instances the YAG laser was used to pump a wavelength-tunable dye laser.

In early experiments with the probes, second-harmonic generation was exclusively used, predominantly with the wavelength combination of 532 nm incident to generate a 266 nm signal. It was determined early on that for this wavelength combination the ocean surface presented an anomalously large optical nonlinearity. This was especially true when surfactant was absent from the surface. (The origin of this anomalously large signal from a clean ocean surface is still a mystery. This effect may be explained in part by an in-plane anisotropy of the water surface, which will be discussed later in this section.) *Ex situ* measurements of the ocean water surface tension as a function of time and the correlated time-averaged second-harmonic signal intensity are presented in Figure 15.7 (Frysinger *et al.*, 1992). The experiments of the Figure 15.7 were obtained during the SAXON-CLT experiment. Samples for surface tension measurements were automatically drawn by a surface skimmer floating within the Chesapeake Light Tower support structure. These samples were to some degree influenced by the breaking wave action and turbulence generated by the tower support structure. On the other hand, second harmonic measurements were performed *in situ* at a location some 9.5 metres outside of the tower support structure, where the ocean surface remained relatively uninfluenced by the structural supports extending into the ocean. Even with the disrupting influence of the tower on the surface films, a correlation between the ocean surface tension measurements and the SHG signal was consistently present for experiments performed on separate days. This rough correlation is evident in the experiments of Figure 15.7. The corresponding time series of the second harmonic signal are presented in Figure 15.8. Here, the passage of surfactant bands or slicks is associated with the decreased second-harmonic signal. It was determined from the simultaneously detected surface laser reflection at 532 nm that the decreases in second-harmonic intensity were not the direct result of wave dampening phenomena altering the signal collected by the detection system. In fact, the resulting second-harmonic signal decrease was the result of a weaker nonlinear optical surface in the presence of natural organic surfactants.

Further evidence that surface slicks were associated with surfactants was obtained by probing with a different wavelength combination (Frysinger, 1992; Korenowski, Frysinger and Asher, 1993). Exploiting the sum-frequency technique, a sum-frequency signal was obtained from the ocean surface at 213 nm, while employing incident laser wavelengths of 355 nm and 532 nm. The results of this experiment are shown in Figure 15.9. In Figure 15.9 the increases in the sum-frequency signal are associated with the increasing surfactant concentrations. The experiment exploited the fact that organic surfactants show strong electronic resonances in the deep ultraviolet, where the response of water remains flat.

Additional information is also obtained from the concentration

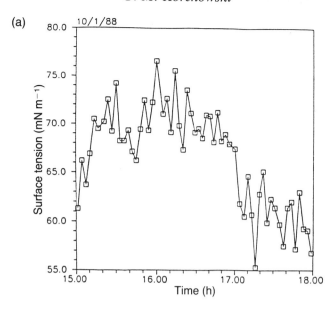

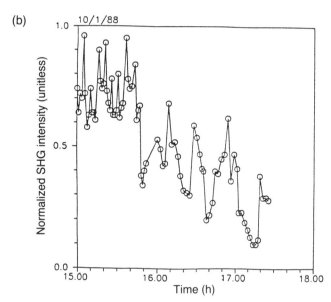

Figure 15.7. Ocean surface tension measurements (a) and corresponding reflected averaged SHG measurements, 532 nm incident light and 266 nm signal detected (b). These results were obtained during the same time period during the SAXON-CLT experiment. The wind-speed was 2.0–3.0 m s^{-1} (0.7 m significant wave height, *swh*). Decreasing surface tension and decreasing SHG signal at this wavelength arrangement both imply increasing surface organic concentration on the ocean surface. (Frysinger *et al.*, 1992.)

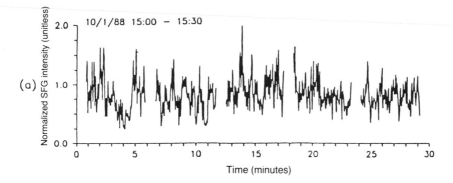

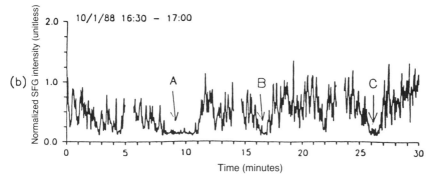

Figure 15.8. The corresponding times series of the reflected SHG shown in Figure 15.7. The passage of slicks through the sampling area is indicated by dramatic decreases in the SHG signal and is marked as A, B, and C in the figure. (Frysinger *et. al.*, 1992).

gradient of organic material observed in the experiment. The surfactant was associated with a banded slick and largest concentrations of the surfactant were consistently found on the upwind edge of these bands. Although little was known about the bulk dynamics during the experiment, one can speculate that the wind aided in spreading the surfactant on the downwind edge and compressing the surfactant on the upwind side. It is also interesting to note that the natural surfactant in these particular banded slicks appears to act more like a monolayer material than a multilayer film. However, it is highly unlikely that it is a monolayer film in total character. (A multilayer film would not show a concentration gradient but would look more like a step function because it does not spread. The topmost surface-layer concentration would remain constant.)

Recent laboratory experiments with these nonlinear probes indicate

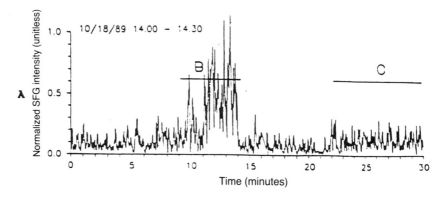

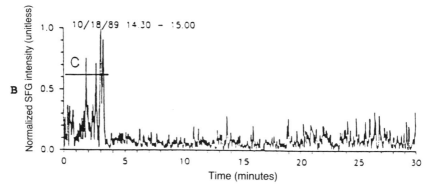

Figure 15.9. The time series of the reflected SFG196 signal taken during SLIX-89. The signal was detected at 213 nm. The choice of wavelength was made to exploit the UV resonance of organics versus the flat response of water in this wavelength region. Here, the banded slicks on the ocean surface are denoted by increases in the SFG signal labelled B and C. (Korenowski, Frysinger and Asher, 1993.)

that the ocean surface may be even more complex than originally believed (Judd and Korenowski, pers. commun). The spatial dependence of the reflected second-harmonic signal implies that there is a large-scale anisotropy in the plane of the air–water interface. Figure 15.10 shows the results from a recent experiment on a clean water surface. In the experiment, the angle of incidence of the pump laser light was varied from normal incidence to near-grazing incidence, while the detected second-harmonic beam was detected at the appropriate reflection angle (matching the incidence angle). Also varied in the experiment are the polarization of the incident pump laser light and the polarization of the emitted second-harmonic signal. Of particular interest is the S-polarized incident laser light and S-polarized detected second-harmonic signal experiment shown in Figure 5.10. (S-polarized light corresponds to the electric field of the light being perpendicular to the plane of incidence and detection. P-polarized light corresponds to the electric field of the light being parallel to or in the plane of incidence and detection.) In the electric-dipole approximation, the only mechanism which can account for an S-polarized second-harmonic signal for an S-polarized pump laser source is the existence of an in-plane macroscopic anisotropy in the surface layer of water molecules. The existence of a maximum in this signal at normal incidence also supports this conclusion. Experiments performed with a surfactant-covered surface exhibit no signal under the S-polarized pump and S-polarized detection conditions indicating a random or isotropic in-plane orientation of the surfactant molecules at the interface. Under other polarization conditions, the signal from the surfactant exhibits a signal minimum at near-normal incidence and detection angle. An isotropic surface layer exhibits a signal minimum in the second-harmonic signal at normal incidence. The results displayed in the figure representing this preliminary laboratory study imply that there is a net orientation of the water molecules within the interfacial plane on the timescale of the laser pulse (8 nanoseconds). The hypothesis of a structured water surface is not a new one. It has long been proposed that a water surface is ice-like in structural character (Drost-Hansen, 1965). A second-harmonic study by Goh *et al.* (1988) and a more recent infrared vibrational sum frequency experiment by Du *et al.* (1993) have probed the surface orientation of water at the air–water interface. These studies, performed at a fixed angle of incidence, focused on determining the net orientation of the water molecule at the interface with emphasis on the orientation of the water molecule relative to the surface normal. The experiment of Du *et al.* (1993) provides the more refined and definitive look at the orientation of water using the vibrationally resonant sum frequency generation to look at the O–H stretching vibration. What is new from the preliminary experiment described above is the dependence of the second-harmonic signal on the angle of incidence and the inferred in-plane anisotropy for water. That one can measure for the first time these surface structures and extend such microscopic measurements to *in situ* studies on the ocean surface opens an exciting new field of microscopic structural studies.

Reflected second-harmonic and sum-frequency generation are not the only possible probes for studying the ocean microlayer. The large number of possible nonlinear optical processes possible via second-, third-, and higher-order nonlinear optical wave mixing offer almost a limitless

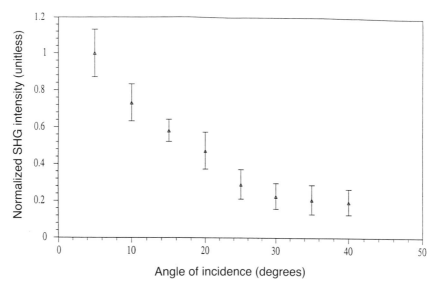

Figure 15.10. The angular dependence of the second-harmonic signal (532 nm incident laser light and 266 nm detected second-harmonic signal) from a clean water surface. The incident laser light and detected second-harmonic signal are both S-polarized (the electric field of the light perpendicular to the plane of incidence and reflection). The existence of this signal under these polarization conditions implies an in-plane anisotropy in the structure of the water surface on the time scale of the laser pulse (8 nanoseconds). The maximum exhibited by the signal at normal incidence also supports this conclusion. (From Judd and Korenowski, pers. commun.)

number of spectroscopic and optical combinations for probing the ocean microlayer. Only the imagination of the researcher limits the number of possibilities. The following example is an illustration of the possibilities.

Recently, it was proposed that surface optical rectification may also be a highly complementary *in situ* surface probe to second-harmonic and sum-frequency generation (Korenowski *et al.*, 1993). In the proposed method, a wavelength-tunable short-pulsed laser beam would be incident on the ocean surface. The light pulse, in addition to producing a reflected second-harmonic signal at $\omega = \omega_1 + \omega_1$, produces a DC polarization of the surface ($O = \omega_1 - \omega_1$). This surface polarization is produced only when the light pulse is present. It turns on and off during the temporal duration of the laser pulse. The energy stored in the DC surface polarization is radiated from the surface with a frequency spread given by the Fourier transform of the laser pulse temporal profile. For example, if the laser pulse is in the picosecond range of duration, the rectification signal will be centered in the gigahertz region. As long as the laser temporal

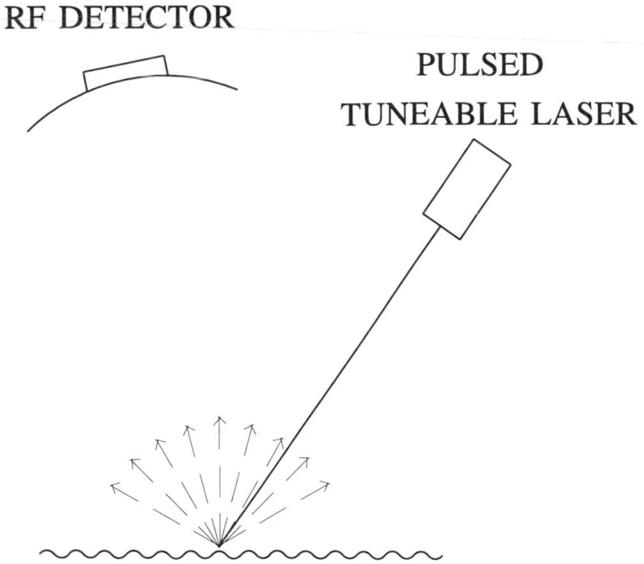

RF DETECTOR

PULSED
TUNEABLE LASER

Figure 15.11. A proposed surface optical rectification probe for ocean surface spectroscopy. (From Korenowski, Frysinger and Asher, 1993.)

pulse length remains the same, the signal will always be centered in the same wavelength region of the microwave. As the laser wavelength is tuned in and out of molecular resonances, such as vibrational or electronic resonances, the intensity of the rectification signal would vary but not its wavelength. In this manner, the method would be used as a remote sensing spectroscopy. A schematic diagram of the proposed apparatus is shown in Figure 15.11. As stated above, it is only the imagination of the research scientist which limits the number of nonlinear optical probes which can be developed for *in situ* microlayer and other ocean studies.

Conclusions

With renewed interest in the ocean microlayer and interfacial ocean surface processes, there is increasing incentive to develop techniques to study the ocean surface *in situ* and in a nonintrusive manner. Laser technology and laser spectroscopy promises to provide these tools to the ocean scientist.

In the laboratory, LIF methods are providing the scientist with information on mass transport across the air–water interface, concentration fluctuations at the interface, and information about turbulence in the upper few millimetres of the water surface. Laser technology employed in devices like the scanning laser slope gauges and advances in nonlinear surface spectroscopy have emerged not only as excellent laboratory probes of the air–water interface but are easily extended for use *in situ*. The scanning laser slope gauge provides estimates of the capillary–gravity wave spectra on the ocean surface and the ability to study these spectra as a function of ocean surface conditions and composition. Reflected second-harmonic and sum-frequency generation spectroscopy provides methods of studying the chemical composition, concentrations, and structure at the ocean surface. The further development of these techniques and development of new *in situ* laser methods offers the ocean scientist the first opportunity to characterize *in situ* the physico-chemical properties and processes of the ocean microlayer.

Acknowledgements

The author would like to thank the Office of Naval Research (grant N0001490J1537) for their support during the preparation of this manuscript.

References

Apel, J. R. (1987). *Principles of Ocean Physics*. London: Academic Press.

Asher, W. E. and Pankow, J. F. (1986). The interaction of mechanically generated turbulence and interfacial films with a liquid phase controlled gas/liquid process. *Tellus*, **38B**, 305–18.

Asher, W. E. and Pankow, J. F. (1989). Direct observation of concentration fluctuations close to a gas–liquid interface. *Chem. Engng. Sci.*, **44**, 1451–5.

Asher, W. E. and Pankow, J. F. (1990). The effect of surface films on concentration fluctuations close to a gas/liquid interface. In *Air–Water Mass Transfer Second International Symposium*, ed. S. C. William and J. S. Gulliver, pp. 68–80, Minneapolis, MN: ASCE.

Asher, W. E. and Pankow, J. F. (1991). Prediction of gas/water mass transport coefficients by a surface renewal model. *Environ. Sci. Technol.*, **25**, 1294–300.

Asher, W. E., Frysinger, G. S. and Korenowski, G. M. (1988). Reflected optical second-harmonic generation in the study of naturally occurring organic films at the ocean surface. *J. Geophys. Res.*, **93 C4**, 6955–7.

Barnoski, A. A., Frysinger, G. S., Gaines, G. L. Jr and Korenowski, G. M. (1994). Macroscopic and microscopic surface characterization of a Vitamin K1 monolayer on water. *Surface Colloids*, **A 88**, 123–8.

Barnoski, A. A. Gaines, G. L., Jr and Korenowski, G. M. (1995). The molecular arrangement of a Vitamin K1 monolayer at the air–water interface. *Surface Colloids*, **A 94**, 59–65.

Bock, E. J. and Frew, N. M. (1993). Static and dynamic response of natural multicomponent oceanic surface films to compression and dilation: laboratory and field observations. *J. Geophys. Res.*, **98 C4**, 14 599–617.

Bock, E. J. and Hara, T. (1992). Optical measurements of ripples using a scanning laser slope gauge part II: data analysis and interpretation from a laboratory wave tank. In *Optics of the Air–Sea Interface: Theory and Measurement*, ed. L. Estep, **1749**, pp. 272–82. San Diego, CA: SPIE.

Bock, E. J. and Hara, T. (1995). Optical measurements of capillary–gravity wave spectra using a scanning laser slope gauge. *J. Atmos. Oceanic Technol.*, **12**, 395–403.

Drost-Hansen, W. (1965). Aqueous methods of study and structural properties. *Ind. Engng. Chem.*, **57**, 18–37.

Du, Q., Superfine, R., Freysz, E. and Shen, Y. R. (1993). Vibrational spectroscopy of water at the vapor/water interface. *Phys. Rev. Lett.*, **70**, 2313–16.

Frysinger, G. S. (1992). Nonlinear optical spectroscopy of air/water interface chemistry. PhD thesis, Rensselaer Polytechnic Institute, Troy, NY.

Frysinger, G. S., Asher, W. E., Korenowski, G. M., Barger, W. R., Klusty, M. A., Frew, N. M. and Nelson, R. K. (1992). Study of ocean slicks by nonlinear laser processes. 1. Second-harmonic generation. *J. Geophys. Res.*, **97 C4**, 5253–71.

Frysinger, G. S., Barnoski, A. A., Gaines, G. L. Jr and Korenowski, G. M. (1994). Molecular arrangement of a poly(acrylonitrile-co-vinylpyridine) monolayer on water. *Langmuir*, **10**, 2277–80.

Goh, M. C., Hicks, J. M., Kemnitz, K., Pinto, G. R., Bhattacharyya, K. and Eisenthal, K. B. (1988). Absolute orientation of water molecules at the neat water surface. *J. Phys. Chem.*, **92**, 5074–5.

Hara, T., Bock, E. J. and Lyzenga, D. (1994). In situ measurements of capillary–gravity wave spectra using a scanning laser slope gauge and microwave radars. *J. Geophys. Res.*, **99 C6**, 12 593–602.

Jähne, B. (1990). From mean fluxes to a detailed experimental investigation of the gas transfer process. In *Air–Water Mass Transfer Second International Symposium*, ed. S. C. William and J. S. Gulliver, pp. 582–92. Minneapolis, MN: ASCE.

Jähne, B. (1993). Imaging of gas transfer across gas–liquid interfaces. In *Imaging in Transport Processes*, ed. S. Sideman and K. Hijikata, pp. 247–56. New York: Begell House.

Korenowski, G. M., Frysinger, G. S., Asher, W. E., Barger, W. R. and Klusty, M. A. (1989). Laser based nonlinear optical measurement of organic surfactant concentration variations at the air/sea interface. *Proceedings of the International Geoscience and Remote Sensing Symposium*, Vol. 3, pp. 1506–9, New York: IEEE.

Korenowski, G. M., Frysinger, G. S. and Asher, W. E. (1993). Noninvasive probing of the ocean surface using laser-based nonlinear optical methods. *Photogramm. Engng. Remote Sensing*, **59**, 363–69.

Li, Q., Zhao, M., Tang, S., Sun, S. and Wu, J. (1993). Two-dimensional scanning laser slope gauge: measurements of ocean ripple structures. *Appl. Opt.*, **32**, 4590–7.

Martinsen, R. J. and Bock, E. J. (1992). Optical measurement of ripples using a scanning laser slope gauge. Part I: Instrumentation and preliminary results. In *Optics of the Air–Sea Interface: Theory and Measurement*, ed. L. Estep, **1749**, pp. 258–71. San Diego, CA: SPIE.

Münsterer, T. and Jähne, B. (1993). A fluorescence technique to measure concentration profiles in the aqueous mass boundary layer. In *Symposium on the Air–Sea Interface Proceedings*, June 24–30, ed. M. Donelan, Marseilles, France.

Münsterer, T. and Jähne, B. (1994). A LIF technique for the measurement of concentration profiles in the aqueous mass boundary layer. In *Seventh International Symposium on Applications of Laser Techniques to Fluid Mechanics*, II, 29.41–5, Lisbon, Portugal.

Pankow, J. F., Asher, W. E. and List, E. J. (1984). Carbon dioxide transfer at the gas/water interface as a function of system turbulence. In *Gas-transfer at Water Surfaces*, ed. W. Brutsaert and G. H. Jirka, pp. 101–13. Hingham, MA: D. Reidel Publishers.

Shen, Y. R. (1984). *The Principles of Nonlinear Optics*. New York: John Wiley.

Shen, Y. R. (1989). Surface properties probed by second-harmonic generation and sum-frequency generation. *Nature*, **337**, 519–25.

Wei, Y. and Wu, J. (1992). In situ measurements of surface tension, wave damping, and wind properties modified by natural films. *J. Geophys. Res.*, **97 C4**, 5307–13.

Wolff, L. M., Liu, Z-C. and Hanratty, T. J. (1990). A fluorescence technique to measure concentration gradients near an interface. In *Air–Water Mass Transfer Second International Symposium*, ed. S. C. William and J. S. Gulliver, pp. 210–18, Minneapolis, MN: ASCE.

16

Remote sensing of the sea-surface microlayer

IAN ROBINSON

Abstract

Remote sensing methods, primarily from satellites, can contribute to the study of the sea-surface microlayer in several ways. An overview is given of the ocean parameters which remote sensing techniques can measure, and the spatial and temporal sampling capabilities of sensors which are useful for microlayer studies are described and explained. Infrared sensors measure the temperature of the sea-surface thermal skin, but to define the difference between radiometric and in-water measurements of temperature requires an understanding of the physical processes which control the thermal structure of the surface microlayer. The skin temperature deviation has an effect on the interpretation of global datasets of sea-surface temperature. Satellite images are also able to identify circumstances where local meteorological conditions introduce heterogeneity into the surface microlayer temperature. Imaging radars are very effective in measuring sea-surface roughness and detecting its spatial variability. The effects, on the radar backscatter cross section, of surface wind fields, waves, swell, slicks and other dynamical features enable radars to image a variety of physical processes which affect the surface microlayer. By combining data from more than one remote sensing technique it is proposed that air–sea fluxes may be estimated, and also that surface microlayer processes can be related to the wider spatial context of mesoscale variability. A better understanding of surface microlayer processes will improve the oceanographic analysis and application of satellite data, but further progress requires new instruments, including accurate shipboard infrared radiometers and instruments to measure surface roughness directly.

Introduction

Since heat, light, water and momentum enter the ocean through its boundary with the atmosphere, as do most of the gases dissolved in the sea, the sea-surface microlayer is the subject of intense study by marine physicists and chemists. In the wider context of 'earth system science', the ocean surface is the boundary between the atmosphere and the ocean,

471

and therefore acts to mediate the diverse air–sea exchange processes which help to control climate and its variations, both globally and regionally.

While the ocean 'surface' can be defined in fluid mechanical terms as a boundary between two fluids, it is a more elusive entity when it comes to studying the physical and chemical processes occurring at the surface. For this it is necessary to define the 'surface' as the thin microlayer of water which is found at the surface. The 1 μm scale of this layer which defines a surface as a visible discontinuity is still very large compared with the size of the water and other molecules of which it consists. However, this is no well-defined membrane such as might be considered in other surface sciences since the fluid particles making up the sea-surface microlayer may be constantly exchanging with others from within the bulk of the fluid. Making measurements in the sea-surface microlayer is therefore not an easy observational task, since the intrusion of the measuring or sampling device may be expected to change the conditions in an indeterminate way. This is the context in which methods of remote sensing have potential importance, since by definition these make non-contact measurements.

This chapter examines the role which remote sensing from satellites and aircraft can play in the study of the sea-surface microlayer. A limitation of satellite oceanography has always been recognized in the inability of remote sensors to observe much, if at all, below the sea surface. In our present context this weakness becomes a strength, and so it is surprising that the two fields of study have not interacted to a greater extent. Remote sensing methods offer a further advantage of being able to capture instantaneous image datasets of the two-dimensional spatial distribution of surface parameters at a resolution which spans two or three orders of magnitude of length scale.

These two themes underlie the rest of this chapter. After an overview of the achieved and potential capabilities of remote sensing for studying the sea-surface microlayer I examine processes in which remote sensing methods can make a major contribution – the thermal skin layer of the ocean, surface roughness modulations and the distribution of surface slicks. The potential for further developing remote sensing applications is discussed – including the possibility of estimating air–sea gas fluxes by combining data from several sensors and other datasets, and the exploitation of the unique spatial sampling capabilities of satellites.

Although prepared specifically for a sea-surface microlayer workshop, this chapter is intended not only to inform the interdisciplinary microlayer community about what remote sensing methods have to offer, but also to demonstrate that the reverse information flow is equally important. That is, the successful development of many of the methods of satellite oceanography depends upon an improved understanding of processes in the sea-surface microlayer. There is need for a two-way flow of ideas and information between the two communities, for collaboration with *in situ* experiments and for the cooperative development of new field instrumentation. Studies of the microlayer are also recognized to be necessary for developing the application of remote sensing to oceanography in general.

Sea-surface parameters observed by remote sensing

Ocean observations have been made from space for over 20 years. General introductions to the methods of satellite oceanography have been provided by Maull (1985), Saltzman (1985), Stewart (1985) and Robinson (1985). Since remote sensing relies on electromagnetic radiation communicating information about the sea through the atmosphere to the remote detector, the only parameters that can be measured directly are ocean colour, temperature, surface roughness and the slope of the surface, although a wide variety of other ocean measurements can be derived from these primary observables. The methods are summarized in Table 16.1. Here, we restrict consideration to those measurement techniques which can make a direct or indirect contribution to the study of the sea-surface microlayer.

Temperature

The thermal emission of the sea surface provides a direct way of determining its temperature from aircraft or satellites. Passive measurements of electromagnetic radiation in the thermal infrared and the microwave parts of the spectrum are used. The 10–12.5 µm waveband is most effective, since the thermal emission is at its strongest, and the measured radiance is sensitive to changes in temperature. Although the Planck relationship between temperature and emission is nonlinear in this waveband, it is not difficult to calibrate the radiance in terms of black-body temperature. The Advanced Very-high Resolution Radiometer (AVHRR) flown on the NOAA series of meteorological satellites is widely used and has a measurement sensitivity of about 0.2 °C (Lauritson *et al.*, 1979). A sensitivity of 0.1 °C is now achievable using the Along-

Table 16.1 *Methods of ocean remote sensing*

Method	Primary measurement	Derived parameters	Utility for microlayer studies
Infrared radiometry	High-resolution brightness temperature images	Sea-surface skin temperature	Detailed spatial view of skin temperature variability patterns
Microwave radiometry	Coarse-resolution brightness temperature	Surface temperature Surface wind Salinity (potentially)	Too coarse to contribute to process studies Applications to global monitoring of air–sea fluxes
Ocean colour sensors	Images of water-leaving spectral radiance (colour)	Near-surface chlorophyll and suspended sediment Evidence of whitecaps and foam	Future potential for imaging spectrometers to detect surface material
Radar scatterometer	Radar backscatter from sea-surface (coarse resolution)	Wind speed and direction	Applications to global monitoring of air–sea fluxes
Imaging radar	High resolution images of radar backscatter	High resolution variability of wind speed	Detailed mapping of surface slicks and modulations of surface roughness due to wind, swell, internal waves, etc
Altimetry	Point measurements of sea-surface height and slope	Geostrophic current variability Sea state	Sea-state measurements indirectly useful for global estimates of surface fluxes

Track Scanning Radiometer on the ERS-1 Satellite (Delderfield *et al.*, 1986). Corrections (see for example McMillin, 1975; Deschamps and Phulpin, 1980; Llewellyn-Jones *et al.*, 1984, McMillin and Crosby, 1984) are made for the atmospheric attenuation of the radiance, using the additional information gained from two wavebands (10.3–11.3 µm and 11.5–12.5 µm) within the main window and/or a further spectral window at 3.7 µm. The latter is only useful at night when there is no reflected solar energy at this wavelength. To improve atmospheric correction further, the ATSR views each sea area twice, once from the nadir and once obliquely. By this means sea-surface temperature (SST) measurements can be derived with an absolute accuracy of around 0.1 K.

Passive microwave sensors are much less sensitive for measuring temperature, although they have the merit of being able to penetrate cloud, which renders infrared detectors useless. Although there is a linear relation between microwave black-body radiation and temperature, the microwave emissivity of sea water also varies with other factors including viewing angle and surface roughness and hence depends on the sea state (Swift, 1980). In general, microwave sensors are not yet a practical option for observing the sea-surface microlayer, although they will make a contribution when improved sensitivity and resolutions are achieved. Whereas the radiation emitted in the thermal infrared comes from the water molecules within a few micrometres of the surface, truly a skin temperature, the emitted microwaves come from the topmost few millimetres, depending on the wavelength used. A lower-frequency detector would observe temperatures from a deeper layer than a higher-frequency detector. Were it possible to intercalibrate two detectors to sufficiently high accuracy, there is the potential for measuring the near-surface thermal gradients and hence heat flow though the surface (McAlister and McLeish, 1970), although an accurate operational method of doing so has yet to be achieved.

Material within the surface microlayer can alter the emissivity. For example, in the infrared a surface film of oil reduces the emissivity, and reduces the apparent brightness temperature below the true temperature. However, because the oil may reduce the heat flux through the surface and absorb more solar radiation, causing a slight increase in the skin temperature, this is an uncertain method for detecting surface films with scientific accuracy, although it is used as one of a range of techniques for operational monitoring of oil spills from aircraft (Hurford and Tookey, 1986). Alpers *et al.* (1982) demonstrated the potential for using microwave radiometers to study surface films.

Surface roughness

Surface roughness is observed remotely by a variety of means. The most readily understood method is that which uses visible light reflected from the surface, since this is familiar to our experience using human vision. Specular reflection of sunlight can be used to measure the mean slope of the surface (Cox and Munk, 1954). Close to the specular point, the increase of roughness reduces the brightness of reflection whereas, for a point on the surface distant from the specular point, roughness increases the number of facets reflecting the sun and hence the brightness increases. However, the use of solar reflection does not lend itself to quantitative observations because of the need to know the viewing geometry relative to the solar direction at each point on a sun-glitter image. The method requires open skies without cloud cover and yields useful data for only a limited range of viewing angles. Qualitatively, the approach is more useful, since given favourable viewing geometry from an aircraft, ship's mast, clifftop or spacecraft it is possible to detect subtle spatial patterns of differing surface roughness.

The photographs from the US Space Shuttle made by Scully-Power (1986) are striking examples of the effectiveness of this method. The sun-glitter patterns reveal the way in which the distribution of surface roughness is controlled by the underlying dynamical structures in the upper ocean at various length scales, e.g. wind-rows, Langmuir circulation patterns and mesoscale eddies. Features appear as lines of rougher water or slicks of smoother water – sometimes brighter, sometimes darker than the surrounding surface. It is not easy to derive quantitative measures from sun-glitter photographs but they can provide a useful complement to *in situ* observations of the surface microlayer by demonstrating the occurrence and spatial patchiness of slicks.

Active microwave sensors, radars, offer a more systematic way of measuring and detecting patterns in the sea-surface roughness. Within the resolution cell (which is dictated by the type of radar and its operating parameters) the radar backscatter is a function of the mean surface slope, as controlled by long-wavelength surface waves, and the surface 'roughness', as defined by surface fluctuations at length scales shorter than the resolution cell and generally longer than or comparable with the radar wavelength. For near-nadir incidence, an image of radar backscatter is similar to sun-glitter patterns near the specular point. The greater the roughness the weaker the backscatter.

The radars most relevant for studying microlayer processes are oblique viewing imaging radars which rely on weak backscattering, which increases with the surface roughness. A generally held assumption is that Bragg scattering is the dominant mechanism controlling the backscatter from oblique radars (Valenzuela, 1978). In this model, the ripples which significantly affect the radar backscatter are those which have a wavelength $\lambda_B = \lambda_r/2 \sin \theta$, where λ_r is the radar wavelength and θ is the incidence angle.

However, despite its wide acceptance as a working hypothesis, there are theoretical and empirical difficulties with applying Bragg resonance to backscatter from a rough surface. Bragg theory is derived from the interaction of electromagnetic waves with a regular lattice structure, whereas the ocean surface has no regular structure. λ_B is typically a few centimetres for X- or C-band radar, and it is assumed that the radar selectively responds to such waves. However, in the ocean the short Bragg waves ride on the back of longer waves whose amplitudes are at least an order of magnitude greater than the Bragg scale length. Their effect is to displace the Bragg waves randomly relative to the incident radar field, potentially interfering with the Bragg resonance mechanism. There has been no experimental demonstration that in these conditions the radar responds to a selective waveband of surface waves. Indeed, the fact that the backscatter goes on increasing with increasing wind speed even when the wave spectrum has saturated at the Bragg wavelength implies that the radar is just as sensitive to the longer waves as to the shorter. The integral solution of the backscatter equations achieved by Holliday *et al.* (1986) seeks to take this into account.

Multiple scattering by microwaves encountering a rough surface can also affect the radar response and the polarization state of the microwaves. Experimental evidence (Bartsch *et al.*, 1987) suggests that VV radars (emitting and detecting vertically polarized microwaves) are better able to detect changes in surface roughness than HH (horizontally polarized) radars, and this is consistent with the experience of the ERS-1 radar. However, cross-polarization radars may be able to image some surface roughness features more clearly than using a single polarization emitter and detector.

There is a family of different active microwave sensors which are used for a variety of purposes in oceanography. *Scatterometers* view obliquely to provide a measure of the average roughness over sea areas with dimensions of many kilometres. Whilst they can make a contribution to microlayer observations in that they measure wind speed and direction, they cannot detect patterns of surface roughness. Similarly *radar alti-*

meters, which are very effective in measuring the significant wave height averaged over a footprint with a linear dimension of around 10 km, cannot detect subtle changes in roughness. In contrast, *imaging radars* can generate detailed images of the spatial distribution of surface roughness, and potentially can make a unique contribution to studies of the sea-surface microlayer, to be discussed later.

Ocean colour

Ocean colour is measured using visible wavelength radiometers. The primary application of these sensors is to determine the concentration of material in the upper ocean which absorbs or scatters light (Robinson, 1990). Visible waveband radiometers are the only sensor type which directly detect information from below the sea surface. The ocean colour is influenced most by the near-surface waters, but some of the photons reaching the sensor will have scattered from the depth to which solar illumination penetrates (30–60 m in clear water). It is the combination of backscatter and absorption of different wavelengths of visible light in the ocean which controls the spectral mix, the colour, of the water-leaving radiance. The data from scanning ocean colour sensors, particularly the ratio of green/blue radiance, can be interpreted in terms of the upper ocean chlorophyll content (Gordon and Morel, 1983) if primary production is the dominant factor affecting the colour. For coastal waters the presence of land-derived dissolved organic substances or suspended sediment make analysis of satellite ocean colour data more difficult.

Probably the most useful contribution which ocean colour remote sensing can make to sea-surface microlayer studies is to provide a synoptic spatial overview of the distribution of phytoplankton biomass in the upper layer of the ocean (e.g. Abbot & Zion, 1987). This is indirectly relevant since phytoplankton are often assumed to be the potential source for biogenic organic compounds found in the surface microlayer. When eventually an ocean colour scanner is in orbit at the same time as an imaging radar, it will be interesting to correlate globally the occurrence of surface slicks with the distribution of phytoplankton blooms, in order to explore the validity of this assumption.

Direct observation from satellites of the colour of the surface microlayer is more difficult because it is difficult to distinguish between light scattered from within the microlayer and that backscattered from the water column zone below. Enhanced reflectance in the wavebands at the red and near-infrared end of the spectrum indicates reflective material close to the surface, since these wavebands are otherwise strongly absorbed as soon as the light penetrates deeper into the water. Scanning spectrometers which derive image data in each of 80 or more narrow spectral wavebands (Gower & Borstad, 1990) may in future provide a way of directly detecting coloured material in the surface layer.

Sampling length and timescales of remote sensing observations

The sampling capabilities of remote sensing methods enable them to provide otherwise unobtainable data from a unique perspective, and hence to cast new light on outstanding scientific problems. On the other hand, the inflexibility of those sampling capabilities (especially from satellites) can also impose severe limitations on their applicability. Figure 16.1 summarizes the space time sampling capabilities of typical remote

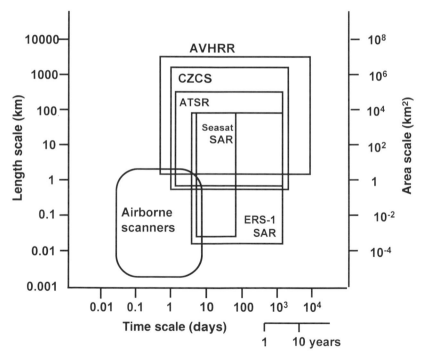

Figure 16.1. Sampling capabilities of remote sensing instruments. Each instrument is represented by a box, the left boundary is the shortest revisit time, the right boundary is the timespan over which data have been collected, the lower boundary is the spatial resolution of the detector and the upper boundary is the largest length scale for a synoptic image. The limitation imposed by cloud cover has been ignored. The sensor abbreviations are explained in Table 16.2.

Table 16.2 *Remote sensing instruments useful for observing the sea surface microlayer*

Sensor	Satellite	Measurement relevant to microlayer studies	Spatial resolution	Size of image (km)	Shortest revisit interval (cloud permitting)
AVHRR – advanced very high resolution radiometer	NOAA series (1978–present)	Infrared radiation (three wavebands)	1.1 km	2400	12 hours
ATSR – along-track scanning radiometer	ERS-1 (1991–present)	Infrared radiation (three wavebands and two directions)	1 km	500	3 days 12 hours for crossovers
C-Band SAR – synthetic aperture radar	ERS-1 (1991–present)	Normalized radar backscatter cross section	12.5 m (cell size) 30 m (resolution)	100	3 days to 17 days depending on orbit
L-Band SAR	Seasat (1978)	Microwave backscatter	25 m	100	3 days
CZCS – coastal zone colour scanner	Nimbus-7 (1978–1986)	Visible radiance (four wavebands)	850 m	1500	2 days
Various visible and infrared sensors; synthetic and real aperture radars	Various aircraft	Visible and infrared radiance Microwave backscatter	Variable 5 m–50 m	Variable 1–10	Variable, minutes to days

sensing platform–sensor combinations identified in Table 16.2, considered to be useful to studies of the surface microlayer. We are concerned here primarily with imaging sensors which yield datasets covering wide areas almost instantaneously. Their usefulness should be considered in the light of the space–time sampling requirements of experiments to observe the surface microlayer.

Spatial scales

The spatial sampling performance is determined by the spatial resolution of the detector, the size of the image cell or pixel, and the maximum width of image which can be captured nearly instantaneously in a single overpass. Most sensors are designed to sample efficiently by arranging that the spatial resolution of the detector approximately corresponds to the area scanned in the spatial sampling interval. Each data point in the image corresponds to an ocean surface parameter, which is integrated over a small area of sea (the instantaneous field of view) contiguous with similar areas which contribute to the value recorded for adjacent pixels. In this respect, remotely sensed data are better matched for comparison with two-dimensional models of ocean processes than *in situ* measurements, which are made at individual points not necessarily representative of their surroundings.

Sensor datasets can be categorized into four classes. *Coarse-resolution* data with a cell size of 50–100 km come from scatterometers on polar orbiting satellites, giving a measure of average surface roughness and the wind speed and direction. Sea-surface temperature at 50 km resolution is derived by integrating higher resolution data; a temperature resolution better than 0.1 °C is achieved at the expense of spatial resolution. Such data have no direct value in detailed studies of the sea-surface microlayer, but they can provide the background context within which to interpret local studies. They also provide the simplest route to worldwide datasets from which the global integration of air–sea interaction processes can be evaluated.

Medium resolution sensors on polar orbiting satellites yield data having a resolution of about 1 km and synoptic coverage over distances of 1000 km or more. Several thermal infrared and ocean colour sensors fall into this category. Their direct use for detailed studies of sea-surface microlayer processes is limited, but their ability to reveal the spatial variability of the sea-surface temperature and patchiness of primary production over length scales down to a few kilometres is of considerable value and will be discussed later.

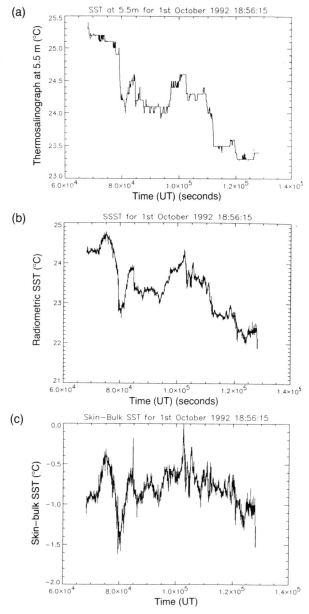

Figure 16.2. An example of a transect of measurements made along the track of the *R.R.S. James Clark Ross* on 1–2 October, 1992, at 20° S, 38° W off Brazil in the South Atlantic Ocean. (a) Sea-surface temperature (bulk) measured by thermosalinograph sampling at 5.5 m depth. (b) Sea-surface temperature (skin) measured by a ship-board infrared radiometer. (c) The skin temperature deviation δT (the difference between the two traces in a and b).

Finer resolutions of 30 m or better are achieved by *high-resolution scanners* on polar-orbiting satellites. To stay within manageable data rates, such high resolutions demand that the swath scanned by the sensor is much smaller, typically less than 200 km wide. Colour sensors are capable of detecting the patchiness of surface material, if it has a visible optical signature. Measurements from high-resolution thermal infrared sensors have so far not achieved radiometric resolution better than 1 °C, and are of little use to microlayer studies at present. A similar resolution of 25–30 m is obtained using synthetic aperture radars (SAR). Both the Seasat SAR, in 1978, and since 1991 the ERS-1 SAR, have provided a completely new view of the ocean surface and the distribution of roughness elements. SAR images are acquired irrespective of cloud cover, and extend over a field of view 100 km wide. As discussed later, they have a significant contribution to make to surface microlayer studies. Future SAR missions are being prepared, promising continuity of coverage for future experiments. Until data-relay satellites are used, SAR is limited by its high data rate to recovering images only from locations viewed when the satellite is within line-of-sight of a receiving station. As more receiving stations are established around the world this becomes less restrictive, but at present most of the available archived data are from US or European marginal seas.

Higher-resolution images which can detect patchiness on a scale of 1 m to 10 m require the use of *airborne scanners*. For visible and infrared detectors, their radiometric resolution is similar to satellite scanners, although airborne infrared sensors do not yet match the thermal stability of satellite instruments. The swath width of airborne scanners is correspondingly reduced to a few kilometres. The finest spatial resolution will be achieved using 'remote' scanning sensors mounted close to the surface on ships or buoys, and in future these may provide a means of observing the instantaneous patchiness of the surface microlayer properties on length scales of centimetres to metres.

Temporal sampling scales

To oceanographers used to making time series of *in situ* measurements at high sampling frequencies, the capabilities of remote sensing may seem very poor. A polar orbiting satellite normally makes about 14 orbits per day fixed in the same plane relative to the sun, while the sensor paints out a swath on the ground or ocean surface which migrates westward as the earth rotates underneath. For wide swath *coarse- and medium-resolution sensors*, the whole earth is covered once every 1 to 3 days, with a daytime pass, and similarly for a nightime pass. The *high-resolution sensors* have a narrower swath and earth coverage is achieved only once every 16 to 30 days. There is thus a trade-off between spatial and temporal resolution. The sampling interval is further extended for visible and infrared sensors because cloud cover often prevents image acquisition.

It must be accepted, therefore, that satellite remote sensing does not have the ability to monitor processes on a timescale of less than a few days. What it does achieve is a near-instantaneous view of detailed spatial patterns without the problem of space–time aliasing encountered by ship sampling. At high latitudes the swaths of adjacent orbits overlap considerably, and successive images are obtained about 100 minutes apart. To improve further on the time sampling capabilities, aircraft must be used. These can revisit the same place far more frequently during a limited period, although it is difficult on each occasion to reproduce exactly similar viewing conditions of altitude, attitude, orientation and solar illumination.

The other advantage of satellite observations, which is sometimes overlooked, is their capacity to continue making self-consistent observations year after year for several decades. Although the sensors may be replaced every few years with new satellites, the stability of instrument calibrations means that remote sensing measurements can be used for monitoring and detecting small changes in surface ocean properties over long periods of time.

Infrared measurements of ocean surface temperature

The thermal skin effect

Infrared radiometry detects the temperature of a layer of water within a few micrometres of the sea surface. It has been realized for a number of years that there are often strong thermal gradients in the topmost millimetre or less of the ocean (see for example: Woodcock and Stommel, 1947; Ewing and McAlister, 1960; Grassl, 1976 and reviews by Saunders, 1973; Katsaros, 1980 and Robinson *et al.*, 1984). These occur when the net heat flux through the surface (normally but not always from sea to atmosphere) flows through the surface microlayer, in which the heat transport is controlled by molecular conduction processes, turbulent eddy convection being inhibited by the stabilizing proximity of the air–sea boundary. Although the laminar sublayer and associated molecular conduction layer are very thin, the flux is such that a temperature deviation of typically 0.2 °C can occur between the radiometrically measured temperature of the skin and the temperature a millimetre below. Normally, the skin is cooler than the water below in conditions of less than 100% relative humidity. The stronger the heat flux, the greater the temperature deviation, but since sub-surface turbulent stirring close to the skin is promoted by shear stress on the surface the effect of increased wind is to reduce the thermal skin effect. Saunders (1967) sought to parameterize this by expressing the temperature deviation, δT, of the skin below the sub-skin temperature, as

$$\delta T = \frac{\lambda Q_N \nu}{k U^*} \qquad (16.1)$$

where Q_N is the heat flux, ν is the kinematic viscosity and k is the molecular conductivity of the sea water, and U^* is the friction velocity of the surface wind. However, attempts to derive the proportionality constant λ have produced inconclusive results and there is as yet no experimental justification for such a simple expression. However, as a simple representation of the important physical processes, the linear expression is useful. For example it indicates that the overall effect of wind on δT is moderated because Q_N also increases with U^* through the latent heat flux. Observations have shown that δT can reach up to 0.6 °C in the sea, and much greater in laboratory studies. Under conditions where the atmosphere is warmer than the sea, and high humidity reduces the latent heat flux, a skin which is 0.3 °C warmer than the water below has been detected.

For calm or very light wind conditions, the physical process promoting turbulent heat exchange is the gravitational instability of the cool surface layer. Under these conditions, Saunders (1967) proposed an expression for δT based on free convection:

$$\delta T = A \left[\frac{Q \varkappa \nu}{k g \alpha} \right]^{3/4} \qquad (16.2)$$

where $\varkappa$ is the thermal diffusivity of sea water, α is the coefficient of thermal expansion and g is the acceleration due to gravity.

A number of other approaches to modelling δT in terms of measurable air–sea interaction parameters such as wind speed, air temperature, humidity, etc. have been presented *inter alia* by Hasse (1971), Grassl (1976), Katsaros *et al.* (1977), Paulson and Simpson (1981) and Eifler (1992). Recently, Soloviev and Schluessel (1994) have developed a renewal model which seeks to parameterize both the temperature dif-

ference and air–sea gas transfer across the surface microlayer in relation
to the different physical processes which cause the skin layer to be
replaced under different wind-speed regimes.

Radiometric observations of the thermal skin

Progress in understanding the thermal skin effect is hampered by the
difficulty of measuring it. *In situ* measurements using contact thermo-
meters cannot be made confidently inside the skin layer, without the
presence of the probe altering the temperature. Mammen and von Bosse
(1990) sought to avoid this problem with a high frequency thermistor
which 'popped-up' through the surface. The best way to make sure that
the true surface is being sampled is to measure the skin temperature
radiometrically in the infrared. However, to be able to measure δT also
requires a measurement of the below-skin temperature, which must come
from an alternative temperature sensor. It is very difficult to intercalibrate the two independent measurements to the accuracy, better than
0.1 °C, which is needed to quantify the skin temperature deviation.
Radiometric measurements also require a knowledge of the surface
emissivity and a correction for reflected downward radiation, which
varies with the viewing angle (Masuda *et al.*, 1988). Most systematic
attempts to measure the skin effect (e.g. Schluessel *et al.*, 1987) have
intercalibrated between the radiometer and an in-water thermistor by
using the radiometer to view a vigorously stirred tank, whose bulk
temperature is measured by the thermistor. Other attempts to measure
δT directly have stirred the actual sea surface in the field of view of the
radiometer for comparison with the unstirred surface. The problem with
either approach is that it must be assumed that stirring completely breaks
down the thermal skin. This is a tenuous assumption because it is known
that the surface skin can reform within seconds.

   Where attempts have been made to measure the skin temperature and
its variation over length scales of centimetres to metres, small-scale
structures with amplitudes comparable to δT itself have been found.
Observations by Jessup (1992) using a small-scale imaging radiometer on
the FLIP research platform have revealed that breaking waves can leave
behind patches of water with a temperature change of 0.3–0.4 °C, which
persist for approximately 1 s. In the same series of experiments more
persistent patches having changes in temperature of about 0.5 °C were
found in low wind conditions, and may be associated with surface slicks.
It appears that the more carefully the sea-surface skin temperature is

observed, the more small-scale variability is found with a magnitude in the range 0.1–0.5 °C. There is evidently a need to make more detailed observations with imaging radiometers over spatial scales of 10 mm to 10 m, in order to understand more about the processes which control the thermal skin, and hence to be able to parameterize and predict it in terms of typical 'bulk' meteorological and oceanographic parameters.

The most useful experimental studies of the thermal skin have been those which have measured radiometric temperature along transects from vessels underway or from aircraft. Schluessel *et al.* (1990) measured the spatial variability of the skin temperature and sought to correlate it with the variability of the bulk temperature. They found correlations greater than 0.75 for long-wavelength variability (>80 km), but much reduced for shorter wavelengths. C. J. Donlon (pers. commun.) has recently made similar measurements in a north–south transect through the Atlantic from 50 ° N to 50 ° S. Figure 16.2 is an example of a section of this transect, showing the variability of skin and bulk temperature and other measurements over a distance of about 350 km.

Calibration of global SST measurements

A better understanding of the physics of the thermal microlayer of the sea surface is required if satellite-derived measurements of sea-surface temperature are to contribute effectively to monitoring changes in SST as evidence of climate change. The spatial distribution of SST is potentially a useful indicator for climate change, and polar orbiting satellites provide an effective way of measuring it. Recently developed sensors such as the ATSR flown on ERS-1 are potentially capable, after atmospheric correction, of measuring the average SST over a 50×50 km area to an accuracy of better than 0.1 °C (Mutlow *et al.*, 1994). However, to turn this into a useful measurement requires a sound understanding of thermal processes in the surface microlayer.

Because it is the skin temperature which is measured from satellites, it is necessary for two reasons to be able to relate the skin temperature to the 'bulk' temperature as measured typically at 1 to 3 m depth from ships. Firstly, to calibrate and validate the atmospheric correction algorithms requires as many *in situ* measurements as possible, coincident with satellite overpasses. Ideally, the atmospheric corrections should be based on skin measurements but, given the scarcity of reliable shipborne radiometers, it is necessary to make use of the many bulk SST observations which are available, corrected for the skin temperature deviation.

Secondly, if satellite measurements of SST are to be compared with historical data from *in situ* observations, either the former should be reduced to bulk temperature, or the latter to skin temperature. For both reasons, it would be desirable to have a reliable model or parameterization which could predict δT from a knowledge of other measurable air–sea parameters. Equations 16.1 and 16.2 have not proved to be sufficiently robust, and there are as yet insufficient test data with which to validate the models proposed by the other authors mentioned above. Consequently, this remains a problem of considerable importance which has not yet been solved satisfactorily.

It should be noted that current SST databases such as NOAA's MCSST product (McClain *et al.*, 1985; Brown *et al.*, 1993), which rely on a global network of buoy measurements of temperature, avoid the issue by incorporating the skin temperature deviation into the atmospheric correction. The algorithms are tuned to match the measured bulk temperatures and thus, while the offset due to the mean δT is included in the correction, the result can never be more accurate than the variance in δT. These issues are discussed further by Wick *et al.*, (1992).

There is another reason why the daytime skin temperature may differ by several degrees from the temperature which is measured from ships at a depth of 2–5 m and which is taken as the temperature of the upper mixed layer. This is the occurrence of the diurnal thermocline which forms in calm sunny conditions and which may raise the surface temperature by 1 to 5 °C above the measured bulk temperature. The most reliable way of eliminating this factor is by basing the satellite SST measurements on night-time images only, because after nightfall it is expected that the surface rapidly cools, overturning takes place and the diurnal thermocline is destroyed.

Evidence of surface processes from infrared images

As a contribution to sea-surface microlayer studies, satellite-derived thermal images can sometimes provide evidence of locally varying air–sea interaction processes which would be difficult to detect in other ways. Certain patterns appear in temperature images which have spatial characteristics more appropriate to atmospheric than oceanic processes. Two examples from UK coastal waters serve to illustrate this.

In Figure 16.3, the brightness temperature distribution in the northeast Atlantic is shown off the Irish Coast. Features defined as zones of water approximately 1 to 2 °C warmer than their surroundings can be seen,

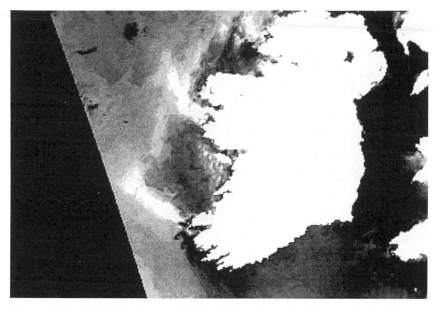

Figure 16.3. AVHRR thermal (10.5–11.5 μm) band 4 image of the Atlantic Ocean off the coast of Ireland, Atlantic Ocean off the coast 1505 h, 24 April 1984. This is presented with higher temperatures brighter and cooler temperatures darker. Note the streaks of warm water spreading northwestwards from several headlands.

apparently spreading northwestwards from prominent headlands and curving gradually towards the north. This was an afternoon image. Although there are no *in situ* observations available to confirm this, it is most likely that the warm patches are regions where the diurnal thermocline has developed preferentially because the wind is reduced and the turbulent stirring is consequently weaker. The zones appear to be associated with wind shadowing behind hills, a coast-induced effect which extends hundreds of kilometres offshore.

Another example, Figure 16.4, shows a wide zone of water several degrees warmer than its surroundings in the centre of the North Sea. Comparison with the subsequent night overpass shows that the warm patch is ephemeral. It reappears the next afternoon but is displaced by tens of kilometres. The suggested explanation is that this also is a diurnal thermocline. Its position in this case corresponds to a ridge of high pressure so it is probably controlled entirely by meteorological factors.

In both these cases, the perspective of satellite remote sensing has detected the spatial extent of a phenomenon which is harder to observe from the close perspective of ship-based measurements. Whilst a diurnal

Figure 16.4. AVHRR thermal (10.5–11.5 μm) band 4 image of the North Sea obtained at 1425 h 28 May 1989. This is presented with higher temperatures brighter and cooler temperatures darker. Note the warm patch stretching from southwest to northeast. The black region to the northwest of the image is cloud.

thermocline is not strictly a surface microlayer phenomenon, its existence is important for the microlayer, and its patchiness is evidence of horizontal variability in air–sea interaction processes.

Radar measurements of surface roughness

Surface roughness is another parameter of relevance to the sea-surface microlayer which can be measured directly using remote sensing. The surface roughness can be used to deduce information about several aspects of the surface microlayer. It serves as an indicator of the mechanical stress being exerted by the wind on the sea surface. Variability of the roughness can be related to the horizontal stretching, shearing and compression of the microlayer by the divergence, shear and convergence of the surface currents associated with long surface waves and a number of other dynamical phenomena. The surface roughness is also affected by the presence of dissolved or dispersed organic material at the surface producing films and slicks. Radar is a particularly useful way of studying any perturbations of the surface microlayer because it is a non-contact measurement, whereas *in situ* instruments disturb the surface by their presence and can interfere with the phenomena being studied. Thus remote sensing can provide unique observations of the surface to obtain measurements which could not be made in any other way.

Surface winds

The spatial distribution of wind fields is detected using oblique-viewing radar. The averaged backscatter from a wide area of sea, as measured by a scatterometer, can be used to obtain the wind speed and direction (see, for example, Jones *et al.*, 1982, for Seasat, and Lecomte and Attema, 1993, for ERS-1). Whereas scatterometer measurements cannot give detailed local information about wind distribution, this can be derived from synthetic aperture radar images. The same processes govern back-scatter from the surface and its dependence on the wind, whether the radars have high or low spatial resolution. Thus, the empirically derived scatterometer model which relates radar backscatter cross-section to wind speed and direction can also be applied to SAR images. The model being used at present for interpreting ERS-1 backscattered radar is CMOD-4 (Stoffelen *et al.*, 1992), derived empirically from scatterometer calibration experiments. It is now possible (Scoon *et al.*, 1995) to calibrate ERS-1 SAR in terms of wind strength, as long as the wind direction is known approximately, enabling detailed measurements to be made of the spatial variability statistics and spectra of the surface wind.

A variety of situations have been observed in which SAR images reveal patterns in the sea-surface roughness which are caused by dynamical structures in the wind field. The simplest case is where the sea surface is sheltered by some orographic feature on the adjacent land, leading to smoother regions of sea surface, which show up as dark patches (low backscatter) on a SAR image. This is clear in Figure 16.5, where the island of St Kilda in the northeast Atlantic shelters an elongated zone of calmer water in its lee. This may be a similar wind phenomenon to those

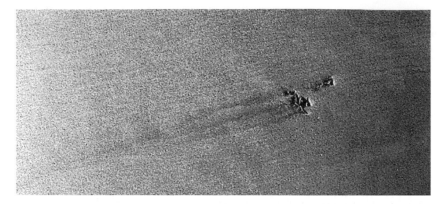

Figure 16.5. ERS-1 SAR image on 13 September 1993 showing the island of St Kilda in the North Atlantic (8.5° W, 57.9° N). Note the dark region (low radar backscatter) in the lee of the island. The width of the image is 100 km.

which can be identified on thermal imagery, as in Figure 16.3. An increasing number of examples are being published of ERS-1 SAR detecting regular atmospheric wavelike phenomena, such as gravity waves (Thomson *et al.*, 1992). Gower (1993) observed atmospheric lee waves generating a regular pattern of bands of higher and lower back-scatter from the sea off Vancouver Island. Vachon *et al.* (1993) documented similar features generated by airflow over the island of Hopen in the Barents Sea. Alpers and Brümmer (1994) have confirmed that the ERS-1 SAR is able to detect the patterns of roughness signature associated with atmospheric boundary layer rolls in the German Bight. The sea-surface signature of atmospheric fronts can also be detected very clearly on some SAR images such as that shown in Figure 16.6, where an occluded front stretches from north to south over the south coast of England, producing a linear, low-backscatter feature across Christchurch Bay.

Waves and swell

Longer wind waves and swell can be observed explicitly by SAR if their wavelength is greater than at least twice the radar resolution cell size. Imaging occurs when the swell modulates the surface roughness in such a way that there is a constant phase relationship between the radar back-scatter and the surface displacement. Under suitable conditions radar images can give the appearance of a photograph of a wave field on the sea surface (Figure 16.7). Three mechanisms for wave imaging have been

Figure 16.6. ERS-1 SAR image on 23 April 1993 showing the effect of an atmospheric front across Bournemouth Bay in the lower left quadrant. The width of the image is 100 km.

identified and are quite well understood (see for example Alpers, 1983, and Valenzuela, 1978). The tilt modulation and velocity bunching mechanisms are related respectively to the orientation of surface facets and the vertical motion of the surface caused by waves. They are not directly concerned with the surface microlayer. The third mechanism, hydrodynamic modulation, is caused by surface convergence and divergence, associated with the orbital motion of the longer waves, compressing and stretching the wave action density (Phillips, 1977) of the short waves, which control the roughness as detected by SAR. This is of more direct relevance to the surface microlayer, since the surface velocity field must also be acting on the material in the surface in a similar way, condensing it or stretching it.

From SAR images which reveal modulations due to long waves and swell it is possible to derive directional wave spectra. Because the strengths of the three swell imaging mechanisms vary with direction and wavelength, the modulation transfer function (MTF) between the surface wave height field and the wavelike perturbations on the corresponding radar backscatter image is directionally dependent and nonlinear (Hasselmann *et al.*, 1985; Brüning *et al.*, 1990). There remains some

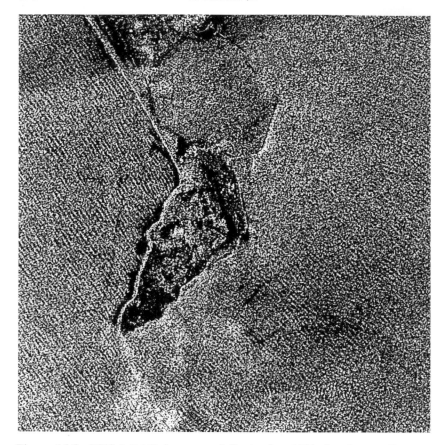

Figure 16.7. ERS-1 SAR image on 9 September 1991 showing swell waves incident on Chesil Beach in southern England and refracting around the headland of Portland Bill. Note that this is at a higher resolution than Figures 16.5 and 16.6, the width of the image being about 12.5 km.

uncertainty about the validity of the proposed models for MTF. For example, Ocampo-Torres and Robinson (1990) pointed out that the MTF could be critically sensitive to assumptions about the directionality of the short wind waves. Assumptions about the high-frequency roughness are hard to test because of the difficulty of measuring the high-frequency end of the wave spectrum independently of the radar method. However, Hasselmann and Hasselmann (1991) and Brüning *et al.* (1993) have made considerable progress in providing a robust method of inversion which permits ocean wave spectra to be derived from SAR image spectra with more confidence.

Surface slicks

The presence of material in a monolayer or thin microlayer at the surface can dampen the short-wavelength ripples which affect radar backscatter (Barger *et al.*, 1970; Alpers and Hühnerfuss, 1989; Tang and Wu, 1992). One mechanism proposed for this is the Marangoni effect (Hühnerfuss *et al.*, 1987; Alpers and Hühnerfuss, 1988). The motion associated with the high-frequency surface waves causes compression and stretching in the surface film, resulting in surface tension gradients. It is suggested that these allow energy to be fed from the gravity–capillary wave into a longitudinal boundary wave known as a Marangoni wave which, being heavily damped, causes rapid attenuation of the wave energy at wave-numbers which couple with the Marangoni waves. The effect is strongest for frequencies in the range 2–15 Hz. Alternatively, Scott (1989) points out that resonating Marangoni waves have not been observed directly and that damping can more simply be attributed to the changes the film makes to the weak boundary layer at the surface. The application of a constraint at the surface, in the form of a dilatational elasticity, increases the dissipation at the surface, where the velocity shear associated with surface waves is the greatest. With increasing surface elasticity the wave damping tends to that which would occur at a rigid wall. Irrespective of which is the appropriate model, it is worth noting that here is a mechanism by which a very thin surface microlayer can exercise a control on the sea surface out of proportion to the quantity of material in the microlayer.

By means of such damping, and also possibly through dielectric effects, films of naturally occurring organic material, or of hydrocarbons of anthropogenic origin, can sometimes be detected on radar images as slicks of lower backscatter than the surrounding sea surface (Vesecky and Stewart, 1982; Onstott and Rufenach, 1992). Slicks will not always be detected, because in very low sea states the whole sea may be too smooth for detectable modulation to be produced by the film, whilst in high seas the radar backscatter will saturate both inside and outside the slick, even if it is not dispersed. It is in light to moderate winds that surface films are likely to be most readily detected by SAR. Higher-frequency radars, whose backscatter is controlled by higher-frequency surface waves, might be expected to be more sensitive to the presence of film damping of the short-wavelength ripples. This appears to be borne out by the C-Band ERS-1 SAR revealing much more detail of slick patterns than the lower frequency L-band SAR of Scasat (Macklin, 1992). Hühnerfuss *et al.*

I. Robinson

Figure 16.8. ERS-1 SAR image showing slicks in the English Channel. The width of this image is 100 km, its height is 200 km.

(1994) have developed further the use of multi-frequency radar with the aim of classifying the material present in slicks.

Figure 16.8 illustrates a variety of slick-like patches within an SAR image. The distinct slick in the central English Channel, whose source is unknown, illustrates two characteristics of slicks on images: they tend to be clearly defined with distinct edges, and they act as tracers of the underlying velocity field. In this case there is clear evidence of horizontal shear in the tidal currents which flow normal to the long axis of the slick patch, resulting in the sinusoidal form of what is assumed to have been initially a linear feature. Ochadlick *et al.* (1992) have made a detailed study of SAR-imaged slicks interacting with surface currents. The well-defined edges of the slick regions on images may be a consequence of a nonlinear relation between the film thickness and its effect on the roughness and hence the radar backscatter. Ermakov *et al.* (1985), (1992) and Demin *et al.* (1985) have studied these processes in some detail. It should be noted that a monolayer or thin film may be present without it affecting the surface ripples and appearing as a slick, and the apparent slick edge may not extend to the true edge of the surface film.

Figure 16.8 reveals, in addition to a number of smaller slicks similar to the one discussed above, another phenomenon often found on SAR images, which is zones of low backscatter very close to the coast. These occur on both sides of the English Channel. Whilst they may simply represent areas where the wind is shielded from the sea by the land, their distinct edges and locations suggest that they may be due to surface films, possibly associated with coastal discharge or runoff from the land, or a concentration of material by bubble processes in the surf zone.

Other dynamical features imaged by SAR

Internal wave fields and mesoscale fronts both generate convergent surface velocity fields which tend to concentrate surface films. This is therefore another mechanism in addition to hydrodynamic surface wave modulation which can cause internal waves to be imaged (Onstott and Rufenach, 1992). Johannessen *et al.* (1993) have studied spiral eddy features imaged by ERS-1 SAR, and concluded that they are caused by surface slicks embedded in the motion of a cyclonic eddy. Similar patterns of the roughness signatures of eddies were detected in sun-glitter photographs of the sea surface observed by Scully-Power (1986). There is, however, a difference between visible (surface-reflective) observations of slicks or roughness and radar observations. The visible images observe

498 *I. Robinson*

foam (caused by breaking waves and collecting at convergence lines), whereas radar does not. The significance of remote sensing observations to the study of the sea-surface microlayer is the way they demonstrate how small-scale, apparently local, surface processes interact with, and are constrained by, mesoscale dynamical processes. Microlayer experiments should therefore not be divorced from an awareness of the wider surface flow field in which they are conducted.

SAR is also effective in imaging the horizontal patterns of surface roughness associated with internal waves. Indeed, SAR has provided the only practical way of observing the horizontal structure of internal wave trains (Apel and Gonzalez, 1983; Alpers, 1985.) For surface signatures of internal waves, the hydrodynamic modulation is the primary imaging

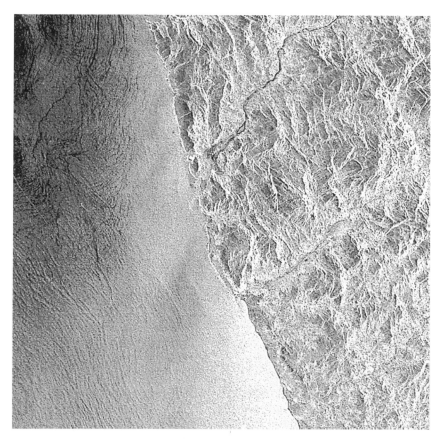

Figure 16.9. ERS-1 image of the Portuguese coast showing internal waves. The image width is 100 km.

mechanism. The surface convergence and divergence of the velocity field in the upper layer of the ocean associated with vertical motions centered on the thermocline interacts with surface waves to produce zones of rougher and smoother surface which correspond to particular phases of the internal wave motion. The presence of surface films provides an alternative mechanism for imaging internal waves in which the convergence leads to thickening of the surface film and the appearance of a smooth slick, but in the divergence zone the roughness is unchanged from the surroundings. Figure 16.9, an ERS-1 SAR image off the coast of Portugal, illustrates both mechanisms operating on similar types of shelf-edge generated internal waves. The hydrodynamic modulation appears to the south and the slicks to the north. For the latter case, it is worth remarking how significantly the sea-surface microlayer appears to be influenced by sub-surface dynamical processes occurring at the thermocline.

Extending remote sensing contributions to sea-surface microlayer studies

Having discussed ways in which remote sensing is already used for measuring aspects of the surface microlayer, we now consider potential developments of the subject, which draw on the unique sampling capabilities of satellite or aircraft sensors. Future new applications of remote sensing are likely to arise from the synergy to be derived from combining observations from several different remote sensing methods and from *in situ* instruments, perhaps in conjunction with numerical models. The synoptic spatial overview available from imaging sensors can contribute a framework for integrating data from diverse sources, and for extrapolating the impact of local process studies to ocean basin scales. Here we consider two areas of promise which still require further development before they can make a reliable contribution to microlayer research: measurement of air–sea fluxes and observations of horizontal turbulence structures.

Air–sea fluxes

Estimates of air–sea fluxes of heat, momentum, water and gas require a knowledge of sea-surface parameters, some of which are measurable by remote sensing. For example, sensible and latent heat fluxes depend on the air–sea temperature difference and the surface wind stress. Evidently, a combination of thermal infrared and radar scatterometer remote sensing could form the basis of heat flux estimates. At present, there are no routine measurements of both which are simultaneous, and to evaluate fluxes it would be necessary to use composite datasets of SST and wind fields, averaged over, for example, one week. However, there is the possibility in future of flying on the same platform an infrared radiometer and wind scatterometer or SAR which would have overlapping swaths, permitting estimates of instantaneous heat fluxes to be evaluated at a resolution of a few kilometres.

Estimates of gas fluxes require, as well as air–sea concentration differences, knowledge of the air–sea transfer velocities, which are a function of wind speed (Liss and Merlivat, 1986). Some global estimates of gas fluxes (e.g., Etcheto and Merlivat, 1988) have used satellite-derived wind speeds. Sea-surface temperature also affects gas transfer through the gas solubility of the surface microlayer. Robertson and Watson (1992) pointed out that if the surface skin temperature rather than the slightly warmer bulk SST is used in gas transfer models, estimates of the surface ocean's capacity as a sink for carbon dioxide is significantly increased. However, there has as yet been no fully integrated use of remotely sensed measurements in estimates of gas transfer. In order fully to exploit remote sensing, it is desirable to re-evaluate the bulk air–sea interaction parameterizations in terms of variables which are measured directly by remote sensing. This may require the fundamental processes to be revisited. For example, parameterizations of transfer velocity are generally based on the ten metre height wind speed, because this is what is routinely measured or inferred. It is likely that transfer velocity is more directly related to the wind stress or friction velocity, possibly even mean-square sea-surface slope (Jähne *et al.*, 1987). In principle, this can be measured directly by radar without the need for the intermediate determination of wind speed. Such an approach should render the parameterization less sensitive to atmospheric stability. However, to proceed in this way will require new empirical determinations of the air–sea interaction formulae.

It will also be important to determine whether such relationships are scale specific. Remote sensing enables the spatial variability of SST and wind stress to be detected at length scales down to a few kilometres. It also permits integrations of air–sea fluxes to be made over thousands of kilometres. It would be inappropriate to apply a nonlinear formula, based on averages over hundreds of kilometres, to detailed spatial structures at shorter length scales, and vice versa. High-resolution SAR images also present the opportunity to refine air–sea flux parameterizations to include explicitly factors such as surface films and breaking waves which have hitherto not been readily quantifiable. If progress is to be made, it will be necessary in future for air–sea gas exchange experiments to have a remote sensing component, and for remote sensing agencies to plan explicitly for suites of sensors to facilitate flux measurements.

Spatial patterns in images

The synoptic overview offered by imaging sensors remains to be fully exploited. It is not only that horizontal variability length scales of surface parameters can now be measured, but also that the patterns revealed in images have significance which awaits fuller explanation. As an example, Figure 16.10 shows the surface temperature structure in the Mediterranean Sea adjacent to the Island of Corsica, detected with the ATSR on ERS-1. To the west, particularly, can be seen narrow (1–3 km wide) parallel patterns of warmer and cooler water, differing by up to 0.5 K. The narrow spacing of these implies that they are controlled by local air–sea interaction processes, and may be caused by material on the surface, or local surface convergence/divergence. However, the wider view reveals also that they seem to be organized in relation to the larger mesoscale eddies associated with the baroclinic dynamics of the region. Images such as this, and at shorter length scales the sun-glitter photographs from the Space Shuttle (Scully-Power, 1986), demonstrate that sea-surface microlayer processes cannot necessarily be studied in isolation from regional dynamics, nor from the physics of what is occurring lower down in the water column both above and below the thermocline. This suggests that models of surface processes need to be nested within wider area dynamical models.

Analysis of the spatial variability may be able to tell us more about the balance between advection and diffusion in the transport of surface tracer parameters. The patchiness and the degree of differentiation of surface patterns may also be able to indicate the strength or weakness of dispersion processes. There is a requirement for new spatial analysis tools to be developed which can express physically useful information derived from the complex patterns which appear on thermal and radar images. Satellites cannot normally deliver rapidly repeated images, but where shore-based radars and aircraft overflights can generate time sequences, a lot can be learned from the evolving patterns about the surface motion, convergence and shear.

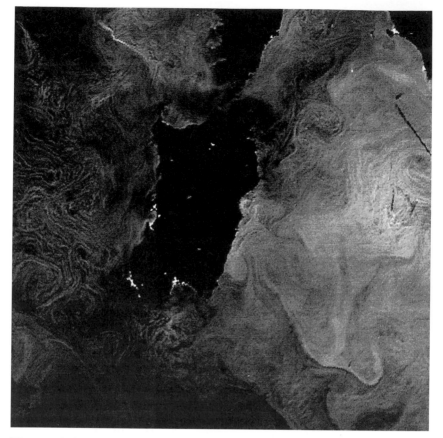

Figure 16.10. ATSR brightness temperature image of the Mediterranean Sea around Corsica, on 13 May 1992 at 2123 UT. This image is 500 km across, and shows warmer temperature as bright tones and cooler as dark.

New in situ *instrumentation*

Ironically, the growing use of remote sensing methods is driving a demand for new *in situ* measurements which are required both to calibrate satellite data, but also to study the near-surface processes which give rise to measurable features on images. The need for reliable shipboard infrared radiometers with an accuracy of 0.1 K has already been mentioned. Imaging thermal sensors deployed from ships are also required in order to study the fine scale patterns of skin temperature. In relation to SARs and scatterometers, there is a strong requirement for new ways of measuring the high-frequency sea-surface roughness, independently of radar if it is to be used to elucidate the mechanisms of microwave backscatter. A number of laser profiling or slope measuring instruments have been used (Hughes *et al.*, 1977, Kwoh and Lake, 1984 and Shemdin and Hwang, 1988) but need further development, and experiments are needed to compare the roughness measured in this way with that inferred from radars. The development of both these types of instrument promises valuable contributions to experiments for observing the sea-surface microlayer.

Conclusion

This chapter has presented a number of ways in which remote sensing is capable of providing information about the sea-surface microlayer which would be impossible, or difficult, to acquire in any other way. Detailed attention has focused on infrared measurement of the ocean thermal skin, and on the detection of small-scale roughness using imaging radars. Suitably coupled, these promise advances in the measurement of surface fluxes, which can be extrapolated from the local to the global scale using the broad-ranging spatial sampling capability characteristic of satellite remote sensing methods. It is to be expected that a new generation of airborne and satellite imaging spectrometers will enable visible wave-band measurements of ocean 'colour' also to contribute to knowledge of the sea surface.

However, useful progress requires a closer collaboration between remote sensing scientists and those studying the surface microlayer. Remotely sensed observations in isolation are of much less value than when acquired in conjunction with complementary *in situ* measurements. Interaction and feedback between researchers is needed in both directions. It is not simply that remote sensing can add additional observational tools to complement those already used in ship-based measurements of the surface microlayer, but that the remote sensing community itself requires improved knowledge and understanding of surface processes in order to develop the oceanographic applications of remote sensors.

Acknowledgements

The author is grateful for helpful discussions with Dr David Woolf and for a number of useful suggestions from the referees. Some of the satellite images and data were processed or provided by Craig Donlon,

José da Silva, Alison Scoon and Miguel Tenorio at the University of Southampton. ERS-1 data used to construct the images in Figures 16.5–16.9 were provided by the European Space Agency. Data for the image displayed in Figure 16.10 were provided by the ATSR Instrument Group at the Rutherford Appleton Laboratory.

References

Abbot, M. R. and Zion, P. M. (1987). Spatial and temporal variability of phytoplankton pigment off Northern California. *J. Geophys. Res.*, **92**, 1745.

Alpers, W. (1983). Imaging ocean surface waves by synthetic aperture radar: a review. In *Satellite Microwave Remote Sensing*, ed. T. D. Allan, pp. 107–119. Chichester: Ellis Horwood.

Alpers, W. (1985). Theory of radar imaging of internal waves. *Nature*, **314**, 245–7.

Alpers, W. and Brümmer, B. (1994). Atmospheric boundary layer rolls observed by the synthetic aperture radar aboard the ERS-1 satellite. *J. Geophys. Res.*, **99**, 12613–22.

Alpers, W. and Hühnerfuss, H. (1988). Radar signatures of oil films floating on the sea surface and the Marangoni effect. *J. Geophys. Res.*, **93**, 3642–8.

Alpers, W. and Hühnerfuss, H. (1989). The damping of ocean waves by surface films: a new look at an old problem. *J. Geophys. Res.*, **94**, 6251–65.

Alpers, W., Blume, H. J. C., Garret, W. D. and Hühnerfuss, H. (1982). The effect of monomolecular surface films on the microwave brightness temperature of the sea surface. *Int. J. Remote Sensing*, **3**, 437–74.

Apel, J. R. and Gonzalez, F. I. (1983). Non-linear features of internal waves off Baja California as observed from the Seasat imaging radar. *J. Geophys. Res.*, **88**, 4459–66.

Barger, W. R., Garrett, W. D., Mollo-Christensen, E. L. and Ruggles, K. W. (1970). Effects of an artificial sea slick upon the atmosphere and the ocean. *J. Appl. Meteorol.*, **9**, 396–400.

Bartsch, N., Grüner, K., Keydel, W. and Witte, F. (1987). Contributions to oil-spill detection and analysis with radar and microwave radiometry: results of the Archimedes II campaign. *IEEE Trans. Geosci. Remote Sensing*, **GE-25**, 677–90.

Brown, J. W., Brown, O. B. and Evans, R. H. (1993). Calibration of advanced very high resolution radiometer infrared channels: a new approach to nonlinear correction. *J. Geophys. Res.*, **98**, 18 257–68.

Brüning, C., Alpers, W. and Hasselmann, K. (1990). Monte Carlo simulation studies of the non-linear imaging of a two-dimensional surface wave field by a synthetic aperture radar. *Int. J. Remote Sensing*, **11**, 1695–727.

Brüning, C., Hasselmann, S., Hasselmann, K., Lehner, S. and Gerling, T. (1993). On the extraction of ocean wave spectra from ERS-1 SAR wave mode image spectra. In *Proceedings of the first ERS-1 Symposium, Space at the Service of our Environment, 4–6 November 1992, Cannes*, pp.747–52. Paris: European Space Agency, SP-359.

Cox, C. and Munk, W. (1954). Measurements of the roughness of sea surface from photographs of the sun's glitter. *J. Opt. Soc. Am.*, **44**, 838–50.

Delderfield, J. *et al.* (1986). The Along Track Scanning Radiometer (ATSR) for ERS-1. In *Instrumentation for Optical Remote Sensing from Space*, ed. J. S. Seeley, J. W. Lear, A. Monfils and S. L. Russak. *Proc. SPIE*, **589**, 114.

Demin, B. T., Ermakov, S. A., Pelinovsky, E. N., Talipova, T. G. and Sheremet'yeva, A. I. (1985). Study of the elastic properties of sea surface-active films (English translation). *Izv. Atmos. Oceanic Phys.*, **21**, 312–16.

Deschamps, P-Y. and Phulpin, T. (1980). Atmospheric correction of infrared measurements of sea surface temperature using channels at 3.7, 11 and 12 µm. *Boundary-layer Meteorol.*, **18**, 131–43.

Eifler, W. (1992). Modelling the skin-bulk temperature difference near the sea–atmosphere interface for remote sensing applications. In *Proc. Central Symposium of the International Space Year Conference, Munich, 30 March – 4 April 1992*, pp. 335–340. Paris: European Space Agency, ESA **SP-341**.

Ermakov, S. A., Panchencko, A. R. and Talipova, T., G. (1985). Damping of high-frequency wind waves by artificial surfactant films (English translation). *Izv. Atmos. Oceanic Phys.*, **21**, 54–8.

Ermakov, S. A., Salashin, S. G. and Panchencko, A. R. (1992). Film slicks on the sea surface and some mechanisms of their formation. *Dynam. Atmos. Oceans*, **16**, 279–304.

Etcheto, J. and Merlivat, L. (1988). Satellite determination of the carbon dioxide exchange coefficient at the ocean–atmosphere interface: a first step. *J. Geophys. Res.*, **93**, 15 669–78.

Ewing, G. and McAlister, E. D. (1960). On the thermal boundary-layer of the ocean. *Science*, **131**, 1374.

Gordon, H. R. and Morel, A. Y. (1983). *Remote Assessment of Ocean Colour for Interpretation of Satellite Visible Imagery. A Review.* Lecture notes on coastal and estuarine studies: 4. New York: Springer-Verlag.

Gower, J. F. R. (1993). Wind and surface features in SAR images: the Canadian program. In *Proceedings of the first ERS-1 Symposium, Space at the Service of our Environment, 4–6 November 1992, Cannes*, pp. 101–6. Paris: European Space Agency, SP-359.

Gower, J. F. R. and Borstad, G. A. (1990). Mapping of phytoplankton by solar stimulated fluorescence using an imaging spectrometer. *Int. J. Remote Sensing*, **11**, 313–20.

Grassl, H. (1976). The dependence of the measured cool skin of the ocean on wind stress and total heat flux. *Boundary-Layer Meteorol.*, **10**, 465–74.

Hasse, L. (1971). The sea surface temperature deviation and the heat flow at the sea–air interface. *Boundary-Layer Meteorol.*, **1**, 368–379.

Hasselmann, K. and Hasselmann, S. (1991). On the non-linear mapping of an ocean wave spectrum into a synthetic aperture radar image spectrum and its inversion. *J. Geophys. Res.*, **96**, 10 713–29

Hasselmann, K., Raney, R. K., Plant, W. J., Alpers, W., Shuchman, R. A., Lyzenga, D. R., Rufenach, C. L. and Tucker, M. J. (1985). Theory of SAR ocean wave imaging, a MARSEN view. *J. Geophys. Res.*, **90**, 4659–86.

Holliday, D., St-Cyr, G. and Woods, N. E. (1986). A radar ocean imaging model for small to moderate incidence angles. *Int. J. Remote Sensing*, **7**, 1809–34.

Hughes, B. A., Grant, H. L. and Chappell, R. W. (1977). A fast response surface wave-slope meter and measured wind–wave moments. *Deep-Sea Res.*, **24**, 1211–23.

Hühnerfuss, H., Walter, W., Lange, P. A. and Alpers, W. (1987). Attenuation of wind waves by monomolecular sea slicks by the Marangoni effect. *J. Geophys. Res.*, **92**, 3961–3.

Hühnerfuss, H., Gericke, A., Alpers, W., Theis, R., Wismann, V. and Lange, P. A. (1994). Classification of sea slicks by multifrequency radar techniques: new chemical insights and their geophysical implications. *J. Geophys. Res.*, **99**, 9835–45.

Hurford, N. and Tookey, D. (1986). A detailed evaluation of the maritime surveillance system for oil slick detection. *Oil Chem. Pollut.*, **3**, 231–44.

Jähne, B., Munnich, K. O., Bosinger, R., Dutzi, A., Huber, W. and Libner, P. (1987). On the parameters influencing air–water gas exchange. *J. Geophys. Res.*, **92**, 1937–49.

Jessup, A. T. (1992). Measurements of small-scale variability of infrared sea surface temperature (abstract), In *EOS Trans. Am. Geophys. Union, 1992 Fall Meeting Supplement*, p. 264.

Johannessen, J. A., Røed, L. P. and Wahl, T. (1993). Eddies detected in ERS-1 SAR images and simulated in reduced gravity model. *Int. J. Remote Sensing*, **14**, 2203–13.

Jones, W. L. *et al*. (1982). The Seasat-A scatterometer: the geophysical evaluation of remotely-sensed wind vectors over the ocean. *J. Geophys. Res.*, **87**, 3297–317.

Katsaros, K., (1980). The aqueous thermal boundary layer. *Boundary-layer Meteorol.*, **18**, 107–127.

Katsaros, K. B., Liu, T., Businger, J. A. and Tillman, J. E. (1977). Heat transport and thermal structure in the interfacial boundary layer measured in an open tank of water in turbulent free convection. *J. Fluid Mech.*, **83**, 311–35.

Kwoh, D. S. W. and Lake, B. M. (1984). A deterministic coherent and dual polarized laboratory study of microwave backscattering from water waves, Part I, Short gravity waves without wind. *IEEE J. Oceanic Eng.*, **OE-9**, 291–308.

Lauritson, L. *et al*. (1979). Data extraction and calibration of TIROS-N/NOAA radiometers. *NOAA Tech Memo. 107*, 73 pp.

Lecomte, P. and Attema, E. P. W. (1993). Calibration and validation of the ERS-1 Wind Scatterometer. In *Proceedings of the first ERS-1 Symposium, Space at the Service of our Environment, 4–6 November 1992, Cannes*, pp. 19–29. Paris: European Space Agency, SP-359.

Liss, P. S. and Merlivat, L. (1986). Air–sea gas exchange rates: introduction and synthesis. In *The Role of Air-Sea Exchange in Geochemical Cycling*, ed. P. Buat-Menard, pp. 113–27. Dordrecht, Holland: Kluwer Academic Publishers.

Llewellyn-Jones, D. T., Minnett, P. J., Saunders, R. W. and Zavody, A. M. (1984). Satellite multichannel infrared measurements of sea surface temperature of the N. E. Atlantic Ocean using AVHRR/2. *Q. J. R. Meteorol. Soc.*, **110**, 613–31.

Macklin, J. T. (1992). The imaging of oil slicks by synthetic aperture radar. *GEC J. Res.*, **10**, 19–28.

Mammen, T. C. and von Bosse, N. (1990). STEP–a temperature profiler for measuring the oceanic thermal boundary layer at the ocean–air interface. *J. Atmos. Oceanic Technol.*, **7**, 312–22.

Masuda, K. Takashima, T. and Takayama, Y. (1988). Emissivity of pure and sea waters for the model sea surface in the infrared window region. *Remote Sensing Environ.*, **24**, 313–29.

Maull, G. (1985). *Introduction to satellite oceanography*. Dordrecht: Martinus Nijhoff.

McAlister, E. D. and McLeish, W. (1970). A radiometer system for measurement of the total heat flux from the sea. *Appl. Opt.*, **9**, 2697–705.

McClain, E. P., Pichel, W. G. and Walton, C. C. (1985). Comparative performance of AVHRR-based multichannel sea surface temperatures. *J. Geophys. Res.*, **90**, 11 587–601.

McMillin, L. M. (1975). Estimation of sea surface temperatures from two infrared window measurements with different obsorption. *J. Geophys. Res.*, **80**, 5113–17.

McMillin, L. M. and Crosby, D. S. (1984). Theory and validation of the multiple window sea surface temperature technique. *J. Geophys. Res.*, **89**, 3655–61.

Mutlow, C. T. *et al.* (1994). The Along Track Scanning Radiometer (ATSR) on ESA's ERS-1 satellite: early results. *J. Geophys. Res.*, **99**, 22 575–88.

Ocampo-Torres, F. J. and Robinson, I. S. (1990). Wind wave directionality effects on the radar imaging of ocean swell. *J. Geophys. Res.*, **95**, 20 347–62.

Ochadlick, A. R., Cho, P. and Evans-Morgis, J. (1992). Synthetic aperture radar observations of currents colocated with slicks. *J. Geophys. Res.*, **97**, 5325–30.

Onstott, R. and Rufenach, C. (1992). Shipboard active and passive microwave measurement of ocean surface slicks off the southern Californian coast. *J. Geophys. Res.*, **97**, 5315–23.

Paulson, C. A. and Simpson, J. J. (1981). The temperature difference across the cool skin of the ocean. *J. Geophys. Res.*, **86**, 11 044–54.

Phillips, O., M. (1977). *The Dynamics of the Upper Ocean*. Cambridge: Cambridge University Press.

Robertson, J. E. and Watson, A. J. (1992). Thermal skin effect of the surface ocean and its implications for CO_2 uptake. *Nature*, **358**, 738–40.

Robinson, I. S. (1985). *Satellite Oceanography*. Chichester: Ellis Horwood, 455pp.

Robinson, I. S. (1990). Remote sensing: information from the colour of the seas. In *Light and Life in the Sea*, ed. P. Herring *et al.*, pp. 19–38. Cambridge: Cambridge University Press.

Robinson, I. S., Wells, N. C. and Charnock, H, (1984). Review article: the sea surface thermal boundary layer and its relevance to the measurement of sea surface temperature by airborne and spaceborne radiometers. *Int. J. Remote Sensing*, **5**, 19–45.

Saltzman, B. (ed.) (1985). *Satellite Oceanic Remote Sensing, Advances in Geophysics*, vol. 27. Orlando, FL: Academic Press. 511pp.

Saunders, P. M. (1967). The temperature at the ocean–air interface. *J. Atmos. Sci.*, **24**, 269–273.

Saunders, P. M. (1973). The skin temperature of the ocean: a review. *Mem. Soc. R. Sci. Liège*, **4**, 93–8.

Schluessel, P., Shin, H-Y., Emery, W. J. and Grassl, H. (1987). Comparison of satellite-derived sea surface temperatures with in situ skin measurements. *J. Geophys. Res.*, **92**, 2859–74.

Schluessel, P., Emery, W. J., Grassl, H. and Mammen, T. (1990). On the bulk-skin temperature difference and its impact on satellite remote sensing of sea surface temperature. *J. Geophys. Res.*, **95**, 13 341–56.

Scoon, A., Robinson, I. S. and Meadows, P. (1995). Validation of an improved calibration for ERS-1 SAR.PRI ocean data, using a wind retrieval algorithm. *Int. J. Remote Sensing*, **17**, 413–18.

Scott, J. (1989). Oil slicks still the waves. *Nature*, **340**, 601.

Scully-Power, P. (1986). Navy oceanographer shuttle observations, Mission Report. *Rep STSS 41-G, NUSC Tech. Doc. 7611*. Newport, RI: Naval Underwater Systems Center.

Shemdin, O. H. and Hwang, P. A. (1988). Comparison of measured and predicted sea surface spectra of short waves. *J. Geophys. Res.*, **93**, 13 883–90.

Soloviev, A. V. and Schluessel, P. (1994). Parameterization of the cool skin of the ocean and of the air–ocean gas transfer on the basis of modelling surface renewal. *J. Phys. Ocean.* **24**, 1339–46.

Stewart, R. (1985). *Methods of Satellite Oceanography*. San Diego: University of California Press.

Stoffelen, A., Anderson, D. L. T. and Woiceshyn, P. M. (1992). ERS-1 Scatterometer calibration and validation activities at ECMWF: B, From radar backscatter characteristics to wind vector solutions. *Proc. of Workshop on ERS-1 Geophysical Validation*, pp. 89–93. Paris: European Space Agency, Publ. WPP-36.

Swift, C. T. (1980). Passive microwave remote sensing. *Boundary-layer Meteorol*, **18**, 25–54.

Tang, S. and Wu, J. (1992). Suppression of wind-generated ripples by natural films: a laboratory study. *J. Geophys. Res.*, **97**, 5301–6.

Thomson, R. E., Vachon, P. W. and Borstad, G. A. (1992). Airborne synthetic aperture radar imagery of atmospheric gravity waves. *J. Geophys. Res.*, **97**, 14 249–57.

Vachon, P. W., Johannessen, O. M. and Johannessen, J. A. (1993). Analysis of atmospheric lee waves near Hopen. In *Proceedings of the second ERS-1 Symposium, Space at the Service of our Environment, 11–14 October, 1993, Hamburg*, pp. 825–8. Paris: European Space Agency, SP-361.

Valenzuela, G. R. (1978). Theories for the interaction of electromagnetic and oceanic waves: a review. *Boundary-layer Meteorol.*, **13**, 61–85.

Vesecky, J. F. and Stewart, R. H. (1982). The observation of ocean surface phenomena using imagery from the Seasat synthetic aperture radar: an assessment. *J. Geophys. Res.*, **87**, 3397–430.

Wick, G. A., Emery, W. J. and Schluessel, P. (1992). A comprehensive comparison between satellite-measured skin and multichannel sea surface temperature, *J. Geophys. Res.*, **97**, 5569–95.

Woodcock, A. H. and Stommel, H. (1947). Temperatures observed near the surface of a fresh water pond at night. *J. Meteorol.*, **4**, 102–3.

Index

Note: figures and tables are indicated by *italic page numbers*

509